Machine Learning for Ecology and Sustainable
Natural Resource Management

Grant R. W. Humphries • Dawn R. Magness
Falk Huettmann

Editors

Machine Learning for Ecology and Sustainable Natural Resource Management

 Springer

Editors
Grant R. W. Humphries
Black Bawks Data Science Ltd
Fort Augustus, Scotland

Falk Huettmann
EWHALE Lab, Biology and
Wildlife Department
Institute of Arctic Biology
University of Alaska-Fairbanks
Fairbanks, AK, USA

Dawn R. Magness
U.S. Fish and Wildlife Service
Kenai National Wildlife Refuge
Soldotna, AK, USA

ISBN 978-3-319-96976-3   ISBN 978-3-319-96978-7   (eBook)
https://doi.org/10.1007/978-3-319-96978-7

Library of Congress Control Number: 2018956275

This Springer imprint is published by the registered company Springer Nature Switzerland AG
The registered company address is: Gewerbestrasse 11, 6330 Cham, Switzerland

This book is dedicated to those who fight for balance and harmony with nature despite the odds being stacked against them

Grant R. W. Humphries

I dedicate this book to those who work towards sustainability, human rights, and environmental justice

Dawn R. Magness

I devote this book to 'everybody' (as things cannot be done any other way these days)

Falk Huettmann

Foreword

The fields of nature conservation and wildlife science have grown and diversified so rapidly over the past 50 years, it is hard to imagine how the speed of change could get even faster – it has. As artificial intelligence and machine learning is increasingly used in our field, change will accelerate. Grant Humphries, Dawn Magness and Falk Huettmann have provided us a book filled with information essential to get any wildlife or conservation biologist up to speed with the concepts and application of machine learning. They cover historical and conceptual information, plus specific case studies and a taste of the future.

Machine learning attempts to extract knowledge from messy data, and it is hard to think of a field where the data is messier. While traditional ecological data analysis and well-designed field experiments will always be essential, we need to find new tools to interpret the flood of environmental data coming from satellites, drones, camera traps, acoustic monitoring devices, tagged animals and more. Machine learning is relatively open-minded about the meaning of this data and the relationships between different kinds of data. While at times this is dangerous, it also liberates us to explore new ideas – such as using data on seabird chick weights to make predictions about the Southern Oscillation Index (Chap. 13).

The future of conservation will require an open mind. We could be using machine learning: for carrying out real-time management of small-scale fisheries, analyzing satellite data to make better landscape change models that inform policy, deploying drones that can deter poachers, and creating ephemeral protected areas for migratory species. As the authors remind us, we have very little time to avert the extinction of most of the world's species and whole-scale reduction in the abundance of even common species. We need every discipline and every tool available to efficiently and effectively solve the world's conservation problems.

As The Nature Conservancy's Chief Scientist, I can say that we will be increasing our investment in using artificial intelligence to make faster and better conservation decisions that deliver outcomes on the ground.

The Chief Scientist of The Nature Conservancy   Hugh Possingham
Arlington, Virginia, USA

Preface[1]

This painting, by Margaret Oliver, depicts Northern Gannet (*Morus bassanus*) and Northern Fulmar (*Fulmarus glacialis*) near an iceberg. Northern Gannet is a species that can be found breeding in the temperate parts of the North Atlantic and as far south as the Gulf of Mexico during the non-breeding season. The Northern Gannet, however, is being found further and further north, with 6 records in Svalbard, well above the Arctic Circle, between 2000 and 2016. This is likely to continue and increase as ecosystem shifts push oceanic biomes further north. Icebergs, like the one shown in this piece, are another complex seascape feature that can be integrated into machine learning models for a holistic understanding of global ecosystems

---

[1] The findings and conclusions in this book are those of the authors and do not necessarily represent the views of the US Fish and Wildlife Service.

We humans are curious animals. While other animals generally live in balance with their respective ecosystems, we almost seem to go out of our way to tip that balance in our favor to the extreme detriment of everything else. If an alien race were to look at us objectively, it is likely they would consider most modern humans a cancer on the planet; growing uncontrollably and using resources without much care, limit, or reason. But within this destructive framework lie some heroes who have at least been able to clearly identify our impacts and are fighting to protect species, habitats, and, ultimately, the human race. Some of them are scientists, environmentalists, conservationists, ecological economists, and others who only want to see balance restored to the planet. This battle is becoming harder to fight as our population expands and we compete more and more for our precious resources. To that end, we are in desperate need for fast decisions in a conservation context. Unfortunately, we often see science efforts get caught up in details like narrow statistical assumptions, mechanisms, and parsimony. While this has been interesting perhaps from a certain academic perspective, it really fails to build a holistic understanding of our world.

In the many years that we (the editors of this book) have been studying the application of machine learning, there have been amazing changes in techniques. I (GH) recall starting my master's degree with a moderate understanding of multivariate statistics but having never used a machine learning algorithm before. Within my first week, I had run my first model using Salford System's machine learning software, TreeNet (an implementation of boosted regression trees). At the time, I did not realize the path I was about to take and spent the next two years studying the method to be able to pull apart and make sense of the science behind the technique. One thing I learned was that machine learning and, subsequently, scripting (in R and Python) was a rabbit hole. The more I did, the deeper I went ("deep learning," you could say), and the more I wanted to know. But what I soon realized was that it was not just that the philosophy was complex and dynamic, but that the methods and tools were changing much faster than I could keep up with. Trying to build an understanding of a complex suite of tools and techniques while having to study ecology (or wildlife biology), take other course work, have a social life, and write was a challenging task. This made me realize something that has become abundantly clear to me over the last decade: there is a major disconnect between ecology and the computational (and quantitative) skills required in today's job market. This disconnect makes it very difficult for applying solutions to complex analytical problems in fast and effective ways (to match the changing pace of our world). We (as ecologists) spend vast amounts of time pouring through obscure and confusing computer "lingo" that we just are not used to, and that slows us down. We have to therefore raise the question: "How do academics (already tasked with writing grants, teaching courses, mentoring students, publishing research, doing admin work, etc...), or government scientists, or other researchers have the time to really understand and apply new techniques?" We (GH, FH, and DM) have all experienced this over our careers to date in various ways, and it is part of the motivation for compiling this book. We want to encourage ecologists at universities and elsewhere to embrace the technological changes that are coming our way; specifically, powerful machine learning algorithms that can be tailored to a number of really interesting scientific questions.

Ideally, this would be done in an open access framework, which would help to make the most meaningful impact on progress. The integration of computational science, machine learning, and data mining into ecology courses at a very basic level has the potential to vastly change how we do science by allowing us to quickly analyze complex and peculiar datasets common in wildlife studies. There is still time to redefine syllabi, performance evaluations, and job descriptions with an outlook and focus on global sustainability.

Machine learning algorithms are being recognized more and more as formidable weapons in natural resource management and wildlife biology. This is because the methods we elaborate on in this book are able to decipher complex relationships between many predictor variables (hundreds or more of the best available) and give meaningful and powerful predictions while allowing us to infer relevant mechanisms and good management action. There is somewhat of a stigma in ecology, however, which we believe stems from a lack of basic teachings when we are students. Firstly, we are not really taught basic data management techniques to help us better organize ourselves digitally, which has direct implications in things like data mining. Secondly, the details of new algorithms (or how to begin deciphering them) are not really taught, which makes them come off as "black boxes." Thirdly, the basic principles of scripting (e.g., R, Python, C++ etc…) are not taught to us and are picked up, if at all, through many frustrating nights in front of online tutorials or trial and error (mostly error!).

There is hope, however, as university statistics courses are more frequently taught in R, which means code can be replicated and explained. But these courses are generally pure statistics courses (though exceptions are arising with quantitative ecology being taught in some universities now). Also, machine learning algorithms and techniques rarely integrated in these courses. Machine learning is an entire suite of tools that cannot be ignored as the evidence for their superiority in many aspects mounts further. They can help lead us toward a holistic understanding of the planet if used correctly and in an ethical framework.

Our book aims to guide you through a series of interesting and new applications of machine learning algorithms in wildlife biology, ecology, and natural resource management. We hope that as you read through these techniques, you get some ideas for your own research and perhaps try a few algorithms yourself. The world is changing rapidly, as are the algorithms we use to model it. Despite the ongoing decay and policy chaos, we see a bright future ahead for machine learning in ecology and our planet.

Fort Augustus, Scotland   Grant R. W. Humphries
Soldotna, AK, USA   Dawn R. Magness
Fairbanks, AK, USA   Falk Huettmann

Acknowledgments

The work in this book stems from a decade of networking and communications with a variety of amazing and talented people, as well as personal experience applying machine learning to complex ecological datasets. A full list of those people could take up a book all on its own. I am indebted to I. Jones, T. Miller, D. Verbyla, D. Atkinson, C. Deal, H. Moller, P. Seddon, J. Overton, G. Nevitt, B. Sydeman, and H. Lynch for direct guidance and supervision over my academic career, allowing me to explore novel uses of machine learning algorithms. A special thanks to my co-editors, without whom this book would not be possible. Many of the projects I have worked on have been inspired by the studies and dedication of many of my peers and lab mates including: J. Lavers, S. Seneviratne, A. Baltensperger, T. Mullet, S. Hazlett, K. Jochum, S. Oppel, M. Lindgren, R. Buxton, M. Savoca, B. Hoover, S. Jamieson, C. Tyson, H. Robinson, M. Garcia-Reyes, C. Youngflesh, M. Schrimpf, M. Lynch, C. Foley, B. Goncalvez, A. Borowicz, B. Weinstein, F. McDuie, and many others. Finally, thanks go to my family – my rocks in this world: Alex, Dylan, and River.

Grant R. W. Humphries

Having worked on the issue worldwide and professionally for over 25 years, for a machine learning book and in the environmental and conservation sciences, by now one should carry a long list of people to thank. Though the reality is – and as the intense machine learning work has shown us globally – the amount of people to thank on that issue remains relatively few. Surprising to us, machine learning is still "just" a narrow aspect in the quantitative sciences and an even smaller group in the governmentally dominated wildlife conservation management field. People, nations, and institutions who "saw the light" and supported that issue were far and few: that is also the Leo Breiman career experience (whom I acknowledge here the most). On the good side, those people that are to be thanked are true visionaries and one must be grateful for their massive progress and for being open-minded. An incomplete but tight list consists of: R. O'Connor, A.W. Diamond, G. Yumin, M. Chunrong, H. Huesong, B. Raymond, C. Fox, P. Paquet, D. Kirk, S. Pitman, J. Evans,

S. Cushman, B. Walther, L. Koever, B. Danis, S. Ickert-Bond, D. Verbyla, J. Morton, G. Heyward, T. Gotthardt, and R. Yoccum. The discussions and support we received from Salford Systems Ltd. – D. Steinberg, L. Solomon, and their team – cannot be valued highly enough for discussing and learning more about "modern" machine learning applications in real life. The institutions, agencies, and individuals who invited us for machine learning workshops and conference sessions worldwide are greatly appreciated also. All my 60+ students in the EWHALE lab with the University of Alaska-Fairbanks (UAF) must be thanked for their efforts working on the issue with us on machine learning and open access, too. And then, there are many big but indirect supporters who just kept an open mind and pursued holistic non-parsimonious views, namely D. Carlson, J. Liu, T. Lock, B. Czech, J. Ausubel, P. Skylstad, B. Welker, T. Williams, H. Possingham, P. Koubbi, H. Griffith, C. De Broyer, A. Bult-Ito, D. Misra, R. Ganguli, P. Layer, B. Norcross, T. Bowyer, K. Hundertmark, E. Bortz, A. Shestopalov, K. Kieland, and the Springer publisher (namely J. Slobodien). Lastly, I thank H. Berrios Alvarez, S. Linke, I. Presse, H. Hera, C. Cambu, L. Luna, and S. Sparks for wider infrastructure support and for plain fun in the world of machine learning and data.

Falk Huettmann

The co-editors would like to jointly thank all the reviewers who helped make this book happen: A. Baltensperger, E. Craig, T. Miewald, B. Weinstein, C. Fox, D. Kirk, D. Karmacharya, A. Bennison, M. Lindgren, M. Garcia-Reyes, V. Morera, S. Pittman, T. Mullet, A. Raya Rey, M. Ryo, and C. Mi.

We greatly appreciate the artwork provided to this book by A. Price, G. Brehm, M. Oliver, and C. Humphries—fantastic artists with hearts for the planet. We would also like to extend our deepest gratitude to Hugh Possingham for writing the foreword to this book. His work on Marxan and related issues has led to protections for some of the most vulnerable places in the world, and we are honored to have his thoughts reflected here. And finally, thanks to Springer for publishing this book and making it available to ecologists and natural resource managers around the world.

Grant, Falk, and Dawn

Photo by Grant Humphries. This photo was taken at over 100m high at Port Charcot in Antarctica showing a colony of gentoo penguins that have integrated rope (possibly left by earlier explorers) into their nest. Even in the remote corners of the world, we see the direct impact of humans on wildlife

Contents

Part I   Introduction

1   **Machine Learning in Wildlife Biology: Algorithms, Data Issues
     and Availability, Workflows, Citizen Science, Code Sharing,
     Metadata and a Brief Historical Perspective**   3
     Grant R. W. Humphries and Falk Huettmann

2   **Use of Machine Learning (ML) for Predicting and Analyzing
     Ecological and 'Presence Only' Data: An Overview
     of Applications and a Good Outlook**   27
     Falk Huettmann, Erica H. Craig, Keiko A. Herrick,
     Andrew P. Baltensperger, Grant R. W. Humphries, David J. Lieske,
     Katharine Miller, Timothy C. Mullet, Steffen Oppel,
     Cynthia Resendiz, Imme Rutzen, Moritz S. Schmid,
     Madan K. Suwal, and Brian D. Young

3   **Boosting, Bagging and Ensembles in the Real World:
     An Overview, some Explanations and a Practical Synthesis
     for Holistic Global Wildlife Conservation Applications
     Based on Machine Learning with Decision Trees**   63
     Falk Huettmann

Part II   Predicting Patterns

4   **From Data Mining with Machine Learning to Inference
     in Diverse and Highly Complex Data: Some Shared Experiences,
     Intellectual Reasoning and Analysis Steps for the Real World
     of Science Applications**   87
     Falk Huettmann

5   **Ensembles of Ensembles: Combining the Predictions
     from Multiple Machine Learning Methods**   109
     David J. Lieske, Moritz S. Schmid, and Matthew Mahoney

6   Machine Learning for Macroscale Ecological Niche Modeling -
     a Multi-Model, Multi-Response Ensemble Technique
     for Tree Species Management Under Climate Change   123
     Anantha M. Prasad

7   Mapping Aboveground Biomass of Trees Using Forest Inventory
     Data and Public Environmental Variables within the Alaskan
     Boreal Forest   141
     Brian D. Young, John Yarie, David Verbyla, Falk Huettmann,
     and F. Stuart Chapin III

Part III   Data Exploration and Hypothesis Generation
           with Machine Learning

8   'Batteries' in Machine Learning: A First Experimental
     Assessment of Inference for Siberian Crane Breeding Grounds
     in the Russian High Arctic Based on 'Shaving' 74 Predictors   163
     Falk Huettmann, Chunrong Mi, and Yumin Guo

9   Landscape Applications of Machine Learning: Comparing
     Random Forests and Logistic Regression in Multi-Scale
     Optimized Predictive Modeling of American Marten
     Occurrence in Northern Idaho, USA   185
     Samuel A. Cushman and Tzeidle N. Wasserman

10   Using Interactions among Species, Landscapes, and Climate
     to Inform Ecological Niche Models: A Case Study of American
     Marten (*Martes americana*) Distribution in Alaska   205
     Andrew P. Baltensperger

11   Advanced Data Mining (Cloning) of Predicted Climate-Scapes
     and Their Variances Assessed with Machine Learning:
     An Example from Southern Alaska Shows Topographical
     Biases and Strong Differences   227
     Falk Huettmann

12   Using TreeNet, a Machine Learning Approach to Better
     Understand Factors that Influence Elevated Blood Lead Levels
     in Wintering Golden Eagles in the Western United States   243
     Erica H. Craig, Tim H. Craig, and Mark R. Fuller

Part IV   Novel Applications of Machine Learning Beyond Species
           Distribution Models

13   Breaking Away from 'Traditional' Uses of Machine Learning:
     A Case Study Linking Sooty Shearwaters (*Ardenna griseus*)
     and Upcoming Changes in the Southern Oscillation Index   263
     Grant R. W. Humphries

14 Image Recognition in Wildlife Applications   285
Dawn R. Magness

**15 Machine Learning Techniques for Quantifying Geographic
Variation in Leach's Storm-Petrel (*Hydrobates leucorhous*)
Vocalizations**   295
Grant R. W. Humphries, Rachel T. Buxton, and Ian L. Jones

Part V   Implementing Machine Learning for Resource Management

**16 Machine Learning for 'Strategic Conservation and Planning':
Patterns, Applications, Thoughts and Urgently Needed Global
Progress for Sustainability**   315
Falk Huettmann

**17 How the Internet Can Know What You Want Before You Do:
Web-Based Machine Learning Applications for Wildlife
Management**   335
Grant R. W. Humphries

**18 Machine Learning and 'The Cloud' for Natural Resource
Applications: Autonomous Online Robots Driving Sustainable
Conservation Management Worldwide?**   353
Grant R. W. Humphries and Falk Huettmann

**19 Assessment of Potential Risks from Renewable Energy
Development and Other Anthropogenic Factors to Wintering
Golden Eagles in the Western United States**   379
Erica H. Craig, Mark R. Fuller, Tim H. Craig, and Falk Huettmann

Part VI   Conclusions

**20 A Perspective on the Future of Machine Learning: Moving Away
from '*Business as Usual*' and Towards a Holistic Approach
of Global Conservation**   411
Grant R. W. Humphries and Falk Huettmann

Index   431

Contributors

Andrew P. Baltensperger National Park Service, Fairbanks, AK, USA

Rachel T. Buxton Department of Fish, Wildlife and Conservation Biology, Colorado State University, Fort Collins, CO, USA

Erica H. Craig Aquila Environmental, Fairbanks, AK, USA

Tim H. Craig Aquila Environmental, Fairbanks, AK, USA

Samuel A. Cushman U.S. Forest Service, Rocky Mountain Research Station, Flagstaff, AZ, USA

Mark R. Fuller Boise State University, Raptor Research Center, Boise, ID, USA

Yumin Guo College of Nature Conservation, Beijing Forestry University, Beijing, China

Keiko A. Herrick EWHALE Lab, Biology and Wildlife Department, Institute of Arctic Biology, University of Alaska-Fairbanks, Fairbanks, AK, USA

Falk Huettmann EWHALE Lab, Biology and Wildlife Department, Institute of Arctic Biology, University of Alaska-Fairbanks, Fairbanks, AK, USA

Grant R. W. Humphries Black Bawks Data Science Ltd., Fort Augustus, Scotland

Ian L. Jones Department of Biology, Memorial University, St. John's, NL, Canada

David J. Lieske Department of Geography and Environment, Mount Allison University, Sackville, NB, Canada

Dawn R. Magness U.S. Fish and Wildlife Service, Kenai National Wildlife Refuge, Soldotna, AK, USA

Matthew Mahoney Department of Geography and Environment, Mount Allison University, Sackville, NB, Canada

Chunrong Mi Institute of Zoology, Chinese Academy of Sciences, Beijing, China
College of Nature Conservation, Beijing Forestry University, Beijing, China

Katharine Miller Auke Bay Laboratories, Alaska Fisheries Science Center, National Marine Fisheries Service, NOAA, Juneau, AK, USA

Timothy C. Mullet EWHALE Lab, Biology and Wildlife Department, Institute of Arctic Biology, University of Alaska-Fairbanks, Fairbanks, AK, USA

Steffen Oppel RSPB Centre for Conservation Science, Royal Society for the Protection of Birds, Cambridge, UK

Anantha M. Prasad Research Ecologist, USDA Forest Service, Northern Research Station, Delaware, OH, USA

Cynthia Resendiz EWHALE Lab, Biology and Wildlife Department, Institute of Arctic Biology, University of Alaska-Fairbanks, Fairbanks, AK, USA

Imme Rutzen EWHALE Lab, Biology and Wildlife Department, Institute of Arctic Biology, University of Alaska-Fairbanks, Fairbanks, AK, USA

Moritz S. Schmid Hatfield Marine Science Center, Oregon State University, Newport, OR, USA

EWHALE Lab, Biology and Wildlife Department, Institute of Arctic Biology, University of Alaska-Fairbanks, Fairbanks, AK, USA

CERC in Remote Sensing of Canada's New Arctic Frontier Université Laval, Québec, Canada

F. Stuart Chapin III Institute of Arctic Biology, University of Alaska Fairbanks, Fairbanks, AK, USA

Madan K. Suwal Department of Geography, University of Bergen, Bergen, Norway

David Verbyla Department of Forest Sciences, University of Alaska Fairbanks, Fairbanks, AK, USA

Tzeidle N. Wasserman School of Forestry, Northern Arizona University, Flagstaff, AZ, USA

John Yarie Department of Forest Sciences, University of Alaska Fairbanks, Fairbanks, AK, USA

Brian D. Young Department of Natural Sciences, Landmark College, Putney, VT, USA

About the Editors

Dr. Grant R. W. Humphries is an ecological data scientist who has worked on a number of marine and terrestrial projects (mostly seabirds) around the world where machine learning tools were critical to solving complex problems. He has over a decade of experience working with machine learning tools and techniques and loves applying them in novel and interesting ways. He is the founder of Black Bawks Data Science Ltd., a small data science company based in the highlands of Scotland, where he works on building interactive, web-based decision support tools that integrate advanced modeling. He is also a penguin counter, traveling to Antarctica every year to collect data for the Antarctic Site Inventory. His spare time is dedicated to music, cooking, and spending time with his two daughters, River and Dylan, and wife, Alex.

Dr. Dawn R. Magness is a landscape ecologist interested in climate change adaptation, landscape planning, ecological services, and spatial modeling. She earned her M.S. in Fish and Wildlife Science at Texas A & M University and her Ph.D. in the interdisciplinary Resilience and Adaptation Program at the University of Alaska, Fairbanks. Her current projects use multiple methods to assess ecosystem vulnerability to inform strategic adaptation planning. She has conducted research on songbirds, flying squirrels, and American marten.

Dr. Falk Huettmann is a "digital naturalist" linking computing and the Internet with natural history research for global conservation and sustainability. He is a Professor in Wildlife Ecology at the University of Alaska Fairbanks (UAF) Biology & Wildlife Department and Institute of Arctic Biology, where he and many international students run the EWHALE lab. In his lab, he pursues biodiversity, land- and seascapes, the atmosphere, global governance, ecological economics, diseases, and new approaches to global sustainability on a pixel-scale. Most of his 200 publications and 7 books are centered on Open Access and Open Source, Geographic Information Systems (GIS), and data mining/machine learning.

Part I
Introduction

Artwork by Grant Humphries and Catherine Humphries

"The world is one big data problem."
– Andrew McAfee

Chapter 1
Machine Learning in Wildlife Biology: Algorithms, Data Issues and Availability, Workflows, Citizen Science, Code Sharing, Metadata and a Brief Historical Perspective

Grant R. W. Humphries and Falk Huettmann

1.1 Introduction

We can all agree that machine learning has come a long way from Alan Turing's original theories in the 1950s. With many new methods and publications arising, we are seeing a new statistical culture beginning to form (for natural resource applications see Hilborn and Mangel 1997; Breiman 2001a; Hochachka et al. 2007; Drew et al. 2011). This new culture has granted us access to a major tool box that is handy to any practitioner trained in machine learning techniques, frequentist statistics, or neither (Hastie et al. 2009; Fernandez-Delgado et al. 2014). Its success across multiple disciplines and its accessibility means that machine learning as a technique should not be ignored. Machine learning methods outperform other algorithms when it comes to predictive power (and inference), elegance, and convenience (e.g., Elith et al. 2006; Mi et al. 2017), particularly when rapid decision making is required, as is the case in natural resource management. New algorithms and methods for looking at complex ecological data are arising fast; unfortunately, faster than most scholars or tenured professors (for example) can keep up with.

The adoption of machine learning in the ecological community has been slow. Beyond a possible discomfort with unknown methods, perhaps this has also been due to lack of communication between natural scientists and the machine learning community, the absence of machine learning in natural sciences education, or

G. R. W. Humphries (✉)
Black Bawks Data Science Ltd., Fort Augustus, Scotland
e-mail: grwhumphries@blackbawks.net

F. Huettmann
EWHALE Lab, Biology and Wildlife Department, Institute of Arctic Biology, University of Alaska-Fairbanks, Fairbanks, AK, USA
e-mail: fhuettmann@alaska.edu

© Springer Nature Switzerland AG 2018
G. R. W. Humphries et al. (eds.), *Machine Learning for Ecology and Sustainable Natural Resource Management*, https://doi.org/10.1007/978-3-319-96978-7_1

because there is little communication of examples of machine learning methods applied to ecological data (Thessen 2016). However, the literature on applications of machine learning in ecology is growing (e.g., Cushman and Huettmann 2010; Crisci et al. 2012; Thessen 2016) and integrating into ecological informatics (i.e. the use of modern computational techniques and algorithms on problems of ecological interest; Petkos 2003). We aim, through this chapter, to introduce some machine learning topics in an ecological context. We hope you keep these in mind throughout this book to help guide you towards a better understanding of potential applications of machine learning in ecology.

1.2 Some Terminology

Machine learning consists of many algorithms and thus it is a complex and evolving field. Often the language can be new and unusual for anyone not familiar with it. Included here in Table 1.1 is a short glossary of commonly used terms/concepts and their brief definitions. Many of the terms (e.g., categorical or continuous data) will be familiar to ecologists, but some may be new. These refreshers and additions to the ecologists' vocabulary are important because our required skill sets are changing; as are the institutions, nations, and publication landscapes that we work within.

Table 1.1 A basic glossary of machine learning terms that can be found in ecology. These terms will appear throughout the book in various places

Term	Definition
AUC (Area under the ROC Curve)	An evaluation metric for classification problems that consider a variety of classification thresholds. Values greater than 0.8 are considered 'good' in ecology.
BACK PROPAGATION	Mostly used in *neural networks* for performing *gradient descent* (see below) on neural networks. Basically, predictive error is calculated backwards through a neural network graph (the representation of the data flow through neural network nodes). See Rumelhart et al. (1986) for a good description.
BAGGING	Also known as bootstrap aggregating, it is an algorithm designed to reduce variance and over-fitting through model averaging. This is the meta-algorithm used in any random forests implementation. Many *decision trees* are grown (a forest) and can be over-learned (see *over-fitting* below). Individual trees might have high variance, but low bias. Averaging the predictions across the trees reduces the variance and thus reduces over-learning across the whole dataset.
BIG DATA	A term that describes large structured or un-structured datasets most commonly found in business, but extends into ecology through image recognition problems, or high resolution spatial modeling.

(continued)

Table 1.1 (continued)

Term	Definition
BOOSTING	Boosting is a meta-algorithm used to minimize loss in a machine learning algorithm through iterative cross-validation. This is the meta-algorithm used in the generalized boosted regression modeling method (i.e., boosted regression trees). At each step, the predictive error is measured by cross validation, which is then used to inform the next step. This removes issues of over-learning because it is constantly testing itself to ensure predictive power does not decrease. The best model is selected by where the error has been minimized the most over the process.
CATEGORICAL VARIABLE	A variable with a discrete qualitative value (e.g., names of cities, or months of the year). Sometimes, *continuous variables* can be broken down into categories for analysis; however, it is very difficult to create a *continuous variable* from a categorical one.
CLASSIFICATION	A category of *supervised learning* that takes an input and predicts a class (i.e. a *categorical variable*). Mostly used for presence/absence modeling in ecology.
CLUSTERING	A method associated with *unsupervised learning* where the inherent groupings in a dataset are learned without any *a priori* input. For example, principal component analysis (PCA) is a type of unsupervised clustering algorithm.
CONTINUOUS VARIABLE	A variable that can have an infinite number of values but within a range (e.g., a person's age is continuous).
DATA MINING	The act of extracting useful information from structured or unstructured data from various sources. Some people use this as an analogy for machine learning, but they are separate, yet related fields. For example, using a machine learning algorithm to automatically download information from the internet could be considered data mining. *Unsupervised learning* could also be considered data mining, and some machine learning algorithms take advantage of this to build models.
DATA SCIENCE	The study of data analysis, algorithmic development and technology to solve analytical problems. Many ecologists actually classify as data scientists (without even knowing it); particularly those with a quantitative background and strong programming skills.
DECISION TREE	A type of *supervised learning* that is mostly used in *classification* problems but can be extended to regression. This is also known in the ecological literature as CART (classification and regression trees) and is the basis for algorithms like generalized boosted regression modeling (gbm), TreeNet, or random forests.
DEEP LEARNING	Deep learning is advanced machine learning using neural nets for a variety of purposes. It uses vast amounts of data but can output highly flexible and realistic results for simultaneous model outputs. Although not used in ecology yet (save for image recognition), there are vast implications for ecosystem modeling.
DEPENDENT VARIABLE	The dependent variable is what is measured and is affected by the *independent variables*. Also known as: **target variable**, or **response variable**.
ENSEMBLE	A merger of the predictions from many models through methods like averaging.

(continued)

Table 1.1 (continued)

Term	Definition
FREQUENTIST STATISTICS	Refers to statistics commonly used in ecology and elsewhere that are geared towards hypothesis testing and p-values. They aim to calculate the probability of an outcome of an event or experiment occurring again under the same conditions.
GRADIENT DESCENT	A technique to minimize the *loss* (e.g., root mean squared error) by computing the best combination of weights and biases in an iterative fashion. This is the basis for *boosting*.
HOLDOUT DATA	A dataset that is held back independently from the dataset used to build the model. Also known as the **testing** or **validation data**.
IMPUTATION	A technique used for handling missing data by filling in the gaps with either statistical metrics (mean or mode) or machine learning. In ecology, nearest neighbor observation imputation is sometimes used.
INDEPENDENT VARIABLE(S)	This is the variable or set of variables that affects the dependent variable. Also known as the **predictor variable**(s), **covariate(s)** or the **explanatory variable(s)**
MACHINE LEARNING	The core of our book; this refers to the techniques involved in dealing with vast and/or complex data by developing and running algorithms that learn without *a priori* knowledge or explicit programming. Machine learning methods are often referred to as black box, but we argue that this is not the case and that any machine learning algorithm is transparent when one takes the time to understand the underlying equations.
LOSS	The measure of how bad a model is as defined by a loss function. In linear regression (for example), the loss function is typically the mean squared error.
NEURAL NETWORK	A model inspired by the human brain and neural connections. A neuron takes multiple input values and outputs a single value that is typically a weighted sum of the inputs. This is then passed on to the next neuron in the series.
OVER-FITTING	This happens when a model matches training data so closely that it cannot make correct predictions to new data. We argue that the term "over-fitting" is often mis-used in a machine learning context. Fitting suggests that an equation is fit to a dataset in order to explain the relationship. However, in machine learning, this sort of fitting does not occur. Patterns in the data are generally learned by minimizing variance in data through splits or data averaging (e.g. regression trees or support vector machines). A more appropriate term would be "**over-learning**" or perhaps "**over-training**" (both used in the machine learning community) where functions describe data poorly and result into a poor prediction. However, algorithms that use back-propagation and cross-validation limit and, in some cases, eliminate over-learning. In species distribution modeling (using binary presence/absence data), this can be measured using *AUC/ROC*.
RECURSIVE PARTITIONING	This is the technique used to create CARTs (i.e., CART is a type of recursive partitioning, but recursive partioning is not CART), and thus the offshoot algorithms (e.g., boosted regression trees and random forests). The method strives to correctly classify data by splitting it into sub-populations based on *independent variables*. Each sub-population can be split an infinite number of times, or until a threshold or criteria is reached. See Strobl et al. (2009) for a comprehensive overview.

(continued)

Table 1.1 (continued)

Term	Definition
SUPERVISED LEARNING	This type of learning occurs when the algorithm consists of a *dependent variable* which is predicted from a series of *independent variables*.
TESTING SET	See *holdout set* above. This is used to generate independent validation of the model. This can be done iteratively (as in *boosting*).
TRAINING SET	This is the set of data that is used to build the model and is usually independent of the *testing set*. Could also be known as **assessment data**.
UNSUPERVISED LEARNING	The goal of this technique is to model the underlying structure of the data in order to learn more about it without having a *dependent variable*. *Clustering* falls into this category

1.3 A Few Paragraphs on the History of Machine Learning

Machine learning algorithms really had their beginning after the second world war when Alan Turing introduced the concept of thinking machines (Turing 1950). He envisioned a world when computers would be able to actively learn, and 'communicate' new information back to users after that learning process. The 'Turing test' (i.e., the test that demonstrates whether a machine is learning) became the basis for the machine learning discipline that would soon develop. Only a few years after this, the first computer learning program, designed to play checkers, was developed, and learned as a game progressed. This was inspired by the Turing test (Samuel 1959).

Through the 1960s, there was a boom of interest in machine learning, with neural networks (e.g. the 'Perceptron') being a topic of great interest (e.g. for enemy/friend recognition in radar images). However, computational power limited the forward momentum of the subject. This, combined with little to no computational training in life sciences, meant that from the 1970s to the 1990s, during the "boom" of ecology, machine learning, although having been around for some time, was not something really accessible to ecologists. Frequentist statistics and hypothesis testing were the primary tools for life scientists, which made sense at the time because machine learning algorithms had yet to develop to a point where they could be useful. The notion of probabilities, essentially index numbers between 0–1, became globally established with linear regression and logistic functions at its core. Being quantitative essentially meant that everything was approximately expressed as a 0 or 1 using regression formulas borrowed from theoretical probability theory. It made for the mainstay of what is referred to as being scientific and also for textbook natural resource management practices (e.g. Silvy 2012). Although we cannot devalue the importance of baseline ecological studies, quantitative practices were moving faster than ecologists were able to keep up with (O'Connor 2000).

Stock markets boomed and technology advanced, and soon we were in an era of computers and the internet through the 1990s. It was during this time when machine learning shifted away from basic pattern learning towards finding algorithms that could predict complex behavior (See Mueller and Massaron 2016). Humans were collecting vast amounts of data, and traditional statistical techniques were not suited to dealing with the problems being faced. More explicitly, the large sample sizes essentially meant that hypothesis testing was too challenging as tiny differences could lead to "significant" results (p value <0.05), and analysis was computationally expensive.

Through the 2000s, things progressed rapidly with computers getting smaller and more powerful. It was during this period when machine learning (primarily through a few select algorithms) was picked up in ecology. Algorithms like classification and regression trees (CART; Stone 1984) were incorporated early on (De'ath and Fabricius 2000). Soon after, these were refined with boosting and bagging methods outside of the life sciences, leading to boosted regression trees (Friedman 2002) and random forests (Breiman 2001b). Currently, many software companies and programming languages compete for better and more efficient algorithms (see Textbox 1.1 for one such implementation).

A short reference list for wildlife-relevant machine learning methods consists of, but is not limited to: Breiman et al. 1984; Verner et al. 1986; Stockwell and Noble 1992; Stockwell 1994; Boston and Stockwell 1995; Fielding 1999; Phillips and Dudik 2008; Hastie et al. 2009; Mueller and Massaron 2016.

Textbox 1.1 Salford Systems: A Brief and Selective Personal History

Dan Steinberg
Salford Systems (A Subsidiary of Minitab, Inc.)
9685 Via Excelencia, Suite 208, San Diego CA 92126

I had the good fortune to be introduced to the work of Leo Breiman and Jerome Friedman in 1985 while I was an Assistant Professor of Economics at the University of California, San Diego (UCSD). A new UC Berkeley Agricultural Economics Ph.D., Richard Carson, had just joined the Economics Department and he was enthusiastically encouraging everyone to look at CART (Classification and Regression Trees) and MARS (Multiple Adaptive Regression Splines), new analytical tools he had studied at Berkeley. He was also touting the marvelous advantages of cross-validation for assessing the true performance of predictive models.

So far as I know I was the only one in the department who took an interest in these radically new methods and I started experimenting mainly with CART. I had already founded Salford Systems several years earlier to market some statistical and econometric software I had developed while I was a graduate student at Harvard and I was looking for some new tools to expand Salford's offerings. In 1990 I met Leo Breiman at a Joint Statistical Meetings of the American Statistical Association. Leo and I spoke about creating a PC-based commercial version of the CART software. With this conversation began a many-year email correspondence with Leo while I slowly worked my way through the 1984 Classification and Regression Tree monograph (by Leo Breiman, Jerome Friedman, Richard Olshen, and Charles Stone 1984). This correspondence, which also included Richard Olshen 1990–1992, helped me write an easy to follow introduction to the methodology and manual to Salford's new version of the CART software. By 1992 we already had a far easier CLI (command line interface) developed using commands such as.

```
USE MYFILE
MODEL TARGET
KEEP X1 X2 X3 X4
LIMIT ATOM = 10
CATEGORY TARGET X1
CART GO
```

with software available for the major commercial UNIX platforms and the IBM PC.

During 1993 this product was marketed as a SYSTAT add-on module but Salford subsequently took over all responsibility for sales and marketing. We had also begun work on a full GUI interface. The first GUI CART with the trademark Salford navigator and drill down into nodes and other model details was released in 1995. At that time, we began extensive consulting work for three major US banks developing a variety of targeted marketing, credit risk, and mortgage default models on what were fairly large data sets for the time. We often dealt with 100,000 records or more, and typically started with 300 to 2000 predictors (and sometimes more!). Our largest training data sets had around three million records by 1997. Being early adopters of the DEC Alpha hardware line, we were already working with a true 64-bit UNIX 20 years ago, and we made the most of the RAM we could fit into our servers. In 1997 one of our banking customers subsidized our purchase of huge external disk storage for our servers: 100 2GB disk drives plus a number of redundant drives for hot swap replacement of failed drives. Our discount price: $225,000! (20 years later in 2017 you can buy this storage in a pair of $15 thumb drives). The cabinet was a beauty to behold, and it really it looked like it belonged on the space shuttle (Fig. 1.1.1).

Fig. 1.1.1 Salford systems data mining conference, March 2004, San Francisco Bottom row: Leo Breiman, Charles Stone, Richard Olshen, Jerome Friedman. Top row: Richard Carson, Dan Steinberg, Nicholas Scott Cardell

Our work for the banks was exciting in what it taught us about model building and analytics, and it provided a fantastic stress test bed for our software. Between 1990 and 1998, we fixed numerous exotic CART bugs that revealed themselves only when dealing with large complex data sets, and we thus developed a truly industrial strength product. One of our 1998 projects for an automobile manufacturer required us to predict which make and model a new car purchaser would select from a menu of some 400 options, and our battle-hardened CART had no difficulty handling our 250,000 row by 200 column data set with a 400 level target. We continually perfected and adapted the software in hundreds of ways, for example, to handle text data and optional user specified penalties on predictors due to missingness, cardinality, or cost of acquisition. The penalties act as inhibitors on the use of the predictors and made for better performing and more intuitively acceptable models. Many other enhancements were guided by our Director of Research and Development, Nicholas Scott Cardell often in close collaboration with Breiman and Friedman during the 1990s.

Also, in early 1995 I began a new round of email exchanges with Breiman about the 'bagger' (bootstrap aggregation of CART trees). Originally, Breiman wanted to cross-validate every tree in order to prune them all to the right size. I thought that the computational burden was excessive and so I argued for using the OOB (out of bag) data for this purpose; Breiman ultimately also rejected cross-validation to find right sized trees in favor of using the largest possible (unpruned) tree and so the OOB data was leveraged to great advantage in his subsequent work on Random Forests. Salford was the first to implement bagging for decision trees in 1996 under the rubric of the COMBINE command.

Around 1999 I decided that our consulting work was hampering our software development and that we would henceforth focus almost exclusively on new software development. The timing was good for this as we began work on MARS in 1999 and in 2001 we tackled Gradient Boosting (TreeNet). We were well aware of Random Forests from its earliest incarnations, but development lagged behind due to lack of resources. Although 2001 experienced the shocks of the internet bubble crash and September 11th, Salford Systems grew throughout the crisis period in part by expanding our sales effort to include Australia and New Zealand. During the period 2001 through 2006 I toured the world at least 10 times giving lectures and presentations wherever I could, focusing mainly on gradient boosting but also on machine learning in general. My stops included universities, academic and commercial conferences, research groups in pharmaceutical companies and banks, as well as organizations such as the European Central Bank. Much to my surprise, we made very little headway persuading people to take gradient boosting seriously. A few large companies did absorb the message though and in consequence profited greatly from being early and well ahead of their competitors.

By 2007 we finally started to see a substantial increase in interest in our software, particularly by the major banks, and 2008 started to shape up as by far our best growth year ever. However, the general financial collapse put a stop to many of the contracts we were in the midst of negotiating and this also led to a contraction in our business. But internet marketing came to our rescue and we engaged in a successful major project to automate the targeted marketing of web ads in an ad network. We made major advances in such targeted marketing leveraging a highly customized version of CART that included extensive within terminal node analysis. Coupled with a dedicated scoring engine we were able to rank 1000 potential ads in 6 milliseconds. The models were automatically rebuilt every 6 h on learning samples exceeding 1 billion impressions.

In 2010 we embarked on another major application adventure building a predictive modeling system for a large chain of Latin American grocery stores. Working with a network of close to 200 stores and 122,000 products our TreeNet models predicted daily and weekly sales for all promoted products. The final system was fully automatic, rebuilding and refining models as required, and reducing typical sales prediction errors by at least 50%.

After this project we returned to our roots and began developing substantial upgrades to our core learning machines with an even sharper focus on model automation and developing Japanese, and Chinese language versions of SPM. By the time this history is published we will have released yet another major upgrade we are calling SPM8.2 and we will be well on the way to SPM9.0 which raises the automation of complex sequences of analyses to our highest level ever. 2017 and 2018 should also bring our distributed learning machines to market allowing users wanting to analyze data spread out over many servers to leverage the power of a cluster of servers working in parallel with all of the features and capabilities of CART, TreeNet, RandomForests and ensemble model combinations. This latest set of developments is one we are especially eager to deliver as it opens up entirely new horizons for our 26-year research adventure and begins our latest chapter as a subsidiary of statistical and quality control software specialist Minitab.

Acknowledgments The full history involves many more events and people than I have had room to mention in this abbreviated note. Richard Olshen was a constant source of technical advice to us and provided a fabulous overview of Leo Breiman's work and character in his paper in Statistical Science. Charles Stone was instrumental in helping our adventure result in real world results as well as always being available to give us sage technical advice. I hope others understand that the Salford story is principally bound up with Breiman, Friedman, Olshen, and Stone and they remain the principal subject of this brief essay.

Further Readings

Breiman L, Friedman J, Olshen R, Stone C (1984) Classification and regression trees. Pacific Grove, Wadsworth
Breiman L (1996) Bagging predictors. Mach Learn 24:123–140
Friedman JH (1991) Multivariate adaptive regression splines (with discussion). Ann Stat 19:1–141
Friedman JH (1999) Stochastic gradient boosting. Statistics Department, Stanford University, Stanford
Friedman JH (1999) Greedy function approximation: a gradient boosting machine. Statistics Department, Stanford University, Stanford
Olshen R (2001) A Conversation with Leo Breiman Stat Sci 16(2):184–198
Steinberg D, Colla P (1995). CART: Tree-structured non-parametric data analysis. Salford Systems, San Diego
Seigel E (2016) Predictive analytics: the power to predict who will click, buy, lie, or die, Wiley Publisher, New York

1.4 Machine Learning in Ecology and Wildife Biology to Date

In this book, we will present several cases of how machine learning can be used effectively in ecological studies. It expands on a growing body of literature demonstrating how machine learning can help ecologists and natural resource managers (e.g., Recknagel 2001; Olden et al. 2008; Crisci et al. 2012; Thessen 2016; Valletta et al. 2017), and picks up after Fielding's seminal book on machine learning in ecology when it was still in its infancy in our field (Fielding 1999). We have a good understanding of the current state of machine learning in ecology thanks to these aforementioned works and can therefore start plotting a course forward in the implementation of these methods.

In the disciplines of natural resource management, ecology, and conservation biology, machine learning remains somewhat absent as evidenced by the lack of discussion on the topic in current textbooks: Primack (2010) for conservation, Miller and Spoolman (2012) for environmental studies, Silvy (2012) for wildlife biology, Gill (2007) for ornithology and even in animal physiology (Moyes and Schulte 2007). Most of the advancements in machine learning in ecology lie in the scientific literature (e.g., Elith et al. 2006; Hochachka et al. 2007; Drew et al. 2011). Thus, prior to graduate studies, machine learning techniques for ecological data are inaccessible to students unless they make a point of learning it on their own or take a specialized course. Beyond this, it is only recently that computing methods have begun to be taught in University courses for ecologists. More and more, "quantitative ecology" courses (usually taught with the R programming language), are becoming commonplace. However, machine learning methods are not always integrated into these courses, and when they are, they are often glossed over with regards to the fine-scale detail of their inner-workings.

This disconnect between machine learning, computing, ecology and science-based conservation management has led to missed opportunities in our opinion at a time when the environment is facing rapid changes and efficient solutions are required. While we press forward with massive resource exploitation for the sake of "modernity" (Alexander 2013), biodiversity is being lost at alarming rates. This is partly because we have yet to adopt precautionary approaches to management, but also possibly because we spend valuable time debating mechanisms and methods instead of combining or comparing all available methods for a full understanding of our ecosystem. This has led us far away from embracing holistic views of natural resource management (Rosales 2008; Sandifer et al. 2015) that may be possible by integrating our thinking with machine learning.

Machine learning can easily be used in a huge array of applications (Hastie et al. 2009; Mueller and Massaron 2016); the method represents a true paradigm shift in statistics, wildlife management and conservation. Data mining, predictions and classifications are amongst its most common applications (Hastie et al. 2009); specifically, when data are complex, not normally distributed (statistically), and 'messy' (as is typical in ecological and environmental data; McArdle 1988; Breiman 2001a). Machine learning also works on big and small datasets, lending to their versatility across many fields. The application of machine learning to spatial and global data carries a specific benefit in that it can help to show the lack of space and resources available on earth (Wackernagel et al. 2002; Humphries and Huettmann 2014). This could potentially resolve these space and resource conflicts with a fast turn-around time ("rapid assessment", Huettmann 2007; Kandel et al. 2015 for an example). Whereas, traditional reductionist views have failed to deliver us successful solutions in this respect (as is obvious with respect to the state of the environment, Rockstroem et al. 2009, Mace et al. 2010). A shift in our strategy for wildlife management is needed.

Currently, machine learning in ecology is mostly restricted to species distribution modeling (SDM; Elith et al. 2006, Guisan and Thuiller 2005, and many others) where georeferenced occurrences or abundances of species are associated with layers of environmental data (e.g., topography, precipitation). Those associations are modeled with algorithms like maxent (Phillips et al. 2006), or random forests (Breiman 2001b) and then predicted to the spatial and temporal extent of those layers of environmental data to determine where species might occur in space (and time). This has been boosted by increases in computing power, allowing us to model more complex data at higher temporal and spatial resolutions (Cushman and Huettmann 2010; Watson et al. 2016). See Textbox 1.2 for a brief example.

Beyond SDMs, machine learning has been increasingly used in species recognition applications. These could be audio recognition exercises (Armitage and Ober 2010; Stowell and Plumbley 2014, Chap. 15), or species recognition from images (Goodwin et al. 2014; Rosa et al. 2016). Machine learning algorithms have also been applied to a lesser degree to animal behavior (e.g., Valletta et al. 2017), and population dynamics modeling (e.g., Recknagel et al. 2002). A review of some of these and other applications can be found in Thessen (2016).

Textbox 1.2 Cheer Pheasant predictions in the Hindu-Kush Himalaya: What does machine learning bring to the table for wilderness regions, endangered species and the future?

Falk Huettmann and co-authors[1]
EWHALE lab, University of Alaska Fairbanks.
e-mail: fhuettmann@alaska.edu

Wilderness regions of this world are precious for mankind, but they often lack protection and support. Still, they are the ultimate goal when it comes to natural resource conservation management. Wilderness is nature at its finest, and it needs to be assessed and managed for future generations. The state of wilderness (untouched habitat by humans) and its global extent reflects how humans manage the Anthropocene. But how does that link with machine learning, so far?

Well, while still at its infancy, machine learning has already been used successfully in wilderness region management worldwide (Table 1.2.1 for overview).

The Cheer Pheasant (*Catreus wallichii*)) is a great example of a rare bird species of global concern living in the wilderness of the western Hindu Kush Himalaya region, at around 3000 m of altitude. Its life history is somewhat known (Garson et al. 1992), but probably quite outdated, and its exact distribution lacks a lot of information, with animals being dispersed over many countries. Besides new threats such as climate change, this species is strongly hunted and poached, and some local cultures see it as a medical cure against certain ailments. The global population of the species is declining (BirdLife International 2017), and its conservation status is likely to decay.

In an ongoing study (Kandel et al. *in prep*) we found that machine learning has helped to compile all publicly available data for this species, and then sparked the

Table 1.2.1 Machine learning applications in wilderness regions and their future

Wilderness Area	Application	Machine learning citation
Alaska (largest holder of protected and wilderness areas in the U.S.)	Spatial information, noise impact	Ohse et al. (2009) Mullet et al. (2015)
Ross Sea (one of the most remote and pristine oceanic areas in the world)	Overview of data available, Marine Protected Area (MPA) decision, biodiversity modeling of Southern Ocean	Reygondeau and Huettmann 2014 Huettmann et al. (2015)
Antarctica	Open access data and models for management	Huettmann and Schmid (2014a, b)
Amazonia	Biodiversity modeling	Buermann et al. (2008)
Hindu-Kush Himalaya	Red Panda habitat prediction	Kandel et al. (2015)
Northern Asia	Crane predictions	Chunrong et al. (2017)
Chinese steppe and farmlands	Future of great bustard subspecies and habitats (niche) predictions	Chunrong et al. (2016)

[1] This work was carried out with the Global Primate Network in Nepal, namely Ganga Ram Regmi, Madan Krishna Suwal, Dikpal Krishna Karmacharya, Kamal Kandel and Sonam Tashi Lama.

interest on the issue of cheer pheasant conservation and put it on the wider agenda. Some progress might be made to help bring this species back using some innovative machine learning techniques (e.g., automated occurrence/density sampling in space/time, holistic species distribution modeling) The findings from this study further support the need for science-based management (Fig. 1.2.1).

It is easy to understand that machine learning provides progress to a subject that stalled for over a decade and where the usual set of quantitative analysis cannot provide much further progress. The projects and analyses shown in Table 1.2.1 do not only provide analytical advances but also emphasize the need to improve data collection and management. Perhaps this is one of the biggest improvements provided by machine learning. But as Table 1.2.2 shows, machine learning has not reached its potential yet and more is to be done, it must, if we want to maintain the biodiversity in the globe and its ecological services.

Last but not least, machine learning can be used for future predictions (e.g. Chunrong et al. 2016, Suwal et al. in review), but also for adaptive management progress (Huettmann 2007). We find that, by now, machine learning has shown its benefits and applicability in conservation and awaits urgently to become the analytical platform of choice, worldwide.

Fig. 1.2.1 Cheer pheasant (*Catreus wallichii;* left) and its habitat (right) in the Himalayan region

Table 1.2.2 A selection of machine learning applications and steps of future relevance for natural resource management applications (for citations and applications see chapters throughout this book and references within). These applications can be applied anywhere around the world and at any temporal or spatial scale pending data availability

Application topic	How done	Why needed
Data mining	Mining of compiled databases	Obtain best available information
Species distribution modeling (SDM)	Species-habitat models	Spatial information as the basis of management
Remote sensing-based habitat classification	Classification algorithm	Detailed habitat management requires 'maps'
Climate models	Climate data numerical models	Climate as the key topic for human well-being
Forecasting	Prediction	Pre-cautionary management
Impact predictions	Prediction	Pre-cautionary management

References

BirdLife International (2017) *Catreus wallichii* (amended version of 2016 assessment). The IUCN Red List of threatened species 2017

Buermann W, Saatchi S, Smith TB, Zutta BR, Chaves JA, Milá B, Graham CH (2008) Predicting species distributions across the Amazonian and Andean regions using remote sensing data. J Biogeog 35:1160–1176

Chunrong M, Huettmann F, Guo Y (2016) Climate envelope predictions indicate an enlarged suitable wintering distribution for Great Bustards (*Otis tarda dybowski*) in China for the twenty-first century. PeerJ 4:e1630

Chunrong M, Huettmann F, Guo Y, Han X, Wen L (2017) Why choose Random Forest to predict rare species distribution with few samples in large undersampled areas? Three Asian crane species models provide supporting evidence. PeerJ 5:e2849

Garson PJ, Young I, Kaul R (1992) Ecology and conservation of the Cheer pheasant (*Catreus wallichii*): Studies in the wild and the progress of a reintroduction project. Biol Cons 59:25–35

Huettmann F (2007) Modern adaptive management: adding digital opportunities towards a sustainable world with new values. Forum on public policy: climate change and sustainable development 3:337–342

Huettmann F, Schmid M (2014a) Climate change and predictions on pelagic biodiversity components. In: De Broyer C, Koubbi P (eds), with Griffiths H, Danis B, David B, Grant S, Gutt J, Held C, Hosie G, Huettmann F, Post A, Ropert-Coudert Y, van den Putte A. The CAML/SCAR-MarBIN BIOGEOGRAPHIC ATLAS OF THE SOUTHERN OCEAN. Scientific Committee on Antarctic Research (SCAR), Cambridge, pp 390–396

Huettmann, F, Schmid M (2014b) Publicly available open access data and machine learning model-predictions applied with open source GIS for the entire Antarctic Ocean: a first meta-analysis and synthesis from 53 charismatic species. In: Veress B, Szigethy J (eds) Horizons in earth science research. 11(3):24–34

Huettmann F, Humphries GRW, Schmid M (2015) A first overview of open access digital data for the Ross Sea: complexities, ethics and management opportunities. Special Issue Ross Sea, Hydrobiologia 761:97–119

Kandel K, Suwal MS, Regmi GR, Lama ST, Karmacharya DK, Baral HS, Huettmann F (in prep) Open access and open source distribution and occupancy modelling of the globally threatened cheer pheasant *Catreus wallichii* in Nepal: implications for strategic conservation planning Global Primate Network-Nepal

Kandel K, Huettmann F, Suwal MK, Regmi GR, Nijman V, Nekaris KAI, Lama ST, Thapa A, Sharma HP, Subedi TR (2015) Rapid multi-nation distribution assessment of a charismatic conservation species using open access ensemble model GIS predictions: Red panda (*Ailurus fulgens*) in the Hindu-Kush Himalaya region. Biol Cons 181:150–161

Mullet TC, Gage SH, Morton JM, Huettmann F (2015) Temporal and spatial variation of a winter soundscape in south-central Alaska. Landsc Ecol 31:1117–1137

Ohse B, Huettmann F, Ickert-Bond S, Juday G (2009) Modeling the distribution of white spruce (*Picea glauca*) for Alaska with high accuracy: an open access role-model for predicting tree species in last remaining wilderness areas. Pol Biol 32:1717–1724

Reygondeau D, Huettmann F (2014) Past, present and future state of pelagic habitats in the Antarctic Ocean. In: De Broyer C, Koubbi P (eds), with Griffiths H, Danis B, David B, Grant S, Gutt J, Held C, Hosie G, Huettmann F, Post A, Ropert-Coudert Y, van den Putte A. The CAML/SCAR-MarBIN BIOGEOGRAPHIC ATLAS OF THE SOUTHERN OCEAN. Scientific Committee on Antarctic Research (SCAR), Cambridge, pp 397–403

Suwal MK, Huettmann F, Regmi GR, Vetaas OR (in review) Parapatric subspecies of *Macaca assamensis* show a marginal overlap in the predicted climate niche: Some elaborations for a modern conservation management. Ecol Evol

1.5   Algorithms as a Bottleneck for Wildlife Conservation

If simply judged by the use and publications in the wildlife sciences, some of the major machine learning algorithms might be maxent (Phillips et al. 2006), GARP (Stockwell and Noble 1992; Stockwell 1994; Stockwell 1999) and random forests (Elith et al. 2006), all while linear regressions are still ruling the analytical skill set of most wildlife conservation practitioners. Bayesian analytics is growing but only carries a small share of the market and is rarely used for high-performance predictions (Elith et al. 2006).

But outside of the wildlife and conservation discipline the picture looks very different. Most software companies are using machine learning for business solutions. Market analysis is done with machine learning, so are web searches and social media analyses to great success (see for example, the Cambridge Analytica scandal of 2018 and the 2016 United States general election; arguably won due to the influence of machine learning targeting advertisements to specific groups of people). Thus, the reality is that machine learning as a discipline is much bigger than just maxent, GARP or random forests. The discipline of SDM (Guisan and Zimmermann 2000), widely centered on the maxent algorithm designed for point-and-click usage, has been limited by lack of exploration into alternatives (N.B. open source implementations of maxent are becoming available in the R programming language; Phillips et al. 2017). Some overviews and applications of new analytical approaches in wildlife conservation include: Elith et al. (2006), Jiao et al. (2016), Hochachka et al. (2007), Humphries and Huettmann (2014), Thessen (2016) and Huettmann and Ickert-Bond (2017).

There are easily over 100 machine learning algorithms (including their derivatives in various programming languages; Fernandez-Delgado et al. 2014). Therefore, it is easy to be quite liberal in the choice of the modeling algorithm and use whatever works best (i.e., what gives us the best predictions). We know that each algorithm has its pros and cons, and ideally, an entire ensemble of models is to be used to help average out any model bias. Here we list several broad categories of machine learning algorithms that have been applied in wildlife biology and ecology (Table 1.2).

1.6   Data Issues and Availability Related to Data Mining and Machine Learning

One of the biggest criticisms that the machine learning community receives is that these algorithms are 'black boxes' (Craig and Huettmann 2008). This suggests that we arbitrarily input data into an algorithm, and then some sort of 'voodoo magic' occurs, and an output is given. This is an unfortunate misnomer and has led to "machine learning" being considered a 'dirty term' in some circles. However, we posit here that we use 'black boxes' all the time (e.g., cars, computers, mobile phones, social media, etc....). The term itself is quite subjective, particularly at a

Table 1.2 A selection of machine learning algorithms and associated descriptions most commonly used in ecology and wildlife biology

Algorithm	Description
MAXIMUM ENTROPY	Maxent is a specific software package that has been used almost exclusively for species distribution modeling, and the various wrappers for this algorithm have been designed as such (Phillips et al. 2006; Phillips et al. 2017). The input tends to be a series of presence-only locations and environmental covariates. Maxent calculates a set of constraints on the environmental covariates and then estimates the probability of presence by measuring uncertainty within the set constraints. Maxent works by 'thinking' in terms of covariate space as opposed to geographic space (i.e. we guess a species occurs in a certain range of values, say temperature, in proportion to the availability of those range of values). A full statistical description can be found in Elith et al. (2011).
CLASSIFICATION AND REGRESSION TREES (CART)	This is the earliest of the tree-based methods (Breiman et al. 1984) and forms the basis for both *boosted regression trees* and *random forests*. CART works by randomly selecting variables and splitting the variable space (e.g., the x-y plane of dependent vs independent variables). The variance of the data on each side of the split is measured and recursive partitioning is applied to determine the split that minimizes the variance of the two data 'clouds'. This split forms a sort of 'if/then' conditional statement (e.g., if a data value of variable x > 2, then go down the left branch, or if the data value of variable x < 2 go down the right branch). A series of these branches form a tree, which can be queried for the most important predictors (independent variables). Predictions to data are made by applying the rule set to new data points.
BOOSTED REGRESSION TREES	This algorithm is derived directly from CART, and is essentially a series of iterated trees, where at each iteration the error is minimized through the application of a loss function (e.g., root mean squared error). The point of this method is to begin with poorly performing variables and trees, and re-fit those using the residuals from the previous model. Predictive performance is measured at each iteration and when performance starts getting worse, the iterating stops. The final tree ends up being a linear combination of all the trees and has the lowest amount of error (Friedman 2002; De'ath 2007). Since many trees are used, it can be seen as a sort of ensemble model (although not in the traditional sense in that there is no model averaging). There is a clever extension and optimization of this approach in some commercial algorithms and can be linked with 'bagging' techniques (see details below). Those latter concepts tend to be among 'the best' available (i.e., make the best predictions and are the easiest to interpret).
RANDOM FORESTS	This method is derived from applying a 'bagging' method to CART (Breiman 2001b) and is probably the most successful (or most popular) machine learning algorithm used in ecology (a list of papers would be too extensive but see Cutler et al. 2007 for an introduction to their use in ecology). It is an ensemble modeling method based on a specific bootstrapping technique. Many large trees (many branches) are constructed by sampling with replacement, and the final model ends up being an average of all those trees (by votes in classification schemes or averaging in regression schemes). Random forests improves on bagging by making the splitting process more efficient through the use of out-of-bag data instead of computationally expensive k-fold cross-validation. Many implementations of random forests exist on varying platforms.

(continued)

Table 1.2 (continued)

Algorithm	Description
GENETIC ALGORITHMS	Genetic algorithms are derived from the process of evolution wherein competing solutions 'evolve' over time until an optimal solution is reached (Holland 1975; Olden et al. 2008; Fernández et al. 2010). These have been used in ecology to a limited degree under the acronym GARP (genetic algorithm for rule set production; Stockwell and Noble 1992). GARP has only really be used in ecological niche modeling to date (Elith et al. 2006; Peterson et al. 2007) but could be extended to other problem sets. Genetic algorithms work by creating a series of random 'solutions', which are then 'mutated'. The best solutions are selected from these and then recombined or re-'mutated'. However, current implementations of genetic algorithms lag behind other machine learning tools and traditional statistical techniques can outperform these due to their tendency to over-learn (over-fit) the data (Olden et al. 2008).
Bayesian machine learning	Based on Bayes' theorem (Laplace 1986; one of the oldest statistical concepts), these methods work by building an understanding of systems through the expression of probabilities and updating of those probabilities with new evidence. In a machine learning context, Bayesian methods can be applied to classification, where the likelihood of membership to each of the classes is calculated for each of our data points, and new data are assigned to the class with the highest likelihood. This method can give good results with few training data (Kotsiantis et al. 2006) as they use "prior" information on the parameters of the model, that help inform the outcome. However, a good ecological knowledge of the system at hand is not always possible, and the choice of prior distributions often require a 'best guess'.
SUPPORT VECTOR MACHINES	This method is not commonly used in ecology, but it has some merit for both classification and regression. The essence of support vector machines lies in x-y planes, where data are separated by straight lines. The lines are created by making the margins the largest possible difference between all the points (see Fig. 6 in Thessen 2016). Data are classified or predicted based on which side of the margin they fall into. These are trained in an iterative fashion and can be tuned using other functions in cases where data are 'messy' (as in ecological data). See Kotsiantis et al. (2006) for a detailed description of support vector machines.
ARTIFICIAL NEURAL NETWORKS	This 'machine learning algorithm' is the basis for much of the image and vocal recognition software that exists currently, and also the basis for many of the artificial intelligence algorithms that currently dominate the technological world. With the advent of 'deep learning', neural networks have taken the center stage in the machine learning community but are actually one of the first machine learning methods to gain popularity in the 1970s and again the early 1990 (backward propagation etc). They work by simulating the way the human brain processes information (Recknagel 2001; Hsieh 2009). Input data are taken into a series of 'nodes' in the form of the independent variables. These data are weighted in the links between nodes and then passed to the next level, where information about those data are extracted, weighted, and passed to a next set of nodes. This could be viewed as a sort of 'conveyor belt', except using back-propagation, errors can be corrected. These algorithms are exceptionally difficult to program (though this is changing rapidly) and are very efficient for high dimensional data. See Hagan et al. (2014) for a detailed description of artificial neural networks.

time when a plethora of easy-to-follow tutorials exist for free on the internet (see for instance   https://www.r-bloggers.com/in-depth-introduction-to-machine-learning-in-15-hours-of-expert-videos/ for an in-depth breakdown of machine learning or Hastie et al. 2009).

There is no mystery to machine learning algorithms given enough time to study their inner workings. They have been programmed by people with a thorough understanding of machine learning tools and can be decoded and re-traced. We believe the 'black box' argument is incorrect as an objective a criticism of machine learning. While we appreciate the difficulty in having to decode these methods, particularly without the computational training required to do so (we have all been there, and continue to learn to this day), a fairer thing to say would be "machine learning is a black box to me, and I should strive to learn about it".

Sometimes, the 'black box' argument turns into "where is the code?". This could be turned around to GAMs, GLMs, GLMMs, LMs, and other frequentist methods as well. We put inputs in, we read the and interpret the output, but there are few people who know the exact inner workings of the code (save for those fluent in programming languages). The only difference is that frequentist methods have been around for so long, and that the mathematical equations are taught to us as students, so we claim to understand them better. This would easily change if the inner workings of machine learning were taught the same way. Either way, this is why transparency is key in the sciences. If a machine learning scientist can show the code from input to predictions (from scratch, without using any pre-packaged libraries in R, for example), while someone else comes along and simply uses the 'lm' (linear modeling) function in R, which one is a 'black box'? The point we strive to make here, as stated above, is that the term 'black box' is really not objective. Getting past the notion of the 'black box' as an objective criticism and turning it into a 'transparent box'- or at least a grey box - would greatly help ease ecologists into the use of machine learning algorithms.

Open access code and data are a must for data mining and machine learning work. Not just to get past the stigma of the 'black box', but also for scientific transparency. Science operates (or at least, in our opinion, it should) on the tenet that another scientist should be able to come along and replicate experiments exactly and get the same results. This ensures that our work remains truthful and honest (e.g. Zuckerberg et al. 2011). To do this, the entire workflow (i.e. code and data) must be available and formatted in a way that someone could come along and easily run the analysis. We would recommend that the data sources are adequately referenced and if possible, a datafile (either a database or a flat spreadsheet) with all the pertinent data are provided. This is not typically done, however, and certainly not with well documented metadata. An argument for this has been out of fear of 'scooping', which is not documented and very rare to occur in ecology. In some cases where data could be used by malevolent members of the public to harm a species, there is some argument towards restricting data but making it available to other scientists (Tulloch et al. 2018).

The metadata associated with a dataset is the data that describes its format and workflow (i.e., the description of how data and code are applied to create the output) as well as contact information for those involved in its creation (Huettmann 2015). Although this concept has been around for more than 10 years, metadata is rarely included or mandated upon submission to most journals. Although a somewhat painstaking process to some people, it is another important part of science to ensure transparency (Huettmann 2007). A smart budget for any project should take this, and other data management techniques, into account (i.e., by including salary time for preparing metadata and curating code and data sources).

Data management in and of itself requires a whole separate book, and its importance in ecology is highly under-stated (Zuckerberg et al. 2011; Huettmann 2015). Within our field, there is not a 'best practice' guide, but it is greatly needed. Good data management techniques such as proper computer filing systems, documenting code, documenting spreadsheets, metadata, backing up, redundant hard drives, high quality software and hardware, etc. ensure long-term survival of data and the science we do. It is a form of scientific accounting, which is non-existent in our current system.

1.7   Workflows

With the above-said, entire workbenches (i.e., code, or graphic user interfaces for running machine learning algorithms) exist, which means that if a scientist documents which software version is used (and the exact settings) and provides their dataset, the work can be repeated (even if we do not have the underlying code). A classic workbench for species distribution modeling is OpenModeler (Sutton et al. 2007), which uses data from the Global Biodiversity Information Facility (GBIF; Flemons et al. 2007). However, there are others out there (e.g., Maxent, Salford's predictive modeling suite, or AZURE with Microsoft), some of which are 100% free. However, as we have stated, ensuring all the relevant settings are reported is incredibly important for various reasons, e.g. reputation, science accuracy, liability and sustainability. We would advocate more for the use of coding whenever possible.

When using large datasets or doing computationally heavy calculations, tasks are frequently farmed out to distributed networks (the cloud), high performance computers (super computers), or clusters (groups of local computers working together). When it comes to workflows, this is where things start to become very complex. The cloud is amorphous as it is (Chap. 18), and how it exactly splits computations across multiple machines is somewhat mysterious to non-developers. This provides an interesting conundrum for us as ecologists because aside from saying "this process was run on a super computer", we are generally at a loss as to the computing details. Furthermore, most supercomputers require specialized access, therefore meaning that the analysis in question is not easily repeatable. If computations are sent to the cloud, for example, the exact method for doing this should be described (e.g., if a cloud service that creates a virtual machine is used, the details of that virtual machine are important).

1.8 Citizen Science

One of the more common problems in ecology is what to do when data are scarce. We argue that we should be looking for any evidence we can get, from anywhere we can get it. While citizen science gets widely criticized for its various design issues according to classic science, the reality is that it floods web portals (e.g., eBird and GBIF). The amount of information in these portals easily classifies as big data, and we, as ecologists, are not prepared (quantitatively or philosophically) to deal with the quantity of data now presented to us. Citizen science data is extremely complex and widely under-analyzed, so much so that the use of machine learning analyses (e.g. Artificial Neural Networks) becomes almost a requirement in order to handle the challenges in these datasets (*sensu* Huettmann and Ickert-Bond 2017).

The critics of citizen science are slowly realizing that the collection of data by the general public does not take away from traditional scientific methods, but actually adds to them (see the hashtag #CougarOrNot on Twitter for an example, a 'game' by M LaRue for getting people to identify cougars from cameras by popular vote and expert opinion). This is important to note because funding issues are becoming increasingly problematic with the rise of Nationalism in developed countries and other problems. We need to leverage as many data sources as possible for more efficient progress of conservation efforts. Although citizen science data has its challenges, there is good information to be extracted from it, and a machine learning framework can be very helpful in providing good generalizations. The need for techniques to handle big data (from citizen science for example) is one of many things that is guiding us as ecologists towards new quantitative methods for understanding complex ecosystems.

1.9 A Great Future could be around the Corner, Waiting for you Online, and in the Wilderness of this World

With that, we would like to invite the reader to browse through the provocative chapters that we present, and celebrate machine learning opportunities as applied to species and habitat data. We hope the reader finds these examples and techniques as exciting as we do and applies them to advance natural resource management and sustainability world-wide. As this field is developing quickly we welcome any feedback and updates in the light of decaying wilderness and global habitats. As we aim to do in this book and its subsequent chapters, we encourage everybody to engage and interact with us and the wider global community on the topic of machine learning, especially towards applications for global wildlife conservation management and associated human well-being.

Acknowledgements We would like to thank all the proponents of machine learning algorithms and their use in ecology. The list of people to thank for the discussions in this chapter is too extensive, but without them, our work would not be possible. Special thanks to V Morera and M Garcia Reyes for reviewing this chapter and providing insightful comments.

References

Alexander JC (2013) The dark side of modernity. Polity Publishers, New York

Armitage DW, Ober HK (2010) A comparison of supervised learning techniques in the classification of bat echolocation calls. Ecol Info 5(6):465–473

Boston AN, Stockwell DRB (1995) Interactive species distribution reporting, mapping and modelling using the world wide web. Computer Networks and ISDN Systems 28:231–228

Breiman L (2001a) Statistical modeling: the two cultures (with comments and a rejoinder by the author). Stat Sci 16:199–231

Breiman L (2001b) Random forests. Mach Learn 45:5–32

Breiman L, Friedman J, Stone CJ, Olshen RA (1984) Classification and regression trees. Taylor & Francis, New York

Craig E, Huettmann F (2008) Using "blackbox" algorithms such as Tree Net and random forests for data-mining and for finding meaningful patterns, relationships and outliers in complex ecological data: an overview, an example using golden eagle satellite data and an outlook for a promising future, Chapter IV. In: Wang H-f (ed) Intelligent data analysis: developing new methodologies through pattern discovery and recovery. IGI Global, Hershey, pp 65–83

Crisci C, Ghattas B, Perera G (2012) A review of supervised machine learning algorithms and their applications to ecological data. Ecol Model 240:113–122

Cushman SA, Huettmann F (2010) Spatial complexity, informatics, and wildlife conservation. Springer, Tokyo

Cutler DR, Edwards TC, Beard KH, Cutler A, Hess KT, Gibson J, Lawler JJ (2007) Random forests for classification in ecology. Ecology 88(11):2783–2792

De'ath G, Fabricius K (2000) Classification and regression trees: a powerful yet simple technique for ecological data analysis. Ecology 81(11):3178–3192

De'ath G (2007) Boosted trees for ecological modeling and prediction. Ecology 88(1):243–251

Drew CA, Wiersma Y, Huettmann F (2011) Predictive species and habitat modeling in landscape ecology. Springer, New York

Elith J, Graham CH, Anderson RP, Dudík M, Ferrier S, Guisan A, Hijmans RJ, Huettmann F, Leathwick JR, Lehmann A, Li J, Lohmann LG, Loiselle BA, Manion G, Moritz C, Nakamura M, Nakazawa Y, Overton JMM, Peterson AT, Phillips SJ, Richardson K, Scachetti-Pereira R, Schapire RE, Soberón-Mainero J, Williams S, Zimmermann NE (2006) Novel methods improve prediction of species' distributions from occurrence data. Ecography 29:129–151

Elith J, Phillips SJ, Hastie T, Dudík M, Chee YE, Yates CJ (2011) A statistical explanation of MaxEnt for ecologists. Divers Distrib 17(1):43–57

Fernández A, García S, Luengo J, Bernadó-Mansilla E, Herrera F (2010) Genetics-based machine learning for rule induction: state of the art, taxonomy, and comparative study. IEEE Trans Evol Comput 14(6):913–941

Fernandez-Delgado M, Cernades E, Barro S, Amorim D (2014) Do we need hundreds of classifiers to solve real world classification problems? J Mach Learn Res 15:3133–3181

Fielding A (1999) Machine learning methods for ecological applications. Springer, New York

Flemons P, Guralnick R, Krieger J, Ranipeta A, Neufeld D (2007) A web-based GIS tool for exploring the world's biodiversity: the global biodiversity information facility mapping and analysis portal application (GBIF-MAPA). Ecol Inform 2(1):49–60

Friedman JH (2002) Stochastic gradient boosting. Comp Stat & Data Anal 38:367–378

Gill FB (2007) Ornithology, Third edn. W. H. Freeman & Co., New York

Goodwin JD, North EW, Thompson CM (2014) Evaluating and improving a semi-automated image analysis technique for identifying bivalve larvae. Limnol Oceanogr Methods 12(8):548–562

Guisan A, Thuiller W (2005) Predicting species distribution offering more than simple habitat models. Ecol Lett 10:993–1009

Guisan A, Zimmermann NE (2000) Predictive habitat distribution models in ecology. Ecol Model 135:147–186

Hagan MT, Demuth HB, Beale MH, Jesus Od (2014) Neural network design. Martin Hagan, 1012 pp

Hastie T, Tibshirani R, Friedman J (2009) The elements of statistical learning: data mining, inference, and prediction. Springer Series in Statistics

Hilborn R, Mangel M (1997) The ecological detective: confronting models with data. Princeton University Press, Princeton

Hochachka W, Caruana R, Fink D, Munson A, Riedewald M, Sorokina D, Kelling S (2007) Data mining for discovery of pattern and process in ecological systems. J Wildl Manag 71:2427–2437

Holland J (1975) Adaptation in natural and artificial systems: an introductory analysis with applications to biology, control, and artificial intelligence. MIT Press, Cambridge, 211 pp

Hsieh W (2009) Machine learning methods in the environmental sciences. Cambridge University Press, Cambridge, 349 pp

Huettmann F (2007) Modern adaptive management: adding digital opportunities towards a sustainable world with new values. Forum Public Policy: Clim Chang Sustain Dev 3:337–342

Huettmann F (2015) On the relevance and moral impediment of digital data management, data sharing, and public open access and open source code in (tropical) research: the Rio convention revisited towards mega science and best professional research practices. In: Huettmann F (ed) Central American biodiversity: conservation, ecology, and a sustainable future. Springer, New York, pp 391–418

Huettmann F, Ickert-Bond S (2017) On open access, data mining and plant conservation in the circumpolar north with an online data example of the Herbarium, University of Alaska Museum of the North. Arc Sci

Humphries GRW, Huettmann F (2014) Putting models to a good use: a rapid assessment of Arctic seabird biodiversity indicates potential conflicts with shipping lanes and human activity. Divers Distrib 20(4):478–490

Jiao S, Huettmann F, Guoc Y, Li X, Ouyang Y (2016) Advanced long-term bird banding and climate data mining in spring confirm passerine population declines for the northeast Chinese-Russian flyway. Glob Planet Chang 144:17–33

Johan Rockström, Will Steffen, Kevin Noone, Åsa Persson, F. Stuart Chapin, Eric F. Lambin, Timothy M. Lenton, Marten Scheffer, Carl Folke, Hans Joachim Schellnhuber, Björn Nykvist, Cynthia A. de Wit, Terry Hughes, Sander van der Leeuw, Henning Rodhe, Sverker Sörlin, Peter K. Snyder, Robert Costanza, Uno Svedin, Malin Falkenmark, Louise Karlberg, Robert W. Corell, Victoria J. Fabry, James Hansen, Brian Walker, Diana Liverman, Katherine Richardson, Paul Crutzen, Jonathan A. Foley, (2009) A safe operating space for humanity. Nature 461 (7263):472–475

Kandel K, Huettmann F, Suwal MK, Regmi GR, Nijman V, Nekaris KAI, Lama ST, Thapa A, Sharma HP, Subedi TR (2015) Rapid multi-nation distribution assessment of a charismatic conservation species using open access ensemble model GIS predictions: red panda (*Ailurus fulgens*) in the Hindu-Kush Himalaya region. Biol Conserv 181:150–161

Kotsiantis SB, Zaharakis ID, Pintelas PE (2006) Machine learning: a review of classification and combining techniques. Artif Intell Rev 26(3):159–190

Laplace PS (1986) Memoir on the probability of the causes of events. Stat Sci 1(3):364–378

Mace G, Cramer W, Diaz S, Faith DP, Larigauderie A, Le Prestre P, Palmer M, Perrings C, Scholes RJ, Walpole M, Walter BA, Watson JEM, Mooney HA (2010) Biodiversity targets after 2010. Environ Sustain 2:3–8

McArdle BH (1988) The structural relationship: regression in biology. Can J Zool 66(11):2329–2339

Mi C, Huettmann F, Guo Y, Han X, Wen L (2017) Why to choose random forest to predict rare species distribution with few samples in large undersampled areas? Three Asian crane species models provide supporting evidence. PeerJ 5:e2849

Miller GT, Spoolman SE (2012) Living in the environment. Brooks/Cole Publishers, New York

Moyes CD, Schulte PM (2007) Principles of animal physiology, 2nd edn. Pearson Publishers

Mueller JP, Massaron L (2016) Machine learning for dummies. John Wiley & Sons

O'Connor RJ (2000) Why ecology lags behind biology. Scientist 14(Part 20):35 16 Oct 2000

Olden JD, Lawler JJ, Poff NL (2008) Machine learning methods without tears: a primer for ecologists. Q Rev Biol 83(2):171–193

Peterson AT, Papeş M, Eaton M (2007) Transferability and model evaluation in ecological niche modeling: a comparison of GARP and Maxent. Ecography 30(4):550–560

Petkos G (2003) Applying machine learning techniques to ecological data. M.Sc Dissertation. University of Edinburgh

Phillips SJ, Dudik M (2008) Modelling of species distributions with Maxent: new extensions and a comprehensive evaluation. Ecography 31:161–175

Phillips SA, Anderson RP, Schapire RE (2006) Maximum entropy modeling of species geographic distributions. Ecol Model 190:231–259

Phillips SJ, Anderson RP, Dudík M, Schapire RE, Blair ME (2017) Opening the black box: an open-source release of Maxent. Ecography 40:887–893

Primack R (2010) Essentials of conservation biology. Fifth, Sinauer Associates Inc

Recknagel F (2001) Applications of machine learning to ecological modelling. Ecol Model 146(1–3):303–310

Recknagel F, Bobbin J, Whigham P, Wilson H (2002) Comparative application of artificial neural networks and genetic algorithms for multivariate time-series modelling of algal blooms in freshwater lakes. J Hydroinf 4(2):125–133

Rockström J, Steffen W, Noone K, Åsa P, Stuart Chapin F, Lambin EF, Lenton TM, Scheffer M, Folke C, Schellnhuber HJ, Nykvist B, de Wit CA, Hughes T, van der Leeuw S, Rodhe H, Sörlin S, Snyder PK, Costanza R, Svedin U, Falkenmark M, Karlberg L, Corell RW, Fabry VJ, Hansen J, Walker B, Liverman D, Richardson K, Crutzen P, Jonathan A. Foley (2009) A safe operating space for humanity. Nature 461(7263):472–475

Rosa D, Isabel M, Marques AT, Palminha G, Costa H, Mascarenhas M, Fonseca C, Bernardino J (2016) Classification success of six machine learning algorithms in radar ornithology. Ibis 158(1):28–42

Rosales J (2008) Economic growth, climate change, biodiversity loss: distributive justice for the global north and south. Conserv Biol 22:1409–1417

Rumelhart DE, Hinton GE, Williams RJ (1986) Learning representations by back-propagating errors. Nature 323(6088):533–536

Samuel AL (1959) Some studies in machine learning using the game of checkers. IBM J Res Dev 3(3):210–229

Sandifer PA, Sutton-Grier AE, Ward BP (2015) Exploring connections among nature, biodiversity, ecosystem services, and human health and well-being: opportunities to enhance health and biodiversity conservation. Ecosyst Serv 12:1–15

Silvy NJ (2012) The wildlife techniques manual: research & management. 2 volumes. The Johns Hopkins University Press; Seventh edition

Stockwell DRB (1994) Genetic Algorithm for Rule-set Production (GARP), ERIN WWW Server http://www.erin.gov.au/general/biodiv_model/ERIN/GARP/home.html

Stockwell DRB (1999) The GARP modelling system: problems and solutions to automated spatial prediction. Int J Geogr Inf Sci 3:143–158

Stockwell DRB, Noble IR (1992) Induction of sets of rules from animal distribution data: a robust and informative method of data analysis. Math Comput Simul 33:385–390

Stone CJ (1984) Classification and regression trees, vol 8. Wadsworth International Group, pp 452–456

Stowell D, Plumbley MD (2014) Automatic large-scale classification of bird sounds is strongly improved by unsupervised feature learning. PeerJ 2:e488

Strobl C, Malley J, Tutz G (2009) An introduction to recursive partitioning: rationale, application, and characteristics of classification and regression trees, bagging, and random forests. Psychol Methods 14(4):323

Sutton T, De Giovanni R, Siqueira MF (2007) Introducing open modeller-a fundamental niche modelling framework. OSGeo J 1(1)

Thessen A (2016) Adoption of machine learning techniques in ecology and earth science. One Ecosyst 1:e8

Tulloch AIT, Auerbach N, Avery-Gomm S, Bayraktarov E, Butt N, Dickman CR, Ehmke G, Fisher DO, Grantham H, Holden MH, Lavery TH, Leseberg NP, Nicholls M, O'Connor J, Roberson L, Smyth AK, Stone Z, Tulloch V, Turak E, Wardle GM, Watson JEM (2018) A decision tree for assessing the risks and benefits of publishing biodiversity data. Nat Ecol Evol 2(8):1209–1217

Turing AM (1950) Computing machinery and intelligence. Mind 59(236):433

Valletta JJ, Torney C, Kings M, Thornton A, Madden J (2017) Applications of machine learning in animal behaviour studies. Anim Behav 124:203–220

Verner J, Morrison ML, Ralph CJ (1986) Wildlife 2000: modeling habitat relationships of terrestrial vertebrates. University of Wisconsin Press, Madison

Wackernagel M, Schulz NB, Deumling D, Linares AC, Jenkins M, Kapos V, Monfreda C, Loh J, Myers N, Norgaard R, Randers J (2002) Tracking the ecological overshoot of the human economy. PNAS 99:9266–9271

Watson JE, Darling ES, Venter O, Maron M, Walston J, Possingham HP, Dudley N, Hockings M, Barnes M, Brooks TM (2016) Bolder science needed now for protected areas. Conserv Biol 30(2):243–248

Zuckerberg B, Huettmann F, Friar J (2011) Proper data management as a scientific foundation for reliable species distribution modeling. Chapter 3. In: Drew CA, Wiersma Y, Huettmann F (eds) Predictive species and habitat modeling in landscape ecology. Springer, New York, pp 45–70

Chapter 2
Use of Machine Learning (ML) for Predicting and Analyzing Ecological and 'Presence Only' Data: An Overview of Applications and a Good Outlook

Falk Huettmann, Erica H. Craig, Keiko A. Herrick, Andrew P. Baltensperger, Grant R. W. Humphries, David J. Lieske, Katharine Miller, Timothy C. Mullet, Steffen Oppel, Cynthia Resendiz, Imme Rutzen, Moritz S. Schmid, Madan K. Suwal, and Brian D. Young

> *"…There is such a thing as being too late. This is no time for apathy or complacency…"*
>
> Martin Luther King Jr

2.1 Introduction

Over a decade ago, Leo Breiman (2001a) wrote: *"There are two cultures in the use of statistical modeling to reach conclusions from data. One assumes that the data are generated by a given stochastic data model. The other uses algorithmic models and treats the data mechanism as unknown. The statistical community has been committed to the almost exclusive use of data models. This commitment has led to irrelevant theory, questionable conclusions, and has kept statisticians from working on a large range of interesting current problems. Algorithmic modeling, both in theory and practice, has developed rapidly in fields outside statistics."*

F. Huettmann (✉) · K. A. Herrick · T. C. Mullet · C. Resendiz · I. Rutzen
EWHALE Lab, Biology and Wildlife Department, Institute of Arctic Biology,
University of Alaska-Fairbanks, Fairbanks, AK, USA
e-mail: fhuettmann@alaska.edu

E. H. Craig
Aquila Environmental, Fairbanks, AK, USA

A. P. Baltensperger
National Park Service, Fairbanks, AK, USA

G. R. W. Humphries
Black Bawks Data Science Ltd., Fort Augustus, Scotland

D. J. Lieske
Department of Geography and Environment, Mount Allison University, Sackville, NB, Canada

© Springer Nature Switzerland AG 2018   27
G. R. W. Humphries et al. (eds.), *Machine Learning for Ecology and Sustainable Natural Resource Management*, https://doi.org/10.1007/978-3-319-96978-7_2

Understanding the complex relationships between animal species and their habitats, and classifying and predicting the responses of species to existing or novel environmental conditions is one of the primary challenges in ecology and conservation (Wilson 1998; Mac Nally 2000; Strogatz 2001). Machine learning (ML) is based on the principle that computers ('the machine') are effective tools for detecting patterns in data and making predictions based on those patterns (Hastie et al. 2009; Strobl et al. 2009). ML consists of many (over 100) algorithms (Fernandez-Delgado et al. 2014), which are even more powerful when 'ensembled' (Hastie et al. 2009). The human brain is challenged to grasp the complexities of ecological systems; it can hardly compete with modern computers and ML algorithms for gaining insight into the 1000's of dimensions these systems encompass. If the learning process using data from ecological systems is successful (and well tested), the recognized patterns can be generalized and used for classification, prediction, subsequent inference and extrapolation of complex data. These are critical components for achieving science-based conservation management (see Figs. 2.1 and 2.2 for an example and management schema). This approach can be applied to virtually any data and it eliminates the need to specify, *a priori*, generally untested and potentially biased assumptions regarding the underlying statistical distribution of the data (Breiman 2001a); as a consequence, 'self-fulfilling prophecies' are avoided by design (compare also with Kéry and Schaub 2012). In spite of the fact that available data may lack a traditional research design, ML algorithms can be used to model and provide insight into the complex, nonlinear relationships that are typical of real ecological systems. This presents a paradigm shift affecting not only data treatment and analysis (Breiman 2001a; Hastie et al. 2009; Huettmann 2005, 2007a), but also monitoring schemes (Magness et al. 2008), specifically the understanding and management of natural resources (Huettmann 2007b) and the way institutions carry out

K. Miller
Auke Bay Laboratories, Alaska Fisheries Science Center, National Marine Fisheries Service, NOAA, Juneau, AK, USA

S. Oppel
RSPB Centre for Conservation Science, Royal Society for the Protection of Birds, Cambridge, UK

M. S. Schmid
Hatfield Marine Science Center, Oregon State University, Newport, OR, USA

EWHALE Lab, Biology and Wildlife Department, Institute of Arctic Biology, University of Alaska-Fairbanks, Fairbanks, AK, USA

CERC in Remote Sensing of Canada's New Arctic Frontier Université Laval, Québec, Canada

M. K. Suwal
Department of Geography, University of Bergen, Bergen, Norway

B. D. Young
Department of Natural Science, Landmark College, Putney, VT, USA

State of Alaska Division of Forestry, Fairbanks, AK, USA

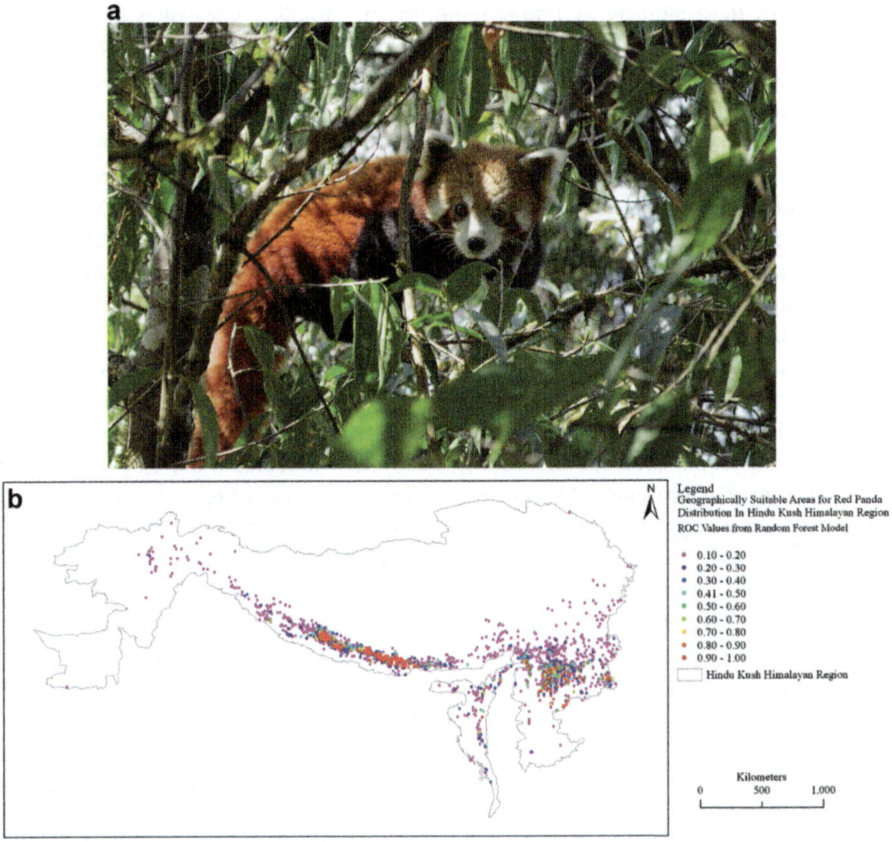

a

b

Legend
Geographically Suitable Areas for Red Panda
Distribution In Hindu Kush Himalayan Region
ROC Values from Random Forest Model

- 0.10 - 0.20
- 0.20 - 0.30
- 0.30 - 0.40
- 0.41 - 0.50
- 0.50 - 0.60
- 0.60 - 0.70
- 0.70 - 0.80
- 0.80 - 0.90
- 0.90 - 1.00

Hindu Kush Himalayan Region

Kilometers
0   500   1,000

Fig. 2.1 (**a**) Photo of the globally endangered Red Panda (*Ailurus fulgens*; taken by S. Tashi Lama/Global Primate Network-Nepal during 2009 in the Choyatar Community Forest in Eastern Nepal at an elevation of ~2400 m asl.). (**b**) Machine Learning predictions (RandomForest ensemble predictions of Red Panda ecological niche in the Hindu-Kush Himalaya region; Kamel et al. in review)

Fig. 2.2 Flow Chart of machine learning methods, and how it can be used in wildlife, biodiversity and habitat analysis, using 'presence only' data and others

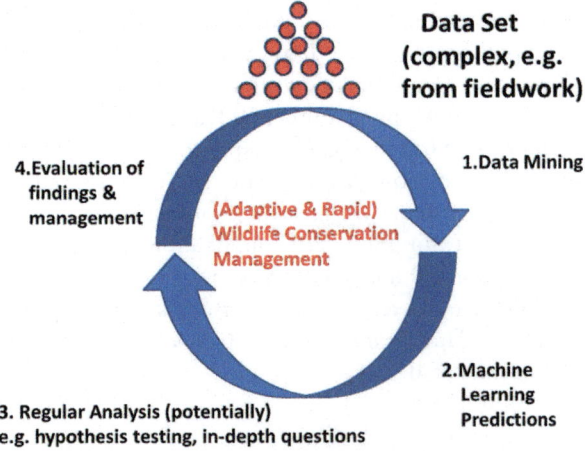

Data Set
(complex, e.g.
from fieldwork)

4.Evaluation of
findings &
management

(Adaptive & Rapid)
Wildlife Conservation
Management

1.Data Mining

2.Machine
Learning
Predictions

3. Regular Analysis (potentially)
e.g. hypothesis testing, in-depth questions

and support these endeavors (Huettmann 2012). This is a very relevant topic for sustainability. Arguably, if one considers the very poor state of the world's biodiversity and habitats (e.g., Mace et al. 2010; Huettmann 2012) it comes as an ethical requirement to take advantage of the objectivity and power presented by ML approaches. ML is 'best available' science! ML approaches have been successfully applied for decades in disciplines such as genetics, medicine, engineering industry, finance, and in some environmental sciences (Bureau et al. 2005; Cooper et al. 1997; Cutler et al. 2007; Dhar 1998; Galindo and Tamayo 2000; Goldberg and Holland 1988; Kononenko 2001; Kubat et al. 1998; Lee et al. 1996; Reich and Barai 1999; Rosten and Drummond 2006; Shipp et al. 2002), but less so in the wildlife and animal sciences, including behavioral research (e.g. primatology).

Habitat-species/biodiversity relationship modeling is one of the fastest growing sub-disciplines in ecology, as judged by rising citations of such publications (e.g. Elith et al. 2006). However, wildlife and ecology disciplines currently still choose to rely heavily on stochastic data models, with their associated use of p-values and the concept of parsimony (Akaike Information Criterion [AIC]), to describe and quantify complex systems (Mac Nally 2000; Mogie 2004; Whittingham et al. 2006; Breiman 2001a). This has translated to the virtually dogmatic application of AIC for inference (e.g. Johnson 1999; Anderson et al. 2000, 2001; Anderson and Burnham 2002; see Guthery et al. 2001 and Stephens et al. 2007 for a discussion). These traditional analyses generally avoid ML or use it in a constrained fashion (Braun 2005; Hochachka et al. 2007) and almost universally advocate the theoretical development of biologically plausible and constrained statistical data models prior to analysis (Burnham and Anderson 2002) before the actual confirmatory test has been completed. It resembles an analysis where the outcome is already known before it was tested. Arguably, this approach to modeling requires a level of *a priori* knowledge that is typically absent in real world ecology studies, particularly for broad scale analyses. In practice, many real-world situations involve huge datasets where the number of plausible models is not easy to identify, widely unknown or is too large to enumerate, and the prior information required to support the formulation of appropriate statistical data models is unavailable or weakly understood (Hochachka et al. 2007). Even with extensively studied species, previously unknown relationships may be identified that change the theoretical framework and interactions, and which require completely new models and explanations to be developed. The power of ML applications is that they can extract and infer the relevant signals and relationships from complex data without any prior knowledge of the nature and shape of those relationships (Breiman 2001a; Hochachka et al. 2007). This trait makes ML techniques particularly applicable for data exploration and in a time of increased availability of powerful computers (Cushman and Huettmann 2010), e.g. 'cloud-computing', and with an ever growing supply of data from global online databases. At a minimum, ML applications can intelligently guide the preliminary exploration and analysis of vast amounts of data, and they do so in less time and with greater efficiency than hitherto possible using traditional statistical approaches (see e.g., Huettmann 2007a; Hochachka et al. 2007; Kampichler et al. 2010; Hochachka et al. 2012).

2.2 Popular and Widely Available Machine Learning Techniques

ML is actually a generic term for a broad array of currently over 20 major algorithms that use different approaches, alone or in combination, to extract information from data (Fernandez-Delgado et al. 2014 for a wider technical overview of algorithm software code). In ecology, some of the most widely used ML techniques include: Classification and Regression Trees (CARTs; De'ath 2002), random forests (Cutler et al. 2007; Hochachka et al. 2007); boosted regression trees (De'ath 2007; Elith et al. 2008 [the powerful version following Friedman 2002 is also known as stochastic gradient boosting and marketed commercially as TreeNet by Salford Systems Inc.]), Maximum entropy (Maxent; Phillips et al. 2004, 2006; Baldwin 2009; Elith et al. 2011) as well as artificial neural networks, genetic algorithms, ecological niche factor analysis (ENFA); and support-vector machines (see Hsieh 2009 and Hegel et al. 2010 for overview). There are a number of excellent introductory texts and resources that explain the basic underlying principles and advantages of these and other ML techniques, and we refer interested readers to these texts rather than provide more detail here (Fielding 1999; Recknagel 2001; Olden et al. 2008; Hsieh 2009; Hegel et al. 2010).

Rapid expansion in knowledge and computing power has led to the evolution of ML algorithms. For example, both, boosted regression trees and random forests are powerful modifications and extensions of classification and regression tree (CART) analysis (De'ath and Fabricius 2000; De'ath 2002). They work by combining hundreds or thousands of individual trees in a single model (the "forest"). Development of more powerful algorithms is ongoing, and many techniques have an active development community (see for instance randomForest package and documentation in R, https://cran.r-project.org/web/packages/randomForest/randomForest.pdf and en.wikipedia.org/wiki/Random_forest). For example, progressive development and extension of the random forests approach (Table 2.1) includes a conditional inference framework for improved variable selection (implemented in the R package 'party'; Hothorn et al. 2006; Strobl et al. 2008), or extensions for survival analysis (Random Survival Forest, Ishwaran et al. 2008). Of specific note should be the fact that the settings, fine-tuning and implementation details of the software code can be key drivers for the performance of algorithms, beyond the initial algorithm group they come from. ML algorithms are used for imputations and are beginning to be incorporated into latent state variable models to account for the imperfect detection process that is pervasive in most ecological field data and for an analysis (Hutchinson et al. 2011). Several widely used statistics programs now include ML implementations (e.g. Statistica, SPSS, SAS, MINITAB), and many new ML methods are freely available and accessible in the R computing environment (R Core Team 2016). Two of the ML approaches that are frequently used for predicting the distribution of species from opportunistic sightings or museum collections are

Table 2.1 Overview of some codes and different implementations for RandomForests

Citation of code	Name of code	Software platform of code	Details of code
Breiman and Cutler	RandomForest	FORTRAN	The orginal RF code
Salford Systems Ltd.	RandomForest	Windows and UNIX	An improved version of the RF code and for commercial, industrial and research use
Liaw and Wiener	RandomForest	R	An early implementation of the original code
Crookston and Finley	YAIMPUTE	R	An extension of RF code for imputations
Hothorn, Hornik, and Zeileis	Party	R	An extension of RF code for more accurate variable importance metrics for categorical predictors
Thuiller et al. (2009)	BIOMOD 1 & 2	R	An ensemble of model algorithms, which includes RandomForest but also many other machine learning algorithms

For more details, please see text, and also check with en.wikipedia.org/wiki/Random_forest

Maxent and Ecological Niche Factor Analysis (ENFA) (e.g. via the Biomapper algorithm: www2.unil.ch/biomapper/). These two techniques are 'presence-only' or 'location only' modeling techniques that require only species presence records, but no confirmed absence locations as input data (Maxent, e.g., creates them internally). The recent establishment of machine learning platforms like TensorFlow, which can be easily accessed via R packages such as 'keras', 'greta' or 'tfestimators' has facilitated applications to automatically process monitoring data (Bruijning et al. 2018; Weinstein 2018) or detect wildlife crimes (Di Minin et al. 2018).

We expect to see continued progress in the development and enhancement of ML algorithms over the coming years for addressing a wide range of problems. The application of several popular ML methods to biological data indicates that ML can be regarded as a serious and valid addition to standard statistical analysis methods. The adoption of ML is nothing short of a paradigm shift in how we manage the earth and its science.

2.3 Applications of Machine Learning in Wildlife Biology

With the rapid development and evolution of increasing numbers of ML algorithms worldwide (Hastie et al. 2009; Fernandez-Delgado et al. 2014), ML sits at the forefront of science for analysis, inference and interpretation of massive datasets. It makes for a 'deep' and intellectual science subject affecting the globe, society and governance (Huettmann 2007b). Beyond the application and management issues, it often involves how to grasp and embrace complex algorithms and data, as well as advanced computational applications. This may well be one reason that ML has not yet been more widely embraced by ecologists and managers (Hochachka et al. 2007; Olden et al. 2008; Drew et al. 2011). Some of the resistance to its wider adoption may further stem from an educational and cultural bias, and the fact that the ML approach to classification/prediction is unfamiliar and, hence, gets confused (Mac Nally 2000) or perceived as a mysterious unexplored 'black box'. However, as interest in ML increases, and is being taught in university programs, user-friendly software and code are becoming more widely available and more easily applied to biological problems (see for instance Elith et al. 2008; Olden et al. 2008; Elith et al. 2011). To our knowledge, there are three principal applications of ML in ecology: (1) data exploration and 'data mining' (i.e., the extraction of signals and relationships from data sets and characterization of data structure); (2) identification of important variables, classification of patterns, and prediction of patterns and classes beyond the data used to construct the model (training data); and (3) as a precursor or input into more detailed mathematical analyses or modeling exercises and specific hypothesis testing. ML can be used for just one or for all of these applications, and it can also be used in combination with more traditional analyses including algorithm comparisons (Elith et al. 2006; Ritter 2007), variable selection (Murphy et al. 2010; Hervias et al. 2013; Buechley and Şekercioğlu 2016), determining the number of target groups or behaviours (Guilford et al. 2009; Oppel et al. 2011), as an aid in scientific hypothesis generation (Stephens et al. 2007), or as part of an improvement loop in a workflow when new data and analysis options become available (similar to adaptive management; Huettmann 2007b).

For the biologist, data exploration, identification of important variables, and prediction are probably the most appealing uses of ML, and likely sufficient for most applications (e.g. see citations in Elith et al. 2006, and in Drew et al. 2011). More quantitatively-inclined users who intend to pursue mathematical models (e.g., resource selection functions or RSFs, Manly et al. 2002; see Venables and Ripley 2002; Hastie et al. 2009 for other applications) will also benefit from the ML insights regarding underlying patterns in the data (Ritter 2007 for an example). Further, ML offers great applications for mixed models, random effects, and even specifically, for data cloning (Jiao et al. 2016; Buston and Elith 2011). Advanced users will want to take advantage of flexible platforms such as R, SAS or Matlab in order to program custom-algorithms. These advanced applications allow the running of multiple simulations, assessment of the sensitivity of models to variation in particular parameters, and the linkage of data between model output and spatial (Geographic Information System [GIS]) or non-spatial databases.

Many datasets, particularly those collected for global (or large scale) analyses, originate all, or in part, from incidental records, or by citizen scientists; they rarely satisfy the prerequisites of a valid research design such as the ones mandated and outlined by Manly et al. (2002), Braun (2005) and Garton et al. (2005). Because of widely implemented imputation algorithms for missing data (Crookston and Finley 2008 for details and R code), and lower sensitivity to outliers and data gaps (Craig and Huettmann 2008), ML can handle these suboptimal data better than classical statistical methods. This then allows for the exploration and analysis of data that may provide insights that might otherwise remain unknown and unanalyzed (see Ritter 2007 for caribou, Huettmann et al. 2011 for seabirds). ML is a great tool to explore data perceived as 'marginal' for their information content. This ability to utilize all available data may be vital for helping inform decision makers and guide future research using 'best available data' in a rapidly changing world.

The growing literature involving the application of ML methods indicates that it is becoming a rather valuable tool for wildlife science and management (Hochachka et al. 2007; Drew et al. 2011); its use in wildlife biology, biodiversity science, and conservation management come from projects worldwide and they cover diverse applications (see Table 2.2 for examples). Currently, 'presence-only' models are probably the most widely known case examples of ML in the biological sciences. The predictive capabilities of Maxent and other ML algorithms mentioned here perform rather well when compared to more traditional modeling approaches (Elith et al. 2006; Pittmann and Huettmann 2006; Buchanan et al. 2011; Hardy et al. 2011; Oppel et al. 2012). Recent evaluations and some misunderstandings of the generic use of ML for modeling presence only data by frequentist mindsets (Royle et al. 2012; Yackulic et al. 2012) have generated rebuttals and lively discussion in the literature regarding the appropriate use and statistical interpretation of such models (Hastie and Fithian 2013; Phillips and Elith 2013). These interactions also exposed many misperceptions and misunderstandings about the use of ML in ecology. Overall, ML applications are in line with Braun (2005) and Stephens et al. (2007) because they provide a better understanding of the data and issues to be studied and addressed; they are the latest and 'best available' tools for inference and generalization (Breiman 2001a), truly based on 'best available' science.

2.4 Strengths and Some Described Weaknesses of Machine Learning

Three main strengths of ML algorithms at hand are i) the ability to model complex, nonlinear relationships without having to make the *a priori* assumptions that frequently constrained parametric approaches, ii) the ability to simultaneously use and evaluate large or messy data sets, and iii) to achieve this very quickly. ML algorithms overcome several problems that are frequently encountered in ecology and which are difficult to solve with conventional parametric data models. Because the structure of most ML algorithms automatically incorporates statistical interactions between all predictor variables, there is no need to specify and interpret complex interactions *a priori*. Variable interactions are incorporated because of the use of 'recursive partitioning', as employed in CART-based methods. That way one also retains the ability to model and detect those interactions if they exist in the data. In addition, most ML algorithms can cope with missing data by either ignoring or imputing missing values. These algorithms are often less

Table 2.2 Applications of machine learning algorithm (RandomForests as an example)

Name of application	Citation	Details of application
'Presence/location only' modeling for wildlife research and management	Baldwin (2009)	This publication presents a classic prediction method for 'presence only' data, based on Maxent, and it reviews numerous applications and for wildlife research and management.
	Edrén et al. (2010)	This paper uses satellite tracking data to model the distribution of a species. Naturally, tracking data are presence-only, as they only provide information on where the tagged animals were.
Modeling a continuous response variable	Yen et al. (2004)	This publication provides predicted population estimates for a seabird of conservation concern in the study area, based on relative abundance indeces.
	Wei et al. (2011)	This publication presents the first global prediction of benthos biomass distribution.
	Humphries (2010)	This M.Sc. thesis, and its publications (in review, in press) presents global predictions of Dimethylsulfat (DMS; a climate change gas) and how it links with storm petrels for their olfactorial seascapes.
Modeling resource selection function design	Popp et al. (2007)	This study models and infers the resource selection of howler monkeys, but not based on a proper research design for habitat preferences according to Manly et al. (2002).
Modeling species distribution via GIS without a proper resource selection function design	Huettmann et al. (2011)	This study models and infers the resource selection of circumpolar arctic seabirds, based on compiled presence only data, and not based on a proper research design for habitat preferences according to Manly et al. (2002).
Multispecies modeling	Lawler et al. (2006, 2011)	This two studies show the use and assessment of climate change models, and when model predicting many species at once.
	Miller et al. (2014)	This study uses many benthos species in complex estuaries of Southeast Alaska and with many predictors
Open source and open access modeling	Ohse et al. (2009)	This publication uses compiled, open access presence only data for white spruce predictions in Alaska and makes them available and for further use.
	Young (2012)	This publication uses compiled, open access forest inventory data for boreal forest coverage and species diversity predictions in Alaska and makes them available and for further use.
	Booms et al. (2009)	This publication model predicts the distribution of a high profile species, based on open access nest data and open source code.

(continued)

Table 2.2 (continued)

Name of application	Citation	Details of application
Remote sensing and landscape forestry application	Evans et al. (2011)	A terrestrial application for vegetated landscapes, using an advanced RandomForest application and predictors.
Landscape clustering	Murphy et al. (2012b)	This application shows how regionalized climate data can be clustered into 'Cliomes' using RandomForest and other approaches, and for future predictions.
Rapid assessment	Kandel et al. (2015)	This study makes use of open access 'presence only' data for an endangered species (red panda) for the entire Hindu Kush-Himalaya (HKH) region, covering 11 countries. While this model is perhaps not perfect, it represents the first, valid and rapid model for this species of global conservation concern.
Ensemble models	Jones-Farrand et al. (2011), Hardy et al. (2011), Oppel et al. (2012)	Ensemble models are becoming popular, and here an approach is shown for commercial fisheries and seabird species using 'averaged' ensemble models across machine learning algorithms.
Forecasting and backcasting	Murphy et al. (2012a,b)	Climate and Ecological Niche Forecasting for 210 using regionalized climate data for Alaska and northern Canada.
	Wickert et al. (2010)	Backcasting and re-assembling the historical distribution of the ecological niche for White Storks in Easten Prussia for 1930s.
	Booms et al. (2011)	Back- and Forecasting of 200 years for Peregerine Falcons and their prey items in space and time for Alaska.
Alternative comparisons of statistical approaches	Ritter (2007)	This study assembled for the first time Caribou data, and modeled them for a comparison of GLMs with CARTs.
	Yen et al. (2004), Elith et al. (2006), Oppel et al. (2012), Fox et al. (2017), Mi et al. (2017)	These studies all compare species-habitat correlations across many methods, e.g. correlational, p-value, parsimonial and machine learning analysis.
Experimental use of machine learning	Wisz et al. (2008)	This study tests via modeling the impacts of sample size on model prediction performance.
	Elith et al. (2006), Hardy et al. (2011)	Comparisons are made among machine learning algorithms and for multispecies analysis (one species at a time).
Data mining	Craig and Huettmann (2008)	This model assesses data quality issues and in telemetry data and for filtering. It uses simulations and finds no relevant performance loss when data carry noise of up to 30%.
Decision-making process and habitat selection	Oppel et al. (2009a, b)	This study quantifies how animals make decisions, mechanistically, and using satellite telemetry data.

(continued)

Table 2.2 (continued)

Name of application	Citation	Details of application
Evolutionary ecology	Buston and Elith (2011)	This study uses serially autocorrelated data of reproductive success and introduces a ML approach analogous to random effects in conventional generalised linear mixed model analyses.
Machine learning and research design optimization and assessment	Magness et al. (2008)	This study assesses (and simulates) monitoring designs and how to optimize them with a focus on predictions for management questions.
	Lawler et al. (2011)	This study reviews and assesses research designs and how to optimize them with predictions in mind.
Machine learning with direct management feed-in and applications	Magness et al. (2008), Miller et al. (2015), Han et al. (2018)	The obtained results are to inform directly the management of natural resources.

sensitive to outliers in the data, and some can even detect outliers, as well as track the major signals of interest.

Further, ML performs very well for mimicking and classifying existing patterns. In our experience, valuable time can be saved when using ML because of the constraints imposed by standard frequentist approaches, which are often demanding in terms of time, statistical expertise and *a priori* knowledge. The necessary frequentist steps of formulating valid statistical data models that attempt to approximate the complexity of ecological systems, determining whether model assumptions are met, testing null hypotheses, and evaluating which set of models receives the most support can be facilitated by the information provided through ML or even avoided. With ML, more attention can be paid to data exploration, investigation of the relevant interpretation and ramifications of results, and improving research designs (Hochachka et al. 2007; Huettmann et al. 2007b; see Magness et al. 2008 for monitoring programs). This allows investigators to concentrate more on the questions that need to be addressed, focus on management and good policy outcomes, consider how a study should be improved, and evaluate what needs to be done to advance actual conservation management and science as a whole (Stephens et al. 2007). It allows for providing a truly science-based leadership of the global society. From our experience, the traditional quantitative analysis instead often gets stuck for months assessing the adequacy of modeling methodology, model fits and how to resolve modeling efforts in a parsimonious fashion based on negotiations and compromises rather than focusing on the underlying ecological questions of interest.

ML can be used as a 'black box'(i.e. a machine that delivers output through an ecological correlation that the user does not necessarily understand yet), as well as a mathematical tool at various levels of performance and transparency (turning the black box into a 'grey' box or making it fully transparent, if wanted). This makes ML a great platform for introducing undergraduate and high school students to statistical analysis with increments in understanding the entire process. With ML tools, virtually any decision based on empirical data can be quantified and generally assessed, allowing for objective decision-making. Lastly, ML encourages scientists to debate the performance of hypothesis testing, linear models and model selection in the context of wildlife management and ecology. Considering the current state of global biodiversity and the shortage of effective science-based management schemes (Mace et al. 2010), a well-thought-out ML application can quickly contribute to urgently needed progress. Who would need or want those tools?

Of course, different ML algorithms have varying strengths and weaknesses. It can become a 'science' to pick, optimize and employ the right algorithm for the application, but rapid assessment methods and ensembles can help (Elith et al. 2006; Hardy et al. 2011; Kandel et al. 2015). Ultimately, ML is just 'a tool' to extract the main patterns in data, to find outliers and for predictions and testing. But it is the best one to do the job. Usually, ML cannot overcome the basic challenge of identifying meaningful and appropriate scientific questions and, like any tool, it can be misused and misapplied. ML needs an ethical context (see Huettmann et al. 2011 for Ecological Economics context of ML applications) and approaches have been criticized for several reasons; here we outline the most important criticisms we are aware of and offer solutions for minimizing or resolving these problems:

> **Overfitting** (i.e., the model provides an excellent fit to one data set with many predictors, but the modeled relationships are not transferrable in time and space and to other data sets): Depending on the quality of the algorithm used, most ML algorithms incorporate internal cross-validation that is designed to reduce the problem of overfitting ('over specialization'): Maxent, random forests, and boosted regression trees (TreeNet) all allow the user to specify a meaningful subset of the data that is withheld from the actual model building process (test data), against which the model predictions are tested. ML, especially tree-based algorithms such as random forests (bagging) and boosted regression trees, can be implemented to provide virtually a 'full fit' to the data, that fits tree splits at every single data point. Despite the widely misunderstood claim of 'overfitting' and its definition these 'full fits' are not necessarily bad (Why would one exclude valid data points?). Instead they describe the data exactly and present a starting point to quantify and to learn the patterns in the data for generalization. They can easily be pruned back to achieve better transferability, if justifiable. These 'full fit' models also can be relaxed towards 'stumps' (trees with fewer nodes and higher sample sizes per node) to investigate the additive nature of the data, interactions, and to draw more coarse and generalized conclusions. The use of stumps and their explorations within the analysis can be necessary when the signal in the data is weak and singular, as is often the case with 'messy' data, when ecology is complex, or when data are collected without a good research design (e.g., 'presence only' data sets compiled from incidental field observations), or when data contain errors (as is often the case in social science data, or when data loggers provide faulty measurements). However, steps as described here help to avoid overfit models in such cases.

i. The other overfitting situation, where data are fit beyond the actual content and data themselves, resulting in non-data fits and artifacts (Venables and Ripley 2002; Elith et al. 2008) is quite rare in good tree-based algorithms and in well-done software implementations. The criticism of overfitting often relates to a poor algorithm, a poor implementation, or odd use of ML. Overfitting often exposes poor use and understanding of the ML statistics and employment. Whereas, such things can easily be checked by simply looking at response variables and the model accuracy metric (e.g., available as a running tally in bagging methods such as RandomForest, Stochastic Gradient Boosting and Boosted Regression Trees. This type of overfitting can be avoided, or at least highly minimized, by the underlying optimization algorithm itself, which suggests a tree cut-off that maximizes the predictive performance in relation to tree complexity and predictor numbers to be reduced. It is here where very good (usually commercial, e.g., TreeNet by Salford, Inc.) algorithm implementations eventually outcompete. In some cases, we actually have found that this optimization method in tree-based methods undervalues the benefit of high-dimensional data. Often, it is even set to be too conservative and parsimonious, i.e., it excessively penalizes more complicated but potentially useful models. The best hedging against overfitting is the independent evaluation of model predictions with external real-world test data that were not incorporated in the model ('confront models with data').

ii. **"Machine learning is not parsimonious"**: The principle of parsimony aims to balance the complexity of a model with its explanatory performance. It penalizes models that are so complex that they are likely to fit just a single data set, and thus would be biased or spurious (Burnham and Anderson 2002; Braun 2005; Fig. 2.3 for a visualized example). This principle was adopted in ecology because in conventional statistical data models, an excessive number of parameters can potentially lead to a so-called overfit model that would be a poor representation of general ecological relationships and predictors. Burnham and Anderson (2002) used this argument a lot to promote model selection via AIC, which can also lead to overfit, bias, being uninformative and a ritual (Arnold 2010; Guthery et al. 2005; Guthery 2008; Galipaud et al. 2014; Brewer et al. 2016). While it is generally assumed to be undesirable to collect and process data that may not have large or direct relevance to the research question (e.g., due to financial costs), it is generally not necessary to constrain the number of variables in ML models. That is because these techniques are specifically designed to handle 'many' predictors (e.g., 10 to 100's), noisy datasets, and to optimize results within these conditions. Being non-parametric, ML does not rely much on pre-determined parameters. The underlying algorithms automatically ignore variables that do not explain any variation in the data (usually expressed by model prediction performance metrics from the confusion matrix and ROC; Boyce et al. 2002; Archer and Kimes 2008; Strobl et al. 2008). Because the predictive accuracy of ML models tends to be more or less insensitive to the number of variables specified, there is no real need to adopt and to worry about the principle of parsimony in ML applications (Breiman 2001a, b; Burnham and Anderson 2002; Quinn and Keough 2004). Usually, the more relevant variables/predictors that can be included in the model the better, as it can aid signal detection, information gain and lead to better predictions and generalizations; all relevant features to explain events are included. However, this may initially result in diffuse answers when many different variables affect a population of interest and no key 'drivers' can be identified, which is a reflection of the ecological complexity of the real world

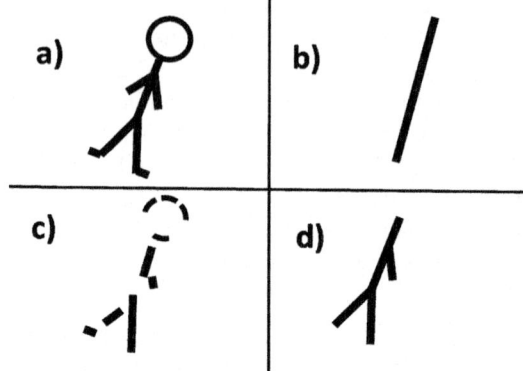

Fig. 2.3 Schema of binary model data, e.g. 'presence only' vs. random, modeled with many predictors: (**a**) Pattern of training data, and when they can be perfectly modeled with high ROCs, (**b**) the same data as in a) but modeled with a linear regression, (**c**) the same data as in a) but modeled with an incomplete set of just a few predictors, and (**d**) the same data as in a) but modeled with a parsimonious paradigm (=bias)

(e.g. Oppel et al. 2017). But a focus on predictions (not model fit) and inference from that (Breiman 2001a) resolves that issue. ML predictive modeling is in stark contrast to the AIC metric and approach where just the major predictors get included and all else are to be fully excluded. The latter approach can not only be very labor intensive, but we actually perceive such concepts as exclusive and not-informative; they may be useful perhaps to *test* well-justified hypotheses, but they are not suitable to *generate* hypotheses, new information and tend not to generalize well for a valid inference.

iii. "**Machine learning results do not produce a significance metric**": ML approaches are not specifically designed to test statistical hypotheses, applications to use ML are much bigger. Therefore ML applications cannot and should not really generate true probabilities (e.g., *p*-values) as to whether the data conforms to certain underlying theories, distributions and pre-determined hypotheses (Murphy and Winkler 1992; Mac Nally 2000). Instead, the purpose of ML simply is to isolate the major patterns in the data at hand and to develop an algorithm to best describe and predict them. This is not a small feat, and it allows data analysis and exploration beyond pre-specified assumptions, which can result in major achievements. Identifying a classic null-hypothesis prior to prediction is not necessary before implementing a ML model. Provided the research objectives involve prediction, there is no need at all to adhere to pre-defined statistical theory such as expressed in Zar (2009). Although this procedure described by Zar (2009) presents a constraint of the ML power, many ML methods actually produce a pseudo-r^2 metric, similar to the variance explained in a linear model. This pseudo-r^2 can be useful for ranking similar models generated by the same ML algorithm and can be particularly helpful in the preliminary stages of model development (e.g., comparing effects of spatial or temporal variation in the data on model performance and for best inference). While most of the authors here do not recommend this evaluation as a final step (for instance, the R 'party' package applies significance concepts to trees; Hothorn et al. 2006), in general, we highly recommend testing the predictive performance of ML models using independent test datasets as suggested by (Breiman 2001a; Elith et al. 2006; Fielding and Bell 1997; Hernandez et al. 2006; Manel et al. 2001) rather than focusing on metrics from frequency statistics. Ideally, those tests and comparable lines of evidence for a match (model vs reality) should be repeated several times, all independent of each other. This helps to gain confirmation and trust in the findings within a transparent framework overall. With that approach comes a shift in the research design, which can then ignore most (parametric) requirements brought by frequency statistics (Magness et al. 2008) but which asks to collect both testing and assessment data in the field. The two-fold approach to research design is a new feature we promote. For binary applications (e.g. presence or absence of species), predictive accuracy can easily be assessed using sensitivity and specificity metrics and the area under the receiver-operated characteristic curve (AUC), a widely used performance criterion (Elith and Leathwick 2009, see also Pearce and Ferrier 2000, as well as, Murphy and Winkler 1992). For models involving continuous response measures, metrics such as the root mean squared error (de Smith et al. 2007) can be employed to assess precision. We believe that for many ecological questions these are the preferred, most relevant, and informative predictive performance metrics rather than p- and AIC-values. However, projects modeling the future still tend to lack a good 'truth' for comparison, regardless of the modeling approach that is used (Huettmann and Gottschalk 2011).

iv. **Potential problems related to autocorrelation in space and time**: Autocorrelation is a common property of virtually all ecological data (Quinn and Keough 2004; Fortin et al. 2010); it is an inherent part of nature. But for many frequentist statisticians it is considered to be an undesirable 'nuisance'. Autocorrelation is perceived a problem factor that detracts from the linear regression assumption of independence, introduces non-random error and biases the estimation of presumed symmetric confidence intervals (Dormann et al. 2007). For instance, when traditional linear models are used in spatial data analysis (Manly et al. 2002), autocorrelation is perceived to be a problem requiring correction prior to subsequent analysis (e.g., through point processes). Here we argue that this is an overly simplistic view that disregards the underlying ecological reality (Betts et al. 2009 for a review of Dormann et al. 2007). Organisms not only respond to the dynamic spatial and temporal distributions of their resources, but this interaction is an inherent part of living systems; organisms also respond to each other through social behavior (e.g., territoriality and spacing mechanisms). Life has a social foundation in order to exist. For a true assessment, environmental factors do not affect an organism in an experimental isolation from other factors. For this reason autocorrelation is a natural part of the evolution of virtually any biological system, and can be used to the advantage of improving ecological understanding (Braun 2005; Fortin et al. 2010). Tree-based ML methods can actually benefit from autocorrelation in the data as their reliance on binary recursive partitioning means that such structure can be used to better explain the patterns (De'ath and Fabricius 2000), and they can also be structured to flexibly model variations in the response variable across time and space (Fink et al. 2010; Hochachka et al. 2012). We illustrate this property using a case example (see below). While novel methods exist to account for spatial and temporal autocorrelation (Hothorn et al. 2011), ML emphasizes predictive accuracy. Thus it is the patterns, and the ability to predict those patterns, that are most important. Meaningful inferences can be drawn from such predictions (Breiman 2001a), and appropriate conservation management actions taken (Ritter 2007 for an example). Hijmans (2012) suggested calibrating predictions with a null model that uses only spatial locations as predictor variables as a method for dealing with autocorrelation. It should be noted here that not explicitly accounting for autocorrelation can potentially lead to some variance inflation and overestimation of predictive accuracy. But with ML approaches, the inclusion of many predictors tends to result in models where autocorrelation naturally becomes less relevant or widely disappears (Young 2012 for forest plots). In many large-scale applications, such error is likely to be rather small but case studies in linear models have reported instances where ~15% of the variance is involved. Still, this should not overly deter detection of the 'main signal' in robust ML applications which are known to buffer fuzzy data through the covariates (Craig and Huettmann 2008).

v. **Potential Pseudoreplication**: When multiple observations of the same individual contribute heavily to a dataset (e.g. as in animal tracking studies), these data are not strictly independent or representative for the population and can interfere with inference. However, many ML methods do not really require independence for predictions (e.g random forests, boosted regression trees and ensembles). The use of mixed effects models, which can include for instance a random term to overcome the non-independence of individual data points, has become very prominent in the traditional linear analyses of such data types (Bolker et al. 2009; Gillies et al. 2006). To our knowledge, while there are no explicit options for including random effects in ML algorithms, there are no theoretical reasons why individuals could not be identified in ML models through the inclusion of a variable containing unique identifiers. For conventional frequentist models, relying on this type of "fixed effect" approach would normally be far too costly in terms of the number of dummy parameters required, degrees of freedom, and for low variances, but this poses no real problem for ML methods. ML offers a real solution to this otherwise prohibitive problem, because through a user-specified division of the dataset into training and testing subsets for cross-validation, which is inherent in many ML algorithms, the natural structure of the data can be retained and the lack of independence of data points within individuals somewhat alleviated (Buston and Elith 2011). Another effective way to test for model bias due to pseudoreplication, is to apply model results to subsets of the larger dataset (e.g., tracking data for each individual animal). Model accuracy when applied to each individual provides a measure of whether the model adequately represents all individuals in the sample population and is useful for inference. Poor model performance for any individual indicates a model biased by pseudoreplication and good model performance on all data subsets, a relatively unbiased model suitable for generalization.

vi. **Poorly defined response index**: Most users of frequentist functions expect the modeled response variable output to follow a Bernoulli distribution on a 0 to 1 scale, matching pre-conceived assumptions. Instead, many ML approaches can sometimes produce an open-ended and asymmetric 'index of response' with a relative scale between slightly negative and higher positive numbers (which can be converted to a 0 to 1 scale). ML provides a predicted response index for both categorical and continuous response variables because that is simply what the algorithm can produce as a machine-optimized compromise to the data provided. The algorithms are confronted with the data structure and from there they produce the best technical and mathematically possible fit and classification solution with the lowest error and best predictive performance metric. These predictions usually result in highly accurate classifications, but the generated index is not necessarily equivalent or proportional to the true probability of a binary outcome (Breiman 2001a, b; Royle et al. 2012; Yackulic et al. 2012; Murphy and Winkler 1992 for overview). This is simply because no probability theory is involved in ML. Also, some data do not contain the essential information to estimate probabilities (Phillips and Elith 2013; Hastie and Fithian 2013). That is not usually a problem because such cases still tend to be very reliable for classifications and predictions (e.g. to identify hotspots and coldspots and to gauge the relative difference between them). If wanted, such offsets ('shrinkage') can simply be regressed back, based on the known training cases, or it can be tried to fit probability curves. In most cases, we found no reasons for doing that, because most managers are content using an accurate index.

vii. **Lack of parameter estimates ("regression coefficients")**: As outlined above, ML methods do not rely on parameters to specify a model, and thus, naturally, the model output will not produce any parameter estimates. The specific outputs of the different ML methods vary. However, the quantified relationship between response and predictor variables can be captured quantitatively and presented visually in partial dependence plots, which show the partial contribution of a given variable considering all other variables at their measured values. In tree-based models for example, the binary partitioning rules at every node in every tree underlying the entire model can be captured by a set of very sophisticated and precise 'rule sets' (often expressed in a digital and binary format). Because many ML algorithms involve some random features (e.g. drawing a random subset of data for building the model, drawing a random number of predictor variables for a specific split), the detailed rule sets can vary slightly between different model runs. Extensions exist to run multiple bootstrap iterations of ML algorithms to achieve stable and repeatable results (Murphy et al. 2010).

viii. **"Machine learning algorithms are difficult to understand and visualize"**: While the underlying algorithms of many ML approaches are indeed rather complex, the model output can generally be readily accessed and visualized. For example, Maxent provides output files that plot the relationship of each predictor variable with the response (Elith et al. 2011). Likewise, results of tree-based methods can be visualized using partial dependence plots (Elith et al. 2008, see also the case example below), and most ML approaches include a default output that summarizes the predictive ability of the model. There are good methods to track and visualize the relationships and predictions obtained by ML. When ensemble models are used, describing 'the tree' is overly simplistic and potentially biased, and therefore not meaningful and best avoided. The whole concept of ML rests on fully utilizing multi-dimensional complexity, which is best grasped and expressed through the predictions, and subsequent inference. The same approach is described by frequentist statisticians for visualizing complicated conventional models, for example, those involving interaction terms (Harrell 2001). These predictions and the behavior of the algorithm in experimental settings can be visualized though, and tends to be rather informative (Elith et al. 2005) for inference (this follows the approach by Breiman 2001a: predictions for inference).

ix. **"Tree-based models cannot show linear relationships"**: Tree-based models with discrete cut-offs at each node within a tree will essentially fit step-wise functions (the size of the individual step usually is a function of the tree size; with stumps being very coarse). However, these stepwise functions closely approximate smooth linear functions for large samples and a large number of trees (Ishwaran 2007). Tree-based models will therefore readily approximate most strictly linear relationships if they exist, especially when 'deep trees' with many nodes and small sample sizes per split are grown with recursive partitioning (De'ath and Fabricius 2000; Strobl et al. 2009). However, for extremely small sample sizes ($n < 20$), as with any other statistical test, care should be exercised when interpreting the step-wise relationships between predictor and response variables. Still, ML presents a major advance given that predictions and inference can be drawn from such sample sizes!

2.5   A Case Example

In the following example we make use of a simulated dataset in order to illustrate some of the issues previously discussed. To proceed we created a region of 21 x 21 unit dimensions, resulting in $N = 441$ locations, and assumed that the density of our hypothetical species corresponds to the combined influence of three habitat variables: x1, x2, and x3. For this "perfectly known" system, average species response is defined based on the following linear predictor:

$$lp = 5.5 + 0.001x_1 - 0.92x_2 + 3.4x_3^{2.5}$$

For each of the 441 locations, density was calculated as random normal draws from the linear predictor, allowing for random variation in the response. Variable x_1 was generated based on random draws from a Gaussian distribution, x_2 based on random draws from a Poisson distribution, and x_3 from a Gaussian random field with a spherical spatial covariance structure (using function *grf* in the geoR package, Ribiero and Diggle 2013). As can be seen from the linear predictor equation and its associated parameters, variable x_1 is scarcely influential (though slightly positive in effect), x_2 much more influential (though negative in effect), and x_3 most influential of all. To render this example less trivial, we included a power term of 2.5 to introduce a measure of non-linearity in the species response.

Figure 2.4 illustrates the species' response under the influence of these three hypothetical habitat factors. Figures 2.5, 2.6, 2.7 indicate the spatial pattern of the

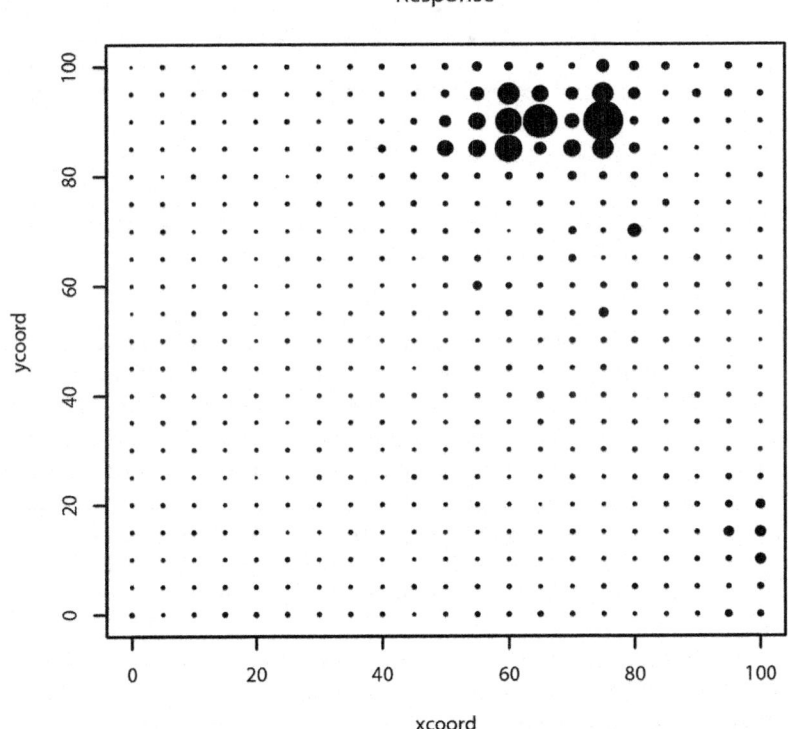

Fig. 2.4 A visualization of the response variable with respect to location in an exhaustively sampled landscape, i.e., when the response value is known at every location. The strong positive spatial autocorrelation is apparent (size of points), resulting in a clear cluster of high magnitude values in the north-east quadrant species distribution model (SDM) using a conventional generalized linear model (GLM; McCullagh and Neder 1989)

Unimportant Factor (x1)

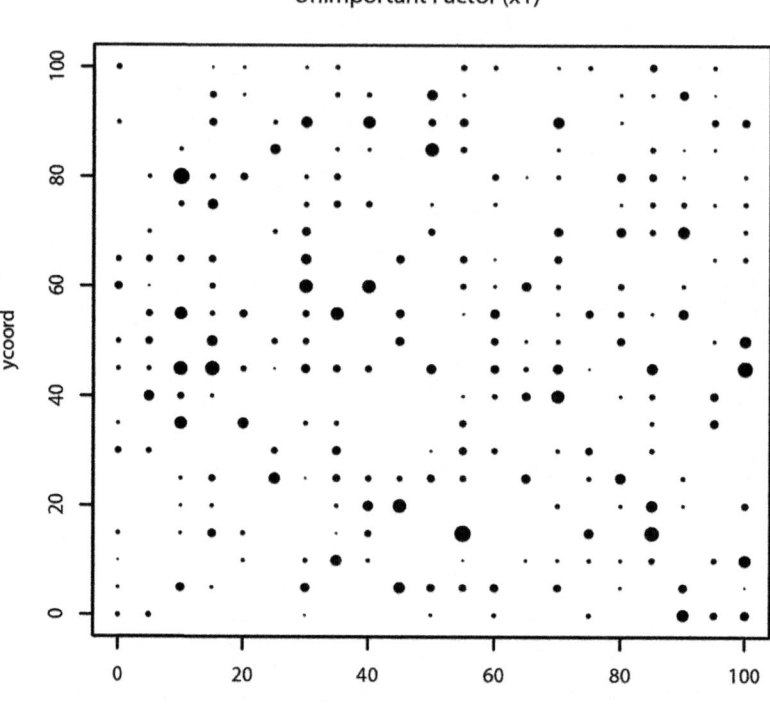

Fig. 2.5 Spatial variation in the first predictor variable ("x1"), which was assigned a low weight in determining the response values. There is no spatial structure in variable x1

individual predictor variables themselves. We then randomly sampled 20% of the locations in the region ($n = 88$) as part of a hypothetical monitoring program and used that randomly selected set of observations to construct a species distribution model (SDM) using a conventional generalized linear model (GLM) as well as a random forests (RF) model. As an added consideration, we constructed the SDMs using only the information provided by variables x_1 and x_2, assuming that variable x_3 is either unknown or unmeasured. By convention, to us this constitutes a significantly flawed model *a priori* as a highly important habitat factor has been excluded, but this is arguably a common feature with species that are poorly studied.

The linear predictor portion of the GLM model was estimated to be: $19.648 - 2.213x_1 - 1.299x_2$. While the negative effect of variable x_2 can be inferred from this model, the parameter estimate for variable x_1 was significantly biased: the coefficient should have been close to zero. Figure 2.8 displays the prediction surface that results from application of this model to the original $N = 441$ cells, RMSE was 18.59, and Moran's I statistic was 0.53 which confirmed that considerable spatial structure remained in the model errors. A quick comparison of Fig. 2.8 with Fig. 2.4

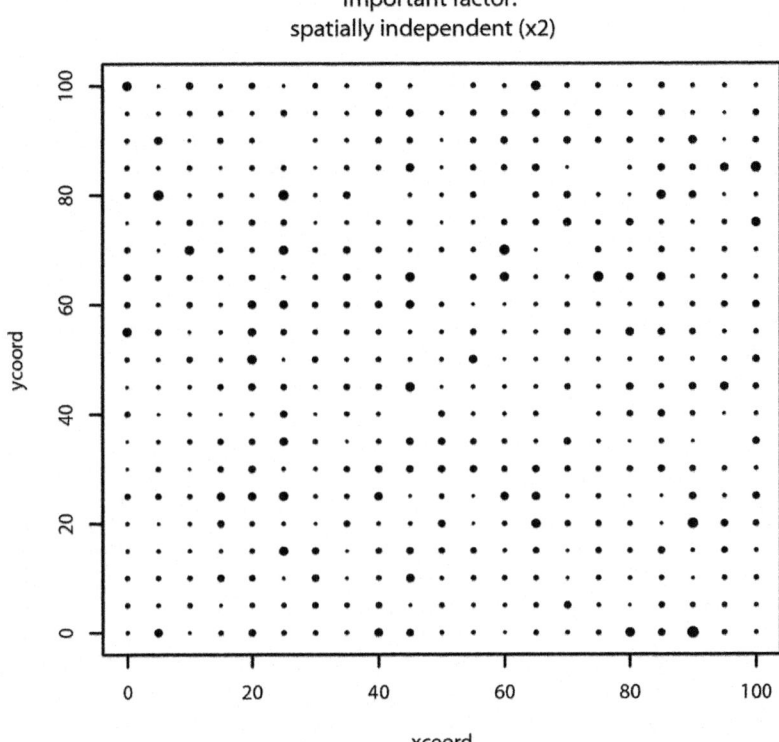

Fig. 2.6 Spatial variation in the second predictor variable ("x2"), which was a moderately important determinant of response values. There is no spatial structure in variable x2

confirms how significant the discrepancy is between the known and predicted species distributions.

As stated above, the RF model does not estimate parameters, but instead constructs multiple regression trees to build an internal rule set for "voting" on the values of new predictions (Liaw and Wiener 2002). Figure 2.9 displays the prediction surface that results from application of this model to the original $N = 441$ cells. RMSE was 19.96, and Moran's I statistic was 0.17, confirming that while some spatial structure in the model errors remained, a considerable portion of spatial structure was captured by RF.

Furthermore, and most importantly, a comparison of Fig. 2.9 and Fig. 2.4 demonstrates how a wildlife manager could still use the RF results to make a meaningful management decision. Based on the GLM prediction surface (Fig. 2.8), the crucial "signal" for the population cluster in the north-east quadrant is invisible, but it can be observed in the RF model (Fig. 2.9). While both of these example models are "naive" in that they suffer from incomplete knowledge of the factors contributing to the distribution of the hypothetical species, the RF approach was more flexible in its model construction, still allowing the crucial signal to be detected.

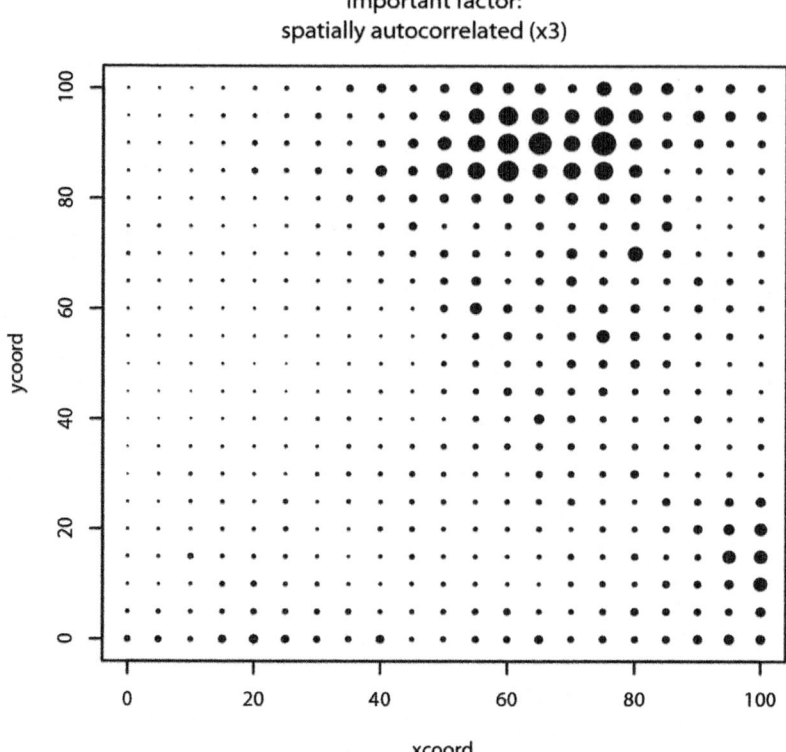

Fig. 2.7 Spatial variation in the third predictor variable ("x3"), which is a very strong influence on response values, exhibits spatial structure, and has a non-linear component (the power value of 2.5)

2.6 Machine Learning in Climate Change Models and Other Complex Applications

ML is especially appropriate for evaluating the complex ecological changes resulting from climate change and its predictors. Both, Maxent and random forests have achieved much positive attention in this regard (e.g. Prasad et al. 2009; www.nrs. fs.fed.us/atlas/, Lawler et al. 2006; see also Murphy et al. 2012a). While the Intergovernmental Panel on Climate Change (IPCC) and most official climate and sea ice modelers have widely ignored ML as a data mining, data cleaning, modeling and clustering method so far, Murphy et al. (2012b) have successfully used random forests for obtaining a distance matrix to cluster climate data into 'cliomes' (climate envelope based biomes) and then project these 100 years into the future. The concept of YAIMPUTE (tree-based imputation; Crookston and Finley 2008; Ellis et al. 2012) was used for climate models, too. Booms et al. (2011) showed similar applications of ML for wildlife-habitat relationships 100 years

Naive GLM Prediction

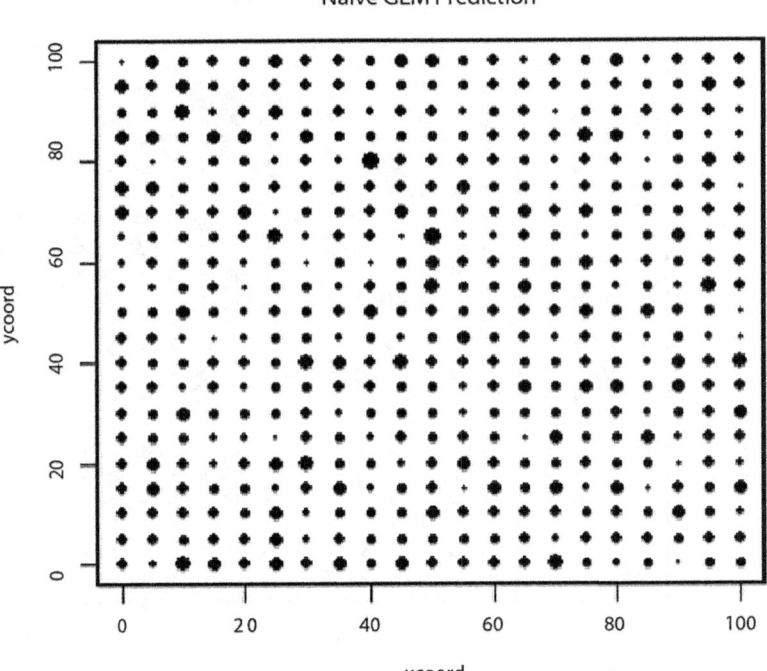

Fig. 2.8 "Naive" GLM. While this model is comparable to the "naive" RF model in terms of RMSE statistic, the underlying spatial structure in the response (stemming from variable "x3") remained undetected. Moran's $I = 0.53$ statistic for the raw residual values confirmed that considerable unexplained spatial structure remains in the model errors

forward as well as 100 years backwards, and Louzao et al. (2013) assessed the influence of climate on albatross distributions using a boosting algorithm (Hothorn et al. 2011). It can be expected that we will continue to see the expanded use of ML for analyses that use multi-species models and online databases. It is a tool of choice that can be extremely useful for climate change and impact studies worldwide (Lawler et al. 2011). The relevance of these approaches can hardly be overstated (Cushman and Huettmann 2010).

2.7 Conclusions: Future Outlook and Topics Awaiting Research and Application for Machine Learning (ML)

ML has existed already for over 30 years but it awaits its conservation and sustainability applications while ML as a discipline keeps rapidly developing. With the advent of cloud computing, global data availability, and increased computational speed, the use and applications of ML are likely to expand (see OpenModeler for an

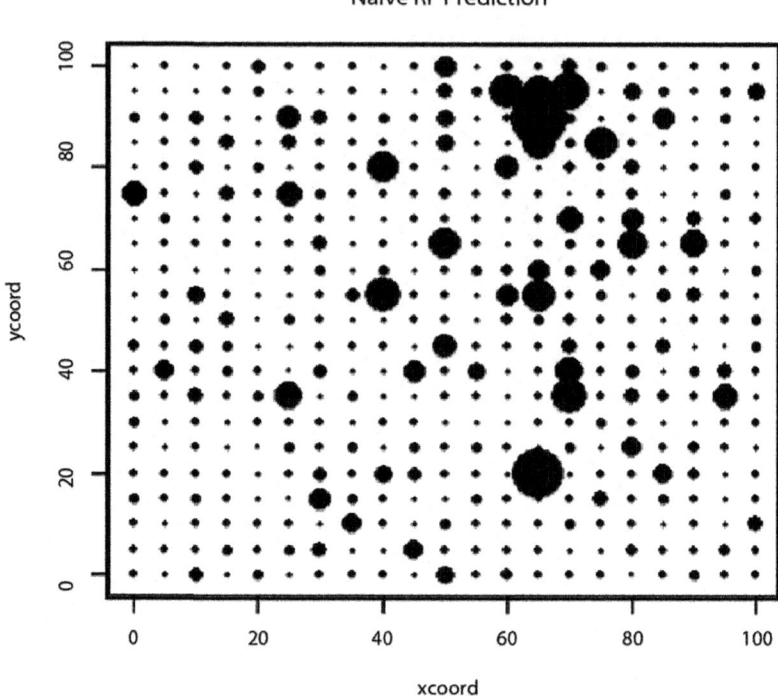

Fig. 2.9 "Naive" RF. While this model is comparable to the "naive" GLM model in terms of RMSE statistic, the underlying spatial structure in the response was well represented. Moran's $I = 0.17$ indicated that some unexplained spatial structure remained, however

example: open-modeling.sourceforge.net/ or TensorFlow: https://www.tensorflow. org/). Application of ML to a wider range of problems will not only lead to ecological and management insights, but will also contribute to a better understanding of the role ML methods can play in data exploration, pattern recognition, and robust prediction (Jean et al. 2016). The use of ML to model presence only data has already received a lot of attention, with interesting applications emerging in the study of disease outbreaks (e.g., Herrick et al. 2014 for a global Avian Influenza model prediction) and for reliably assessing conditions in under-sampled regions (Mullet et al. 2016 for soundscapes). There are many other research possibilities and directions for further development of ML, of which demographic modeling, incorporation of latent variables, ensemble forecasting, tighter integration with GIS software, and the distillation of 'best practices' will yield important insights.

An important field in ecology that has so far not benefited much from ML applications is demographic modeling. Because the analysis of survival and fecundity are at the core of understanding population trajectories, we would welcome further developments that integrate powerful ML algorithms in models examining the survival probability of animals. Random forests models are beginning to be used in survival-based analyses (Ishwaran 2008; see also www.stat. berkeley. edu/~breiman/RandomForests/), but capture-mark-recapture models with random forests do not exist yet. The survival forests that have been developed are not available for population viability analysis because they focus more on identifying important variables affecting survival than on estimating the probability of survival with a good prediction as the main goal. The discipline of demography is widely held back by a software (MARK and its derivatives and practitioners) originating from the 1980s. In ecology, these random survival forests are therefore only useful for 'known-fate' datasets, such as satellite-tracking where the survival state of an animal is known with certainty. We are not aware of any ML applications to mark-recapture data that attempt to distinguish between survival and capture probabilities in a predictive context, yet.

A similar worthy application field for ML is distance sampling and spatially-explicit mark-recapture models, where the model fit (detection curve and function) and the spatial predictions are currently not obtained with ML and ensemble methods. Based on our experience, the use of ML will change, usually improve, many of the currently used estimates and the related wildlife management (Huettmann et al. 2011 and Fox et al. 2017 for pelagic seabirds).

Further, ecological latent variable models are another field that may benefit greatly from the incorporation of ML algorithms. These models recognize that detection of wildlife is always imperfect, and provide a framework to estimate detection probability either from ancillary data (e.g. distance sampling: www. ruwpa.st-and.ac.uk/distance/) or from repeated visits (e.g. Occupancy analysis: Presence software: www.mbr-pwrc.usgs.gov/software/presence.html). Approaches to incorporate boosted regression trees into these latent variable models have been developed (Hutchinson et al. 2011), but we would welcome further research to make ML techniques more widely accessible for models that address the problem of imperfect detection.

In an earlier review, Clemen (1989) pointed out the strengths of combining predictions from different models, particularly when different forecasting models capture different aspects of the information available for prediction. For this reason, ensemble forecasting is an appealing approach to predict ecological responses in the future or at unsurveyed locations. Recent ecological studies (Araujo and New 2007; Buisson et al. 2009; Jones-Farrand et al. 2011; Hardy et al. 2011; Oppel et al. 2012; Kandel et al. 2015) have usually reaffirmed the performance gains associated with ensemble prediction, and they are becoming easier to implement using R packages such as 'ssdm' (Schmitt et al. 2017). As discussed throughout this paper, ML methods should be at the center of any ensemble forecast by virtue of their flexibility in modeling non-linear relationships, as well as their ability to evaluate large, potentially messy data sets. Ensemble models speak to the simple paradigm in ML *"Many weak learners make for a strong learner"*. The typical hurdles facing frequentist techniques, e.g., having to specify a model structure *a priori*, continually determining whether model assumptions are met, and evaluating which of a set of candidate models make "the most sense", are widely avoided with ML. While a blend of methods could be a reasonable way to proceed (see e.g., Hothorn et al. 2006, 2011), we caution that this may also limit the results that can be obtained (as noted by Breiman 2001a), e.g. when poor performers like Linear Models (LMs) are included in the ensemble (as in the default settings of the BIOMOD R package). At the very least, further research is needed in this area.

Despite the impressive advantages of ML methods, they remain underutilized in GIS, modeling, and ecological policy arenas (Huettmann 2007b). In the first and second case, easy access to convenient linkages with GIS software may still be obstructing wider adoption of these tools. Maxent does have a GIS interface (www. cs.princeton.edu/~schapire/maxent/; the same is true for BIOMAPPER), however, easily implementable and available code with direct and adjustable links between ArcGIS and ML (Humphries et al. unpublished for Maxent) remains elusive for random forests (Humphries and Huettmann in prep.). Even if the actual model interface exists, having high-quality GIS layers with high spatial resolution still remains a problem (but see Oppel and Huettmann 2010, and Worldclim and similar public data sets in Herrick et al. 2014 made available worldwide). Regarding the third issue of limited application in the policy arena, we argue that ML techniques may suffer from the same difficulties as the more conventional techniques, i.e., a perception that they are too complicated to understand or based on unreasonable assumptions. The buy-in of mathematical tools is generally not only driven by researchers but by the public at large, students and even more so, whether industry and their attorneys, lawyers and courts are trained and fluent in such methods and actually use them. While future research will help to clarify how research studies can be designed, executed and promoted to capitalize on ML strengths (see for instance Magness et al. 2008; Lawler et al. 2011), the growing number of impressive ML case studies will help to raise awareness of the usefulness of ML techniques and reduce any prejudices that may exist. Future research synthesizing ML 'best practices', as well as relevant ethical issues (Daly 1997; Naess 1997; Czech 2000; Ott 2005), open-minded statistics (Hilborn and Mangel 1997; Strobl et al. 2007; Kelling et al. 2009; Schaub and Kery 2012; Azoulay et al. 2015), data sharing (Bluhm et al. 2010; Huettmann 2011; Zuckerberg et al. 2011) and Open Source, education and outreach, will help to communicate the ways in which ML empirically identifies the key signals in data while simultaneously making no *a priori* assumptions about the data structure.

Ironically, the main 'shortcoming and failure' of ML so far is that it is so widely underutilized and largely unexplored in the natural sciences. It can be argued that this is partially a result of the way philosophy of science, statistics and its relationship to quantitative modeling and with 'nature' is taught and awarded in universities and society. As Breiman (2001a) so eloquently expressed, this inevitably influences all sorts of data-based decision making, ranging from publication policies, peer review, and management policy, to courts of law. We encourage high schools, universities and policy-makers to incorporate ML into their standard lecture and lab material for undergraduate and graduate courses to facilitate a better understanding and more rapid adoption of these powerful approaches, as well as a knowledge of best practices (Hochachka et al. 2012; Cushman and Huettmann 2010; Drew et al. 2011).

In summary, we foresee a significant role and relevance for ML in guiding the sustainable science-based management of global biodiversity. It can become the computational standard and benchmark against which other analytical methods are measured. As stated in Huettmann (2007a, b) and elsewhere, now is the time to make the best use of these efficient tools and to share the associated expertise, methods, and general philosophy globally. We believe that a wider use of these methods will necessarily improve wildlife management frameworks, both locally and globally.

Acknowledgements   This is a shared MS summarizing work efforts from over 2 decades on international projects. FH is grateful to all individuals who were open-minded enough to develop and try machine learning algorithms and to support them. The late R. O'Connor and A.W. Diamond are thanked for introducing us to CARTs early on. J.Liu kindly helped to start a co-authored model session at IALE-U.S. in 2007 on such subjects, published with Springer. Salford Systems Ltd., D.Steinberg and his great team, are specifically thanked for the long collaboration, for ideas and for helpful support using their thoughts and their software in many ways. Most EWHALE students heroically supported machine learning projects, either helping to evaluate the paradigms of statistics, or putting themselves out there for the debate and advancement of conservation science and management with machine learning; finding new knowledge and information. FH is further grateful to S. Linke, L. Strecker, and to the ArcOD project (B. Bluhm), Alaska GAP project (T. Gotthard), SNAP (N. Fresco et al.), Antarctic Biogeography Atlas project (B. Danis, C. Broyer, Philiippi et al.), Red Panda project (G. Regmi, K. Kamal, MS et al), the Chinese Crane and Bustard projects (G. Yumin and students like H. Juang, M. Chunrong, P. Guopanlian), J. Morton, S. Cushman, J. Evans, T. Hegel, J. Ritter, D. Watts, A. Drew, Y. Wiersma, W. Thogmartin, T. Gottschalk, B. Raymond, B. Walther, I. Presse and H. Berrios for general support, publications, replies, and advice regarding machine learning implementations and applications. This is EWHALE publication # 125.

References

Anderson D, Burnham K (2002) Avoiding pitfalls when using information-theoretic methods. J Wildl Manag 66:912–918

Anderson D, Burnham K, Thompson W (2000) Null hypothesis testing: problems, prevalence, and an alternative. J Wildl Manag 64:912–923

Anderson DR, Link WA, Johnson D, Burnham KP (2001) Suggestions for presenting the results of data analysis. USGS Northern Prairie Wildlife Research Center. Paper 227. https://digitalcommons.unl.edu/usgsnpwrc/227

Archer KJ, Kimes RV (2008) Empirical characterization of random forest variable importance measures. Comput Stat Data Anal 52:2249–2260

Araujo M, New B (2007) Ensemble forecasting of species distributions. Trends Ecol Evol 22:42–47

Arnold TW (2010) Uninformative parameters and model selection using Akaike's information criterion. J Wildl Manag 74:1175–1178

Azoulay P, Fons-Rosen C, Zivin JSG (2015) Does science advance one funeral at a time? National Bureau of Economic Research Working Paper Series. No. 21788. http://www.nber.org/papers/w21788

Baldwin RA (2009) Use of maximum entropy modeling in wildlife research. Entropy 11:854–866. https://doi.org/10.3390/e11040854

Betts MG, Ganio L, Huso M, Som N, Huettmann F, Bowman J, Wintle BW (2009) Comment on "Methods to account for spatial autocorrelation in the analysis of species distributional data: a review". Ecography 32:374–378

Bluhm B, Watts D, Huettmann F (2010) Free database availability, metadata and the internet: an example of two high latitude components of the census of marine life. In: Cushman SA, Huettmann F (eds) Spatial complexity, informatics and wildlife conservation. Springer, Tokyo, pp 233–244

Bolker BM, Brooks ME, Clark CJ, Geange SW, Poulsen J, Stevens MHH, White J-SS (2009) Generalized linear mixed models: a practical guide for ecology and evolution. Trends EcolEvol 24:127–135

Booms T, Huettmann F, Schempf P (2009) Gyrfalcon nest distribution in Alaska based on a predictive GIS model. Pol Biol 33:1602–1612

Booms T, Lindgren M, Huettmann F (2011) Linking Alaska's predicted climate, Gyrfalcon, and ptarmigan distributions in space and time: a unique 200-year perspective. In: Watson RT, Cade TJ, Fuller M, Hunt G, Potapov E (eds) Gyrfalcons and ptarmigan in a changing world, vol I. The Peregrine Fund, Boise, pp 177–190

Boyce MS, Vernier PR, Nielsen SE, Schmiegelow FKA (2002) Evaluating resource selection functions. Ecol Model 157:281–300

Braun CE (ed) (2005) Techniques for wildlife investigations and management. The Wildlife Society (TWS), Bethesda

Breiman L (2001a) Statistical modeling: the two cultures (with comments and a rejoinder by the author). Stat Sci 16:199–231

Breiman L (2001b) Random forests. Mach Learn J 45:5–32

Brewer MJ, Butler A, Cooksley SL (2016) The relative performance of AIC, AICC and BIC in the presence of unobserved heterogeneity. Meth Ecol Evol 7:679–692

Bruijning M, Visser MD, Hallmann CA, Jongejans E (2018) Trackdem: automated particle tracking to obtain population counts and size distributions from videos in R. Meth Ecol Evol 9:965–973. https://doi.org/10.1111/2041-210X.12975

Buechley ER, Şekercioğlu ÇH (2016) The avian scavenger crisis: looming extinctions, trophic cascades, and loss of critical ecosystem functions. Biol Conserv 198:220–228

Buisson L, Thuiller W, Casajus N, Sovan L, Grenouillet G (2009) Uncertainty in ensemble forecasting of species distribution. Glob Chang Biol 16:1145–1157. https://doi.org/10.1111/j.1365-2486.2009.02000.x

Bureau A, Dupuis J, Falls K, Lunetta KL, Hayward B, Keith TP, Van Eerdewegh P (2005) Identifying SNPs predictive of phenotype using random forests. Genet Epidemiol 28:171–182

Burnham K, Anderson D (2002) Model selection and multimodel inference: a practical information-theoretic approach. Springer, New York

Buchanan GM, Lachmann L, Tegetmeyer C, Oppel S, Nelson A, Flade M (2011) Identifying the potential wintering sites of the globally threatened Aquatic Warbler *Acrocephalus paludicola* using remote sensing. Ostrich 82:2, 81–85. https://doi.org/10.2989/00306525.2011.603461

Buston PM, Elith J (2011) Determinants of reproductive success in dominant pairs of clownfish: a boosted regression tree analysis. J Anim Ecol 80:528–538

Clemen RT (1989) Combining forecasts: a review and annotated bibliography. Int J Forecast 5:559–583

Craig E, Huettmann F (2008) Using "blackbox" algorithms such as TreeNet and Random Forests for data-mining and for finding meaningful patterns, relationships and outliers in complex ecological data: an overview, an example using golden eagle satellite data and an outlook for a promising future. In: Wang H-f (ed) Intelligent data analysis: developing new methodologies through pattern discovery and recovery. IGI Global, Hershey, pp 65–84

Cooper GF, Aliferis CF, Ambrosino R, Aronis J, Buchanan BG, Caruana R, Fine MJ, Glymour C, Gordon G, Hanusa BH et al (1997) An evaluation of machine-learning methods for predicting pneumonia mortality. Artif Intell Med 9:107–138

Crookston NL, Finley AO (2008) yaImpute: an R package for kNN imputation. J Stat Softw 23:1–14

Cushman S, Huettmann F (eds) (2010) Spatial complexity, informatics and wildlife conservation. Springer, Tokyo

Cutler DR, Edwards TC Jr, Beard KH, Cutler A, Hess KT, Gibson J, Lawler JJ (2007) Random forests for classification in ecology. Ecology 88:2783–2792

Czech B (2000) Shoveling fuel for a runaway train: errant economists, shameful spenders, and a plan to stop them all. University of California Press, Berkeley

Daly H (1997) Beyond growth: the economics of sustainable development. Beacon Press, Boston

Dhar V (1998) Data mining in finance: using counterfactuals to generate knowledge from organizational information systems. Inf Syst 23:423–437

De'ath G, Fabricius K (2000) Classification and regression trees: a powerful yet simple technique for ecological data analysis. Ecology 81:3178–3192. https://doi.org/10.1890/0012-9658 (2000)081[3178:CARTAP]2.0.CO;2

De'ath G (2002) Multivariate regression trees: a new technique for modeling species–environment relationships. Ecology 83:1105–1117. https://doi.org/10.1890/0012-9658(2002)083[1105:MRT ANT]2.0.CO;2

De'ath G (2007) Boosted trees for ecological modeling and prediction. Ecology 88:243–251

Di Minin E, Fink C, Tenkanen H, Hiippala T (2018) Machine learning for tracking illegal wildlife trade on social media. Nat Ecol Evol 2:406–407. https://doi.org/10.1038/s41559-018-0466-x

Dormann CF, McPherson JM, Araújo MB, Bivand R, Bolliger J, Carl G, Davies RG, Hirzel A, Jetz W, Kissling WD (2007) Methods to account for spatial autocorrelation in the analysis of species distributional data: a review. Ecography 30:609–628

Drew CA, Yo W, Huettmann F (eds) (2011) Predictive modeling in landscape ecology. Springer, New York

Edrén SMC, Wisz MS, Teilmann J, Dietz R, Söderkvist J (2010) Modelling spatial patterns in harbour porpoise satellite telemetry data using maximum entropy. Ecography 33:698–708

Elith J, Graham C, NCEAS working group (2006) Novel methods improve prediction of species' distributions from occurrence data. Ecography 29:129–151

Elith J, Ferrier S, Huettmann F, Leathwick J (2005) The evaluation strip: a new and robust method for plotting predicted responses from species distribution models. Ecol Model 186:280–289

Elith J, Leathwick JR, Hastie T (2008) A working guide to boosted regression trees. J Anim Ecol 77:802–813. https://doi.org/10.1111/j.1365-2656.2008.01390.x

Elith J, Leathwick JR (2009) Species distribution models: ecological explanation and prediction across space and time. Ann Rev Ecol Evol Syst 40:677–697

Elith J, Phillips SJ, Hastie T, Dudík M, En Chee Y, Yates CCJ (2011) A statistical explanation of MaxEnt for ecologists. Div Distrib 17:43–57

Ellis N, Smith SJ, Pitcher JR (2012) Gradient forests: calculating importance gradients on physical predictors. Ecology 93(1):156–168. http://www.esajournals.org/doi/abs/10.1890/0012-9658 (2002)083%5B1105:MRTANT%5D2.0.CO%3B2

Evans J, Murphy M, Cushman S, Holden Z (2011) Modeling tree distribution and change using random forests. In: Drew CA, Wiersma Y, Huettmann F (eds) Predictive wildlife and habitat modeling in landscape ecology. Springer Publishers, New York

Fox CH, Huettmann F, Harvey GKA, Morgan KH, Robinson J, Williams R, Paquet PC (2017) Predictions from machine learning ensembles: marine bird distribution and density on Canada's Pacific coast. Mar Ecol Prog Ser 566:199–216

Jones-Farrand DT, Fearer TM, Thogmartin WE, Thompson FR 3rd, Nelson MD, Tirpak JM (2011) Comparison of statistical and theoretical habitat models for conservation planning: the benefit of ensemble prediction. Ecol Appl 21:2269–2282

Fielding AH, Bell JF (1997) A review of methods for the assessment of prediction errors in conservation presence/absence models. Environ Conserv 24:38–49

Fernandez-Delgado M, Cernades E, Barro S, Amorim D (2014) Do we need hundreds of classifiers to solve real world classification problems? J Mach Learn Res 15:3133–3181

Fielding AH (1999) Machine learning methods for ecological applications. Springer, New York

Fink D, Hochachka WM, Zuckerberg B, Winkle DW, Shaby B, Munson MA, Hooker G, Riedewald G, Sheldon D, Kelling S (2010) Spatiotemporal exploratory models for broad-scale survey data. Ecol Appl 20:2131–2147

Fortin M-J, Dale MRT, Bertazzon S (2010) Spatial analysis of wildlife distribution and disease spread. In: Huettmann F, Cushman S (eds) Spatial complexity, informatics, and wildlife conservation. Springer, Tokyo, pp 255–273

Friedman JH (2002) Stochastic gradient boosting. Comp Stat Data Anal 38:367–378

Galindo J, Tamayo P (2000) Credit risk assessment using statistical and machine learning: basic methodology and risk modeling applications. Comput Econ 15:107–143

Galipaud M, Gillingham MAF, David M, Dechaume-Moncharmont F-X (2014) Ecologists overestimate the importance of predictor variables in model averaging: a plea for cautious interpretations. Methods Ecol Evol 5:983–991

Garton EO, Ratti JR, Giudice JH (2005) Research and experimental design. In: Braun CE (ed) Techniques for wildlife investigations and management. The Wildlife Society, Bethesda, pp 43–71

Gillies CS, Hebblewhite M, Nielsen SE, Krawchuk M, Aldridge CL, Frair JL, Saher DJ, Stevens CE, Jerde CL (2006) Application of random effects to the study of resource selection by animals. J Anim Ecol 75:887–898

Goldberg DE, Holland JH (1988) Genetic algorithms and machine learning. Mach Learn 3:95–99

Guilford T, Meade J, Willis J, Phillips RA, Boyle D, Roberts S, Collett M, Freeman R, Perrins, C (2009) Migration and stopover in a small pelagic seabird, the Manx shearwater *Puffinus puffinus*: insights from machine learning. Proc R Soc Lond B Biol Sci: rspb 2008.1577

Guthery FS (2008) Statistical ritual; versus knowledge accrual in wildlife science. J Wildl Manag 72:1872–1875

Guthery FS, Lusk JJ, Peterson MJ (2001) The fall of the null hypothesis: liabilities and opportunities. J Wildl Manag 65:379–384

Guthery FS, Brennan LA, Peterson MJ, Lusk LL (2005) Information theory in wildlife science: critique and viewpoint. J Wildl Manag 69:457–465

Han X, Huettmann F, Guo Y, Mi C, Wen L (2018) Conservation prioritization with machine learning predictions for the black-necked crane *Grus nigricollis*, a flagship species on the Tibetan Plateau for 2070. Glob Environ Chang. https://doi.org/10.1007/s10113-018-1336-4

Hardy SM, Lindgren M, Konakanchi H, Huettmann F (2011) Predicting the distribution and ecological niche of unexploited snow crab (*Chionoecetesopilio*) populations in Alaskan waters: a first open-access ensemble model. Integr Comp Biol 51:608–622. https://doi.org/10.1093/icb/icr102

Harrell FE Jr (2001) Regression modeling strategies: with applications to linear models, logistic regression, and survival analysis. Springer, New York

Hastie T, Tibshirani R, Friedman J (2009) The elements of statistical learning: data mining, inference, and prediction, 2nd edn. Springer, New York

Hastie T, Fithian W (2013) Inference from presence-only data; the ongoing controversy. Ecography 36:864–867

Hegel T, Cushman SA, Evans J, Huettmann F (2010) Chapter 16: Current state of the art for statistical modelling of species distributions. In: Cushman S, Huettmann F (eds) Spatial complexity, informatics and wildlife conservation. Springer, Tokyo, pp 273–312

Hernandez PA, Graham CH, Master LL, Albert D (2006) The effect of sample size and species characteristics on performance of different species distribution modeling methods. Ecography 29:773–785

Herrick KA, Huettmann F, Lindgren MA (2014) A global model of avian influenza prediction in wild birds: the importance of northern regions. Vet Res 44:42. https://doi.org/10.1186/1297-9716-44-42.

Hervías S, Henriques A, Oliveira N, Pipa T, Cowen H, Ramos JA, Nogales M, Geraldes P, Silva C, de Ruiz Ybáñez R, Oppel S (2013) Studying the effects of multiple invasive mammals on Cory's shearwater nest survival. Biol Invasions 15:143–155

Hijmans RJ (2012) Cross-validation of species distribution models: removing spatial sorting bias and calibration with a null model. Ecology 93:679–688

Hilborn R, Mangel M (1997) The ecological detective: confronting models with data. Princeton University Press, Princeton, p 330

Hochachka WE, Caruana R, Fink D, Munson A, Riedewald M, Sorokina D, Kelling S (2007) Data-mining discovery of pattern and process in ecological systems. J Wildl Manag 71:2427–2437. https://doi.org/10.2193/2006-503

Hochachka WM, Fink D, Hutchinson RA, Sheldon D, Wong W-K, Kelling S (2012) Data-intensive science applied to broad-scale citizen science. Trends Ecol Evol 27:130–137

Hothorn T, Hornik K, Zeileis K (2006) Party: a laboratory for recursive part(y)itioning. Available at: http://CRAN.R-project.org/. Accessed 21 Dec 2008

Hothorn T, Müller J, Schröder B, Kneib T, Brandl R (2011) Decomposing environmental, spatial, and spatiotemporal components of species distributions. Ecol Monogr 81:329–347

Hsieh WW (2009) Machine learning methods in the environmental sciences. Cambridge University Press, Cambridge

Humphries G (2010) 'The Ecological Niche of Storm-Petrels in the North Pacific and a Global Model of Dimethylsulfide DMS'. Unpublished M.Sc. thesis. University of Alaska-Fairbanks USA

Huettmann F (2005) Databases and science-based management in the context of wildlife and habitat: towards a certified ISO standard for objective decision-making for the global community by using the internet. J Wildl Manag 69:466–472

Huettmann F (2007a) Constraints, suggested solutions and an outlook towards a new digital culture for the oceans and beyond: experiences from five predictive GIS models that contribute to global management, conservation and study of marine wildlife and habitat. In: VandenBerghe E et al (eds) Proceedings of 'ocean biodiversity informatics': an international conference on marine biodiversity data management Hamburg, Germany, 29 November–1 December, 2004. IOC Workshop Report, 202, VLIZ Special Publication 37, pp. 49–61. www.vliz.be/vmdcdata/imis2/imis.php?module=ref&refid=107201

Huettmann F (2007b) Modern adaptive management: adding digital opportunities towards a sustainable world with new values. Forum Public Policy 3:337–342

Huettmann F (2011) Serving the Global Village through public data sharing as a mandatory paradigm for seabird biologists and managers: why, what, how, and a call for an efficient action plan. Open Ornith J 4:1–11

Huettmann F (2012) Protection of the three poles. Springer, Tokyo

Huettmann F, Gottschalk T (2011) Simplicity, model fit, complexity and uncertainty in spatial prediction models applied over time: we are quite sure, aren't we? In: Drew CA, Wiersma YF, Huettmann F (eds) Predictive species and habitat modeling in landscape ecology, pp 189–208. https://doi.org/10.1007/978-1-4419-7390-0_10

Huettmann F, Artukhin Y, Gilg O, Humphries G (2011) Predictions of 27 Arctic pelagic seabird distributions using public environmental variables, assessed with colony data: a first digital IPY and GBIF open access synthesis platform. Mar Biodivers 41:141–179. https://doi.org/10.1007/s12526-011-0083-2

Hutchinson RA, Liu L-P, Dietterich TG (2011) Incorporating boosted regression trees into ecological latent variable models. In: 25th AAAI conference on artificial intelligence. Association for the Advancement of Artificial Intelligence, San Francisco

Ishwaran H (2007) Variable importance in binary regression trees and forests. Electron J Stat 1:519–537

Ishwaran H, Kogalur UB, Blackstone EH, Lauer MS (2008) Random survival forests. Ann Appl Stat 2:841–860

Jean N, Burke M, Xie M, Davis WM, Lobell DB, Ermon S (2016) Combining satellite imagery and machine learning to predict poverty. Science 353:790–794

Jiao S, Huettmann F, Guo Y, Li Y, Ouyang Y (2016) Advanced long-term bird banding and climate data mining in spring confirm passerine population declines for the Northeast Chinese-Russian flyway. Glob Planet Chang. https://doi.org/10.1016/j.gloplacha.2016.06.015

Johnson DH (1999) The insignificance of statistical significance testing. J Wildl Manag 63:763–772

Kampichler C, Wieland R, Calmé S, Weissenberger H, Arriaga-Weiss S (2010) Classification in conservation biology: a comparison of five machine-learning methods. Eco Inform 5:441–450

Kandel K, Huettmann F, Suwal MK, Regmi RG, Nijman V, Nekaris KAI, Lama ST, Thapa A, Sharma HP, Subedi TR (2015) Rapid multi-nation distribution assessment of a charismatic conservation species using open access ensemble model GIS predictions: red panda (*Ailurus fulgens*) in the Hindu-Kush Himalaya region. Biol Conserv 181:150–161

Kelling S, Hochachka WM, Fink D, Riedewald M, Caruana R, Ballard G, Hooker G (2009) Data-intensive science: a new paradigm for biodiversity studies. Bioscience 59:613–620 www.jstor.org/stable/10.1525/bio.2009.59.7.12

Kéry M, Schaub M (2012) Bayesian population analysis using WinBUGS. Academic Press, Oxford

Kononenko I (2001) Machine learning for medical diagnosis: history, state of the art and perspective. Artif Intell Med 23:89–109

Kubat M, Holte RC, Matwin S (1998) Machine learning for the detection of oil spills in satellite radar images. Mach Learn 30:195–215

Lawler JJ, White D, Neilson RP, Blaustein AR (2006) Predicting climate-induced range-shifts: model differences and model reliability. Glob Chang Biol 12:1568–1584

Lawler JJ, Yo W, Huettmann F (2011) Chapter 5: Designing predictive models for increased utility: using species distribution models for conservation planning, forecasting, and risk assessment. In: Drew CA, Wiersma Y, Huettmann F (eds) Predictive modeling in landscape ecology. Springer, New York, pp 271–290

Lee KC, Han I, Kwon Y (1996) Hybrid neural network models for bankruptcy predictions. Decis Support Syst 18:63–72

Liaw A, Wiener M (2002) Classification and regression by randomforests. R News 2(3):18

Louzao M, Aumont O, Hothorn T, Wiegand T, Weimerskirch H (2013) Foraging in a changing environment: habitat shifts of an oceanic predator over the last half century. Ecography 36:057–067. https://doi.org/10.1111/j.1600-0587.2012.07587.x

Mace G, Cramer W, Diaz S, Faith DP, Larigauderie A, Le Prestre P, Palmer M, Perrings C, Scholes RJ, Walpole M, Walter BA, Watson JEM, Mooney HA (2010) Biodiversity targets after 2010. Environ Sustain 2:3–8

Mac Nally R (2000) Regression and model-building in conservation biology, biogeography and ecology: the distinction between – and reconciliation of – 'predictive' and 'explanatory' models. Biodivers Conserv 6:655–671

Magness DR, Huettmann F, Morton JM (2008) Using random forests to provide predicted species distribution maps as a metric for ecological inventory & monitoring programs. In: Smolinski TG, Milanova MG, Hassanien AE (eds) Applications of computational intelligence in biology: current trends and open problems, studies in computational intelligence, vol 122. Springer, Berlin/Heidelberg, pp 209–229

Manel S, Williams HC, Ormerod SJ (2001) Evaluating presence–absence models in ecology: the need to account for prevalence. J Appl Ecol 38:921–931

Manly BF, McDonald L, Thomas DL, McDonald TL, Erickson WP (2002) Resource selection by animals: statistical design and analysis for field studies. Springer, Dordrecht

McCullagh P, Nelder J (1989) Generalized linear models. Chapman and Hall, London

Mi C, Huettmann F, Guo Y, Han X, Wen L (2017) Why to choose random forest to predict rare species distribution with few samples in large undersampled areas? Three Asian crane species models provide supporting evidence. PeerJ. https://doi.org/10.7717/peerj.2849

Miller K, Huettmann F, Norcross B, Lorenz M (2014) Multivariate random forest models of estuarine-associated fish and invertebrate communities. MEPS 500:159–174

Miller K, Huettmann F, Norcross B (2015) Efficient spatial models for predicting the occurrence of subarctic estuarine-associated fishes: implications for management. Fish Manag Ecol 22:501–517

Mogie M (2004) In support of null hypothesis significance testing. Proc R Soc Lond B 271:S82–S84

Mullet TC, Gage SH, Morton JM, Huettmann F (2016) Temporal and spatial variation of a winter soundscape in Alaska. Landsc Ecol 31:1117–1137

Murphy AH, Winkler RL (1992) Diagnostic verification of probability forecasts. Int J Forecast 7:435–455

Murphy MA, Evans JS, Storfer A (2010) Quantifying *Bufo boreas* connectivity in Yellowstone National Park with landscape genetics. Ecology 91:252–261

Murphy K, Huettmann F, Fresco N, Morton JM (2012a) Connecting Alaska landscapes into the future. U.S. Fish and Wildlife Service, And the University of Alaska. Prepared by the Scenarios Network for Arctic Planning (SNAP). www.snap.uaf.edu/attachments/SNAP-connectivity-2010-complete.pdf

Murphy K, Reynolds J, Whitten E, Fresco N, Lindgren M, Huettmann F (2012b) Predicting future potential climate-biomes for the Yukon, northwest territories, and Alaska: a climate-linked cluster analysis approach to analyzing possible ecological refugia and areas of greatest change. Prepared by the Scenarios Network for Arctic Planning (SNAP) and the EWHALE lab, University of Alaska-Fairbanks on behalf of The Nature Conservancy Canada, Government Northwest Territories. www.snap.uaf.edu/attachments/Cliomes-FINAL.pdf

Næss A (1997) Ecology, community and lifestyle: outline of an ecosophy (trans: D. Rothenberg). Cambridge University Press, Cambridge

Ohse B, Huettmann F, Ickert-Bond S, Juday G (2009) Modeling the distribution of white spruce (*Picea glauca*) for Alaska with high accuracy: an open access role-model for predicting tree species in last remaining wilderness areas. Pol Biol 32:1717–1724

Olden JD, Lawler JJ, Poff NJ (2008) Machine learning without tears: a practical primer for ecologists. Q Rev Biol 83:171–193

Oppel S, Huettmann F (2010) Chapter 8: Using a random forests moedel and public data to predict the distribution of prey for marine wildlife management. In: Cushman S, Huettmann F (eds) Spatial complexity, informatics and wildlife conservation. Springer, Tokyo, pp 151–164

Oppel S, Pain DJ, Lindsell J, Lachmann L, Diop I, Tegetmeyer C, Donald PF, Anderson G, Bowden CGR, Tanneberger F, Flade M (2011) High variation reduces the value of feather stable isotope ratios in identifying new wintering areas for aquatic warblers in West Africa. J Avian Biol 42:342–354

Oppel S, Strobl C, Huettmann F (2009a) Alternative methods to quantify variable importance in ecology. Technical Report Number 65, Department of Statistics, University of Munich, Germany

Oppel S, Powell AN, Dickson DL (2009b) Using an algorithmic model toreveal individually variable movement decisions in a wintering sea duck. J Anim Ecol 78:524–531

Oppel S, Meirinho A, Ramírez I, Gardner B, O'Connell A, Miller PI, Louzao M (2012) Comparison of five modelling techniques to predict the spatial distribution and abundance of seabirds. Biol Conserv 156:94–104

Oppel S et al (2017) Landscape factors affecting territory occupancy and breeding success of Egyptian Vultures on the Balkan Peninsula. J Ornithol 158:443–457

Ott R (2005) Sound truth & corporate myth: the legacy of the Exxon Valdez oil spill. Dragonfly Sisters Press, Cordova

Pearce J, Ferrier S (2000) Evaluating the predictive performance of habitat models developed using logistic regression. Ecol Model 133:225–245

Phillips SJ, Dudík M, Schapire RE (2004) A maximum entropy approach to species distribution modeling. In: Proceedings of the 21st international conference on machine learning. ACM Press, New York, pp 655–662

Phillips SJ, Anderson RP, Schapire RE (2006) Maximum entropy modeling of species geographic distributions. Ecol Model 190:231–259

Phillips SJ, Elith J (2013) On estimating probability of presence from use–availability or presence–background data. Ecology 94:1409–1419

Pittmann S, Huettmann F (2006) Chapter 4: Seabird distribution and diversity. An ecological characterization of the Stellwagen Bank national marine sanctuary region: oceanographic, biogeographic, and contaminants assessment. In: Battista T, Clark R, Pittmann S (eds) Prepared by NCCOS's Biogeography Team in cooperation with the National Marine Sanctuary Program. Silver Spring, MD. NOAA Technical Memorandum NCCOS 45

Popp J, Neubauer D, Huettmann F (2007) Using TreeNet for identifying management thresholds of mantled howling monkeys' habitat preferences on Ometepe Island, Nicaragua, on a tree and home range scale. J Med Biol Sci 1(2):1–14 www.scientificjournals.org/journals2007/articles/1096.pdf

Prasad A, Iverson L, Matthews S, Peters M (2009) Atlases of tree and bird species habitats for current and future climates. Ecol Restor 27:260–263

Quinn G, Keough Q (2004) Experimental design and data analysis for biologists. Cambridge University Press, Cambridge

Core Team R. (2016) R: a language and environment for statistical computing. R foundation for statistical computing. www.r-project.org

Recknagel F (2001) Applications of machine learning to ecological modelling. Ecol Model 146:303–310

Reich Y, Barai SV (1999) Evaluating machine learning models for engineering problems. Artif Intell Eng 13:257–272

Ribiero Jr., P J., Diggle PJ (2013) Package 'geoR'. www.leg.ufpr.br/geoR

Ritter J (2007) Species distribution models for Denali national park and preserve, Alaska. Unpublished M.Sc. thesis, University of Alaska-Fairbanks (UAF), Alaska

Rosten E, Drummond T (2006) Machine learning for high-speed corner detection. In: European conference on computer vision. Springer, pp 430–443

Royle JA, Chandler RB, Yackulic C, J D N (2012) Likelihood analysis of species occurrence probability from presence-only data for modelling species distributions. Methods Ecol Evol 3:545–554

Schaub M, Kery M (2012) Combining information in hierarchical models improves inferences in population ecology and demographic population analyses. Anim Conserv 15:125–126. https://doi.org/10.1111/j.1469-1795.2012.00531.x

Schmitt S, Pouteau R, Justeau D, Boissieu F, Birnbaum P (2017) ssdm: An r package to predict distribution of species richness and composition based on stacked species distribution models. Methods Ecol Evol 8:1795–1803. https://doi.org/10.1111/2041-210X.12841

Shipp MA, Ross KN, Tamayo P, Weng AP, Kutok JL, Aguiar RC, Gaasenbeek M, Angelo M, Reich M, Pinkus GS et al (2002) Diffuse large B-cell lymphoma outcome prediction by gene-expression profiling and supervised machine learning. Nat Med 8:68–74

de Smith MJ, Goodchild MF, Longley PA (2007) Geospatial analysis: a comprehensive guide to principles, techniques, and software tools. Troubadour Publishing, Ltd., Leicester

Stephens PA, Buskirk SW, Hayward GD, Martinez del Rio C (2007) A call for statistical pluralism answered. J Appl Ecol 44:461–463. https://doi.org/10.1111/j.1365-2664.2007.01302.x

Strobl C, Boulesteix A-L, Zeileis A, Hothorn T (2007) Bias in random forests variable importance measures: illustrations, sources and a solution. Research Report Series/Department of Statistics and Mathematics, 40. Department of Statistics and Mathematics, WU Vienna University of Economics and Business, Vienna

Strobl C, Boulesteix A-L, Kneib T, Augustin T, Zeileis A (2008) Conditional variable importance for random forests. Bioinformatics 9:307. https://doi.org/10.1186/1471-2105-9-307

Strobl C, Malley J, Tutz G (2009) An introduction to recursive partitioning: Rationale, application, and characteristics of classification and regression trees, bagging, and random forests. Psychol Methods 14:323–348. https://doi.org/10.1037/a0016973

Strogatz SH (2001) Exploring complex networks. Nature 410:268–276

Thuiller WB, Lafourcade R, Engler J, Araujo MB (2009) BIOMOD a platform for ensemble forecasting of species distributions. Ecography 32:369–373. https://doi.org/10.1111/j.1600-0587.2008.05742.x

Venables WN, Ripley BD (2002) Modern applied statistical analysis, 4th edn. Springer, New York

Wei C et al (15 co-authors) (2011) A global analysis of marine benthos biomass using random forests. Public Libr Sci 5:e15323

Weinstein BG (2018) A computer vision for animal ecology. J Anim Ecol 87:533–545. https://doi.org/10.1111/1365-2656.12780

Wickert C, Wallschlaeger D, Huettmann F (2010) Spatially predictive habitat modeling of a white stork (*Ciconiaciconia*) population in former East Prussia in 1939. Open Ornithol 3:1–12

Wilson EO (1998) Consilience: the unity of knowledge. Alfred A Knopf, Inc., New York

Wisz MS, Hijmans RJ, Peterson AT, Graham CT, Guisan A, NCEAS Predicting Species Distributions Working Group (2008) Effects ofsample size on the performance of species distribution models. Divers Distrib 14:763–773

Whittingham MJ, Stephens PA, Bradbury RB, Freckleton RB (2006) Why do we still use stepwise modelling in ecology and behaviour? J Anim Ecol 75:1182–1189

Yackulic CB, Chandler R, Zipkin EF, Royle JA, Nichols JD, Campbell Grant EH, Veran S (2012) Presence-only modeling using MAXENT: when can we trust the inferences? Methods Ecol Evol 4:236–243

Yen P, Huettmann F, Cooke F (2004) Modeling abundance and distribution of Marbled Murrelets (*Brachyramphusmarmoratus*) using GIS, marine data and advanced multivariate statistics. Ecol Model 171:395–413

Young B (2012) Diversity in the boreal forest of Alaska: distribution and impacts on ecosystem services. Unpublished PhD thesis. University of Alaska-Fairbanks (UAF), Fairbanks

Zar JH (2009) Biostatistical analysis, 5th edn. Prentice Hall, Upper Saddle River

Zuckerberg B, Huettmann F, Frair J (2011) Data management as a scientific foundation for reliable predictive modeling. In: Drew A, Wiersma Y, Huettmann F (eds) Predictive modeling in landscape ecology. Springer, New York

Chapter 3
Boosting, Bagging and Ensembles in the Real World: An Overview, some Explanations and a Practical Synthesis for Holistic Global Wildlife Conservation Applications Based on Machine Learning with Decision Trees

Falk Huettmann

> *"My goal is simple. It is a complete understanding of the universe, why it is as it is and why it exists at all."*
>
> Stephen Hawkins

3.1 Introduction

Machine Learning offers a very different mindset than the traditional analysis (e.g. Zar 2010). It is not based on just one constrained single approach and algorithm (e.g. linear regression, normal distribution and other parametric assumptions; Harrell 2001, Reinhart 2015), but it extends Bayesian approaches (Hobbs and Hooten 2015) and easily consists of over 100 algorithms (Hastie et al. 2009; Ferandez-Delgado et al. 2014). That number of algorithm tools is still growing, and new concepts, implementations and approaches are still developed and open up for many platforms and disciplines (Witten et al. 2011, Aggarwal 2015; see for updates here https://en.wikipedia. org/ wiki/ Machine_learning). This represents a major paradigm shift in the sciences, in quantitative analysis and how to do business on earth, for conservation, policy making and global well-being. The earlier notion of employing just one binding algorithm, parsimony, reductionism and a mandatory use of information-theoretic concepts by the government in charge - as promoted before for resource management (see Verner et al. 1986, Silva 2012 and Zar 2010) - now

F. Huettmann (✉)
EWHALE Lab, Biology and Wildlife Department, Institute of Arctic Biology,
University of Alaska-Fairbanks, Fairbanks, AK, USA
e-mail: fhuettmann@alaska.edu

© Springer Nature Switzerland AG 2018   63
G. R. W. Humphries et al. (eds.), *Machine Learning for Ecology and Sustainable Natural Resource Management*, https://doi.org/10.1007/978-3-319-96978-7_3

fades out quickly, and rightly so (Loftus 1996): the reductionistic approach (Burnham and Anderson 2002) does not perform well for nature and natural resources, and never really has! (Naess 1989; Liu et al. 2018) It violates most what Aldo Leopold and good conservation stands for (Leopold and Meine 2013; http://www.steadystate.org/brian-czech/). From the vast quantitative analysis landscape that we see ahead of us, now most of it is in shambles (Alexander 2013; Cockburn 2013) and so is their conservation (Kurt 1982; Mace et al. 2010). It's the Anthropocene that is causing this disruption in conservation and the decay in quantitative analysis (Smith and Zeder 2013).

For a new approach and an urgently needed re-start (Breiman 2001a) in science-based natural resource conservation management, here I will focus on three dominant tree-based algorithms and their combinations: boosting, bagging and ensembles. While those algorithms are a somewhat arbitrary and convenient choice from all the Machine Learning algorithms available they have not only been already introduced early (O'Connor et al. 1996; Fielding 1999) but also perform very highly (Kampichler et al. 2010, Elith et al. 2006, Chunrong et al. 2017) and can be useful for progressing a wide variety of applications (e.g. Yen et al. 2004, Drew et al. 2011, Kandel et al. 2015; see for instance Reich and Barai 1999 for engineering, Kononenko 2001 for medicine and Dhar 1998 for economics). At minimum, those methods add to the tool box that otherwise is fully discriminated against when using Machine Learning (see Romesburg 1989, Anderson et al. 2000, Burnham and Anderson 2002 or Silva 2012 for examples), or what just employs maxent (Phillips and Dudik 2008) or worse (e.g. maxlike, Merow and Silander 2014; see Oppel et al. 2012 for low performance). As referred to in this text, boosting and bagging are usually based on recursive partitioning and they essentially represent optimized extensions of classification and regression trees (CARTs; Breiman et al. 1984, https://www.salford-systems.com/support/spm-user-guide/help). Based on the paradigm in Machine Learning that 'many weak learners make for a very strong learner' (Schapire 1990, 1992; Schapire and Singer 1999; Hastie et al. 2009), it is probably rather suitable to perceive the methods of boosting and bagging all as ensembles in their own right. Arguably, much is still to be learned about those methods (Mueller and Massaron 2016) and their applications (Hochachka et al. 2007) and potentials. While still growing these methods and situations demand for a new ethics in regard to conservation-based analysis in many of their aspects. However, a maturity was reached that makes it difficult to deny them their participation, if not even their monopoly and a lead in the global management arena of natural resources. Some of those details will be shown next.

3.2   A Quick Refresher on Linear Models (LMs), Parsimony and Classification and Regression Trees (CARTs)

Although using linear models (LMs) or classification and regression trees (CARTs) are not a base-requirement for boosting and bagging as a concept, they tend to play a big role in the overall discussion and application for these methods. For generic concepts and history, these methods are very well treated elsewhere (see for instance Hilborn and Mangel 1997, Venables and Ripley 2002 and MacNally 2000 for LMs, Burnham and Anderson 2002 for parsimony and Breiman et al. 1984, 2001a and De'ath and Fabricius 2000 for CARTs). In the following I will simply refer to the major characteristics of boosting and bagging and put them in context with inference and their applications.

I have heard people making reference to linear models (LMs) as being part of Machine Learning; but one must beg to differ (see Hegel et al. 2010 for overview in wildlife conservation; see Elith et al. 2006 and Oppel et al. 2012 for the performance of LMs). While Machine Learning represents a really big 'family' of algorithms (Ferandez-Delgado et al. 2014), methods that employ binary recursive partitioning tend to be consistently on the forefront of performance (Mueller and Massaron 2016; see Oppel et al. 2012 and Chunrong et al. 2017 for real-world applications). That is not to say that other methods and concepts, e.g. splines, neural networks, support vector machines, entropy or others are less relevant or not powerful; far from it (Hastie et al. 2009, Mueller and Massaron 2016). But here I simply focus on tree (CART)-based methods because they are already widely tested and discussed, are robust, offer good software implementations, and are still widely underused, even mis-understood, in the 'community' of natural resource managers (see for their wide absence in Verner et al. 1986, Romesburg 1989, Silva 2012). CARTs are convenient and readily available models to exemplify the power of Machine Learning overall, including boosting and bagging! In the near-future the concept of Machine Learning will obviously grow further (Aggarwal 2015). It is worthwhile to agree here that the generic concept of linear regression is actually part of some internal and initial CART optimizations (Breiman et al. 1984; De'ath and Fabricius 2000). Secondly, I found that leaving linear regressions 'unconstrained' and applying them without parametric assumptions can, and sometimes does, lead to suitable predictions (but they are usually topped by more sophisticated algorithms; see for instance in the BIOMOD2 package in R). However, this means to apply LMs in the sense of Breiman (2001a; prediction used for inference) unconstrained, but not like suggested by Burnham and Anderson (2002) with parsimony and hypothesis testing. Clearly, the wider failure of LMs really comes to show when they are assumed to be strict parametric tools and when using them with parsimony and for 'inference only' in a reductionistic mindset (as promoted by Manly et al. 2002 and Burnham and Anderson 2002, Johnson et al. 2008, for instance). An even higher level of error is introduced when using them with assumed (but not well-tested) distributions such as poisson and logistic link functions (Keating and Cherry 2004 for performance); those assumptions are very often not well met and just theoretical mathematical ones without a good sense of reality (*sensu* McArdle 1988). Most of those latter LMs applications cannot compete much with the usual Machine Learning applications, e.g. Elith et al. (2006), Oppel et al. (2012), see Fox et al. (2017) for an applied example. Consequently, one will not see them used and published much in modern data mining and prediction projects. Machine Learning dominates there in the wide absence and ignorance of practitioners as shown in their textbooks (Verner et al. 1986; Romesburg 1989; Silva 2012).

3.3   Boosting

3.3.1   What Boosting is in a Nutshell

Boosting refers essentially to a sequence of algorithms, each explaining the left-over variance of the previous model (This is inherent in the notion of '*many weak learners make for a strong learner*' which is described further below in detail). Boosting is essentially a fine-tuned sequence, or chain, of model runs. The more often this concept is applied, the more variance gets 'explained down' each time and eventually the model explains 'most' of the variance in the system (as represented in the initial data cube to be modeled). So a system of high variance becomes explained through the repetitive use of an algorithm to capture and explain it. The name of 'boosting' refers to a solution that must boost the low accuracy of a weak learner to the high accuracy of a strong learner (Schapire 1990). Essentially this concept is maximizing (='boosting') the variance explained using individual steps (Freund and Schapire 1997; see also Breiman 1998 for variations and De'ath 2007 for applications). Similar to deep learning, this method is by now a mainstay in Machine Learning (Hastie et al. 2009; Mueller and Massaron 2016).

The concept of 'boosting' can essentially be applied to any algorithm, it's not just limited to 'binary recursive partitioning' and CARTs as such (see Drucker et al. 1993 for neural networks for instance). Boosting of linear regressions for instance can lead to rather interesting and eye-opening insights, e.g. when compared to the flawed AIC log linear regressions as mandated by federal governmental agencies (Akaike 1974, Burnham and Anderson 2002; see for Guthery et al. 2005 and Arnold 2010 for an assessment). However, boosting with tree-algorithms proved to be very successful as a combination. One reason is that 'trees' break virtually any linear dependencies in the data, as they are otherwise known as a requirement from linear regressions (which makes them part of parametric statistics; Hastie et al. 2009). Being able to break those linearities is due to the fact of binary partitioning and it being 'recursive' (Breiman et al. 1984). Thus, a binary splitting rule can be used again and again throughout the entire 'tree' and dataset. This allows to make good use of the many dimensions in a data set. Correlations in the data can then actually be used for explaining data and for predictions and such type of inference (Breiman 2001a,   https://www.salford-systems.com/news/salford-systems-introduces-cart). Belonging to the group of non-parametric methods (Hastie et al. 2009), this makes for a new concept that many scientists and conservationists still need to wrap their head around and which needs to be taught, e.g. Silva (2012). It opens up many new avenues and also makes existing analysis paths stronger. Further, boosting became 'stochastic boosting' (Friedman 2001) as a specific and very powerful variation. It is stochastic because a random element gets introduced in the data analysis (e.g. Breiman 1998, Friedman 2001) and for what makes the testing and the internal assessment data. That way, the entire data set gets 'ploughed' through for its information and for a meaningful optimum to be found in the data analysis. There are many variations how boosting is implemented, and those fine-tuning settings can

Table 3.1 A selection of fine tuning parameters that make boosting and bagging so powerful, beyond just the underlying algorithm as such[a]

Fine-tuning parameter	How done	Why it matters	Software and reference
Weighting	Puts a well-balanced weight on specific samples, e.g. to predict 'best'.	Allows for imbalanced samples and for 'finding a needle in the hay stack' (=data mining)	SPM8
Random draw	Draw columns and rows in a specific fashion (e.g. weighted)	Allows improved performance of the bootstrap aspect	SPM8, Breiman (1998)
mtry (bagging)	Select a specific number of subsetted predictors	Allows to find the best size of interactions in a complex data cube	SPM8
Prediction metric (ROC, Gains, Entropy, LL Average)	Choose the most powerful metric to express prediction performance	Better optimization for predictions	SPM8
Focus on predictions	Optimize models based on predictive performance (not model fit)	Allows for best possible inference (Breiman 2001a, b)	SPM8

[a]Many of those fine-tunings are the recipe for the great performance of boosting and bagging and thus the guarantee of business success. Those are virtually not shared and thus not really available in R and similar packages and codes

make a major change in the performance overall (see Table 3.1). Arguably, boosting as such is not automatically a guarantee for a good model *per se*, and other settings also play a role and can overrule. However, if the 'boosting machine' is applied right it makes for one of the best solutions for classification, regression and prediction available to mankind. Boosting can indeed create a great 'Learner'.

3.3.2   Short History of 'Boosting'

'Boosting' started with several algorithms such as ADABOOST (Freund and Schapire 1997). A major step forward was presented by Friedman (2001, 2002) leading eventually to the commercial TreeNet implementation with Salford Systems Ltd. (SPM8, https://www.salford-systems.com/). Along this way many other variations of boosting were pursued, developed and exist (e.g. LogitBoost, GentleAdaBoost, CoBoosting, BrownBoosting etc.; Hastie et al. 2009) and more are developed still (see online http://www.boosting.org/) exemplifying the magnitude and depth of this approach. Fielding (1999) is among the first to introduce such methods for conservation and ecology. By now, boosting enjoys a wide set of applications (see for instance Drew et al. 2011 and Cai et al. 2014 for wildlife conservation).

3.3.3 Why is Boosting so Powerful?

The notion of sequential approaches to variances explained is not really new. Linear regressions have been tried, but they are limited in that a linear regression does not bring anything new to the data once employed and when re-employed. Once their variance gets explained down, nothing more can usually be done with linear models (LMs) because they are 'static'. As long as one stays in the classic mindset (Zar 2010) it's a dead end with them. But in tree-based methods new options exist to look at remaining variances each time (that's due to recursive partitioning, and specifically in stochastic approaches). Boosting, in general, does not really focus on the predictive performance for optimization. But if that is achieved (like done in TreeNet; Friedman 2000) then it becomes very sophisticated. The latter is an approach that is fully in line with Leo Breiman (2001a, b; predictions as the goal) and thus it makes it for a very powerful approach to inference, like also done in bagging (see next section, Breiman 2001b).

3.4 Bagging

3.4.1 What Bagging is in a Nutshell

Like with boosting, bagging is usually based on 'trees', binary recursive partitioning. Also, it is a technique that summarizes many 'trees' (which classifies it *per se* as an ensemble; Breiman 2001b). But bagging, as a scheme, the same as it is done in boosting, could be applied with many algorithms. Bagging differs from boosting in two parts: i) it subsets rows as well as columns (similar to bootstrapping), and ii) it has a specific procedure to average out the trees for 'the best' result. Both of these steps are sophisticated and are reasons for why bagging is such a success. The performance sits once more in how those details are really implemented (Table 3.1). Breiman (1996, 2001b) presented a version of bagging, called random forests and which seems to be among the top classifiers, world-wide (Ferandez-Delgado et al. 2014). A base version was released online (https://www.stat.berkeley.edu/~breiman/RandomForests/) and subsequently in R (by A. Liaw https://cran.r-project.org/web/packages/randomForest/randomForest.pdf), and a commercial version was implemented by Salford Systems Ltd. Details are provided below in the next section.

3.4.2 Short History of Bagging

Bagging in 'trees' is credited to the work by Leo Breiman (1996, 2001a, b) and his former PhD student Adele Cutler (https://www.stat.berkeley.edu/~breiman/RandomForests/; see also Cutler et al. 2007). However, previous work is based on

CARTs, by Leo Breiman et al. (1984). Many of the publicly available random forests algorithms use the code presented by Breiman and Cutler (Breiman 2001a, b). Publicly made available packages in R and Python are essentially just wrappers of that code but which leave the relevant questions of fine-tuning and ultimate inference to the coder and user (Table 3.1). This is a big flaw in those public implementations of random forests because most users lack the testing and understanding and thus, it leaves out much of the award-winning performance that is found in random forests. Consequently, several random forests implementations and applications can be found in the literature that are substandard, see Table 3.2 for a discussion.

3.4.3 Why Bagging is so Powerful

Bagging involves drawing random (re-sampling) samples from rows (bootstrapping), which is a relatively simple and old procedure as such (Efron and Tibshirani 1993). Trees get build from each of those subsamples and then summarized. Whereas the real innovative part is in the random draw of columns (=predictors). Users usually did not do that because they pre-selected and then 'hugged' their predictors at all costs and wanted to keep all of them for outermost inference. But instead, when 'bootstrapping predictors' one obtains more robust information from them! That way, random forests 'rarely overfits a model (that means, it always uses a lower amount of predictors than what the data allow). Thus, always just a subset of the predictors is used, and then, the best tree gets inferred from that eventually. All of it is fine-tuned for optimization. There is another 'trick' in bagging, and that is, it gets optimized for the best prediction. Whereas linear regressions, as an example, do not work that way. Their optimization essentially is a) relatively primitive (based on the 'least squares', at best (an approach that is over a century old), and b) based on minimizing the variance (r^2, Zar 2010). Whereas in bagging, an overall optimization on the predictions (=metrics of ROC/AUC; Fielding and Bell 1997) allows for a higher level of generalization, and thus, robust inference (Breiman 2001a). The emphasis is thus put on prediction for inference and that is where the power sits; whereas r^2 values have less meaning and relevance in that discussion (the author knows of models that have a relatively low r^2 but a high ROC/AUC and thus perform rather well when tested with alternative evidence). The often-demanded practice to compute a pseudo- r^2 in Machine Learning should be dismissed because r^2 is derived from a linear concept (Zar 2010) and virtually impossible to mimic in Machine Learning algorithms, e.g. for Neural Networks. It thus should be left alone and just be used for linear regressions (which tend to perform low anyways and thus not a real option) (Textbox 3.1).

Table 3.2 A selection of random forests algorithm implementations and their details (Note that many more are described in Ferandez-Delgado et al. 2014 using C++ and Fortran etc)

Name of the package	Software	Source citation	Performance assessment	Specific feature
Random forests	Fortran	Breiman (2001b)	The original raw code with relevant settings	Raw code
SPM8 Random forests	C++	https://www.salford-systems.com/products/randomforest	An optimized commercial version of the initial Breiman (2001b) algorithm	GUI, weigthing, optimized settings
randomForest	R	Liaw and Wiener (2002) https://cran.r-project.org/web/packages/randomForest/randomForest.pdf	The raw code in R (r port) leaving relevant performance enhancements to the user	regression (added), get tree, combine
Biomod2	R	Thuiller et al. https://cran.r-project.org/web/packages/biomod2/index.html	Use of the raw code from (Breiman ad Cutler via Liaw and Wiener in R). Not making use of the relevant fine-tuning options to show the true performance of random forests	Adds other algorithms in parallel as an ensemble model overall
Scikit learn	python	http://scikit-learn.org/stable/modules/generated/sklearn.ensemble.RandomForestClassifier.html	Similar to the above, emphasize on ensembles	All relevantsettings

Textbox 3.1 On the suggested 'best' use of competing landcover, altitude and climate layers for better inference with Machine Learning: Being inclusive beats being exclusive and parsimony (The 'Kitchen Sink' model tends to win)

There is a wide range of habitat information to pick from these days. Often, those layers are competing with each other to be picked and used. A classic human approach is 'to select the best' and throw out any other layers as inferior and dismiss them from 'science'. But with Machin Learning - boosting and bagging- better approaches now exist and to obtain robust inferences by using all data at hand in an optimized fashion for the benefit of a best possible inference, and subsequent conservation.

The way to approach such problems in Machine Learning is to allow correlations and to understand that some data have strengths in some areas, and weaknesses in other areas of the data space. If tree-based algorithms are used, correlations can be powerful to use, interactions can be addressed, and best options for a given case can be found and used in such a Machine Learning analysis. Binary re-cursive partitioning allows for a new approach, often a better one than done in previous times.

What is proposed here is to keep many layers, similar sounding by name but different in the detail, and let the Machine Learning handle it to find the best prediction, and inference. One may refer to it as 'the kitchen-sink model. That is, all data available get thrown in at first, and if adjustments are to be made, then one may act accordingly, but should do only very reluctantly and always to keep in mind to focus on best predictive metrics, less so r^2. It's the 'smart' algorithm that tries to sort things out and to optimize the best possible prediction for inference.

Like it initially appeared to be with ensembles (Elder 2003) and when *'many weak learners can create a superior learner'* (Friedman 2001, 2002), this is another example where the analysis approach in Machine Learning might at first appear counter-intuitive, and where the human wish to always 'select for the best and for the superior' and 'to narrow things down' so that it all gets very easy fails and makes things less informative. But unless tested and shown, this concept pf narrowing things down is not always the best option, nor is Ecology easy ad simple (Naess 1989) or parsimonious (as demanded claimed by Burnham and Anderson 2002).

Table 3.1.1 shows examples of competing climate layers, elevation and landcover data analysis. Figure 3.1.1 shows a concept on how to run such an analysis, and how to interpret them, and for drawing 'the best possible' inference.

Table 3.1.1 Example of repetitive-content datasets that should be used for SDM, habitat analysis and data mining with Machine Learning for best-available inference of data available

Topic	Data set	Reference	Comment
Climate	IPCC	www.ipcc.ch	Many sources (modeled data) exist within.
	NCAR	climatedataguide.ucar.edu/	A complex and leading 3D climate data set (modeled).
	CRU	www.cru.org	A somewhat coarser but widely used climate dataset from UK.
	Worldclim	www.worldclim.org	A standard by now, See version 2 for update.
	CHELSA	www.chelsa-climate.org	This dataset claims superiority over Worldclim but Suwal et al. (submitted) shows otherwise for mountain areas in Asia.
Landcover	GLC2000	https://ec.europa.eu/jrc/en/ scientific-tool/ global-land-cover	All landcover maps are far from perfect, vary in their legends and are usually not up-to-date (A lag time of +10 years is common).
	Maryland Landcover	http://glcf.umd.edu/research/ portal/geocover/	
	GlobeCover V2.2	https://earth.esa.int/web/ guest/-/ globcover-v22-land-cover- product-now-available-5999	
Elevation[a]	WorldClim	www.worldclim.org	A standard for many applications, e.g. food security and species distribution models (SDMs).
	ETOPO2	https://www.ngdc.noaa.gov/ mgg/global/etopo2.Html	Old elevation data now superseded by ETOPO1.
	ETOPO1	https://www.ngdc.noaa.gov/ mgg/global/global.html	One of the best elevation data sources (also for oceans).
	GMTED2010	https://topotools.cr.usgs.gov/ gmted_viewer/	An elevation data source by USGS.

[a]It must be stated here that elevation is often a core data set in natural resource models, and also the base source for slope, aspect and ruggedness. Clearly, those data sets are far from error-free and thus create many subsequent errors in the inference, especially when 'only' model fits are used without further and alternative evidence tests

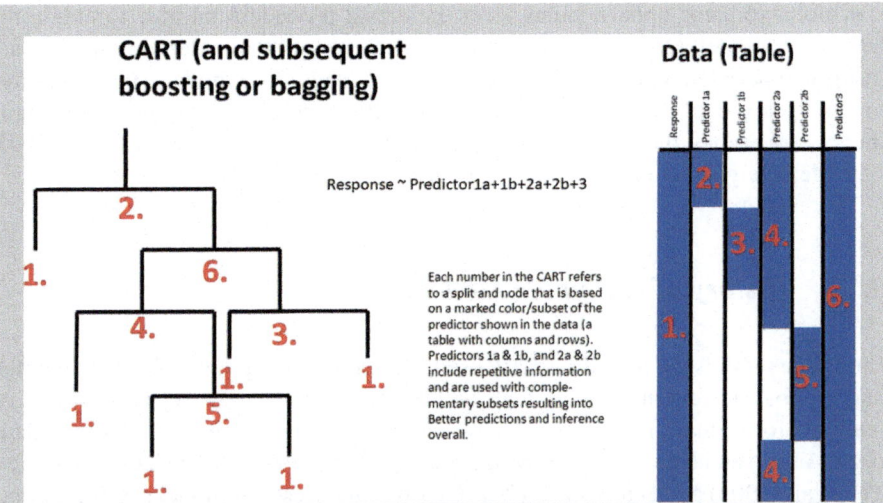

Fig. 3.1.1 Recursive binary partitioning

The concept proposed here works is almost opposite to what is usually done and suggested. So why does that work ? Several answers can be given. As long as tree-based analysis using boosting and bagging is used - as part of Machine Learning - one data point can help to inform and better interpret another data point and towards a better response and model prediction overall. The inference is then done through the prediction itself. Secondly, theory and education has just not caught up with reality.

Whatever the reasons are, it is highly suggested to test those concepts and compare, in order to find best possible solutions and inferences to advance conservation, worldwide. The proof sits in the pudding, clearly.

3.5 Ensemble Models

3.5.1 What is an Ensemble Model?

In generic terms, ensemble models simply consist of many models underneath and combine them into one single output. The actual definition of what ensemble models really are, and how combined into one, is not so precise though. Random forest qualifies as an ensemble model, and in a way, and so can boosting be called an ensemble model (that's because they are built on, and derived from, many models and then summarize into a single model overall; Elder 2003, Hastie et al. 2009). However, in most applications, ensembles consist of many algorithms and can be much more complex than just bagging and boosting. While the actual definition of ensembles remains open-minded, so is the actual procedure on how to obtain the best, final, model. Often in an ensemble model many of the underlying models are just run in parallel, and the best result is selected from those models (instead of just one merged model overall). Development is still going on, but arguably, ensemble models are among the best classifiers and predictors possible by now (Elder 2003; Ferandez-Delgado et al. 2014; Mueller and Massaron 2016).

3.5.2 History of Ensemble Models

Ensemble models are already widely used for many years, but they became popular and a global success in recent climate and financial models. The IPCC is using such methods for climate predictions for over a decade already (http://www.ipcc-data. org/). Araujo and New (2007) provided a summary for species distributions, but also Yen et al. (2004) and Hardy et al. (2011) showed their use for model predictions and natural resource contexts. Biomod2 and SPM8 (details see Table 3.2) are software packages that easily allow to run model algorithms in parallel and to then to select 'the best' result from it (SPM8 extends them to the use of 'batteries'; see Chap. 8 this publication). However, in the Machine Learning discipline, ensemble models are nothing new really and have been used for a long time in various forms and concepts (Hastie et al. 2009; Mueller and Massaron 2016). It should be stressed that random forests, and even many boosting algorithms are essentially ensemble models. By now, ensemble models became a global standard when predictions are the goal.

3.5.3 Why Ensemble Models are so Powerful

According to theory (e.g. Zar 2010 and Burnham and Anderson 2002), model fitting practices and following the paradigm 'garbage in garbage out' (GIGO) ensemble models are not to work well. Their success is therefore commonly referred to as a 'paradox' (Elder 2003) and not so easy to explain when following the classic wisdom of statistics and western science. But ensemble models offer several advantages over the classic approach to just fitting data in a parsimonious fashion.

(i) A key scheme in the success of ensembles might be the diversity of algorithms applied to a data problem. That way, always 'the best' model algorithm gets used and applied for a certain section of the data. It matches the best ability of algorithms in the ensemble with suitable data subsets. A whole plethora of algorithms can be applied in parallel, and only the best models remain and are used, e.g. optimized for predictions. This is a powerful approach because often the exact data details, and the best-possible model suiting the data, remains unknown before the model is actually run. It can be surprising for the practitioner what makes the best model for a dataset. Using ensembles offers many model algorithms to be optimized during the analysis run, just as the data allow. It's therefore inherently non-parametric (Hastie et al. 2009).

(ii) The actual models, optimized in their subsequent own runs, can get averaged and further optimized by a model-weighted model contribution (aka '*ensemble of an ensemble*'), creating a form of synergy that simple and singular model algorithms can usually not achieve.

(iii) Because the 'correct' data fit, nor such models are really known ahead of the analysis, errors and biases caused by a pre-perceived model and algorithm choice (=human expert error) are kept to the bare minimum.

(iv) If the prediction metric is optimized for, it allows for the best prediction as well as for the best inference all in one run, based on 'the best' algorithms employed.

(v) For very complex data and model applications it might well be that statistical theory has not well kept up with those complexities (e.g. Zar 2010), and thus these cases cannot be fit nor explained well simply for those reasons, yet the model performs well regardless of the theory and their institutions and experts lagging behind! The science of chaos and percolation theories are good examples for that (see Forman 1995 for real-world landscapes examples). And more of those cases are likely to exist and to be discovered the more models are run on very diverse data situations. Our statistical theory is widely behind the data complexities in nature, but it must never hold us back for good progress in conservation management!

As a matter of fact, ensemble models are currently the best-possible inference for mankind on the data, if the best algorithms are used underneath and combined appropriately. Arguably, the better and the more meaningful the data are, the better will ensemble models, any model, perform! While the data are widely incomplete (as it is specifically the case in nature and in natural resource management), ensembles remain the analysis path of choice. Ignoring or avoiding them - knowingly - is not only poor science, it actually harms the natural resource. The implications for science and education, as well as for governance and real-life, are tremendous and not even really fathomed yet, which makes for a rather scary situation.

3.6 Model Applications and Inference

3.6.1 Boosting Experiences and Applications

Stochastic gradient boosting is a rather strong concept *per se* and allows for a wide range of applications (see for instance De'ath 2007 for Ecology). I use it daily for rapid assessments, first error checks and pattern detections in data to gauge further analysis. It is specifically used for data mining, variance reduction and pattern detections in data across disciplines. Using it for predictions, and when optimized, becomes really strong for inference. I noticed a few atypical cases where boosting can be erroneous for overfitting (=where a good model cannot be achieved, even when fine-tuned and with good predictors that achieve well in other algorithms). In those cases -rare though - I tend to use 'stumps' (very simple split trees) and cut back on predictors and node depths.

Using the R implementations of boosting often asks for known distributions of data to be analyzed; I found this a question impossible to answer in complex data mining applications beforehand, and thus, I see it as a major drawback in the implication (R). In my own work, primarily with SPM8, I use boosting for instance to explore data and to mine data for their initial patterns. Once those 'signals' in the data are found and become clear, I then follow up with other tools for a test, and whether those patterns can even improve further (often not the case though and thus I remain with the initial analysis!).

3.6.2 Bagging Experiences and Applications

Bagging is popular in several ways. By now, it is among the most widely used high-performance classifier out there (Ferandez-Delgado et al. 2014). It's the method of choice for many remote sensing classification and modeling problems (e.g. Evans and Cushman 2009; Evans et al. 2010; Chunrong et al. 2017; see Cutler et al. 2007 for Ecology). Myself, I see it as the prime prediction machine for many species model applications (Drew et al. 2011; see Yen et al. 2004; Kandel et al. 2015; Chunrong et al. 2017 for a test and application with species of high conservation profile). Using random forests with regression problems can prove less successful for the unexperienced user. Ensembles models are suggested in such cases.

Unfortunately, the vast implications of random forests in order to change, improve and progress science in itself have not been fully realized by the natural resource conservation management community. First steps have been done, but arguably, many large ones are to come still, e.g. to use RandomForest as the base approach for obtaining a robust 'Learner.' One way or another, this will move the conservation management of natural resources into computer-aided decision-making, for instance, beyond also having many other impacts, e.g. job descriptions, publications, ethics, business models and education and institutions. Textbooks may be re-written and are to be extended accordingly.

3.6.3 Ensembles

The global discussion on climate change - arguably the major scheme of our time and certainly for natural resource management as we know it - is driven by a set of ensemble models (IPPC). There is no need to stress the relevance of this approach, and of its underlying algorithms, e.g. Baltensperger and Huettmann (2015). While bagging is on the rise, so are ensemble models even more so (see Elder 2003, for an application: Jiao et al. 2014).

3.6.4 Precautionary, Pro–Active, and Predictive Models for Better Resource Conservation Management

Applications that involve predictions for natural resource management are not only mature established, but for globally relevant ones they now become the norm (Table 3.3 for overview). In addition, the true relevance of those approaches is still not widely embraced, nor is climate change and how to manage natural resources in a predictive fashion (e.g. see in Silva 2012; but see O'Connor et al. 1996, Nielsen et al. 2008 and Chunrong et al. 2016 for examples). As long as man-made climate change is not accepted as a key problem, one cannot go ahead much with a meaningful approach to predictions used for policy. We are thus in true violation of the UN Agreement on Pre-cautionary, Pro-Active Management Principles. One can only be pro-active, in the best available fashion, when being predictive. And best predictions are only possible when using best-available approaches (See Tables 3.2 & 3.4) That is what Leo Breiman (2001a, b) and others offer, but that is widely missing from the textbooks on conservation and natural resource management (Verner et al. 1986; Romesburg 1989; Silva 2012). Table 3.4 shows some disciplines where Machine Learning still needs to be established further for progress.

Table 3.3 A small selection of application fields of CARTs, boosting, bagging and ensembles for wildlife conservation worldwide

Application	Algorithm	Detail of progress achieved	Citation
Climate prediction	Ensembles	Global progress of climate for sustainable decision-making	IPCC (https://www.ipcc.ch)
Species forecast	Ensembles	First-time forecast of the species niche for an endemic subspecies	Lawler et al. (2006, 2011), Hardy et al. (2011), Chunrong et al. (2016)
Human health	CART	First-time connection on a national scale between avian diversity and human health	O'Connor et al. (1996)
	Boosting & Bagging	First-time global model for influenza	Herrick et al. (2013)
Rapid assessment	Boosting & Bagging	First-time model for that species over large areas	Kandel et al. (2015), Regmi et al. (2018)

Table 3.4 A small selection of application fields for Machine Learning-driven wildlife conservation worldwide but currently still left open and not used to the full potential overall

Application	Justification and explanation
Wildlife and landscape ethics	How machines can make 'the best' decision for mankind and for the earth.
Definition of 'the best' decision	Inclusion of human and wildlife welfare as well as global sustainability for the next millennium, e.g. through metrics and feedback.
Definition of 'the best' method and algorithm	Selection of 'the best' approach overall.
Oversight of the process	How implemented in a democratic process, e.g. co-management.
Alternative information to assess quality from machine-driven decisions	Comparison and controls are required.

3.7 A Commonly Heard Criticism and Misunderstanding of Machine Learning, and Characteristics of Man–Made Science and Conservation Driven by Reductionism

Arguably, tree-based methods are highly performing, but also are slow in the implementation with the sciences, specifically when it comes to conservation management of natural resources, and policy worldwide (Table 3.4). So how can it be that the best-available conservation science paradigm has not resulted in a faster use of Machine Learning while we face an unprecedented global conservation crisis? Do we not need better methods, better policy, better actors or all together? Arguably, human systems, the dominating cultures and their institutions are way too slow to acknowledge and to implement good research, and relevant research for progress. This is specifically true for conservation and environmental issues (see Berthold 2016 for a real-world example), which get marginalized in our current scheme of economy and globalization for over five decades (e.g. Czech et al. 2000). Perhaps the current governance has an affiliation with destruction (e.g. Alexander 2013; Cockburn 2013)?

While virtually all relevant industries and online services are heavily driven by Machine Learning and data mining (Mueller and Massaron 2016) -the sciences, specifically conservation of natural resource management as well as the humanities have not made yet much strides to use those tools. In conservation management, we are over 40 years behind, and subsequently lost the battle against the industries that are harming the wilderness thus far. Our legal guidance and Massive long-term flaws hit much the notion that ecology is complex (Naess 1989, and as described already by Aldo Leopold and others 100 years ago; Silva 2012) and thus, any approach of simplification must be biased, missing relevant bits and pieces. The discussion about p-values, hypothesis testing and AIC show that simplification, ignorance and bias rule, all driven by the science view of reductionism (parsimony) and not making good use of available tools (Cushman and Huettmann 2007; Drew et al. 2011). This situation shows the massive flaws of a so-called science-driven management western style-, and when it gets truly confronted with the state of global conservation (Mace et al. 2010). Arguably, a reductionist view just relates to the concept of singling out factors 'at will' and to manage or resolve them (conveniently) by a narrowly targeted effort and done through money and/or law. Instead, *the problem is more complicated than we had thought*, and this approach is not only biased and seriously limited, it is in violation of what Aldo Leopold suggested (Leopold and Meine 2013). It widely failed us and is to be replaced with a more holistic scheme, as expressed in a multivariate framework of thought. Machine Learning can achieve exactly that, and in a fast way. So it must be on the agenda as the method of choice and in an open access and open source fashion.

3.8 Synthesis and Outlook

Machine Learning is a wide field offering hundreds of variations, if not thousands of algorithms and their implementations to tackle a conservation problem in a holistic fashion for good progress. Tree-based methods like boosting, bagging and ensembles are a big part of this approach; and they are known to be extremely powerful, certainly convenient.

What will the future hold for those methods and for the Earth? Arguably, we are all moving with Machine Learning and Data Mining into Artificial Intelligence (AI) and Deep Learning, that's specifically true for robots and associated drone applications and inferences. In conservation and resource management, the institution and people are meant to be at the core of the decision-making process. This is achieved when the people and institutions decide with democratic principles on resource management and conservation planning with the assistance of computing and up-to date decision making tools - including considerate ethics -, instead of using expert's opinion that are not based on the latest up to date tools.

However, the people are assisted by computing and aided through decision-making tools, all while we see increasingly the failure of traditional experts (Perera et al. 2010). It is here where Machine Learning can help in a good way (e.g. Huettmann 2007). While not perfect, boosting and bagging has reached a level of maturity, and accuracies of over 80%, and sometimes way over 95% accuracies have been observed for global models. How much of an accuracy do we really need? It is unlikely that major changes will occur in boosting and bagging any time soon. The methods are there, mature and stable! However, a few things are increasingly developing: a better embedding and workflow, rise in computing power, more applications overall, and an emphasis on decision-making process that has Machine Learning, Deep Learning and Artificial Intelligence, at its core.

Our current conservation problem is not so much defined anymore really by which (single) algorithm and model to choose, but to be on the Machine Learning platform overall, online,with all data freely available, free open-access robust softwares, and then to implement the obtained predictions in a pro-active fashion before more damage occurs on earth and its atmosphere. While the Machine Learning methods are perfected further, the real culprit is by now, policy, governance, the human aspect and its role on sustainable management of the earth and universe. Apart from major ethical questions that's where the biggest effort is to be placed by now for conservation management of natural resources worldwide. Boosting and bagging are here to stay and thus to be embraced for the best-possible global sustainable outcomes.

Acknowledgement I thank Profs R. O'Connor and A.W. (Tony) Diamond for an early workshop on statistics with ACWERN at UNB, Canada introducing me in the late 1990s to tree-based techniques (CART) and multivariate analysis. I thank Dan Steinberg and Salford Systems Ltd. for a workshop with U.S. IALE at Snowbird, Utah, as well as with The Wildlife Society, Alaska Chapter, for a wider debate and introduction of tree-based methods, boosting and bagging. I am indebted to U.S.IALE, the Global Primate Network in Kathmandu, Nepal, Medical University Taipeh, Taiwan, and the Wildlife Institute of India in Dheradun for their workshop promotion and support. Thanks to S. Linke, I. Presse, B. Walter, G. Regmi, M. Suwal, R. Lama, C. Cambu, H. Hera, S. Sparks, Y. Subaru, H. Berrios and the many members of the -EWHALE lab- at UAF for their discussions and partly, support. This is EWHALE lab publication #187.

References

Aggarwal C (2015) Data mining: the textbook. Springer

Akaike H (1974) A new look at the statistical model identification. IEEE Trans Automat Contr AC-19. Institute of Statistical Mathematics, Minato-ku, pp 716–723

Alexander JC (2013) The dark side of modernity. Polity Press, Cambridge

Anderson DR, Burnham KP, Thompson WL (2000) Null hypothesis testing: problems, prevalence, and an alternative. J Wildl Manag 64:912–923

Araujo MB, and New M (2007) Ensemble forecasting of speies distributions. Trends in Ecology and Evolution 22:42–47

Arnold TW (2010) Uninformative parameters and model selection using Akaike's information criterion. J Wildl Manag 74:1175–1178

Baltensperger AP, Huettmann F (2015) Predicted shifts in small mammal distributions and biodiversity in the altered future environment of Alaska: an open access data and Machine Learning. PLoS One. https://doi.org/10.1371/journal.pone.0132054

Berthold P (2016) Mein Leben fuer die Voegel. Kosmos Publisher, Berlin

Breiman L (1996) Bagging predictors. Mach Learn 26:123–140

Breiman L (1998) Arcing classifier (with discussion and a rejoinder by the author). Ann Stat 26(3):801–849. https://doi.org/10.1214/aos/1024691079

Breiman L (2001a) Statistical modeling: the two cultures (with comments and a rejoinder by the author). Stat Sci 16:199–231

Breiman L (2001b) Random forests. Mach Learn 45:5–32

Breiman L, Friedman J, Stone CJ, Olshen RA (1984) Classification and regression trees. CRC Press, Boca Raton

Burnham KP, Anderson DR (2002) Model selection and multimodel inference: a practical information-theoretic approach. Springer, New York

Cai T, Huettmann F, Guo Y (2014) Using stochastic gradient boosting to infer stopover habitat selection and distribution of hooded cranes *Grus monacha* during spring migration in Lindian, Northeast China. PLos ONE 9. https://doi.org/10.1371/journal.pone.0097372

Chunrong M, Huettmann F, Guo Y (2016) Climate envelope predictions indicate an enlarged suitable wintering distribution for great bustards (*Otis tarda dybowski*) in China for the 21st century. PeerJ 4:e1630. https://doi.org/10.7717/peerj.1630

Chunrong M, Huettmann F, Guo Y, Han X, Wen L (2017) Why choose random Forest to predict rare species distribution with few samples in large undersampled areas? Three Asian crane species models provide supporting evidence. PeerJ 5:e2849. https://doi.org/10.7717/peerj.2849

Cockburn A (2013) A colossal wreck: a road trip through political scandal, corruption and American culture. Verso Publishers, New York

Cutler DR, Edwards TC, Beard KH, Cutler A, Hess KT, Gibson J, Lawler JJ (2007) Random forests for classification in ecology. Ecology 88:2783–2792. https://doi.org/10.1890/07-0539.1

Czech B, Krausman PR, Devers PK (2000) Economic associations among causes of species endangerment in the United States. Bioscience 50:593–601

De'ath G (2007) Boosted trees for ecological modeling and prediction. Ecology 88:243–251

De'ath G, Fabricius K (2000) Classification and regression trees: a powerful yet simple technique for ecological data analysis. Ecology 81:3178–3192 https://doi.org/10.1890/0012-9658(2000)081[3178:CARTAP]2.0.CO;2

Dhar V (1998) Data mining in finance: using counterfactuals to generate knowledge from organizational information systems. Inf Syst 23:423–437

Drew CA, Wiersma Y, Huettmann F (eds) (2011). Predictive Species and Habitat Modeling in Landscape Ecology. Springer, New York

Drucker H, Schapire R, Simard P (1993) Boosting performance in neural networks. Int J Pattern Recognit Artif Intell 7:705–771

Efron B, Tibshirani R (1993) An introduction to the bootstrap. Chapman & Hall/CRC Monographs, New York

Elder JF (2003) The generalization paradox of ensembles. J Comput Graph Stat 12:853–864

Elith J, Graham CH, Anderson RP, Dudík M, Ferrier S, Guisan A, Hijmans RJ, Huettmann F, Leathwick JR, Lehmann A, Li J, Lohmann LG, Loiselle BA, Manion G, Moritz C, Nakamura M, Nakazawa Y, Overton J, Peterson AT, Phillips SJ, Richardson K, Scachetti-Pereira R, Schapire RE, Soberón J, Williams S, Wisz MS, Zimmermann NE (2006) Novel methods improve prediction of species' distributions from occurrence data. Ecography 29:129–151

Evans JS, Cushman S (2009) Gradient modeling of conifer species using random forests. Landsc Ecol 24:673. https://doi.org/10.1007/s10980-009-9341-0

Evans JS, Murphy MA, Holden ZA, Cushman SA (2010) Modeling species distribution and change using random forest. Predictive species and habitat modeling in landscape ecology, pp 139–159

Ferandez-Delgado M, Cernadas E, Barrow S, Amorim D (2014) Do we need hundreds of classifiers to solve real world classification problems? J Mach Learn Res 15:3133–3181

Fielding A (1999) Machine learning methods for ecological applications. Springer, Boston

Fielding A, Bell Y (1997) A review of methods for the assessment of prediction errors in conservation presence/absence models. Environ Conserv 24:38–49

Forman RTT (1995) Land mosaics: the ecology of landscapes and regions. Cambridge University Press, Cambridge

Fox CH, Huettmann, F, Harvey GKA, Morgan KH,. Robinson J, Williams R and Paquet PC (2017) Predictions from Machine Learning ensembles: marine bird distribution and density on Canada's Pacific coast. Marine Ecology Progress Series 566:199–216

Freund Y, Schapire RE (1997) A decision-theoretic generalization of on-line learning and an application to boosting. J Comput Syst Sci 55:119–139

Friedman JH (2001) Greedy function approximation: a gradient boosting machine. Ann Stat 29:1189–1232

Friedman JH (2002) Stochastic gradient boosting. Comput Stat Data Anal 38:367–378

Guthery FS, Brennan LA, Peterson MJ, Lusk LL (2005) Information theory in wildlife science: critique and viewpoint. J Wildl Manag 69:457–465

Hardy SM, Lindgren M, Konakanchi H, Huettmann F (2011) Predicting the distribution and ecological niche of unexploited snow crab (*Chionoecetes opilio*) populations in Alaskan waters: a first open-access ensemble model. Integr Comp Biol 51(4):608–622. https://doi.org/10.1093/icb/icr102

Harrell FE Jr (2001) Regression modeling strategies: with applications to linear models, logistic regression, and survival analysis. Springer, New York

Hastie T, Tibshirany R, Friedman J (2009) The elements of statistical learning: data mining, inference, and prediction. Springer Series in Statistics

Hegel TSA, Cushman JE, Huettmann F (2010) Current state of the art for statistical modelling of species distributions. Chapter 16. In: Cushman S, Huettmann F (eds) Spatial complexity, informatics and wildlife conservation. Springer, Tokyo, pp 273–312

Herrick KA, Huettmann F, Lindgren MA (2013) A global model of avian influenza prediction in wild birds: the importance of northern regions. Vet Res. https://doi.org/10.1186/1297-9716-44-42

Hilborn R, Mangel M (1997) The ecological detective: confronting models with data. Princeton University Press, Princeton

Hobbs NT, Hooten M (2015) Bayesian models: a statistical primer for ecologists. University Press, Princeton

Hochachka W, Caruana R, Fink D, Munson A, Riedewald M, Sorokina D, Kelling S (2007) Data mining for discovery of pattern and process in ecological systems. J Wildl Manag 71:2427–2437

Huettmann F (2007) Modern adaptive management: adding digital opportunities towards a sustainable world with new values. Forum on Public Policy: Clim Chang Sustain Dev 3:337–342

Jiao S, Guo Y, Huettmann F, Lei G (2014) Nest-site selection analysis of hooded crane (*Grus monacha*) in northeastern China based on a multivariate ensemble model. Zool Sci 31:430–437

Johnson DS, Thomas DL, Ver Hoef JM, Christ AD (2008) A general framework for the analysis of animal resource selection from telemetry data. Biometrics 64:968–976

Kampichler C, Wieland R, Calmé S, Weissenberger H, Arriaga-Weiss S (2010) Classification in conservation biology: a comparison of five machine-learning methods. Ecol Inform 5:441–450

Kandel K, Huettmann F, Suwal MK, Regmi GR, Nijman V, Nekaris KAI, Lama ST, Thapa A, Sharma HP, Subedi TR (2015) Rapid multi-nation distribution assessment of a charismatic conservation species using open access ensemble model GIS predictions: red panda (*Ailurus fulgens*) in the Hindu-Kush Himalaya region. Biol Conserv 181:150–161

Keating KA, Cherry S (2004) Use and interpretation of logistic regression in habitat- selection studies. Journal of Wildlife Management 68:774–789

Kononenko I (2001) Machine learning for medical diagnosis: history, state of the art and perspective. Artif Intell Med 23:89–109

Kurt F (1982) Naturschutz-illusion. Paul Parey Publisher, Berlin Germany

Lawler JJ, White D, Neilson RP, Blaustein AR (2006) Predicting climate-induced range-shifts: model differences and model reliability. Glob Chang Biol 12:1568–1584

Lawler JJ, Yo W, Huettmann F (2011) Designing predictive models for increased utility: using species distribution models for conservation planning, forecasting, and risk assessment. In: Drew CA, Wiersma Y, Huettmann F (eds) Predictive modeling in landscape ecology. Chapter 5. Springer, New York, pp 271–290

Leopold A, Meine C (2013) A sand county almanac & other writings on conservation and ecology. Library of America, New York

Liaw A, Wiener M (2002) Classification and regression by randomforests. R News 2(3):18

Liu J, Dou Y, Batistella M, Challies E, Conno T, Friis C, DA MJ, Parish E, CL R, Bl BS, Triezenber H, Yang H, Zhao Z, Zimmerer KS, Huettmann F, Treglia M, Basher Z, Chung MG, Herzberger A, Lenschow A, Mechiche-Alami A, Newig A, Roch J, Sun J (2018) Spillover systems in a telecoupled Anthropocene: typology, methods, and governance for global sustainability. Environ Sustain 33:58–69. https://doi.org/10.1016/j.cosust.2018.04.009

Loftus GR (1996) Psychology will be a much better science when we change the way we analyze data. Curr Dir Psychol 5:161–171

Mace G, Cramer W, Diaz S, Faith DP, Larigauderie A, Le Prestre P, Palmer M, Perrings C, Scholes RJ, Walpole M, Walter BA, Watson JEM, Mooney HA (2010) Biodiversity targets after 2010. Environ Sustain 2:3–8

MacNally R (2000) Regression and model-building in conservation biology, biogeography and ecology: the distinction between – and reconciliation of – 'predictive' and 'explanatory' models. Biodivers Conserv 6:655–671

Manly FJ, McDonald LL, Thomas DL, McDonald TL, Erickson WP (2002) Resource selection by animals: statistical design and analysis for field studies, Second edn. Kluwer Academic Publishers, Dordrecht

McArdle (1988) The structural relationship: regression in biology. Can J Zool 66: 2329–2339

Merow C, Silander JA (2014) A comparison of Maxlike and Maxent for modelling species distributions. Methods Ecol Evol 5:215–225

Mueller JP, Massaron L (2016) Machine Learning for dummies. For Dummies Publisher, 435 p

Næss A (1989) Ecology, community and lifestyle: outline of an Ecosophy (trans: Rothenberg D). Cambridge University Press, Cambridge

Nielsen SE, Stenhouse GB, Beyer HL, Huettmann F, Boyce MS (2008) Can natural disturbance-based forestry rescue a declining population of grizzly bears? Biol Conserv 141:2193–2207

O'Connor R, Jones MT, White D, Hunsacker C, Loveland T, Jones B, Preston E (1996) Spatial partitioning of environmental correlates of avian biodiversity in the Conterminuous United States. Biodivers Lett 3:97–110

Oppel S, Meirinho A, Ramírez I, Gardner B, O'Connell AF, Miller PI, Louzao M (2012) Comparison of five modelling techniques to predict the spatial distribution and abundance of seabirds. Biol Conserv 156:94–104

Perera AH, Drew A, Johnson CJ (2010) Expert knowledge and its application in landscape ecology. Springer, New York

Phillips SJ, Dudik M (2008) Modelling of species distributions with Maxent: new extensions and a comprehensive evaluation. Ecography 31:161–175

Regmi GR, Huettmann F, Suwal MK, Nijman V, Nekaris KAI, Kandel K, Sharma N and Coudrat C (2018). First Open Access Ensemble Climate Envelope Predictions of Assamese Macaque Macaca Assamensis in South and South-East Asia: A new role model and assessment of endangered species. Endangered Species Research 36:149–160 https://doi.org/10.3354/esr0088

Reinhart A (2015) Statistics done wrong: The woefully complete guide. No Starch Press. San Francisco

Reich Y, Barai SV (1999) Evaluating Machine Learning models for engineering problems. Artif Intell Eng 13:257–272

Romesburg HC (1989) More on gaining reliable knowledge. J Wildl Manag 53:1177–1180

Schapire RE (1990) The strength of weak learnability (PDF). Machine learning, vol 5. Kluwer Academic Publishers, Boston, pp 197–227. https://doi.org/10.1007/bf00116037

Schapire RE (1992) The design and analysis of efficient learning algorithms. MIT Press, USA

Schapire RE, Singer Y (1999) Improved boosting algorithms using confidence-rated predictors. Machine Learning 37:297–336

Silva NJ (2012) The wildlife techniques manual: research & management. 2 volumes. The Johns Hopkins University Press; Seventh edn

Smith BD, Zeder MD (2013) The onset of the Anthropocene. Anthropocene 4:6–13

Venables WN, Ripley BD (2002) Modern applied statistical analysis, 4th edn. Springer, New York

Verner J, Morrison ML, Ralph CJ (1986) Wildlife 2000. Modeling habitat relationships of terrestrial vertebrates. University of Wisconsin Press, Madison

Witten IH, Frank E, Hall MA (2011) Data mining: practical machine learning tools and techniques, 3rd edn. Morgan Kaufman Publisher, Amsterdam

Yen P, Huettmann F, Cooke F (2004) Modelling abundance and distribution of marbled Murrelets (*Brachyramphus marmoratus*) using GIS, marine data and advanced multivariate statistics. Ecol Model 171:395–413

Zar JH (2010) Biostatistical analysis, 5th edn. Prentice Hall, Upper Saddle River

Part II
Predicting Patterns

Artwork by Catherine Humphries

"Information is the oil of the 21st century, and analytics
is the combustion engine."
– Peter Sondergaard

Chapter 4
From Data Mining with Machine Learning to Inference in Diverse and Highly Complex Data: Some Shared Experiences, Intellectual Reasoning and Analysis Steps for the Real World of Science Applications

Falk Huettmann

> *"The problem posed for biologists by the real world seldom, if ever, have an exact, correct statistical solution. The assumptions of nearly all techniques are violated to a greater or lesser extent by real data."*
>
> McArdle (1988)
>
> *"In a Time of Universal Deceit — Telling the Truth Is a Revolutionary Act."*
>
> Attributed to George Orwell
>
> *"With this method the dangers of parental affection for a favorite theory can be circumvented."*
>
> T. Chamberlin (1890)

4.1 Introduction

For a long time, scientists had limited data and they applied experiments in isolation to test and forward knowledge on a given subject (Salsburg 2001 and Conner 2005 for overview). It's a western tradition that then became applied to modern questions. It is from this perspective that frequentist statistics evolved (Chamberlin 1890; Berkson 1942), its methods became established and promoted (Popper 1945) then widely taught as a 'good practice' (Zar 2010) with many specific statistical tests that could be carried out (e.g. Table 4.1 and references within). It culminated in a study

F. Huettmann (✉)
EWHALE Lab, Biology and Wildlife Department, Institute of Arctic Biology,
University of Alaska-Fairbanks, Fairbanks, AK, USA
e-mail: fhuettmann@alaska.edu

© Springer Nature Switzerland AG 2018
G. R. W. Humphries et al. (eds.), *Machine Learning for Ecology and Sustainable Natural Resource Management*, https://doi.org/10.1007/978-3-319-96978-7_4

Table 4.1 Overview of some basic statistical tests for common data problems

Data situation	Recommended test	Reference	Comments
Normal distribution	Goodness of fit tests (Z scores, Shapiro–Wilk, Kolmogorov–Smirnov Goodness- of-Fit Test, Lilliefors and Anderson–Darling tests)	Filliben (1975), Zar (2010), Razali and Wah (2011)	Nature is virtually never normally distributed. Consequentially, those tests are purely theoretical without any relevant meaning for real life and conservation management applications (McArdle 1988).
Parametric: Test for differences among samples and experiments	Chi-square test (one way, or two way contingency table); widely used	Quinn and Keough (2004), Zar (2010)	Should only be done with a valid hypothesis testing framework. However, this assumes the underlying theory is correct and statistically met. It is possible to use a GLM as well (see below as well). There are many references that question the validity of the threshold (e.g. 0.05%) and others ague to "euthanize" p-values all together (Anderson et al. 2000, Anderson and Burnham 2002, Concato and Hartigan 2016, Stang et al. 2010 for tyranny of p-values).
Regression slope	Linear Regression Model (LM) (e.g. Zar 2010)	Zar (2010), McArdle (1988)	The existence of a slope is often seen as an 'effect', e.g. when compared to a flat line (no slope). However, these details are widely discussed. Usually LMs require a normal distribution of the errors and thus, they are not so realistic for natural processes.
Non-parametric: Test for differences among ordinal samples and experiments	Median Test, Mann- Whitney U-test, Kruskal-Wallis test, Wilcoxon- signed rank test	Venables and Ripley (2002)	There is a wider debate about the power of those tests. Arguably, they are more powerful than the parametric tests, but their predictive performance tends to be poor. And consequently, so is the inference, according to Breiman (2001a).
Interval/ratio data	T-test, ANOVA (F-test)	Thompson (2004)	One of the most frequently used statistical test procedures. However, its predictive performance is rather low, and so is the inference according to Breiman (2001a).

(continued)

Table 4.1 (continued)

Data situation	Recommended test	Reference	Comments
Multiple comparison tests	Bonferroni, Tukey, Scheffe	Quinn and Keough (2004)	There is a wide debate about the validity of those tests. See Rothman (1990) and Perneger (1998) for what's wrong with Bonferroni tests.
Advanced difference tests	MANOVA, ANCOVA, MANCOVA	Hillborn and Mangel (1997)	Those complicated tests rarely meet the real-world assumptions, unless data match the required research design and are free of interactions etc. (Nature is virtually never free of interactions and lacks such a research design)
Normal distribution of errors (Heteroscedasticity)	White test, modified Beusch-Pragan test	Quinn and Keough (2004), Zar (2010)	There should be no heteroscedasticity (i.e., variance of residuals should not increase with values fitted of response variable) and error bias in linear regressions. Often it gets rectified through Box Cox transformation (but which affects the original data for inference)
Confidence Intervals	Confidence intervals and/or standard error	Gardner and Altman (1986), Fidler and Loftus (2009)	Unless confidence intervals are truly assessed with alternative data, they contribute more to unproven claims (so-called confidence trick; Salsburg 2001, Reinhart 2015)
Power of a test and effect size	Alpha and beta levels, simulations	Greenland et al. (2016)	A key question in experimental testing, for sample sizes required and sensitivity setting of a valid inference
Multiple Regressions (many predictors)	Generalized Linear Models (GLM)	Hillborn and Mangel (1997), Venables and Ripley (2002), Quinn and Keough (2004)	The 'workhorse' for many statistical applications and in model selection studies. Usually applied in a logistic and parsimonious setting (Burnham and Anderson 2002, Manly et al. 2002). However, GLMs and GAMs are known to predict poorly (Elith et al. 2006), and they have very strong assumptions
Autocorrelation	Moran's I, Ripley's K	Venables and Ripley (2002)	There is a lot of debate about autocorrelation, e.g. to correct it or use it as a description (Swihart and Stade 1985; Betts et al. 2009).

template used for simplicity (Anderson et al. 2000; Burnham and Anderson 2002). Whereas, similar to good detective work (e.g. Hilborn and Mangel 1997) reflection should still sit at the core of human inquiry and knowledge, specifically for ecology and natural resource management (Naess 1989; Romesburg 1991; Dodds 2001; Stephens et al. 2007; Silva 2012; Stanton-Geddes et al. 2014) especially where modern computing now allows for more than just 'simplicity/parsimony'. Instead, this enables us to carry out powerful predictions (Venables and Ripley 2002) for a more holistic approach (McGarical et al. 2000; Drew et al. 2011) to tackle mankind's problems.

But complex real-world data situations are unknown and difficult for us to comprehend; much can be accomplished within those vast datasets and data cube (McGarical et al. 2000). This is an inherent characteristic of data from nature (McArdle 1988). It applies even more to multivariate problems when many predictors are employed (McGarigal et al. 2000; Drew et al. 2011). Despite many claims made (Zar 2010; see also Anderson and Burnham 2002; Manly et al. 2002; Silva 2012 or even Chamberlin 1890; Berkson 1942 and Popper 1945), there are no good and standard rules to generalize multivariate problems because every data set tends to be different and unique, requiring powerful analytical approaches instead. Reflection is required. Nature is not symmetrical, not linear and not normal distributed; and that reality applies to ecology and natural resources, which has been known and expressed for decades (e.g. Naess 1989; Yoccoz 1991; Dodds 2001). In human medicine and psychology those details are also well known and have been discussed for long time (e.g. Salsburg 1985; Loftus 1996; Lambdin 2012; Rinehart 2015). This is also true in education (Thompson 2004; Ziliak and McCloskey 2009). As early as 1988, McArdle (1988) had already stated clearly that "*in the I-wish-it-were-so land of theoretical statistics …the data cannot be made to conform to the assumptions…*" Thus, violations and surprises can easily occur in biological data sets (Elith et al. 2006; Hastie et al. 2009), and those data defy many other assumptions, theory and untested expectations such as parametric ones (Zar 2010). Making those data cases all equal, parametric, and putting them through the same analysis steps - without much reflection in a study template (as promoted by Burnham and Anderson 2002) - does them no good justice nor the many cases and realities they are representing. The peculiarities of the data records can be easily lost that way. But on the bright side, if one manages to resolve those peculiarities, with accepted methods that lead to a valid generalization, then the inventors of those methods are likely to end up in permanent positions (for life!). This is what data miners are doing for a living (Breiman 2001a, b; Mueller and Massaron 2016) and likely why data science is the 'hottest' job of the early 21st century: it's a profession to turn marginal data into defendable information (i.e., extracting the signal from a very complex set of information). See Chap. 1 in this book for the evolution and historic details of this concept of machine learning.

In the following, I will elaborate on the task of data mining in real life applications and contrast it with the usual approach that is still taught in virtually all universities as the gold standard (Ziliak and McCloskey 2009; Zar 2010; for natural resource management textbook see for instance in Silva 2012) (Fig. 4.1).

Schools of Thought' in the Science Disciplines*

Generic	Example Statistics	Example Economy
School of Thought 1	Frequency Statistics	Neoclassic Economy
School of Thought 2	Bayesian Statistics	Environmental Economy
School of Thought 3	Machine Learning	Ecological Economy

* Commonly more than 3 schools of thought occur, and also based on universities and stakeholders involved. It is nothing unusual to see that those 'schools' have an unequal amount of influence and compete for those roles with fluctuating success. The history of science, and conservation, shows nothing but that, almost worldwide.

Fig. 4.1 Some views of statistics and how an analysis is to be done in a so-called rigorous fashion

4.2   Model Selection with Many Predictors as an Analysis Scheme and as a Major Platform for Statistical Testing, Prediction and Inference

Moving forward from linear regression - described above- is multiple regression; it's old in concept (Chamberlin 1890) and comes in various shapes and forms, e.g. generalized linear models (GLMs), general additive models (GAMs), mixed models and polynomial regressions (e.g. Venables and Ripley 2002). Those are usually perceived as a 'higher art' and conceptually they can interpret deeper complexities in datasets, but many of which those methods cannot really handle by design (parametric, linear).

In mathematics and statistics, model selection is not well resolved; no mutual agreement exists on how to do it, and what methods and metrics to really use (Stephens et al. 2007; Arnold 2010). Mathematical models are not finite (e.g. McArdle 1988) and can potentially contradict each other, even on the same questions and their data (a classic example here is the question whether light consists either of particles or rays; both can be shown mathematically; details found in any Physics textbook). For unresolved model selection ambiguities, this is true numerically, biologically, for management and philosophically (see Stephens et al. 2007 and Strobl et al. 2007 for examples). It must therefore come as a big surprise to see how frequently model selection is taught and presented as a finite answer to a problem. It is usually done with a statistical template (Gigerenzer 2004), as it would be adding up 1-by-1 (where all predictors and their variances act without interactions and no synthesis) and with parsimony being finite and having 'certainty' (Burnham and Anderson 2002). All of this is to be carried without a conceptual and philosophical awareness (Guthery et al. 2001, 2005) without reflection

(Stephens et al. 2007) and often even using stepwise methods (see Whittingham et al. 2006 for a review and critique). It's like a ritual (Gelman 2008) which shuffles all concerns aside (Bandura 2007). Of particular concern must be the statement made in those works about having found 'the best' model and its explanatory variables (Burnham and Anderson 2002). Due to correlations, interactions and repressor variables (unusually and disproportionally dominant variables in the set of predictors that reduce and outcompete other predictors resulting into a skew and bias in model selection, Strobl et al. 2007; Oppel et al. 2009), the exercise of model selection as a scheme can quickly turn into deep philosophy and how to approach science, sampling and its data (see Conner 2005 and Rosales 2008 for real-world implications). I think it should definitely be treated as a philosophical and ethical question for a valid inference and justification of the scientific process. Without such an approach, the concept of valid inference must fail, nor will it ever be scientific, impartial and objective (Popper 1945). As the last reference shows, statistics and quantitative analysis was never to be done without philosophy (Dodds 2001). This can easily be seen in the PhD degree given for such analysis: a Doctorate of Philosophy. Clear guidance and readings exist to follow and to consider - e.g. Dodds (2001), Czech (2002), Bandura (2007), Daley and Farley (2010) - but they are widely ignored (as can easily be seen in the lack of such citations in most statistical and model selection publications in Ecology and Natural Resource Management). In the next section I will show how this conundrum can be addressed, and somewhat resolved with new tools and approaches at hand, namely machine learning and its wider applications (Table 4.2).

4.3   Confront Models with Data: Moving towards an Evidence–Based Analysis

Although rarely done in natural resource management, arguably, all data can be expressed and summarized as a (quantitative) model. Thus, the data then turn into a single formula (which essentially is a set of rules!) that basically 'clones'/mimics the training data, It represents an entire experimental setting, or an ecosystem (for example). This is the act of creating a 'learner'! (Schapire 1990, 1992; Freund and Schapire 1997; Schapire and Singer 1999; Friedman 2002). The beauty here obviously is that now you have captured the data and can express them (i.e., an entire ecological situation) as a formula with metrics to modify them; all done in software on your computer! But while virtually any linear regression application is doing just that, an even better summary can be achieved with non-linear methods provided through machine learning (Elith et al. 2006; Hastie et al. 2009; Mi et al. 2017). Often, this does not work out precisely, and it comes home to an approximation of the data set, and the mismatch is shown as the variance unexplained. Fig. 4.2 shows it in two examples (Fig. 4.3).

Table 4.2 A few methods used to choose 'the best' model (and its predictors) [Arguably, 'the best' is a subjective term, but in modeling it can be presented well when using performance metrics, e.g. Pearce and Ferrier (2000)]

Model selection Method + Ref	Pros + Ref	Cons +Ref	Comments
Visual (eye-balling)	None, but many applications of pure eye-balling exist, e.g. Manly et al. (2002), Johnson and Seip (2008).	Reinhart (2015)	Eye-balling is widely done and even approved of, whereas it is not precise and flips the entire reason and concept of working quantitatively (for more details see Reinhart 2015).
Univariate	It's fast and appears simple (Zar 2010).	Misses reality (McGarical et al. 2000).	Widely applied, although ecology and nature are known to be multi-variate, e.g. Naess (1989), McGarical et al. (2000).
p-values	Shows 'significant' explanations and predictors.	Significant-itis (Chia 1997) to be 'euthanized' (Anderson and Burnham 2002, Burnham and Anderson 2002).	Arguably, still the dominant approach in the sciences, e.g. Quinn and Keough (2004).
AIC (Burnham and Anderson 2002)	Finds 'one' (or the best) strong predictor. It implies due diligence and 'science well done'.	Overfit, biased, poor prediction (Arnold 2010, Guthery et al. 2005, Guthery 2008).	The performance metric of AIC has no biological nor mathemat-ical justify-cation. It is arbitrary.
Stepwise (Venables and Ripley 2002)	Fast; implies an exhaustive approach.	Misses reality Whittingham et al. (2006). Forward and backward approaches are not in good agreement with each other (Venables and Ripley 2002).	This is essentially a first, simplistic and naïve version of data mining but using the wrong (linear) methods.
Bayesian *	Use of priors (informed and uninformed ones), WinBugs algorithm etc. e.g. Hobbs and Hooten (2015)	Automated inference and subjective, e.g. Gelman (2008)	An alternative to frequency statistics and its inference, but poor predictions and such inference overall.
Maxent [a]	Fast, reliable (Elith et al. 2006),	Tends to be point-and-click, low skill and expertise needed; often outperformed (e.g. Mi et al. 2017). For inference discussion see Yackulic et al. (2012)	Among the best prediction, classification and inference tools. A prediction and machine learning approach.

(continued)

Table 4.2 (continued)

Model selection Method + Ref	Pros + Ref	Cons +Ref	Comments
Machine Learning: CART [a]	Fast, alternative to liner regressions; new insights (Breiman et al. 1984); De'ath and Fabricius (2000)	Usually outperformed by boosting and bagging. (Breiman 2001b, Elith et al. 2006).	A strong method from the 1980s. Breaks linearities in data. Often the basis for a great 'Learner.'
Machine Learning: Random Forest (bagging) [a]	Fast, very reliable (Mi et al. 2017), achieving high. Breiman (2001b; Cutler et al. 2007)	Often ignored or not known or favored by investigators and funders. Strobl et al. (2007) reports bias.	Among the best prediction, classification and inference tools, often used in ensembles.
Machine Learning: Boost-ing [a]	Fast, very reliable (Elith et al. 2006).	Often ignored or not known or favored by investigators and funders.	Among the best prediction, classification and inference tools; , often used in ensembles
Machine learning algorithms, e.g. Hastie et al. (2009), Fernandez-Delgado et al. (2014).	Powerful and flexible	Often ignored or not known or favored by investigators and funders.	A large field waiting to be explored further for progress.

[a]While these are concepts and algorithms as such they have inherent metrics and approaches to model selection, and are frequently used to rank, find and compare model selections. Eventually, any model selection is a function of the metric and concept/algorithm used

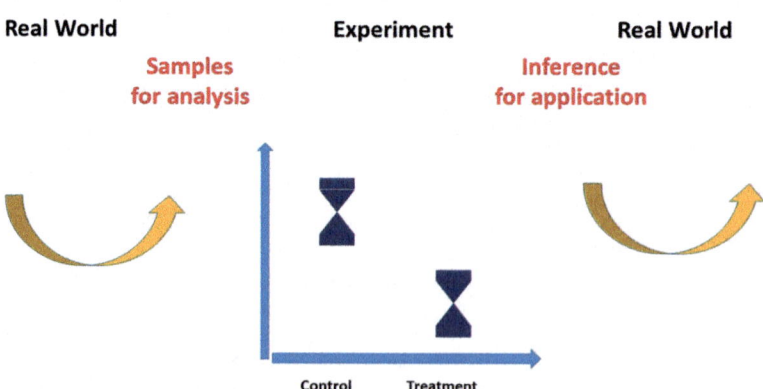

Fig. 4.2 "The Experiment", the real world, sampling and application

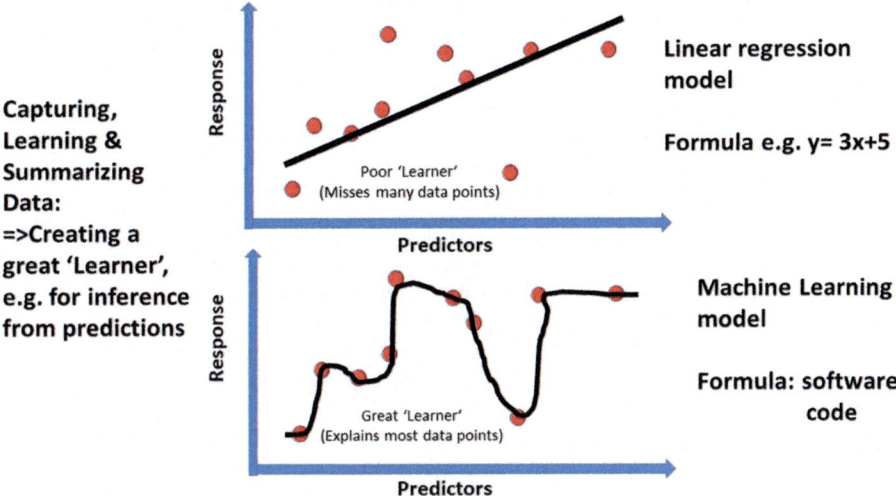

Capturing, Learning & Summarizing Data: =>Creating a great 'Learner', e.g. for inference from predictions

Fig. 4.3 Linear regression data summary vs machine learning summary

In modern terms, that unexplained 'noise' or variance (deviance, depending on the underlying distribution) is captured as the entropy (i.e., unexplained structure, chaos in the data). One may easily become more engaged in this discussion (see Georgescu-Roege 1971 for entropy), but for this section here we simply stick with the notion of a model that is fitting the data, and 'the rest' as unexplained. This 'rest' is a function of what the algorithm is able to capture from the data. Good algorithms capture more, if not almost all the data; whereas bad algorithms do not capture the data well, and others might add explanations that are not 'true to the data'. The latter case is generically referred to as overfitting, but which machine learning algorithms can optimize (see Hastie et al. 2009 and Mueller and Massaron 2016 for identifying local optima and best fitting approaches for best-possible generalizations from 'a learner'). These components are the essence of good algorithms and why there is such a chase in finding 'the best' ones (Fernando-Delgado et al. 2014). Now, 'the best' is usually explained as fitting the data, that is a mainstay of traditional forms of statistics (Hillborn and Mangel 1997; Zar 2010). But beyond the training data, better seems to be fitting the data for the best possible prediction, which only then allows for best-possible inference and generalization (*sensu* Breiman 2001a, b). The focus on best-predictions allows us to generalize from real-world situations, instead of just fitting in a narrow and traditional way some data onto a pre-defined theoretical model assumption creating bias (McArdle 1988; Zar 2010). Machine learning leads the way in this regard, whereas linear regression is shown to be less powerful and usually is widely dismissed if it comes to ecological multivariate interpretations (McGarical et al. 2000, for tests and applications see Elith et al. 2006; Oppel et al. 2012) and high-performance applications such as predictions (Fox et al. 2017; Han et al. 2018). When re-arranging, combining, optimizing or linking machine learning algorithms with each other (i.e., ensembles), one can even get higher performance

In ecology, nature and the universe
1 + 1 is usually not 2 !

Examples: - light (particles or rays)
                - time (can be 'bended', affected by gravity and is relative)
                - genetics (impossible to all understand for humans)
                - diseases and health (a widely non-quantitative discipline)
                - number of species (nobody has ever all counted them)
                - many populations, their individuals and dynamics,
                  e.g. fish, plankton, insects or fungi
                - predator-prey relationships (widely not understood)
                -habitat dynamics (e.g. percolation theory)

=> How to effectively study, quantify and manage ecology
and natural resource for sustainability ?

Fig. 4.4 Some simplistic and pragmatic reasoning like 1+1 = 2 cannot really be applied to ecology and natural resource management questions

(Araujo and New 2007, or boosting or bagging (essentially a specific form of ensembles; Hastie et al. 2009). Ensembles offer great opportunities due to being less biased due to their ability to make good use of many predictors across algorithms in a powerful way of computing (Elder 2003) (Fig. 4.4).

But any model that can capture data in 'good' terms is helpful for predictions, and we then tend to use it. Again, the prediction is the focus, not the algorithm *per se*. And why not? One would be ill-advised to ignore such an algorithm that provides progress.

Therefore, any model, any algorithm that provides progress is used, and others are usually dismissed (unless one finds value and insight for a comparison, let's say). What really matters now is to show that the evidence provided is true and generalizable. That is usually done by confronting the model with other data, alternative data and evidence (Hilborn and Mangel 1997). If those match, two lines of evidence are similar and overall progress is provided. It is very difficult to reject such a case and such logic due to its evidence. Ideally, more evidence is located and the model gets compared with it. Multiple lines of evidence is ideal and widely done, see for instance Huettmann et al. (2011) with 5 lines of evidence and Kandel et al. (2015) with 7 lines of evidence. All of those lines of independent evidence are in good agreement with each other supporting the prediction and thus generalization, and the method overall outlined here. The model should be used and interpreted then, but only then. This approach nicely follows a meta-analysis scheme (Schmidt and Hunter 2014) and proves to be rather successful for synthesis. That's because all existing and available information is used and interpreted. This means that no other evidence is left to show different results than those presented, and hence the argument is difficult to reject because it is based on 'best available' science. If that is all carried out in an open access framework, this form of reasoning is not only repeatable (Savalei and Dunn 2015) and transparent, but it also provides its products used for latest reasoning available as building blocks for people to use and

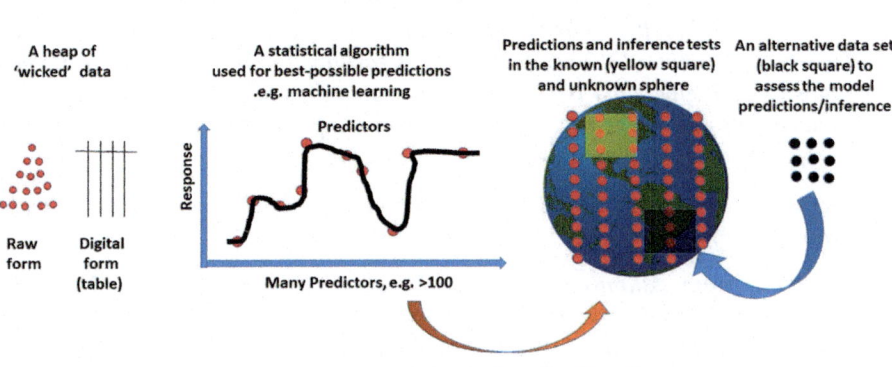

Fig. 4.5 Inference through a prediction machine (Learner)

to work from (Zuckerberg et al. 2011; Greenland 2012). Often, this consists of the best-available data and maps, let's say. That way, science truly did its due diligence and decisions are made from it. While well-known (e.g. Huettmann 2007), unfortunately this concept is not applied much, and the prime bottleneck still sits in the policy arena (Magness et al. 2008, 2011)! (Fig. 4.5)

4.4   A Real–World Data and Analysis Workflow Example

In the following I present a typical task, a common data problem and job description one might come across and encounter as a scientist in the Natural Resource arena. Also, it's common business for data miners and machine learning professionals:

4.4.1   Pre–Assumption (as Experienced in Real World Examples)

A data set that does not follow normal distribution, where errors are not distributed as you wish, samples are not really representative nor independent, but due to communication errors you have many of those types of (non-parameteric) samples …

Suggestion 1: Map the Data in Time and Space

Justification: Before any analysis can start one needs to see what the data are, and what they look like, (i.e., what shape and form they have).

Suggestion 2: Mine the data through in full detail (i.e., data exploration)

Justification: This first data mining exercise is less about an analysis and a quantitative understanding, but rather to get familiar with the data format and to see and find where the (technical) 'speed bumps' are. Learning how to handle data obtained, and how to do it efficiently is a key procedure and for effective data analysis (and associated time and budgeting to get the project completed). Often, those steps need to be repeated and the more one can do it, probably the better.

Suggestion 3: Understand the data, each column (i.e., covariate) and each combination you can test and assess

Justification: Many data are not clean and need to be understood first. Often what looks clean is not clean and needs a test and run first; this applies to columns and their content. That is especially the case with large data sets, and where not all content can be overlooked right away but shows as an odd result instead.

Suggestion 4: Talk to the initial data collector and read all documentation available

Justification: Ideally, such a communication should occur when the data are handed over. However, communication with the data are more effective after one is 'into' the dataset and knows them intimately. Arguably, a good head start on data can be helpful, but most data miners can figure out data sets on their own first. The real crucial content questions come afterwards and continue once the entire data set is analyzed.

Suggestion 5: Address all NAs and data errors.

Justification: According to the public wisdom, wrong data tends to produce wrong results. Cleaning data -technically and scientifically - is an essential but big task; there are different philosophies involved in how to approach it. It can determine the success and failure of such projects. Arguably, one wants 'good' data to run models. But 'life' is not always clean, and an alternative but effective and often not so problematic approach is to initially ignore the errors - use the raw data - and run the analysis and compare the findings subsequently. In machine learning, 'majority voting' can for instance overcome data errors. And different model runs from data cleaned at several levels should be done for comparison. If that cannot be done on the entire data set, it's worthwhile to run it on a subset of the data, or on a certain percentage of the data. The goal here obviously is to find patterns in the data, explain them, and how it all relates to NAs and errors, and what the impacts are on 'cleaning', e.g. to avoid running a self-fulfilling hypothesis with no new Information coming from the data analysis.

Suggestion 6: Predict the data onto themselves (create a great 'learner')

Justification: This step is an essential piece in understanding the inference, the data and the content. Usually, the more the training data can be 'mimicked' the better.

It allows to generalize beyond the data. An essential question is what algorithm to use to predict data onto themselves, what methods to use to describe outliers, and how to measure the variance. While these questions are covered elsewhere (e.g. Fernandez-Delgado et al. 2014), it is suggested to employ different algorithms for a comparison. I highly suggest creating a great 'Learner' (as per Breiman 2001a, b; Friedman 2002) instead of a model fitting mindset (Zar 2010).

Suggestion 7: Re-run the models without taking out all NAs and data errors (Step 4).

Justification: As mentioned in step 4, most data carry errors, and many cannot really be fixed in large data sets, especially when the data carry a legacy and with many people involved. However, to get a sensitivity test done on the real impacts of cleaning efforts, we suggest running a raw data model, and to see how bad the errors and bugs really are. We found that to be usually a very informative task to do, also defining where to put time and efforts for a cleanup of data and how this presents an effective gain. Doing such runs helps to set priorities. In large data sets, this is not a trivial task because based on our experience ~10% of errors tend to be part of the process in large data holdings and where many owners occur; fuzziness becomes part of the game.

Suggestion 8 Think very hard about hypothesis and whether and what it brings to the discussion and to new information

Justification: No doubt, conceptually, hypothesis testing can provide progress and if done well and applied where suitable (Cushman and Huettmann 2010; Reinhart 2015). Hypotheses make for a great narrative, appear very sound, and are very appealing to communicate. But it applies primarily to experimental and theoretical settings. However, for nature and reality it shows us different (McArdle 1988; Reinhart 2015). In complex multivariate situations of nature, those theoretical concepts are very difficult to apply so that all statistical assumptions are correctly met. Alternatives exist, and we outline a good step forward (see below).This matters essentially for any applied conservation management question.

Suggestion 9: In complex multiple regressions and with many predictors, hypothesis probably do not bring much progress, and there are other, better ways to inference.

Justification: Traditionally, the widely used approaches in such cases are p-values, AIC, and perhaps Bayesian approaches. All of them are known to fail on various grounds (as published by Anderson et al. 2000; Guthery et al. 2005; 2008; Whittingham et al. 2006 etc.) Authors like Breiman (2001a, b) and Hastie et al. (2009) showed already powerful ways forward. See also Fig. 4.4, Strobl et al. 2007 and Drew et al. 2011. The key here is not to apply just a techno-fix but to break out of existing limitations and to embrace new and exciting ways to discover knowledge. Fernandez-Delgado et al. (2014) shows options to explore, however there are many more options and new ones exist and found online.

Suggestion 10 Explore any statistical tools and approaches you can get hold of and know and test them in parallel.

Justification: There are many ways to carry out an analysis and to get to a conclusion. Many software platforms exist and should be tested and run in parallel. This is not only insightful for the own learning but presents a 'test' and how robust current findings and knowledge are. Even more so, many algorithms have different versions and software implementations. For instance, a random generator in one software easily differs from another software which can affect outcomes. Even for simple and well-known implementations like ordinary least square regressions many ways exist to make it happen numerically. And not many agree with each other when benchmarked (Sawitzki 1994a, b). Looking then at disciplines like Demography and ones that use Distance Sampling, and Resource Selection Functions (RSFs) the diversity of tools used remains extremely narrow and not allowing diversity to be tested and to evolve. One should consider that much discussion exist already on dubious inference in those disciplines (Yoccoz 1991; Rexstad et al. 1988; Stephens et al. 2007; Arnold 2010).

Suggestion 11: Leo Breiman (2001a, b) offers a valid philosophy, an insightful and robust approach to be used and investigated by all means.

Justification: As we promote in this chapter, the works by Breiman (2001a, b) are very thorough, deep and progressive. That holds to this very day, more options exist and are developing. While progress is not well made in the field of natural resource management, by now, this work is well published and offers good software tools to be employed. It has reached a great foundation to work from. A key concept here is to assess and model data (as outlined in steps above), create a prediction, and have alternative data handy to assess the prediction for performance, and then infer accordingly. Ideally it requires a research design with training and testing data. That will help to overcome many of the problems described here.

Suggestion 12: Exposing all steps, raw data, raw code and details as metadata and 'in the open' supports the buy-in and transparency by the users and the public.

Justification: Since science is to be repeatable and transparent, this suggestion should not come as a surprise, but we find, it often is fully ignored. As Carlson (2011, 2013) and others found, data and project files are not shared, even less so, the actual underlying code (despite the wide use of R, Distance Sampling and MARK software packages easily allowing for it). Making code publicly available allows for feedback and improvements, and it helps others in their work. It should be seen as a collegial task and for serving the wider public good and being a scientist. For any of this, good metadata formats exist (Huettmann 2005, 2009).

4.5 Real World Tools and Minimum Approaches to Start Data Mining and Predictions

In this section, some basic tools are shown that can be easily used by an investigator who wants to pursue a data mining approach. Those are listed in a good sequence and order:

- import data into some of your favorite software packages, e.g. work sheet, SQL database, open source and commercial tools
- map coordinates in a worksheet or any other tool that can plot x and y. Ideally, map the data in a Geographic Information System (GIS) such as QGIS, R or similar.
- run a pairs plot (e.g. in R; https://www.r-project.org/)
- run a data summary (e.g. in R)
- run a data summary (e.g. in SPM Salford Systems Ltd; https://www.salford-systems.com/)
- run varclust (in R)
- repeat your analysis in several statistical analysis platform to assess their performance and outcomes. Easy software packages could be, but are not limited to, R, S+, SPSS, SAS, MATLAB and partly, Open Office, MS EXCEL (but see McCullough and Wilson 1999 for bench- marking)
- in a data table setting: run a prediction in a lm (in R), glm (in R), machine learning (R) and SPM and compare the outcome and what is learned.
- for autocorrelation, and with spatial data, SDMs and ecological niche models, or in any analysis where proxy-predictors are used, run 're-cursive partitioning' (binary classification and regression trees) as they can overcome traditional auto-correlation problems, are robust on those issues, and tend to give a less inflated variance
- in an ecological niche analysis: run a 'global kitchen sink' model of all data, and predict it onto itself and to a lattice point grid
- re-run the latter analysis and steps with a smaller subset to compare.

4.6 A Set of Commonly Heard Criticisms and Comments for these Data Mining Steps and How to Answer them from a Machine Learning Aspect for Best inferences

Argument: *"Data analysis is to be hypothesis driven and uninformed by knowledge of what the data hold!"*

Reply: Not really; many ways exist to carry out science and to get to the solution of a problem. Data mining, and entering the actual analysis well informed is a key step for a good inference and part of textbooks and best professional practice. Data are to be pre-screened and assessed rapidly for those reasons.

Argument: *"This is data dredging!"*
Reply: No, it is data mining, based on latest science, and with much theory, justification and reasoning. It's a good thing and a professional requirement. Actually there is nothing wrong with data dredging *per se*, and if done in a data mining and prediction framework. This provides reliable information not to be dismissed either way.

Argument: *"You are just looking for a pattern!"*
Reply: Yes, exactly. Finding robust patterns in complex data is essential information and highly needed (Hochachka et al. 2007). It's creating a 'Learner'. We try to find the robust signal in complex data, in any data. It is the obtained signal that can be described, predicted and elaborated on further.

Argument: *"This is all new to me and I am not trained on this"*
Reply: Sure, it's time to get at this new method to progress science. Unless shown otherwise, there are no other good solutions we know of to get at best-possible predictions and for inference. Being fluent on research methods is a job requirement (one must not hide behind ignorance or methods from decades ago, as trained back then. Science moves[1])

Argument: *"My funder, supervisor and committee told me to do the other, old method"*
Reply: Suggest to review your understanding of science, progress, ethics, and assess analytical alternatives for scientific progress. Ask funder, supervisor and committee why still entrenched in old methods while new ones exist to produce new insights as progress, as per aims of science (Popper 1945; Fryell and Caughley 2014; Silva 2012)

Argument: *"I tested it, and it flipped my expected results. So I did not continue it further"*
Reply: Suggest you may report yourself to the ethics and science integrity department.

Argument: *"It's just too much work and my deadline is coming up"*
Reply: Suggest you may report yourself to the ethics and science integrity department, again.

---

[1] Changes in the current science concept - as run by the western world - are known to take app. 20 years or longer (e.g. Connner 2005). While this is way too long for relevant and effective conservation management, it's often caused by generic inertia, institutional setup and tenure + funding concepts. In reality, most changes in science are only achieved when earlier researchers retire and new minds and new generations enter the scene. Arguably, many of the 'new/fresh' minds got trained by earlier minds and thus, just little changes truly occur. Examples of this pattern can be found in most western and science nations, e.g. in Academies of Science, many established Journals and Nobel Prize committees.

4.7 On 'Best Professional Practices', Professional Bias, Ignorance, Misconduct, Professional Societies, Education and Culture: What is a Lie and Punishable Intent when underlying Methods Problems are well-known but are ignored?

This elaboration has presented and outlined the many problems that are found in the sciences, across disciplines and practitioners, specifically in quantitative approaches and where inference is supposed to be most precise. Nowadays where 'everything is digital', there is little meaningful resistance to working open access and open source (e.g. Zuckerberg et al. 2011; Greenland 2012). Those problems are far from new (Popper 1945), but why do they remain unsolved, and are still educated and promoted by so-called professional societies and found in their internationally (!) peer-reviewed publications, with editors, reviewers and publishers pushing those concepts? Why is there no drive for improvement while conservation management and natural resources are decaying, globally, and climate change is on the rise?

This question might require a multi-dimensional analysis, and is beyond the scope of this text. However, a few facts should be captured here:

- frequency statistics is widely criticized for its validity
- alternatives to frequency statistics and its institutions are known for well over 30 years
- to this very day frequency statistics has the lion-share of all research publications
- university education, experts and global leadership by governments still center around frequency statistics
- science-based ecology and natural resource management centers around frequency statistics
- -the majority of data and analysis codes used in the sciences are widely not shared nor documented with metadata
- professional societies do virtually nothing to change this situation

Virtually all of those 'facts' hint towards a crisis in the sciences, as we know them. It also shows a crisis in institutions, education, leadership, for ecology and natural resource management, as for the globe and human-wellbeing and society overall.

From a legal perspective, in the western world, if one knows that a certain practice has a bad outcome, this is punishable and thus should be avoided. It might equal an 'intent' and therefore is a serious violation in western law, policy and ethics. This remains the paradox of our time why there is so little progress and change to make it better, for ourselves and future generations.

4.8 Conclusion

Complex data warrant many approaches to extract the strongest and relevant signal. Eventually, it's not a widely used or accepted analysis template that counts but the evidence that convinces. As part of a good science paradigm, every data record and information tends to matter and carries relevant information. A simple template approach - as promoted with p-values (Zar 2010) or AIC (Anderson and Burnham 2002) - is not so informative (Arnold 2010) nor even defendable and meaningful. It's 'me too' research (O'Connor 2000) and lacks reflection and debate, and thus is no good progress. While model selection is not well resolved at all, here it is shown how this can be addressed in the framework of Leo Breiman (2001a, b), with machine learning, with (spatial) predictions and when comparing the predictions with several alternative lines of evidence. While pragmatic and parsimonious approaches do not solve the initial problem but create new ones, I believe that here is presented a good and holistic way forward and it helps conservation management and can progress the wider global well-being. Although the methods of machine learning are relatively simple to run in a free and readily available R software and with other publicly available packages, for over 20 years it urgently awaits its wider implementation for the wider benefit of natural resource management. Here I present some first and powerful steps to reach progress instead of ignoring and marginalizing it.

References

Anderson DR, Burnham KP (2002) Avoiding pitfalls when using information-theoretic methods. J Wildl Manag 66:912–918

Anderson DR, Burnham KP, Thompson WL (2000) Null hypothesis testing: problems, prevalence, and an alternative. J Wildl Manag 64:912–923

Araujo M, New B (2007) Ensemble forecasting of species distributions. Trends Ecol Evol 22:42–47.

Arnold TW (2010) Uninformative parameters and model selection using Akaike's information criterion. J Wildl Manage 74:1175–1178

Bandura A (2007) Impending ecological sustainability through selective moral disengagement. Int J Innov Sustain Dev:8–35

Berkson J (1942) Tests of significance considered as evidence. J Am Stat Assoc 37:325–335

Betts MG, Ganio L, Huso M, Som N, Huettmann F, Bowman J, Wintle WA (2009) Comment on "methods to account for spatial autocorrelation in the analysis of species distributional data: a review". Ecography 32:374–378

Breiman L (2001a) Statistical modeling: the two cultures (with comments and a rejoinder by the author). Stat Sci 16:199–231

Breiman L (2001b) Random forests. Mach Learn 45:5–32

Breiman L, Friedman J, Stone CJ, Olshen RA (1984) Classification and regression trees. CRC press, Boca Raton

Burnham KP, Anderson DR (2002) Model selection and multimodel inference: a practical information-theoretic approach. springer, New York

Carlson D. (2011). A lesson in sharing. Nature. 469: 293.

Carlson D (2013) Reading and thinking about international polar years: five recent books. Polar Res 32:1–7. https://doi.org/10.3402/polar.v32i0.20789CODATA

Chamberlin T (1890) The method of multiple working hypothesis. Reprinted 1965. Science 148:754–759

Chia KS (1997) "Significant-itis"—an obsession with the P-value. Scand J Work Environ Health 23:152–154

Concato J, Hartigan JA (2016) P values: from suggestion to superstition. J Investig Med 64:1166–1171

Conner CD (2005) A People's history of science: miners, midwives and "low Mechanicks". Nation books, New York

Cushman, S. and F. Huettmann. (2010) Spatial Complexity, Informatics and Wildlife Conservation. Springer Tokyo, Japan. 448 p.

Cutler DR, Edwards TC Jr, Beard KH, Cutler A, Hess KT, Gibson J, Lawler JJ (2007) Random forests for classification in ecology. Ecology 88:2783–2792

Czech B (2002) Shoveling fuel for a runaway train: errant economists, shameful spenders, and a plan to stop them all. University of California Press, Berkeley

Daly H, Farley J (2010) Ecological economics. Principles and applications, 2nd edn. Island Press, New York

De'ath G, Fabricius K (2000) Classification and regression trees: a powerful yet simple technique for ecological data analysis. Ecology 81:3178–3192 10.1890/0012-9658(2000)081[3178:CARTAP]2.0.CO;2

Dodds DG (2001) Philosophy and practice of wildlife management, 3rd edn. Krieger Publications, New York

Drew CA, Yo W, Huettmann F (eds) (2011) Predictive modeling in landscape ecology. Springer, New York.

Elith J, Graham C, NCEAS working group (2006) Novel methods improve prediction of species' distributions from occurrence data. Ecography 29:129–151.

Elder JF (2003) The generalization paradox of ensembles. J Comput Graph Stat 12:853–864

Fernandez-Delgado M, Cernadas E, Barro S, Dinan A (2014) Do we need hundreds of classifiers to solve real world classification problems. J Mach Learn 15:3133–3181

Fidler F, Loftus GR (2009) Why figures with error bars should replace p values: some conceptual arguments and empirical demonstrations. J Psychol 217:27–37. https://faculty.washington.edu/gloftus/Downloads/Fidler.Loftus.pdf

Filliben JJ (1975) The probability plot correlation coefficient test for normality. Technometrics. Am Soc Qual 17:111–117

Fox CH, Huettmann F, Harvey GKA, Morgan KH, Robinson J, Williams R, Paquet PC (2017) Predictions from machine learning ensembles: marine bird distribution and density on Canada's Pacific coast. Marine Ecology Progress Series 566:199–216.

Freund Y, Schapire RE (1997) A decision-theoretic generalization of on-line learning and an application to boosting. J Comput Syst Sci 55:119–139

Friedman JH (2002) Stochastic gradient boosting. Comput Stat Data Anal 38:367–378

Fryell JM, Caughley G (2014) Wildlife ecology, conservation, and management. Wiley Blackwell, Brisbane

Gardner MA, Altman DG (1986) Confidence intervals rather than P values: estimation rather than hypothesis testing. Br Med J 292:746–750

Gelman A (2008) Objections to bayesian statistics. Bayesian Anal 3:445–450

Georgescu-Roegen N (1971) The entropy law and the economic process. Harvard University Press, Cambridge, MA

Gigerenzer G (2004) Mindless statistics. J Socio Econ 33:587–606

Greenland S (2012) Transparency and disclosure, neutrality and balance: shared values or just shared words? J Epidemiol Community Health 66:967–970

Greenland S, Senn SK, Rothman KJ, Carlin JB, Poole C, Goodman SN, Altman DG (2016) Statistical tests, P values, confidence intervals, and power: a guide to misinterpretations. Eur J Epidemiol 31:337–350

Guthery FS (2008) Statistical ritual; versus knowledge accrual in wildlife science. J Wildl Manag 72:1872–1875

Guthery FS, Lusk JJ, Peterson MJ (2001) The fall of the null hypothesis: liabilities and opportunities. J Wildl Manag 65:379–384

Guthery FS, Brennan LA, Peterson MJ, Lusk LL (2005) Information theory in wildlife science: critique and viewpoint. J Wildl Manag 69:457–465

Han X, Huettmann F, Guo Y, Mi C, Wen L (2018) Conservation prioritization with machine learning predictions for the black-necked crane *Grus nigricollis*, a flagship species on the Tibetan plateau for 2070. Glob Environ Chang. https://doi.org/10.1007/s10113-018-1336-4

Hastie T, Tibshirani R, Friedman J (2009) The elements of statistical learning: data mining, inference, and prediction, 2nd edn. Springer, New York

Hilborn R, Mangel M (1997) The ecological detective: confronting models with data. Princeton University Press, Princeton, p 330

Hobbs NT, Hooten M (2015) Bayesian models: a statistical primer for ecologists. Princeton University Press, Princeton

Hochachka W, Caruana R, Fink D, Munson A, Riedewald M, Sorokina D, Kelling S (2007) Data mining for discovery of pattern and process in ecological systems. J Wildl Manag 71:2427–2437

Huettmann F (2005) Databases and science-based management in the context of wildlife and habitat: towards a certified ISO standard for objective decision-making for the global community by using the internet. J Wildl Manag 69:466–472

Huettmann F (2007) Modern adaptive management: adding digital opportunities towards a sustainable world with new values. Forum Public Policy Clim Chang Sustain Dev 3:337–342

Huettmann F (2009) The global need for, and appreciation of, high-quality metadata in biodiversity work. In: Spehn E, Koerner C (eds) Data mining for global trends in mountain biodiversity. CRC Press, Taylor & Francis, pp 25–28

Huettmann F, Artukhin Y, Gilg O, Humphries G (2011) Predictions of 27 Arctic pelagic seabird distributions using public environmental variables, assessed with colony data: a first digital IPY and GBIF open access synthesis platform. Mar Biodivers 41:141–179. https://doi.org/10.1007/s12526-011-0083-2

Johnson CJ, Seip DR (2008) Relationship between resource selection, distribution, and abundance: a test with implications to theory and conservation. Popul Ecol 50:145–157

Kandel K, Huettmann F, Suwal MK, Regmi GR, Nijman V, Nekaris KAI, Lama ST, Thapa A, Sharma HP, Subedi TR (2015) Rapid multi-nation distribution assessment of a charismatic conservation species using open access ensemble model GIS predictions: red panda (*Ailurus fulgens*) in the Hindu-Kush Himalaya region. Biol Conserv 181:150–161

Lambdin C (2012) Significance tests as sorcery: science is empirical—significance tests are not. Theory Psychol 22:67–90

Loftus GR (1996) Psychology will be a much better science when we change the way we analyze data. Curr Dir Psychol 5:161–171

Magness DR, Huettmann F, Morton JM (2008) Using random forests to provide predicted species distribution maps as a metric for ecological inventory & monitoring programs. pp 209–229. In: Smolinski TG, Milanova MG, Hassanien A-E (eds) Applications of computational intelligence in biology: current trends and open problems. Studies in computational intelligence, vol 122. Springer-Verlag, Berlin/Heidelberg, p 428

Magness DR, Morton JM, Huettmann F, Chapin FS III, McGuire AD (2011) A climate-change adaptation framework to reduce continental-scale vulnerability across conservation reserves. Ecosphere 2:art112. https://doi.org/10.1890/ES11-00200.1

Manly FJ, McDonald LL, Thomas DL, McDonald TL, Erickson WP (2002) Resource selection by animals: statistical design and analysis for field studies, 2nd edn. Kluwer Academic Publishers, Dordrecht

McArdle (1988) The structural relationship: regression in biology. Can J Zool 66:2329–2339

McCullough BC, Wilson B (1999) On the accuracy of statistical procedures in Microsoft Excel 97. Comput Stat Data Anal 31:27–37

McGarical K, Cushman S, Stafford S (2000) Multivariate statistics for wildlife and ecology research. Springer, New York

Mi C, Huettmann F, Guo Y, Han X, Wen L (2017) Why to choose random Forest to predict rare species distribution with few samples in large undersampled areas? Three Asian crane species models provide supporting evidence. Peerj. https://doi.org/10.7717/peerj.2849

Mueller JP,Massaron L (2016) Machine learning for dummies. For Dummies Publisher, 435 p

Næss A (1989) Ecology, community and lifestyle: outline of an ecosophy (trans: Rothenberg D). Cambridge: Cambridge University Press

O'Connor R (2000) Why ecology lags behind biology. The Scientist 14:35–36

Oppel S, Strobl C, Huettmann F (2009). Alternative methods to quantify variable importance in ecology. Technical report number 65, Department of Statistics. University of Munich

Oppel S, Meirinho A, Ramírez I, Gardner B, O'Connell A, Miller PI, Louzao M (2012) Comparison of five modelling techniques to predict the spatial distribution and abundance of seabirds. Biol Conserv 156:94–104.

Pearce J, Ferrier S (2000) Evaluating the predictive performance of habitat models developed using logistic regression. Ecol Model 133:225–245

Perneger TV (1998) What's wrong with Bonferroni adjustments. Br Med J 316:1236–1238

Popper K (1945) The open society and its enemies. Princeton University Press, Princeton/Oxford

Quinn G, Keough M (2004) Experimental design and data analysis for biologists. Cambridge University Press, Cambridge

Razali N, Wah YB (2011) Power comparisons of Shapiro–Wilk, Kolmogorov–Smirnov, Lilliefors and Anderson–darling tests. J Stat Model Anal 2:21–33

Reinhart A (2015) Statistics done wrong: the woefully complete guide. No Starch Press, San Francisco

Rexstad EA, Miller D, Flather C, Anderson A, Hupp J, Anderson JR (1988) Questionable multivariate statistical inference in wildlife and community studies. J Wildl Manag 52:794–798

Romesburg HC (1981) Wildlife science: gaining reliable knowledge. J Wildl Manag 45:293–313

Romesburg HC (1991) On improving natural resources and environmental sciences. J Wildl Manag 55:744–756

Rosales J (2008) Economic growth, climate change, biodiversity loss: distributive justice for the global north and south. Conserv Biol 22:1409–1417

Rothman KJ (1990) No adjustments are needed for multiple comparisons. Epidemiology 1:43–46

Salsburg DS (1985) The religion of statistics as practiced in medical journals. Am Stat 39:220–223

Salsburg D (2001) The lady tasting tea: how statistics revolutionized science in the twentieth century. W. H. Freeman and Company, New York

Savalei V, Dunn E (2015) Is the call to abandon p-values the red herring of the replicability crisis? Front Psychol. https://doi.org/10.3389/fpsyg.2015.00245

Sawitzki G (1994a) Testing numerical reliability of data analysis systems. Comput Statist Data Anal 18:269–286

Sawitzki G (1994b) Report on the reliability of data analysis systems. Comput Statist Data Anal (SSN) 18:289–301

Schapire RE (1990) The strength of weak learnability. Machine learning, vol 5. Kluwer Academic Publishers, Boston, MA, pp 197–227. https://doi.org/10.1007/bf00116037

Schapire RE (1992) The design and analysis of efficient learning algorithms. MIT Press, Cambridge, MA

Schapire RE, Singer Y (1999) Improved boosting algorithms using confidence-rated predictors. Mach Learn 37:297–336

Schmidt FL, Hunter JE (2014) Methods of meta-analysis: correcting error and bias in research findings, 3rd edn. Sage Publisher, Thousand Oaks

Silva NJ (2012) The wildlife techniques manual: research & management, vol 2, Seventh edn. The Johns Hopkins University Press

Stang A, Poole C, Kuss O (2010) The ongoing tyranny of statistical significance testing in biomedical research. Eur J Epidemiol 25:225–230

Stanton-Geddes J, Gomes De Freitas C, de Sales Dambros C (2014) In defense of P values: comment on the statistical methods actually used by ecologists. Ecology 95:637–642

Stephens PA, Buskirk SW, Hayward GW, Martinez Del Rio C (2007) A call for statistical pluralism answered. J Appl Ecol 44:461–463. https://doi.org/10.1111/j.1365-2664.2007.01302.x

Strobl C, Boulesteix A-L, Zeileis A, Hothorn T (2007) Bias in random forest variable importance measures: illustrations, sources and a solution. Bioinformatics 8:25. https://doi.org/10.1186/1471-2105-8-25

Swihart R, Slade N (1985) Testing for independence in observations of animal movements. Ecology 66:1176–1184

Thompson B (2004) The "significance" crisis in psychology and education. J Soc Econ 33:607–613

Venables WN, Ripley BD (2002) Modern applied statistical analysis, 4th edn. Springer, New York

Whittingham MJ, Stephens PA, Bradbury RB, Freckleton RP (2006) Why do we still use stepwise modelling in ecology and behaviour? J Anim Ecol 75:1182–1189

Yackulic CB, Chandler R, Zipkin EF, Royle JA, Nichols JD, Campbell Grant EH, Veran S (2012) Presence-only modeling using MAXENT: when can we trust the inferences? Meth Ecol Evol 4:236–243

Yoccoz NG (1991) Use, overuse, and misuse of significance tests in evolutionary biology and ecology. Bull Ecol Soc Am 72:106–111

Zar JH (2010) Biostatistical analysis, 5th edn. Prentice Hall, Upper Saddle River

Ziliak ST, McCloskey DN (2009) The cult of statistical significance. Section on statistical education, pp 2302–2319

Zuckerberg, B, F. Huettmann and J. Friar (2011) Proper Data Management as a Scientific Foundation for Reliable Species Distribution Modeling. In C.A. Drew, Y. Wiersma and F. Huettmann (eds). Predictive Species and Habitat Modeling in Landscape Ecology. Springer, New York. Pp 45–70

Chapter 5
Ensembles of Ensembles: Combining the Predictions from Multiple Machine Learning Methods

David J. Lieske, Moritz S. Schmid, and Matthew Mahoney

"Predictive learning is an important aspect of data mining ... However, it is seldom known in advance which procedure will perform best or even well for any given problem."

Hastie et al. 2001: 312

5.1 Introduction

The quotation at the start of this article underscores two problems that perpetually plague predictive modeling: (1) with so many available model methods/algorithms, which is the most appropriate for the problem at hand? and (2) what is the best way to make use of predictive models, i.e., achieve high predictive accuracy while at the same time minimizing the impact of overfitting (loss of generalizability)?

With regards to the difficulty in determining the most appropriate prediction algorithm, many comparative analyses of performance have been conducted, but they typically draw different conclusions (e.g., Franklin 1995; Elith and Burgman 2002; Caruana and Niculescu-Mizil 2006; Elith et al. 2006; Moisen et al. 2006; Elith and Graham 2009; Marmion et al. 2009a). This has resulted in a somewhat confusing body of evidence. Diversity in performance is at least partially attribut-

D. J. Lieske (✉) · M. Mahoney
Department of Geography and Environment, Mount Allison University, Sackville, NB, Canada
e-mail: dlieske@mta.ca; matthew.mahoney@canada.ca

M. S. Schmid
Hatfield Marine Science Center, Oregon State University, Newport, OR, USA
e-mail: Schmidm@oregonstate.edu

© Springer Nature Switzerland AG 2018
G. R. W. Humphries et al. (eds.), *Machine Learning for Ecology and Sustainable Natural Resource Management*, https://doi.org/10.1007/978-3-319-96978-7_5

able to differences in the choice of tuning parameters (e.g., degree of freedom of smoothing functions, specification of interaction terms) and the fact that predictions from, for example, tree-based methods, rely on some form of random partitioning. Because of this, tree-based classifiers exhibit a degree of stochastic variation in performance, though this is reduced by constructing and aggregating multiple trees such as done by bagging, etc. (see James et al. 2013: 316). According to Domingos (2012), the number of algorithms to choose from numbers in the thousands, with hundreds added each year, rendering the identification and selection of a single "best" algorithm a provisional choice at best. At the end of the day the question of which method is most generally appropriate remains elusive, though it should be noted that tree-based machine-learning (ML) techniques performed quite well in the comparative studies previously cited.

Random forests (RF, Breiman 2001) and boosted regression trees (BRT, Friedman 2001) are prominent tree-based techniques which employ binary recursive partitioning to generate predictions. They do not require the *a priori* specification of a functional form for the predictors, they are very flexible in handling missing data, and they can capture complex, non-linear interactions (De'ath and Fabricius 2000; Harrell 2001; Cutler et al. 2007; Hochachka et al. 2007). In fact, James et al. (2013, 299) describe boosting as "one of the most powerful learning ideas introduced in the last 10 years". For this reason they are first-choice modeling methods for a wide range of applications.

Tree-based ML techniques, however, can be vulnerable to the second common problem limiting the generalizability of predictive models: overfitting (Harrell 2001; Caruana and Niculescu-Mizil 2006, and Elith et al. 2008). Harrell (2001, 60) succinctly defined the problem, and articulated its consequence: "When a model is fitted that is too complex, that is it [sic] has too many free parameters to estimate for the amount of information in the data, the worth of the model (e.g., R^2) will be exaggerated and future observed values will not agree with predicted values. In this situation, *overfitting* is said to be present, and some of the findings of the analysis come from fitting noise or finding spurious assocations between X and Y''. The question becomes one of how to make maximum use of the most powerful ML techniques while, at the same time, reducing the potential impact of overfitting so as to achieve the most generally applicable model predictions. We argue that the forecasting literature has already offered a solution to this problem.

Prediction, or forecasting, is a common procedure in many applied fields, and has resulted in the general observation that the combination of multiple predictions leads to increased accuracy (see Clemen 1989for a dated though extensive review across many fields). The combined prediction is usually referred to as an "ensemble" or "consensus"prediction. Such a solution is pragmatic, easy to implement, and makes it possible to capitalize on the way in which different algorithms capture different aspects of the information available for prediction (Clemen 1989). As a side effect, this approach offers a practical way to produce predictions that balance predictive accuracy with generalizability, while at the same time reducing the reliance of predictions on a single technique. The utility of this type of approach is made clear when one considers how many different methods/algorithms are in existence (Fernández-Delgado et al. 2014).

So, how does an ensemble or consensus approach offer a solution to the problems discussed here? According to Clemen (1989), merely averaging predictions can dramatically improve overall predictive accuracy - the so called "concensus" or "committee averaging"technique (Heikkinen et al. 2006; Araújo and New 2007; Das et al. 2008). Other aggregation methods are possible, though Marmion et al. (2009b) determined that other aggregation measures (e.g., the median) performed about the same or worse than committee averaging. Depending upon the choice of prediction aggregation method, weightings can be introduced to dampen the influence of overly-optimistic methods in the final ensemble prediction. Lieske et al. (2014), for instance, employed a weighted mean of three different algorithms (negative binomial regression, hurdle models, and RF), where the weighting was determined by the r^2 values of each algorithm corrected for over-fitting. A similar method was employed by Oppel et al. (2012) and Renner et al. (2013).

We illustrate an ensemble prediction approach using an example data set involving satellite-derived marine vessel traffic information, and applied two techniques based on classification and regression trees: random forests (RF) and boosted regression trees (BRT). We also combined the RF and BRT predictions to produce an ensemble (ENS) prediction. Along the way we also considered the influence of the key tuning parameters governing the behaviour of RF and BRT algorithms.

5.2 Methods

5.2.1 Data Set

Marine traffic data was obtained for the Scotian shelf offshore region of Nova Scotia, originating as hourly automatic identification system (AIS) tracking positions recorded between June 22, 2014 and June 22, 2015. All vessels large than 500 tons are required to transmit AIS signals, although many smaller vessels are also equipped with the technology to support rescue in an emergency response situation. An automated procedure was employed to construct vessel tracks from point location data when gaps between successive positions were less than 4 h in duration, otherwise vessel locations were modelled as isolated positions not part of a continuous cruise (Brawn 2016). For the purposes of this study, analysis was limited to vessels of type "fishing", and for the third quarters (Q3) of the 2014–2015 period. Using a 20-km grid cell size, which corresponded approximately to the 90th percentile of nearest-neighbour distance for points on a cruise track, line (scLineDensity) and point (scPointDensity) densities were calculated and scaled relative to the maximum density over the entire surface (Fig. 5.1). As complete vessel tracks were deemed more informative than solitary point positions, the two vessel density measures were averaged to form a single measure called "scCombined" with a weighting of 80% assigned to line information and 20% to point information.

Environmental data consisted of physical and oceanographic variables with the potential to identify areas of high marine productivity and, therefore, areas likely to

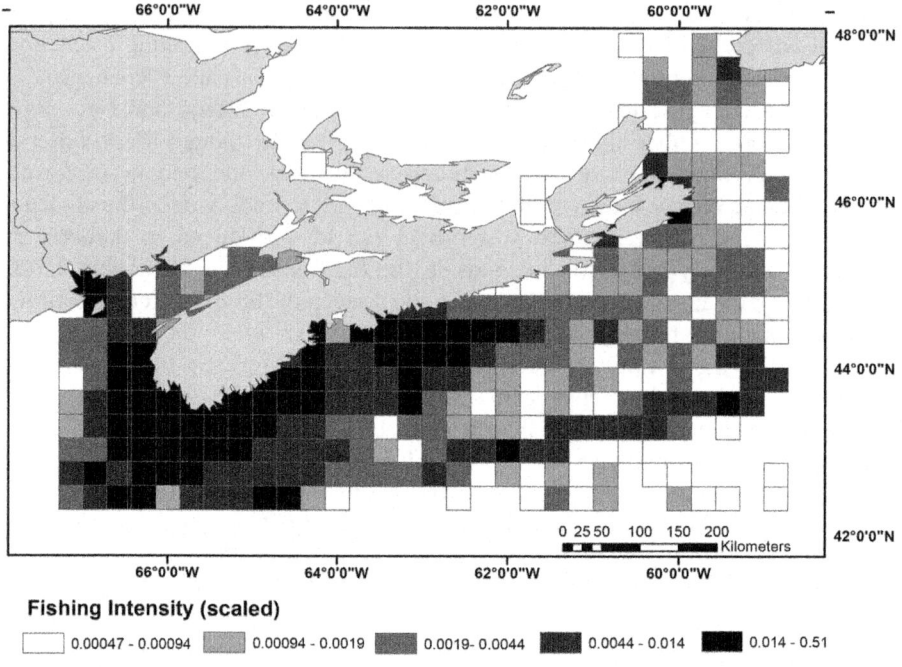

Fig. 5.1 Relative intensity of fishing-vessel traffic for the Scotian shelf region, based on automatic identification system (AIS) tracking data for the third quarters of the period 22 June 2014 to 22 June 2015

be exploited by fishing vessels (Table 5.1). This data was combined within a geodatabase and aggregated to the level of 20-km grid cells to maintain consistency with the vessel traffic response data.

5.2.2   Accuracy Assessment

Model construction was conducted 500 times for each combination of tuning parameter (see Tables 5.2, 5.3). For each of the 500 iterations, we randomly selected 50% of the data for model training purposes, and passed this on to the BRT and RF algorithms. Mean squared error (MSE) was employed to cross-validate predictive accuracy using the withheld testing data. It should be noted, however, that the RF method has its own internal procedure for randomly selecting a portion of the data for model training, and reserving a portion for model testing (the so-called "OOB", or out-of-bag portion of the data; see *Modeling Algorithms*). The data and R-code developed for this study is online accessible at https://doi.org/10.5281/zenodo.1318352 (Lieske et al. 2018) and serves as a documentation of methods and results.

Table 5.1 Summary of the thirteen predictor variables used in the analysis, as well as the data sources

Layer group	Resolution/ scale	Source	Description
CHLA_xxxx	4 km, monthly	MODIS[a]	Concentration of the photosynthetic pigment Chlorophyll a (mg m^{-2}), where *xxxx* indicates the month and year of summary: July 2014, 2015; August 2014, 2015; September 2014.
SST_xxxx	4 km, monthly	MODIS[a]	Sea surface temperature derived from long-wave (11–12 μm) thermal radiation, where *xxxx* indicates the month and year of summary: July 2014, 2015; August 2014, 2015; September 2014.
WIND_yyyy	$1^0 \times 1^0$	QuikSCAT[b]	Surface wind (m s^{-1}), August 1999–October 2009, where *yyyy* indicates the month averaged over the entire time series: July, August, and September.
DISTCOAST	4 km	GML[c]	Distance to coast (km).
DISTPORT	4 km	GML[c]	Distance to closest marina and/or port (km).
DEPTH	4 km	GML[c]	Seadepth (m), derived from ETOPO2v2 2006 product[d]
RUGGED	3 nearest neighbours	GML[c]	Seafloor ruggedness, derived using Benthic Terrain Modeler Extension[e]

[a]NASA OceanColor Web (http://oceancolor.gsfc.nasa.gov/cms/), downloaded 1 September 2016.
[b]QuickSCAT (https://podaac.jpl.nasa.gov/QuikSCAT), downloaded 1 September 2016.
[c]Mount Allison University Geospatial Modelling Lab (GML, http://arcgis.mta.ca).
[d]National Geophysical Data Center. 2006. 2-minute gridded global relief data (ETOPO2) v2. National Geophysical Data Center, NOAA.
[e]Wright, D.J., M. Pendleton, J. Boulware, S. Walbridge, B. Gerlt, D. Eslinger, D. Sampson, and E. Huntley. 2012. ArcGIS Benthic Terrain Modeler (BTM), v. 3.0, Environmental Systems Research Institute, NOAA Coastal Services Center, Massachussetts Office of Coastal Zone Management. https://www.arcgis.com/home/item.html?id=b0d0be66fd33440d97e8c83d220e7926, downloaded 8 April 2016.

Table 5.2 Key tuning parameters for the random forests (RF) method, as implemented in the *R* package of Liaw and Wiener (2002)

Random forests[a]			
Parameter	Software-specific implementation	Typical Values	Comment
Number of iterations, T	`ntree`	5,000[b]	
Size of predictor subset, m	`mtry`	Classification: $p^{1/2}$; regression: $p/3$	$m = p$ (the number of predictors) equivalent to bagging.
Number of observations in terminal node	`nodesize`	Default: 5 for regression, 1 for classification	

[a]Package `randomForest` (Liaw and Wiener 2002); Note: p = the number of predictor variables (covariates).
[b]Breiman (2002)

Table 5.3 Key tuning parameters for the boosted regression tree (BRT) method, as implemented in the *R* Package of Ridgeway (2012)

Boosted regression tree[a]			
Parameter	Software-specific implementation	Typical Values	Comment
Number of iterations, T	`n.trees`	3,000 to 10,000	
Shrinkage (learning rate), λ	`shrinkage`	0.01 to 0.001	Elith et al. (2008) recommend $\downarrow\lambda$ with $\uparrow K$
Number of splits (depth of each tree), K	`interaction.depth`	1 to number of variables in dataset	
Subsampling rate, p	`bag.fraction`	0.5 (recommended)	Controls the level of stochasticity in model selection.

[a]package gbm (Ridgeway 2012)

5.2.3 Modeling Algorithms

Random forests (RF, Breiman 2001) is, in itself, a form of ensemble technique that generates continuous-value predictions by averaging the expectations from multiple regression trees. Stochasticity is introduced into the learning procedure by: (1) bootstrapping 63% of the data (with replacement) for training purposes, reserving the remainder for model testing (referred to as the "out of bag" results, or OOB), and (2) randomly selecting a subset of the predictor variables at each step, thereby "decorrelating" the trees and ensuring that the resulting average is less variable and of higher predictive accuracy (Hegel et al. 2010; James et al. 2013). RF was implemented using the `randomForest` *R* package of Liaw and Wiener (2002). Important tuning parameters for this package are described in Table 5.2. It should be noted that other implementations are available for RF, for example, Salford Systems' (2016) `RandomForests` package. Our analysis focused on the more readily-available open source version of Liaw and Wiener (2002), but we acknowledge that different software implementations may yield different results from those reported here.

The boosted regression tree algorithm (BRT, Friedman 2001) can be described as a "slow learning" technique where predictions are constructed additively, sequentially, and incrementally (Elith et al. 2008; James et al. 2013). Data is not bootstrap sampled nor is the response variable directly modelled, rather the attention of the model-fitting procedure is focused on the residual or unexplained variation (James et al. 2013). The weight of each tree in the final prediction is controlled by the learning rate parameter (λ, Table 5.3), which ensures slow improvements in the model in the areas of the response "space" where predictions are poor (James et al. 2013). BRT was implemented using the gbm *R* package of Ridgeway (2012), though it should be noted that alternative commercial boosting packages are available (e.g., Salford Systems 2016).

Tuning parameter settings (Tables 5.2 and 5.3) were varied systematically for both algorithms. For RF, the size of the predictor subset (m, Table 5.2) was allowed to vary from 1 to the total number of available predictor variables. Simultaneously, the number of iterations varied from 500 to 8000 (T, Table 5.2). In the case of BRT, the learning (or shrinkage) rate (λ) took on a value of either 0.01 or 0.001; interaction depth (or the number of splits, K) was allowed to vary from 1 to the total number of available predictor variables; and the number of iterations was varied from 500 to 8000 (T, Table 5.3).

Ensemble (ENS) predictions were generated, at each iteration of model construction, by calculating a weighted average of the predictions from RF and BRT. The weightings were based on model performance using the cross-validation data, and were defined as the inverse of the MSE (MSE^{-1}).

5.3   Results and Discussion

Figures 5.2–5.3 present some interesting consequences of varying the combinations of tuning parameters when applied to the fishing traffic dataset. Comparing performance for BRT, RF, and ENS when a faster "burn in" (shrinkage, or λ) rate of 0.01 was defined for BRT yielded some clear patterns (Fig. 5.2). First, RF seemed largely unaffected by the number of iterations/trees constructed, and yielded a consistent error pattern. Mean error rate was $0.000836 \pm SE\ 3.58 \times 10^{-6}$ across all combinations of predictor variable size (m) and numbers of trees. BRT, on the other hand, yielded less accurate predictions with larger numbers of trees, particularly beyond about 2000 trees. Overall BRT mean error rate was $0.000863 \pm SE\ 5.52 \times 10^{-6}$. Predictions from the ENS method showed a relatively consistent error pattern across varying numbers of trees, similar to RF, and yielded the lowest overall mean error rate of $0.000818 \pm SE\ 2.467 \times 10^{-6}$. Variation in the number of variables used in a single iteration – the tree depth parameter of BRT or the size of predictor subset parameter of RF – had a particularly strong impact on RF error rates. RF performance degraded linearly with increasing predictor subset size. BRT and ENS algorithms yielded the lowest cross-validated error rates when tree depth was set to 1, but after an initial rise at a tree depth = 2, performance was consistent across higher depth settings. Use of the function `tuneRF` confirms the cross-validated accuracy assessment (Fig. 5.4) in that for this data set, lower `mtry` settings yielded the lowest error rate.

Comparing performance for BRT, RF, and ENS using a slower "burn in" (shrinkage) rate of 0.001 showed consistently superior performance by BRT with a mean error rate of $0.000758 \pm SE\ 3.24 \times 10^{-6}$ (Fig. 5.3). ENS predictive accuracy closely tracked that of BRT, especially for larger numbers of trees (≥ 3000). While exhibiting a slightly higher mean error rate of 0.000772, there was greater consistency in ENS performance as evidenced by the lower standard error of 2.34×10^{-6}. Choice of the number of trees used to train BRT models was less influential on BRT accuracy under this lower shrinkage setting.

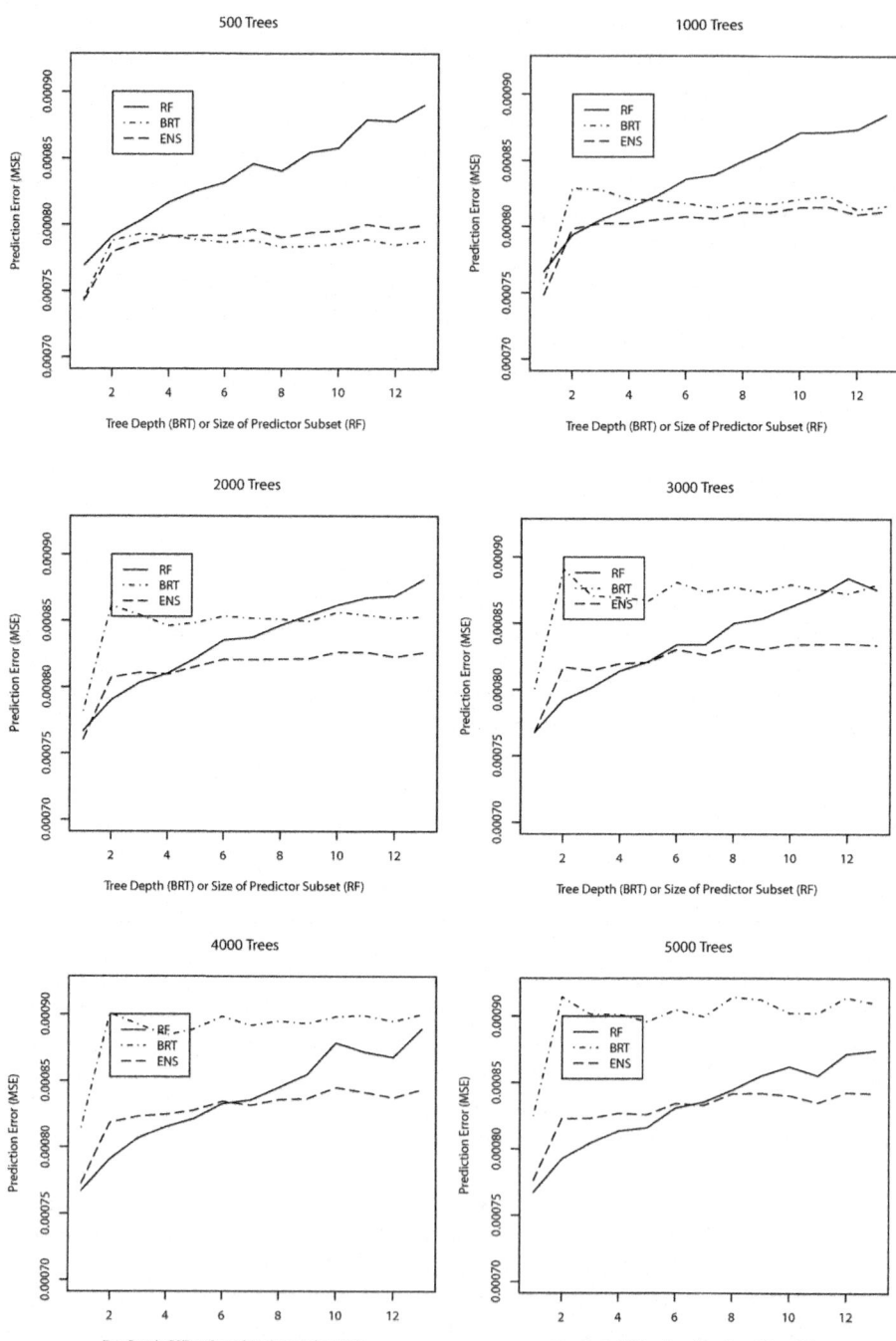

Fig. 5.2 Comparative performance of random forests (RF), boosted regression tree (BRT), and ensemble (ENS) predictions, as measured using cross-validated mean square prediction error (MSE). Tree size was allowed to vary from 500 to 8000 trees ($T = 8000$ not shown), and tree depth/size of predictor subset was set from 1 to 13. In the case of BRT, a shrinkage (λ) value of 0.01 was used

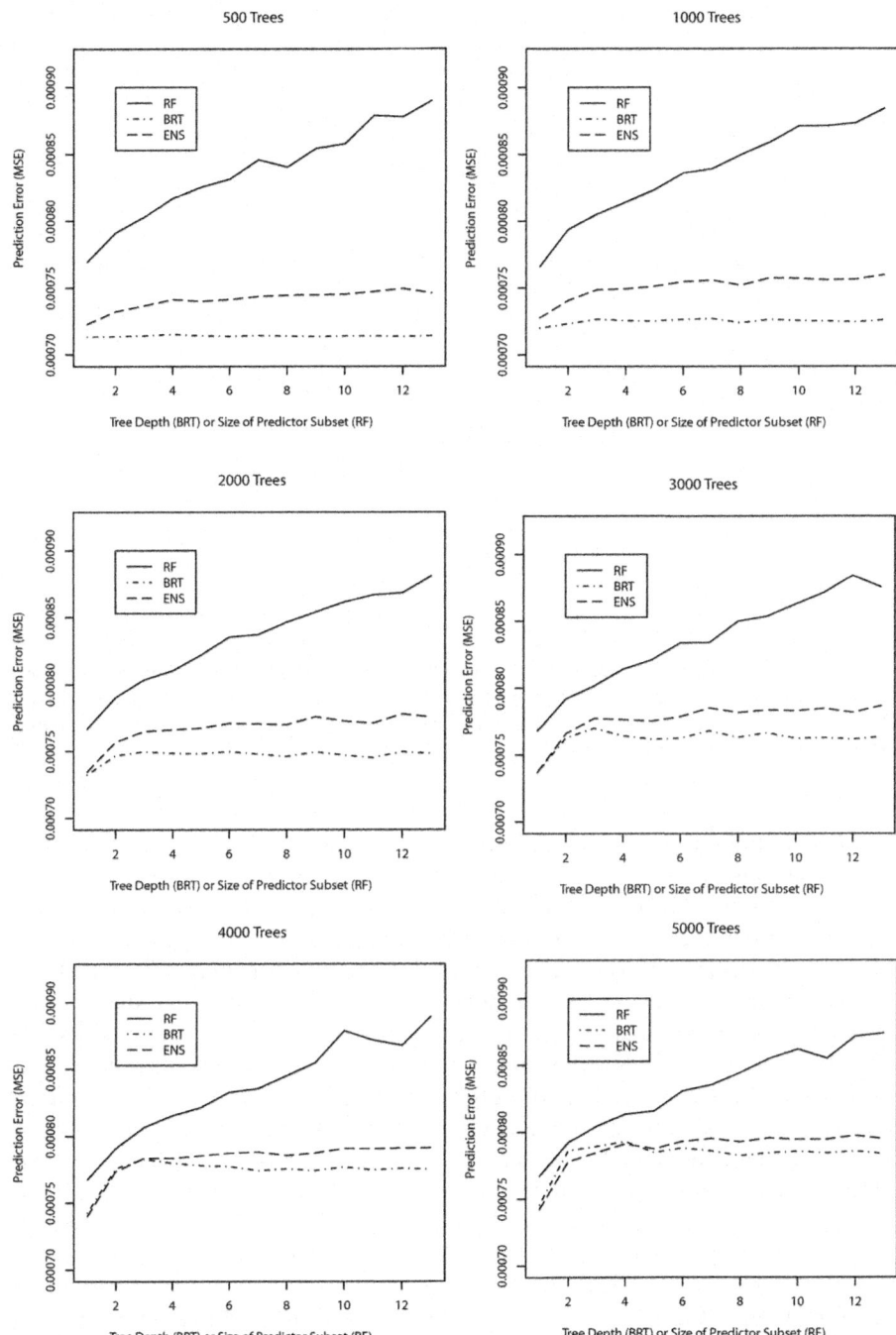

Fig. 5.3 Comparative performance of random forests (RF), boosted regression tree (BRT), and ensemble (ENS) predictions, as measured using cross-validated mean square prediction error (MSE). Tree size was allowed to vary from 500 to 8000 trees ($T = 8000$ not shown), and tree depth/ size of predictor subset was set from 1 to 13. In the case of BRT, a shrinkage (λ) value of 0.001 was used

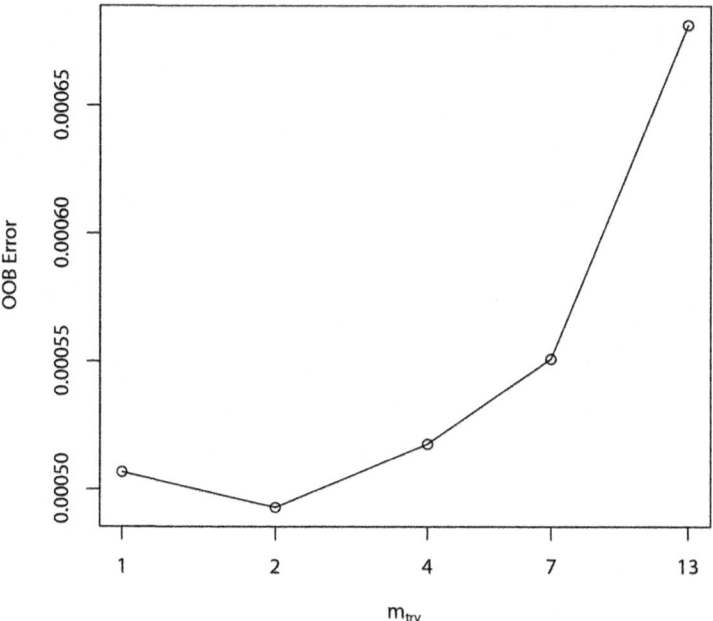

Fig. 5.4 Results for random forests out-of-bag (OOB) error assessment, as a function of the size of the predictor subset (*m*, or mtry in package randomForest of Liaw and Wiener 2002). A commonly cited decision rule is based on the square-root of the number of predictor variables which, in this case, would be 3.6 (~4)

All algorithms experienced highest cross-validated accuracies when lower numbers of covariates were specified - either in terms of the number of splits permissible to BRT trees (K) or the size of the predictor subset (m) for RF. Other authors have reported similar findings. For example, while Prasad et al. 2006 employed a MARS algorithm (Friedman 1991), they reported limitations in the portability of MARS predictions to future climate when more interaction terms were introduced in the model training stage, describing them as more "wild" (Prasad et al. 2006: 197). The BRT and RF equivalent to increasing interaction terms is by increasing m (RF) or K (BRT). Elith et al. (2008, 807) suggest that larger data sets benefit more from this increased complexity, but we demonstrate that this can come at the price of increased prediction bias. As a final note, James et al. (2013, 320) argued that lower values of m are warranted when there is strong correlation amongst the covariates.

The effect of varying the numbers of constructed trees (T) differed depending upon the algorithm; in the case of BRT, cross-validated accuracy tended to be optimal for intermediate numbers of trees (i.e., $T \leq 3000$) whereas it seemed less important for RF provided that m was set to one. By virtue of the additive nature in which BRT fits trees to increasingly smaller portions of residual variation (James et al. 2013), the judicious choice of parameter settings seems particularly important in order to prevent overfitting.

Our results demonstrate that for at least part of the "parameter space", ensemble predictions yielded the lowest prediction error and, at worst, tended to track BRT performance with less variation in performance. In general, variation in the number of trees used to train RF models seemed unimportant, but for this particular dataset, cross-validated error was lowest when tree depth was set to low values. We propose that the strengths of the two be merged using an "ensemble of ensembles" such as we have done here. The impact of producing an "ensemble of ensembles" was three-fold: improved predictive accuracy, particularly when comparing ensemble predictions to those based on RF; somewhat of a dampening of variation in performance from iteration-to-iteration; and lower prediction bias under a range of conditions, as evidenced by the narrowing of the gap between optimistic and crossvalidated accuracy assessments.

A pertinent question to ask is why, despite a growing body of applications of ML methods in the ecohydrological (Peters et al. 2007), marine (Leathwick et al. 2006; Pinkerton et al. 2010; Huettmann et al. 2011; Huettmann and Schmid 2015; Schmid et al. 2016) and terrestrial (Cutler et al. 2007; Jiao et al. 2014; Mi et al. 2014; Baltensperger and Huettmann 2015) ecological literature, are ML techniques not more widely used by ecologists? It may be a result of less familiarity (Olden et al. 2008), or perhaps they are perceived as "black boxes" that are harder to interpret (Elith et al. 2008). We hope that these examples, along with our analysis, will continue to make a case for routine use of ML methods such as BRT and RF under the proviso that idiosyncrasies of particular data sets may render it difficult to determine, in advance, the optimal set of tuning parameters to guide the model building process.

Given the complexity of real-world ecological problems, and the difficulty in assessing, *a priori*, appropriate model structures to test or make predictions means that ML methods should be routinely used. They can be used to explore patterns, evaluate the impact of different predictor variables, and provide predictions in a standalone form or as part of an ensemble of other predictions. They are ideally suited for "mining" large, complex data sets, especially when little prior knowledge about the system exists (Hochachka et al. 2007). Our results demonstrate that the combination of predictions from multiple algorithms, to form ensemble or consensus predictions, is straightforward to implement and can result in higher predictive accuracy than the results from single algorithms alone. Ensemble predictions have the added benefit of reducing the reliance on single techniques and allowing a wider range of potentially useful algorithms to be employed.

References

Araújo MB, New M (2007) Ensemble forecasting of species distributions. Trends Ecol Evol 22:42–47

Baltensperger A, Huettmann F (2015) Predictive spatial niche and biodiversity hotspot models for small mammal communities in Alaska: applying machine-learning to conservation planning. Landsc Ecol 30:681–697

Brawn C (2016) Marine traffic risk density in Atlantic Canada. Unpubl. Report

Breiman L (2001) Random forests. Mach Learn 45:5–32

Breiman L (2002) Manual on setting up, using, and understanding random forests v3.1. https://www.stat.berkeley.edu/~breiman/Using_random_forests_V3.1.pdf. Accessed 31 July 2015

Caruana R, Niculescu-Mizil A (2006) Proceedings of the 23rd international conference on machine learning, Pittsburgh, PA

Clemen RT (1989) Combining forecasts: a review and annotated bibliography. Int J Forecast 5:559–583

Cutler DR, Edwards TC Jr, Beard JH, Cutler A, Hess KT, Gibson J, Lawler JL (2007) Random forests for classification in ecology. Ecology 88:2783–2792

Das SK, Chen S, Deasy JO, Zhou S, Yin F-F, Marks LB (2008) Combining multiple models to generate consensus: application to radition-induced pneumonitis prediction. Med Phys 35:5098–5109

De'ath G, Fabricius KE (2000) Classification and regression trees: a powerful yet simple technique for ecological data analysis. Ecology 81:3178–3192

Domingos P (2012) A few useful things to know about machine learning. Commun ACM 59:78–87

Elith J, Burgman M (2002) Predictions and their validation: rare plants in the central highlands, Victoria, Australia. In: Scott JM, Heglund PJ, Morrison ML, Raphael MG, Wall WA, Samson FB (eds) Predicting species occurrences: issues of accuracy and scale. Island Press, Covelo, pp 303–314

Elith J, Graham CH (2009) Do they? How do they? WHY do they differ? On finding reasons for differing performances of species distribution models. Ecography 32:66–77

Elith J, Graham CH, Anderson RP, Dudik M, Ferrier S, Guisan A, Hijmans RJ, Huettmann F, Leathwick JR, Lehmann A, Li J, Lohmann LG, Loiselle RA, Manion G, Moritz C, Nakamura M, Nakazawa Y, Overton JM, Peterson AT, Phillips SJ, Richardson K, Scachetti-Pereira R, Schapiro RE, Soberón J, Williams S, Wisz MS, Zimmermann NE (2006) Novel methods improve prediction of species' distributions from occurrence data. Ecography 29:129–151

Elith J, Leathwick JR, Hastie T (2008) A working guide to boosted regression trees. J Anim Ecol 77:802–813

Fernández-Delgado M, Cernadas E, Barro S, Amorim D (2014) Do we need hundreds of classifiers to solve real-world classification problems? J Mach Learn Res 15:3133–3181

Franklin J (1995) Predictive vegetation mapping: geographic modelling of biospatial patterns in relation to environmental gradients. Prog Phys Geogr 19:474–499

Friedman JH (1991) Multivariate adaptive regression splines. Ann Stat 19:1–67

Friedman JH (2001) Greedy function approximation: a gradient boosting machine. Ann Stat 29:1189–1232

Harrell FE (2001) Regression modeling strategies: with applications to linear models, logistic regression, and survival analysis. Springer, New York

Hastie T, Tibshirani R, Friedman J (2001) The elements of statistical learning: data mining, inference, and prediction. Springer, New York

Hegel TM, Cushman SA, Evans J, Huettmann F (2010) Current state of the art for statistical modelling of species distributions. In: Cushman SA, Huettmann F (eds) Spatial complexity, informatics, and wildlife conservation. Springer, New York, pp 273–312

Heikkinen RK, Luoto M, Araújo MB, Virkkala R, Thuiller W, Sykes MT (2006) Methods and uncertainties in bioclimatic envelope modelling under climate change. Prog Phys Geogr 30:751–777

Hochachka WM, Caruana R, Fink D, Munson A, Riedewald M, Sorokina D, Kelling S (2007) Data-mining discovery of pattern and process in ecological systems. J Wildl Manag 71:2427–2437

Huettmann F, Artukhin Y, Gilg O, Humphries G (2011) Predictions of 27 Arctic pelagic seabird distributions using public environmental variables, assessed with colony data: a first digital IPY and GBIF open access synthesis platform. Mar Biodivers 41:141–179

Huettmann F, Schmid M (2015) Climate change predictions of pelagic biodiversity components. In: De Broyer C, Koubbi P, Griffiths HJ, Raymond B, Udekem d'Acoz C, Van de Putte AP, Danis B, David B, Grant S, Gutt J, Held C, Hosie G, Huettmann F, Post A, Ropert-Coudert Y (eds) Biogeographic Atlas of the Southern Ocean. Scientific Committee on Antarctic Research, Cambridge, pp 390–396

James G, Witten D, Hastie T, Tibshirani R (2013) An introduction to statistical learning: with applications in R. Springer, New York

Jiao S, Guo Y, Huettmann F, Lei G (2014) Nest-site selection analysis of Hooded Crane (*Grus monacha*) in northeastern China based on a multivariate ensemble model. Zool Sci 31:430–437

Leathwick JR, Elith J, Francis MP, Hastie T, Taylor P (2006) Variation in demersal fish species richness in the oceans surrounding New Zealand: an analysis using boosted regression trees. Mar Ecol Prog Ser 321:267–281

Liaw A, Wiener M (2002) Classification and regression by Random forest. R News 2(3):18–22

Lieske DJ, Fifield DA, Gjerdrum C (2014) Maps, models, and marine vulnerability: assessing the community distribution of seabirds at-sea. Biol Conserv 172:15–28

Lieske DJ, Schmid M, Mahoney M (2018) Data and analysis script. https://doi.org/10.5281/zenodo.1318352

Marmion M, Luoto M, Heikkinen RK, Thuiller W (2009a) The performance of state-of-the-art modelling techniques depends on geographical distribution of species. Ecol Model 220:3512–3520

Marmion M, Parviainen M, Luoto M, Heikkinen RK, Thuiller W (2009b) Evaluation of concensus methods in predictive species distribution modelling. Divers Distrib 15:59–69

Mi C, Huettmann F, Guo Y (2014) Obtaining the best possible predictions of habitat selection for wintering Great Bustards in Cangzhou, Hebei Province with rapid machine learning analysis. Chin Sci Bull. Published online: https://doi.org/10.1007/s11434-014-0445-9

Moisen GG, Freeman EA, Blackard JA, Frescino TS, Zimmermann NE, Edwards TC (2006) Predicting tree species presence and basal area in Utah: a comparison of stochastic gradient boosting, generalized additive models, and tree-based methods. Ecol Model 199:176–187

Olden JD, Lawler JJ, Poff NL (2008) Machine learning methods without tears: a primer for ecologists. Q Rev Biol 83:171–193

Oppel S, Meirinho A, Ramírez I, Gardner B, O'Connell AF, Miller PI, Louzao M (2012) Comparison of five modelling techniques to predict the spatial distribution and abundance of seabirds. Biol Conserv 156:94–104

Peters J, De Baets B, Verhoest MEC, Samson R, Degroeve S, De Becker P, Huybrechts W (2007) Random forests as a tool for ecohydrological distribution modelling. Ecol Model 207:304–318

Pinkerton MH, Smith ANH, Raymond B, Hosie GW, Sharp B, Leathwick JR, Bradford-Grieve JM (2010) Spatial and seasonal distribution of adult *Oithona similis* in the southern ocean: predictions using boosted regression trees. Deep-Sea Res I 57:469–485

Prasad AM, Iverson LR, Liaw A (2006) Newer classification and regression tree techniques: bagging and random forests for ecological prediction. Ecosystems 9:181–199

Renner M, Parrish JK, Piatt JF, Kuletz KJ, Edwards AE, Hunt GL Jr (2013) Modeled distribution and abundance of a pelagic seabird reveal trends in relation to fisheries. Mar Ecol Prog Ser 484:259–277

Ridgeway G (2012) Generalized boosted models: a guide to the gbm package. URL: http://gradientboostedmodels.googlecode.com/git/gbm/inst/doc/gbm.pdf (downloaded January 8, 2016)

Salford Systems, Inc. (2016) https://www.salford-systems.com/

Schmid MS, Aubry C, Grigor J, Fortier L (2016) The LOKI underwater imaging system and an automatic identification model for the detection of zooplankton taxa in the Arctic Ocean. Methods Oceanogr In pres

Chapter 6
Machine Learning for Macroscale Ecological Niche Modeling - a Multi-Model, Multi-Response Ensemble Technique for Tree Species Management Under Climate Change

Anantha M. Prasad

6.1 Introduction

Machine learning has come a long way in recent decades due to huge increases in computing power and the availability of robust public platforms for statistical analysis (e.g., R Core Team 2016). Machine learning techniques have benefited from advances in statistical learning and vice versa (Hastie et al. 2009; Slavakis et al. 2014), resulting in impressive applications of big data in imaging, astronomy, medicine, finance and to a lesser extent in ecology (Van Horn and Toga 2014; Zhang and Zhao 2015; Belle et al. 2015; Hussain and Prieto 2016; Hampton et al. 2013). A healthy relationship with computer science and engineering has invigorated the field even more, resulting in a variety of techniques suitable for diverse applications. One successful and frequently used method is ensemble learning, where learning algorithms independently construct a set of classifiers or regression-estimates and classify or regress newer data points by either taking a weighted vote (classifiers) or an average (regression) of their predictions (Zhou 2012).

A majority of the ensemble learning problems deal with classification due to the binary, or in some cases multinomial, response that is of interest. However, in the field of ecology, and especially in tree species abundance modeling, we have access to continuous data thanks to the Forest Inventory Analysis (FIA) in the United States (Woudenberg et al. 2010) that lends itself to a regression approach. Valuable information can be lost if the continuous data are classified *a priori* into classes. Therefore, it is best to solve the problem in a regression context, and classify the results later to retain most of the information in the response. I will choose the regression approach for this reason and also to highlight this less used aspect of statistical learning.

A. M. Prasad (✉)
Research Ecologist, USDA Forest Service, Northern Research Station, Delaware, OH, USA
e-mail: aprasad@fs.fed.us

© Springer Nature Switzerland AG 2018
G. R. W. Humphries et al. (eds.), *Machine Learning for Ecology and Sustainable Natural Resource Management*, https://doi.org/10.1007/978-3-319-96978-7_6

Modeling the abundance response of trees under current and future climates is an exercise fraught with assumptions and uncertainties due to the dynamic nature of the species' range boundaries. We are essentially capturing a slice in the eco-evolutionary history of the species and trying to project it into future climatic space as forecast by the general circulation models (GCMs; McGuffie and Henderson-Sellers 2014). Of the many uncertainties, the non-equilibrium nature of the tree species (they could still be expanding their ranges and not yet have achieved climatic equilibria) (Garcia-Valdes et al. 2013), and inability to capture biotic interactions (Belmaker et al. 2015) are cited most often. These limitations, however, are often due to the scale of analysis; a macroscale analysis will typically include biotic interactions as an emergent phenomenon. Only finer scale analysis can deal with biotic interactions in a more fundamental way. However, the question of species non-equilibrium also affects macroscale studies because of the historical nature of eco-evolutionary processes and can be addressed to some extent by comparing various studies as slices in time (Prasad 2015).

Of the many techniques that have emerged in recent years (Iverson et al. 2016), ensemble techniques based on decision trees have become the most popular among ecologists modeling niche related phenomena (Galelli and Castelletti 2013; Hill et al. 2017; Vincenzia et al. 2011). The transition from more parametric analysis like generalized linear and additive models (glm, gam and shrinkage based regression) to decision tree based techniques has to do mainly with the nature of ecological systems. They tend to be high dimensional and nonlinear with many embedded interactions; all of which are handled well by decision tree based techniques (Guisan et al. 2002; Guisan and Thuiller 2005). Hence a multitude of techniques have evolved, each appropriate for a subset of problems and dealing mostly with various shortcomings arising from more conventional decision-tree based techniques like bagging, randomized trees and boosting (Elith et al. 2010).

As datasets have become larger and easier to acquire (large scale inventories, digital elevation models, satellite imagery, demographic financial data, to name a few) with a corresponding increase in computing power, there has been a movement away from more parametric forms of analysis towards computationally intensive machine learning, such as non-parametric methods that are flexible and data-driven. While older constraints based on limited data and computing power have relaxed, newer ones have emerged because the analysis has moved more into the "prediction" space (e.g., models that overfit because of non-optimal variance-bias ratio). These newer challenges are being addressed via increasingly sophisticated algorithms that combine flexible models with resampling, permuting, shrinkage and regularization techniques (Tibshirani 1996; Zou and Hastie 2005; Hastie et al. 2009).

The focus of this chapter is to show how to tackle these issues when modeling the abundance of tree species at a macroscale (20 km resolution) in the eastern United States (where we have sufficiently large predictor and response data), and also, how to address the problems of model reliability and prediction confidence while interpreting the results. Towards this goal, I develop a multi-response, multi-model ensemble technique that addresses problems of bias, variance and output noise – resulting in more reliable prediction.

6.2   Controlling Bias and Variance

Some ecological projects are fortunate to have large amounts of data at their disposal while other studies fall into the category of designed experiments where data collection can be cumbersome and costly. Large Data projects are typically those that use datasets that are large and complex, of fairly coarse resolution, and are already available (e.g., remotely sensed topography and land-use, climate, soils, national forest inventory plots and bird surveys). Niche based analyses of these data lend themselves well to statistical machine learning techniques, unlike studies that require formal experimental design, which may be more appropriate for parametric statistical analyses. The existence of Large Data begs for a data-driven approach with complex and flexible models that capture nonlinearities and interactions well and can screen out less important predictors. However, this flexibility can result in overfitting and attendant variance; the models may fit the training data well, but not generalize well to newer prediction space (Domingos 2012; Merow et al. 2014). In statistical terms, these models have low bias (good) but high variance (not good). If bias is too high, the models are less likely to fit the underlying data (think straight line fitting curvilinear data), but if we lower bias too much, we risk overfitting and increased variance, making the models poor predictors of newer data (Dieterich and Kong 1995). To understand this a little better, imagine that we are training a flexible model with a data set that yields low training mean square error (MSE). If we use this same model with data set aside for testing, the test MSE will be much higher because it is picking up too many patterns associated with random noise (Hastie et al. 2009). A less flexible model (say a linear model) would have showed lower MSE with the test data even though the training MSE would be higher than the flexible model because it approximates nonlinearity with a linear fit. The quest in statistical learning is to optimize models to achieve a favorable bias-variance ratio, i.e., to simultaneously achieve low bias and low variance (Hastie et al. 2009).

6.3   Ensemble Learning Via Decision Trees

The basic idea of ensemble learning is to construct a mapping function y = F(x), based on the training data $\{(x_1,y_1), \ldots\ldots, (x_n,y_n)\}$, where

$$F(x) = a_o + \sum_{m=1}^{M} a_m f_m(x)$$

Where M is the size of the ensemble and $\{fm(x)\}$ is an ensemble of functions called base learners (Friedman and Popescu 2008). The base learners are chosen from a function class of predictor variables and can vary with the ensemble methods used. An algorithmic procedure is specified to pick functions and also to obtain linear combination of the parameters $\{am\}0 M$ based on the minimization of some cost function. This procedure generalizes the framework of ensemble learning to include algorithms like bagging, Random Forests, boosting, RuleFit etc.

The fundamental component of all ensemble learning algorithms that use "ensemble of decision trees" algorithms is the individual decision tree (Breiman et al. 1984). Decision tree is a recursive partitioning algorithm that partitions the response into subsets (left and right child nodes) based on splitting rules of the form $x_j < k$, where x_j is the splitting variable (predictor) and k is the splitting value. The left node gets all the observations (response) that satisfy the splitting rule and the right node gets the rest. The algorithm evaluates all possible splitting rules (for all the predictors) based on the response and selects the one that minimizes a statistical criterion (usually lowest MSE for regression). The observations in the resulting left and right nodes are again subject to the same partitioning scheme, and this goes on recursively until a stopping rule is satisfied (usually, minimum number of observations in the node, or the maximum depth of the tree or some other cost parameter). The end result of the recursive partitioning procedure is a decision tree with splitting rules and fitted values for terminal nodes (for regression, the average of the observations that fall into the terminal node).

Decision trees are intuitive, easy to interpret, capture nonlinearities and interactions very well and are very useful for high dimensional data. These properties make them very attractive for many ecological problems that exhibit these behaviors (Loh 2011; Rokach and Maimon 2015; Iverson and Prasad 1998). However, individual decision trees exhibit high variance and have poor prediction ability. Yet, they are very good building blocks in an ensemble setting where they can be used to build more complex models to achieve good variance bias tradeoffs (Dietterich 2000).

6.4 Ensemble Models

6.4.1 Bagging, Random Forest and Extreme Random Forests

Bagging is a way of reducing variance of decision trees via bootstrapping and aggregation of an ensemble of trees (Breiman 1996). In bagging, a number of decision trees are grown without pruning with a bootstrapped sample (sampling with replacement) and the resulting prediction rules averaged. It is based on the principle that if a single regressor has high variance, an aggregated regressor has smaller variance than the original one (Breiman 1996).

Random forests (RF) is a modification of bagging by taking a step further and randomizing even the predictor space. If along with the bootstrap sample, the predictors are also sampled randomly at each node and the results averaged, it results in further reducing variance (Prasad et al. 2006). This is the technique used in RF (randomForest package in R), where both datasets and predictors are perturbed to slightly increase the independence of each tree and then averaged to reduce variance (Breiman 2001). In RF, because a random subset of predictors are chosen at each split, many dominant predictors may not be present to define a split. This results in more local features defining the split instead of the dominant ones. When a large number of such trees are averaged, this can result in good balance between bias and variance and result in extremely reliable predictions. Another innovation in RF is that instead of computationally costly cross-validation or a separate test set to get unbiased error estimates, the observations not used in the training sample (usually one-third of the observations in the bootstrap sample), called "out-of-bag" (OOB), are used to obtain forecasts from the tree fitted to the remaining two-thirds (Liaw and Wiener 2002).

Extremely randomized trees (ERF) takes RF one step further in randomization (extraTrees package in R). While RF chooses the 'best' split at each node, ERF creates p splits randomly (i.e., independently of the response variable, p being the subset of predictors randomly chosen in each node) and then the split with the best gain (MSE for regression) is chosen. The rationale for ERF is that by randomizing the selection of split, the variance is reduced even further compared to the RF. However, ERF typically uses the entire learning sample instead of the boot-strapped sample to grow the trees in order to reduce bias (Geurts et al. 2006). Bias reduction becomes more important with this form of extreme randomization, because randomization increases bias when the splits are chosen independent of the response (Galelli and Castelletti 2013). ERF can be useful as a robust predictor after initially screening for irrelevant predictors. For example we can use RF to select a parsimonious, but ecologically meaningful set of predictors, and then use this set to predict with ERF.

6.4.2 Boosting Decision Trees

Boosting is a method of iteratively converting weak learners to stronger ones (in our case, using decision trees). Boosting initially builds a base learner after examining the data and then reweights observations that have higher errors. Stochastic gradient boosting (gbm package in R) is a form of optimization algorithm of a loss function with added tools to reduce variance by shrinkage and stochasticity (Ridgeway 1999; Friedman 2002). It optimizes a loss function over function space (as opposed to parameter space in ordinary regression problems) by estimating gradient directions of steepest descent (negative partial derivatives of the loss function called the pseudo-residuals) such that each iteration learns from previous errors (pseudo-residuals) and improves on them

$$F_m(x) = F_{m-1}(x) + v \cdot \gamma_m h_m(x)$$

At every stage of gradient boosting $1 < m \leq M$, the weak model Fm is slowly converted to a stronger one by improving on the previous iteration Fm-1 by adding an estimator. The value hm(x) is the decision tree (at the m-th step) with J terminal nodes (the tree partitions the predictor space into J disjoint regions). The goal is to minimize γm as a loss function (typically mean square error for regression), which has its own separate value for each of the J terminal nodes. The depth of the trees (i.e., the number of terminal nodes) J, defines the level of interaction and usually works best between 4 and 8. The shrinkage parameter ν ($0 < \nu \leq 1$) controls the learning rate of the boosting algorithm. If the number of boosting iterations (number of trees grown) is too large, it can lead to overfitting - ν therefore is usually chosen via cross-validation after finding the shrinkage parameter (values between 0.01 to 0.001 works best). In addition, the base learner, instead of using the entire training set, randomly subsamples without replacement (usually set to 50% of the training set), which adds stochasticity and leads to increased accuracy (Friedman 2002).

There is another slightly different approach to boosting that differs in the way the objective function is optimized with separate terms for training loss and regularization (Friedman 2001) called xgboost (Chen and Guestrin 2016). This method (xgboost package in R) differs from gbm in the way regularization is implemented when boosting, improving on its ability to control overfitting. It also handles tree pruning differently; gbm would stop splitting a node if it encounters a negative loss while xgboost splits to the maximum depth specified and then prunes the tree backwards to remove splits with no positive gain. Although boosting with carefully selected parameters can outperform RF, it can overfit noisy datasets due to the iterative learning process and has to be used with caution, or by using algorithms that automatically control overfitting with internal mechanisms (Opitz and Maclin 1999; Hastie et al. 2009).

6.4.3 RuleFit

RuleFit also uses decision tree ensembles to derive rules - however, these rules are used to fit regularized linear models in a flexible way that captures interactions (Friedman and Popescu 2008). It is similar to stochastic gradient boosting in that it combines base learners (decision tree rules) via a memory function with shrinkage to form a strong predictor. A large number of trees are generated from random subsets of the data and numerous rules assembled from a specified subset of terminal nodes. The predictor variables from these nodes allow for the estimation of linear functions where in addition to the rule-based base learner, linear basis functions are included in the predictive model. This is a useful feature because linearity from decision trees are hard to approximate. The large number of rules formed in the rule-generation phase, along with the linear basis functions are then minimized using regularized regression using lasso penalty (Tibshirani 1996; Zou and Hastie 2005). In regularized regression (ridge, lasso or elastic net) an additional penalty is imposed on the coefficients while minimizing the loss function. The final ensemble formed by regularized regression, results in rules, variables and linear coefficients sorted by importance. In contrast with other ensemble methods, RuleFit outputs coefficients in addition to prediction rules, which can be interpreted as regular linear coefficients.

6.5   Multiple Abundances – Habitat Suitability

The response, which in our case is an assessment of the habitat quality of white oak, is typically a measure of species abundance as reflected by its dominance and density (McNaughton and Wolf 1970). Dominance and density together capture many aspects of habitat quality. The measure that we used traditionally (Iverson et al. 2008; Prasad et al. 2016) is the importance value (IV) which captures the relative abundance weighted by other species present in the FIA plot (Woudenberg et al. 2010) as follows for each species X in a FIA plot:

$$IV(x) = \frac{50 * BA(x)}{\sum_{i=1}^{N} BA(i)} + \frac{50 * NS(x)}{\sum_{i=1}^{N} NS(i)}$$

BA is basal area, NS is number of stems (summed for overstory and understory trees) and N is the total number of species in the plot. This measure, which is a blend of dominance and density, reflects the biotic pressure that accounts for the interaction with other species and hence can reflect the realized niche better.

Another measure of species abundance that is proposed here is called mature average diameter (MAD). This dominance measure is derived by averaging the mean diameter of all trees of the target species in the plot after discounting the contribution of juveniles; juveniles are considered ephemeral because their contribution is negligible for this application. Juveniles were defined as: (min (avgdia) + q1(avgdia))/2; where avgdia is the average diameter, min is the minimum and q1 is the first quartile average diameter of all the FIA plots with white oak. This measure of dominance captures the absolute abundance of the species in contrast to the relative importance value (IV).

To capture the density of the species better, I propose another measure of abundance, mature species density (MNT), as the total number of trees of the species in the plot after discounting the juveniles. This measure of abundance denotes how well the species has colonized a site.

All three forms of abundance measures (IV, MAD, and MNT) in FIA plots were aggregated to 20 km cells and scaled from 0–100 (Fig. 6.1). They reflect different

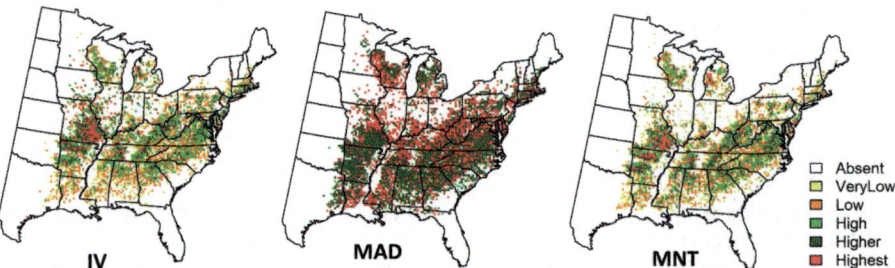

Fig. 6.1   The current maps of abundance for white oak - the importance value (IV), mature average diameter (MAD) and mature number of trees (MNT) per FIA plot aggregated to 20 km cells. The abundance values have been reclassified in the legend for illustrative purposes

aspects of habitat quality and should be modelled separately, with the overall effect spatially summarized similar to the multi-stage ensemble models (Anderson et al. 2012). I expect this approach to provide a better estimate of how the species would respond to climate change at a macro-scale compared to a single measure of abundance. Plurality of outputs and methods are important in gauging the overall response of the species, which has a complex nonlinear relationship with the environment under changing climates (Bowman et al. 2015).

6.6   Explanatory Variables (Predictors)

The explanatory variables represented a blend of climate, soil and topographic variables that were deemed most ecologically relevant after repeated tests (Table 6.1). For sources and other details, refer to Prasad et al. (2016). The current climate data are for the period 1981–2010 (Daly et al. 2008), and the future climate is Hadley Global Environment Model [HAD, Jones et al. 2011] for the greenhouse concentration pathway of RCP 8.5 (Representative Concentration Pathways; Moss et al. 2008) which represents the high emission future scenario (Meinshausen et al. 2011). The future RCP 8.5 climate scenario represents equilibrium conditions of the general circulation model (GCM; McGuffie and Henderson-Sellers 2014) for approximately 2100.

Table 6.1 The explanatory variables (predictors) used in the five models for white oak. These are a parsimonious set of ecologically relevant variables screen selected after repeated modeling

Climate	
tjan	Mean January temperature (°C)
tmaysep	Mean May–September temperature (°C)
pmaysep	May–September precipitation (mm)
gsai	Growing season aridity index (ratio of May–September precipitation by May–September evapotranspiration index)
Elevation	
elvmax	Maximum elevation (m)
elvsd	Elevation standard deviation
Soil	
clay	Percent clay (< 0.002 mm)
om	Organic matter content (% by weight)
ph	Soil pH
sieve10	Percent passing sieve no. 10 (coarse)
sieve200	Percent passing sieve no. 200 (fine)

Climate: Data for the period 1981–2010 from (PRISM Climate Group), GCM data from NEX-DCP30 (Thrasher et al. 2013).
Elevation: From the NASA's Shuttle Radar Topography Mission provided at a resolution of 3" (Guth 2006). We calculated the maximum value and standard deviation at 10 and 20 km² grids.
Soil: From Natural Resource Conservation Service's County Soil Survey Geographic (SSURGO) database (NRCS 2009). Data was processed by (Peters et al. 2013) and aggregated to 10 and 20 km² grids

6.7 Multi-Model Ensemble Approach

To achieve good bias-variance tradeoff, I used an 'ensemble-of-trees' via aggregation, randomization, boosting (randomForest, extraTrees, gbm, xgboost packages in R) and the ruleFit module (http://statweb.stanford.edu/~jhf/R_RuleFit.html). All these five approaches have their strengths and weaknesses depending on the training set. RandomForest and extraTrees have the least number of parameters to manipulate but cannot outperform the carefully tuned gbm and xgboost models. The gbm and xgboost algorithms, however, have more parameters to manipulate although the default settings often perform well. RuleFit in addition to robust prediction, gives linear coefficients and rule-sets. Multi-model ensemble approaches have been used where prediction uncertainty needs to be stabilized to yield more robust predictions (Jones and Cheung 2015; Martre et al. 2015). For the multi-model approach to work well, the models should be based on a similar framework (in this case decision trees) but should adopt structurally different approaches so that the final ensemble averages these heterogeneous approaches (Tebaldi and Knutti 2007). My approach consists of combining the five models (ensemble of models) to obtain two types of predictions: a) where output of all models are averaged (AVGMOD), and b) where they are averaged but only those cells common to these five models (an AND operation) make it to the final model (CAVGMOD). This procedure treats these models as a committee of experts and uses their average and common averaged prediction, improving prediction of single models by averaging out the errors. The overall thrust of the predictions are better captured by this approach for future climates. For this to work most effectively, the parameters for each of these five models need to be optimized via a repeated cross-validation approach in order to obtain a model with the most favorable bias-variance ratio. To do this, I used the caret package in R and repeated the ten-fold cross-validation, five times and chose the parameters with the lowest error (Kuhn 2008).

The multi-model, multi-response ensemble approach for the high emission future climate is illustrated for white oak using the three measures of abundance (IV, MAD and MNT) for the average model (AVGMOD), and the common average model (CAVGMOD) (Fig. 6.2). The CAVGMOD retains all the important habitats, while smoothing out the lower abundance values compared to AVGMOD and is therefore preferred in situations where reducing noise is desirable.

6.8 Results and Interpretation

One of the main goals while modeling future climate habitats of tree species is the need to gauge both model reliability and prediction confidence.

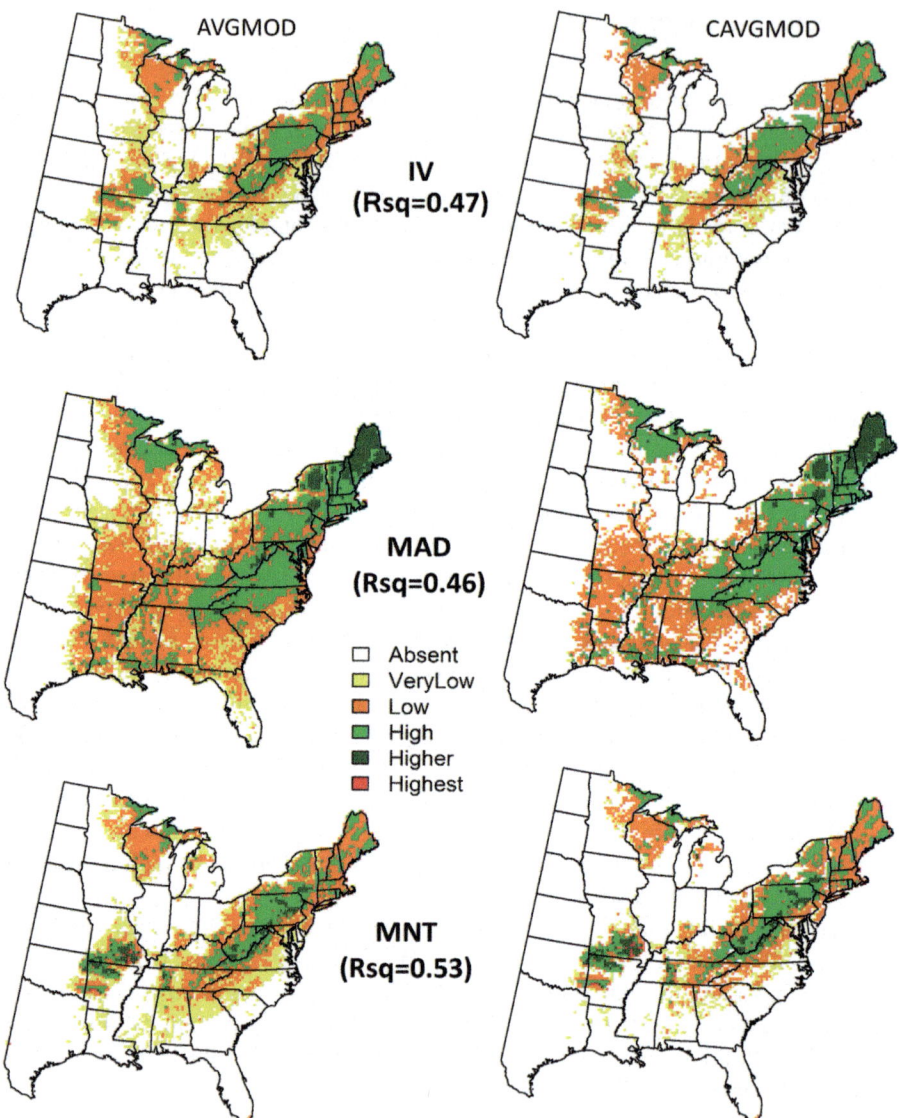

Fig. 6.2 The multi-model predictions for the three responses (importance value (IV), mature average diameter (MAD) and mature number of trees (MNT)) for the future harsh (Hadley, RCP 8.5) climate scenario for white oak. The AVGMOD is the average response across the five models, the CAVGMOD is the average response across the five models restricted to values common to all models. The abundance values have been reclassified in the legend for illustrative purposes

6.8.1 Model Reliability

Model reliability, which measures how well the models fit the data, reflects the vagaries of the training data, depending on whether the tree species is habitat specific, sparse, or a generalist. The sparser species have poor fit due to lack of training data and generally have poor model reliability. The habitat specific trees have the best model fit due to a better correlation with the environmental variables, with higher confidence in future predicted habitats. The model fit of generalists can vary depending on how widely and sparsely the species are distributed spatially. These species-specific vagaries affecting model reliability can be roughly measured via R-square-like measures via OOB, cross-validation or through a separate training and test dataset. For example, the R-square for the IV response of the RF model for the habitat-specific loblolly pine (*Pinus taeda*) was 0.79. In comparison, the R-square measure for our generalist species example of white oak (for the five models and three responses) averaged ~ 0.47.

6.8.2 Prediction Confidence

Even for species with good model reliability, the spatial configuration of the habitat quality in the predicted output (as measured via abundance values) can vary. For example, in Fig. 6.2, the classes 1–3 and 4–7 figure prominently even in CAVGMOD, and are of lower habitat quality than the higher classes. Because we can take advantage of the continuous distribution via regression models (after rescaling the abundances to values between 0 and 100), we have the ability to interpret the predicted habitats in terms of "prediction confidence" by reclassifying the results. The multi-model ensemble method helps mitigate the effects of spurious model artifacts (what can be termed "fuzzy values") at the low end of the abundance spectrum. The CAVGMOD approach further helps us identify only those prediction signals that have been strong in all five of the model predictions. Further, continuous predictions do not lend themselves to easy interpretation. Therefore, reclassifying them with the purpose of identifying the core regions where we have the highest confidence (based on abundance values) becomes useful for interpretation.

6.8.3 Combined Habitat Quality and Prediction Confidence

Using the CAVGMOD approach, we can average the predicted abundances of IV, MAD and MNT to capture the important future habitats as reflected by these three aspects of abundance and then reclassify the output to highlight the prediction confidence of the averaged response (Fig. 6.3). I have classified the future habitats to

Fig. 6.3 The average of
the three predictions
(importance value (IV),
mature average diameter
(MAD) and mature number
of trees (MNT)) for
CVAGMOD (Fig. 6.2),
with values common to the
three predictions for white
oak

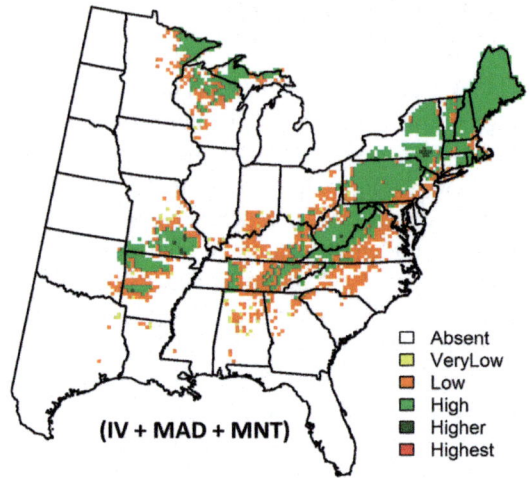

five confidence zones based on the predicted abundance: (1) Very low (1–3);
(2) Low (4–7); (3) High (8–15); (4) Higher (16–25); (5) Highest (26–100). Class 1
(Very low) would include many model artifacts (for example values close to zero
that were regressed as 1–3) that are of dubious habitats that can be discarded as
unreliable. Class 2 (Low) may also contain some regions with dubious habitats and
some with low habitat suitability and should be treated with caution. Confidence in
the habitat suitability classes increase steadily from Class 3 onwards (High, Higher
and Highest).

Compared to the three CAVGMOD responses (Fig. 6.2), the single combined
response (Fig. 6.3) highlights those areas (High and Higher classes) where we have
the most confidence in the habitat quality of future habitats based on all three aspects
of the abundances. For white oak, these areas (green and dark green) are predominantly in the north-east, north-central and south-central regions.

6.8.4 Predictor Importance

The importance of the predictors for each of the responses (IV, MAD and MNT)
varied among the five models for white oak, although the first three were similar for
all five models. These were recorded and averaged across the five models for the
three responses (Table 6.2). For IV and MAD, the three most important variables
are ph, tmaysep and tjan (Table 6.1), which explain 47.5% (IV) and 48.2% (MAD)
of the variation for white oak. For MNT, the order varies with sieve10 and clay
becoming important, but the same three variables (ph, tmaysep and tjan) still explain
40.1% of the variation. The predictor importance of the final combined response of
the multi-model ensemble is the average for the three individual responses (IV,
MAD and MNT) (Table 6.3). Again, the three most important variables (ph, tmaysep and tjan) explain 46.5% of the total variation. Because white oak is a generalist

species occupying a vast swath of the eastern US, ph captures variation from east to west, while tjan and tmaysep are more important in capturing the north-south variation, and hence figures prominently in the final response.

6.9 Discussion

The main goal of the multi-model, multi-response approach developed here is to produce more reliable and ecologically interpretable models that can be used to help decision makers in managing tree species (Bell and Schlaepfer 2016). Tree species ranges are dynamic by nature and the additional impact of anthropogenic climate change makes it harder to predict distribution for future climates irrespective of the

Table 6.2 The predictor importance of white oak averaged across the five models for importance value (IV), mature average diameter (MAD) and mature number of trees (MNT). The Percent Gain reflects the proportion of variance explained by the variable

IV		MAD		MNT	
Variables	Percent gain	Variables	Percent gain	Variables	Percent gain
ph	16.7	ph	22.1	ph	17.2
tmaysep	16.6	tmaysep	16.5	sieve10	12.9
tjan	14.2	tjan	10.6	tmaysep	12.3
sieve10	10.2	elvmax	7.3	clay	10.8
clay	7.9	pmaysep	7.0	tjan	10.6
gsai	6.5	om	6.9	sieve200	8.1
elvmax	6.1	sieve10	6.7	elvsd	7.7
pmaysep	6.1	elvsd	6.4	pmaysep	5.6
om	5.5	gsai	6.3	om	5.5
elvsd	5.3	sieve200	6.0	gsai	5.3
sieve200	5.0	clay	4.2	elvmax	4.0

Table 6.3 The average predictor importance of the five models for white oak averaged across the three responses (IV, MAD and MNT in Table 6.2) and sorted by the Percent Gain

AVG	
Variables	Percent gain
ph	18.7
tmaysep	15.1
tjan	11.8
sieve10	9.9
clay	7.6
elvsd	6.5
sieve200	6.4
pmaysep	6.2
gsai	6.0
om	6.0
elvmax	5.8

modeling approaches used (Zurell et al. 2016). However, managers need to be able to target specific areas for facilitating species conservation and other multiple-use management objectives. The first step in accomplishing these goals is to explore where the most probable future suitable habitats will occur. The multi-model, multi-response approach addresses the inherent complexity in tree species response in a systematic and statistically defensible manner. It also provides maps of regions where we have high confidence in the future suitable habitats for tree species that exhibit good model reliability (Hannemann et al. 2015). The tree species that exhibit high model reliability are typically species that are habitat specific, although generalists like white oak can also be adequately modelled. The tree species that typically have poor model reliability are those that are sparse (both closely and widely distributed), which for eco-evolutionary and biogeographic reasons have not extended their range. Models for these species should be treated with caution because their habitats are difficult to predict with environmental variables; biogeographic and eco-evolutionary variables are not easy to incorporate without extensive Gene X Environment studies.

The multi-model, multi-response model I present as an example, demonstrates that suitable future habitats for white oak are most likely to be in the north-east, north-central and south-central regions of the eastern United States (Fig. 6.3). This type of information is important for resource managers dealing with uncertainty and mandates to incorporate climate change in their management portfolios. While suitable habitats lack information on the likelihood of colonization, these can be assessed at a later stage via dispersal models (Prasad et al. 2016). However, to assess the probability of establishment of colonized sites involves finer scale process-based models that account for biotic interactions.

Another challenge when modeling tree species habitats under current and future climates lies in the transfer of ecological space (the niche of the species) to eco-geographic space (the mapped niche), which results in spatial autocorrelation effects. The problem of spatial autocorrelation can become acute with conventional parametric techniques and, while less problematic with non-parametric statistical learning methods, can still manifest in residual errors (Hawkins 2012; Kühn and Dormann 2012). In this study, there was negligible global residual spatial autocorrelation, although local ones were present. However in niche-based spatial modeling, some residual spatially auto-correlated errors have to be tolerated, and interpreted with caution. The alternative is extremely complex, autoregressive, parametric models that in many cases defeat the purpose of a more flexible modeling approach (Merow et al. 2014).

6.10   Conclusion

Predicting habitat quality is the first stage in the analysis of future distribution of tree species because dispersal and site-specific constraints will prevent colonization and establishment in all available suitable habitats (Prasad et al. 2016). Predicting these suitable habitats using robust modeling techniques is the essential first step and I present the multi-model and multi-response ensemble technique as a method for modeling tree species dynamics for better management under changing climates.

Acknowledgements The author would like to thank Louis Iverson for his comprehensive review and also two anonymous reviewers for their valuable suggestions. Thanks to the Northern Research Station, USDA Forest Service, for funding.

References

Anderson BJ, Chiarucci A, Williamson M (2012) How differences in plant abundance measures produce different species-abundance distributions. Methods Ecol Evol 3:783–786

Bell DM, Schlaepfer DR (2016) On the dangers of model complexity without ecological justification in species distribution modelling. Ecol Model 330:50–59

Belle A, Thiagarajan R, Soroushmehr SMR, Navidi F, Beard DA, Najarian K (2015) Big data analytics in healthcare. BioMed Res Int 370194, 16. doi:https://doi.org/10.1155/2015/370194

Belmaker J, Zarnetske P, Tuanmu M-N, Zonneveld S, Record S, Strecker A, Beaudrot L (2015) Empirical evidence for the scale dependence of biotic interactions. Glob Ecol Biogeogr 24:750–761

Bowman DM, Perry GLW, Marston JB (2015) Feedbacks and landscape-level vegetation dynamics. Trends Ecol Evol 30:255–260

Breiman L (1996) Bagging predictors. Mach Learn 24:123–140

Breiman L (2001) Random forests. Mach Learn 45:5–32

Breiman L, Friedman J, Stone CJ, Olshen RA (1984) Classification and regression trees. CRC press, Boca Raton

Chen T, Guestrin C (2016) XGBoost: reliable large-scale tree boosting system. ar Xiv: 1603.02754 [cs. LG]. http://arxiv.org/pdf/1603.02754v1

Daly C, Halbleib M, Smith JI, Gibson WP, Doggett MK, Taylor GH, Curtis J, Pasteris PP (2008) Physiographically sensitive mapping of climatological temperature and precipitation across the conterminous United States. Int J Climatol 28:2031–2064

Dietterich TG (2000) An experimental comparison of three methods for constructing ensembles of decision trees. Mach Learn 40:139–157

Dietterich TG, Kong EB (1995) Machine learning bias, statistical bias, and statistical variance of decision tree algorithms. Mach Learn 255:0–13

Domingos P (2012) A few useful things to know about machine learning. Commun ACM 55(10):78–87

Elith J, Kearney M, Phillips S (2010) The art of modelling range-shifting species. Methods Ecol Evol 1:330–342

Friedman JH (2001) Greedy function approximation: a gradient boosting machine. Ann Stat 29:1189–1232

Friedman JH (2002) Stochastic gradient boosting. Comput Stat Data Anal 38:367–378

Friedman JH, Popescu BE (2008) Predictive learning via rule ensembles. Ann Appl Stat 2:916–954

Galelli S, Castelletti A (2013) Assessing the predictive capability of randomized tree-based ensembles in streamflow modelling. Hydrol Earth Syst Sci 17:2669–2684

Garcia-Valdes R, Zavala MA, Araujo MB, Purves DW (2013) Chasing a moving target: projecting climate change-induced shifts in non-equilibrial tree species distributions. J Ecol 101:441–453

Geurts P, Ernst D, Wehenkel L (2006) Extremely randomized trees. Mach Learn 63:3–42

Guisan A, Edwards TC Jr, Hastie T (2002) Generalized linear and generalized additive models in studies of species distributions: setting the scene. Ecol Model 157:89–100

Guisan A, Thuiller W (2005) Predicting species distribution: offering more than simple habitat models. Ecol Lett 8:993–1009

Guth PL (2006) Geomorphometry from SRTM: Comparison to NED. Photogramm Eng Remote Sens 72:269–277

Hampton SE, Strasser CA, Tewksbury JJ, Gram WK, Budden AE, Batcheller AL, Duke CS, Porter JH (2013) Big data and future of ecology. Front Ecol Environ 11:156–162

Hannemann H, Willis KJ, Macias-Fauria M (2015) The devil is in the detail: unstable response functions in species distribution models challenge bulk ensemble modelling. Glob Ecol Biogeogr 25:26–35

Hastie T, Tibshirani R, Friedman J (2009) The elements of statistical learning, 2nd edn. Springer Science, New York

Hawkins BA (2012) Eight (and a half) deadly sins of spatial analysis. J Biogeogr 39:1–9

Hill L, Hector A, Hemery G, Smart S, Tanadini M, Brown N (2017) Abundance distributions for tree species in Great Britain: a two-stage approach to modeling abundance using species distribution modeling and random forest. Ecol Evol 7:1043–1056

Hussain K, Prieto E (2016) Big data in the finance and insurance sectors. In: Cavanillas JM et al (eds) New horizons for a data-driven economy. Springer Open. https://doi.org/10.1007/978-3-319-21569-3

Iverson LR, Prasad AM (1998) Predicting abundance of 80 tree species following climate change in the eastern United States. Ecol Monogr 68:465–485

Iverson LR, Prasad AM, Matthews SN, Peters M (2008) Estimating potential habitat for 134 eastern US tree species under six climate scenarios. For Ecol Manag 254:390–406

Iverson LR, Thompson FR, Matthews S, Peters M, Prasad AM, Dijak WD, Fraser J, Wang WJ, Hanberry B, He H, Janowiak M, Butler P, Brandt L, Swanston C (2016) Multi-model comparison on the effects of climate change on tree species in the eastern U.S.: results from an enhanced niche model and process-based ecosystem and landscape models. Landsc Ecol. https://doi.org/10.1007/s10980-016-0404-8

Jones MC, Cheung WWL (2015) Multi-model ensemble projections of climate change effects on global marine biodiversity. ICES J Mar Sci 72:741–752

Jones CD, Hughes JK, Bellouin N, Hardiman SC, Jones GS, Knight J, Liddicoat S, O'Connor FM, Andres RJ, Bell C, Boo K-O, Bozzo A, Butchart N, Cadule P, Corbin KD, Doutriaux-Boucher M, Friedlingstein P, Gornall J, Gray L, Halloran PR, Hurtt G, Ingram WJ, Lamarque J-F, Law RM, Meinshausen M, Osprey S, Palin EJ, Parsons Chini L, Raddatz T, Sanderson MG, Sellar AA, Schurer A, Valdes P, Wood N, Woodward S, Yoshioka M, Zerroukat M (2011) The HadGEM2-ES implementation of CMIP5 centennial simulations. Geosci Model Dev 4:543–570

Kühn I, Dormann CF (2012) Less than eight (and a half) mis- conceptions of spatial analysis. J Biogeogr 39:995–998

Kuhn M (2008) Building predictive models in R using the caret package. J Stat Softw 28:1–26

Liaw A, Wiener M (2002) Classification and regression by random forest. R News 2:18–22

Loh W-Y (2011) Classification and regression trees. WIREs Data Min Knowl Discovery 1:14–23. https://doi.org/10.1002/widm.8

Martre P, Wallach D, Asseng S, Ewert F, Boote KJ, Ruane AC, Peter J, Cammarano D, Hatfield JL, Rosenzweig C, Aggarwal PK, Angulo C, Basso B, Bertuzzi P (2015) Multimodel ensembles of wheat growth: many models are better than one. Glob Chang Biol 21:911–925

McGuffie K, Henderson-Sellers A (2014) A climate modelling primer, 4th edn. Wiley, p 456. isbn:978-1-119-94336-5

McNaughton SJ, Wolf LL (1970) Dominance and the niche in ecological systems. Science 167:131–139

Meinshausen M, Smith SJ, Calvin K, Daniel JS, Kainuma MLT, Lamarque JF, Matsumoto K, Montzka SA, Raper SCB, Riahi K, Thomson A, Velders GJM, van Vuuren DPP (2011) The RCP greenhouse gas concentrations and their extensions from 1765 to 2300. Clim Chang 109:213–241

Merow C, Smith MJ, Edwards TC Jr, Guisan A, McMahon SM, Normand S, Thuiller W, Wuest RO, Zimmermann NE, Elith J (2014) What do we gain from simplicity versus complexity in species distribution models? Ecography 37:1267–1281

Moss R, Babiker M, Brinkman S, Calvo E, Carter T et al (2008) Towards new scenarios for analysis of emissions, climate change, impacts, and response strategies. Intergovernmental Panel on Climate Change, Geneva, p 132 http://www.aimes.ucar.edu/docs/IPCC.meetingreport.final.pdf

NRCS (Natural Resources Conservation Service) (2009) Soil Survey Geographic (SSURGO). Available at https://datagateway.nrcs.usda.gov/. Accessed between August 2009 and November 2010

Opitz D, Maclin R (1999) Popular ensemble methods: an empirical study. J Artif Intell Res 11:169–198

Peters MP, Iverson LR, Prasad AM, Matthews SN (2013) Integrating fine-scale soil data into species distribution models: preparing soil survey geographic (SSURGO) data from multiple counties. US Department of Agriculture, Forest Service, Northern Research Station, Newtown Square, p 70

Prasad AM (2015) Macroscale intraspecific variation and environmental heterogeneity: analysis of cold and warm zone abundance, mortality, and regeneration distributions of four eastern US tree species. Ecol Evol 5:5033–5048

Prasad AM, Iverson LR, Liaw A (2006) Newer classification and regression tree techniques: bagging and random forests for ecological prediction. Ecosystems 9:181–199

Prasad AM, Iverson LR, Matthews SN, Peters MP (2016) A multistage decision support framework to guide tree species management under climate change via habitat suitability and colonization models, and a knowledge-based scoring system. Landsc Ecol. https://doi.org/10.1007/s10980-016-0369-7

PRISM Climate Group. Oregon State University, http://prism.oregonstate.edu

Ridgeway G (1999) The state of boosting. Comput Sci Stat 31:172–181

Rokach L, Maimon O (2015) Data mining with decision trees - theory and applications, 2nd edn. World Scientific

R Core Team (2016) R: A language and environment for statistical computing. R Foundation for Statistical Computing, Vienna URL https://www.R-project.org/

Slavakis K, Giannakis GB, Mateos M (2014) Modeling and optimization for big data analytics. IEEE Signal Process Mag 5:18–31

Tebaldi C, Knutti R (2007) The use of the multi-model ensemble in probabilistic climate projections. Phil Trans R Soc A 365:2053–2075

Thrasher B, Xiong J, Wang W, Melton F, Michaelis A, Nemani R (2013) Downscaled climate projections suitable for resource management. Trans Am Geophys Union 94:321–323

Tibshirani R (1996) Regression shrinkage and selection via the Lasso Robert Tibshirani. J R Stat Soc Ser B Stat Methodol 58:267–288

Van Horn JD, Toga AW (2014) Human neuroimaging as a "big data" science. Brain Imaging Behav 8:323–331. https://doi.org/10.1007/s11682-013-9255-y

Vincenzia S, Zucchettab M, Franzoib P, Pellizzato M, Pranovib F, De Leo GA, Torricelli P (2011) Application of a random Forest algorithm to predict spatial distribution of the potential yield of Ruditapes philippinarum in the Venice lagoon, Italy. Ecol Model 222:1471–1478

Woudenberg SW, Conkling BL, O'Connell BM, LaPoint EB, Turner JA, Waddell KL (2010) The forest inventory and analysis database: database description and User's manual version 4.0 for phase 2. General Technical Report RMRS-GTR-245, USDA Forest Service, Rocky Mountain Research Station, Fort Collins, Colorado, 336 p

Zhang Y, Zhao Y (2015) Astronomy in the big data era. Data Sci J 14:11. https://doi.org/10.5334/dsj-2015-011

Zhou ZH (2012) Ensemble methods: foundations and algorithms. CRC press, Boca Raton

Zou H, Hastie T (2005) Regularization and variable selection via the elastic net. R Stat Soc Ser B Stat Methodol 67:301–320

Zurell D, Thuiller W, Pagel J, Cabral JS, Münkemüller T, Gravel D, Dullinger S, Normand S, Schiffers KH, Moore KA, Zimmermann NE (2016) Benchmarking novel approaches for modelling species range dynamics. Glob Chang Biol 22:2651–2664

Chapter 7
Mapping Aboveground Biomass of Trees Using Forest Inventory Data and Public Environmental Variables within the Alaskan Boreal Forest

Brian D. Young, John Yarie, David Verbyla, Falk Huettmann, and F. Stuart Chapin III

7.1 Introduction

Forest biomass, the aboveground dry mass portion of live trees within a given area (Bonnor 1985), is of interest for both ecological and economic reasons. Forest soils and biomass hold most of the carbon in the Earth's terrestrial biomes (Houghton 2005) and significantly contribute to the overall global carbon exchange (Schimel et al. 2001). Within the boreal forest, the largest terrestrial biome, little is known about the quantity of woody biomass at spatial scales useful to forest practitioners (~ 1m^2 to 3.0 km^2) (Niemelä 1999; O'Neill et al. 1997). In previous investigations of aboveground forest biomass in the boreal region (see for instance Blackard et al. 2008; Botkin and Simpson 1990; Harrell et al. 1995; Yarie and Billings 2002; Yarie and Mead 1982), the spatial scales have either been rather course (Botkin and Simpson 1990; Yarie and Billings 2002) or the predictions lacked precision due to limited ground-truthing.

Within the boreal region, an increased interest in biomass as a possible fuel source has led to the need for a clearer understanding of the quantities of biomass that are available at an operational scale (Fresco 2006; GAO 2005; Loeffler et al. 2010).

B. D. Young (✉)
Department of Natural Sciences, Landmark College, Putney, VT, USA
e-mail: brianyoung@landmark.edu

J. Yarie · D. Verbyla
Department of Forest Sciences, University of Alaska Fairbanks, Fairbanks, AK, USA

F. Huettmann
EWHALE Lab, Biology and Wildlife Department, Institute of Arctic Biology, University of Alaska-Fairbanks, Fairbanks, AK, USA

F. Stuart Chapin III
Institute of Arctic Biology, University of Alaska Fairbanks, Fairbanks, AK, USA

© Springer Nature Switzerland AG 2018   141
G. R. W. Humphries et al. (eds.), *Machine Learning for Ecology and Sustainable Natural Resource Management*, https://doi.org/10.1007/978-3-319-96978-7_7

As of 2015, nine wood biomass energy facilities have been built in Interior Alaska with another ten under construction, and more than eleven are in design or feasibility status (AEA 2015). Energy demands for woody biomass are both expanding total forest harvest and changing the conventional forest management paradigm from production of large-dimension white spruce timber to a slowly expanding harvest of other species and size classes.

The boreal forest of Alaska extends from the Bering Sea on the west to the Canadian border in the east and is bounded in the north by the Brooks Range and in the south by the Chugach and Coastal mountains (Fig. 7.1), covering an area of nearly 500,000 km². The Alaskan boreal forest consists of a mosaic of two general forest types, mixed poplar/birch and mixed spruce (Ruefenacht et al. 2008; Viereck

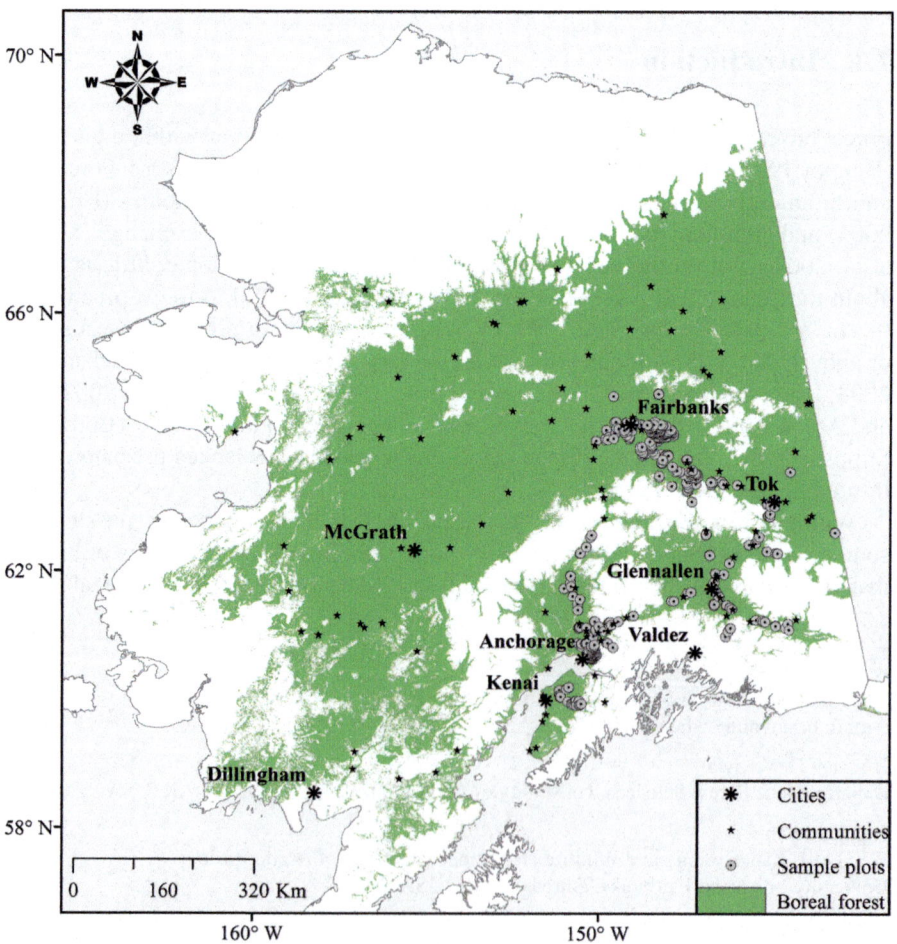

Fig. 7.1 Geographic distribution of the 694 Sample Plots (in circles) within the Alaskan boreal forest (Ruefenacht et al. 2008)

and Little 2007; Young et al. 2011) and primarily contains seven tree species. White spruce (*Picea glauca*) and black spruce (*P. marianana*) are the predominant conifers and two poplars (*Populus tremuloides* and *P. balsamifera*), two birches (*Betula neoalaska* and *B. kenaica*), and tamarack (*Larix laricina*) represent the deciduous species. Significant variation in tree growth occurs due to local differences in topography, soil type, the biota, the successional state, and climate conditions (Chapin III et al. 2006; Liang 2010; Lloyd and Fastie 2002; van Cleve et al. 1983; Wilmking and Juday 2005) which are collectively referred to as state factors (Major 1951). These variations in tree growth can lead to vastly different amounts of aboveground forest biomass depending on site differences in state factors. Biomass models that incorporate as many of the state factors as possible may yield an enhanced predictive ability for this ecologically and economically important forest attribute (Cutler et al. 2007; Grossmann et al. 2010).

The combination of forest inventory data with remote sensing data from both aerial and satellite formats have been previously employed in mapping woody biomes across broad spatial scales (see for instance Fassnacht et al. 2006; Iverson and Prasad 2001; McRoberts et al. 2008; Ruefenacht et al. 2008). In the United States, the inventory data is typically from the Forest Inventory and Analysis (FIA) program of the USDA Forest Service which is uniformly distributed across the landscape, expect in interior Alaska where FIA data is still lacking. Spatial interpolation techniques that combine forest-inventory and remote-sensing data in regions with either sparse or non-uniformly distributed inventory data have been employed to predict the geographical distribution of various forest attributes (Liang and Zhou 2010; Parmentier et al. 2011; Young et al. 2011). The use of machine learning to estimate aboveground forest biomass at the mesoscale from forest inventory and remotely sensed data in remote regions has, however, received less attention.

Maps depicting spatially explicit estimates of forest biomass are valuable for planning and monitoring (Drew et al. 2011). Creating such maps typically employs predictive spatial modeling techniques where the parameters of interest are obtained from inventory data and then related to remotely mapped attributes (see Austin 2002; Cushman and Huettmann 2010; Cushman and McKelvey 2009; Ferrier et al. 2002; Franklin 1995; Guisan and Thuiller 2005). Several statistical approaches have been used to create predictive maps, with non-parametric approaches tending to yield better results than parametric approaches, (Drew et al. 2011; Prasad et al. 2006). Spatial autocorrelation, a general property of most ecological attributes (Legendre 1993), is an additional issue to address, especially in large-extent forest studies (Young et al. 2011). When unaccounted for in a traditional approach, spatial autocorrelation may affect statistical model predictions because it violates the assumption of independence on which most standard statistical procedures rely (Legendre 1993). Thus, non-parametric models that account for, or are tolerant of, spatial autocorrelation and noisy data, e.g. when employing 'recursive partitioning' could be generally useful in assessing the spatial patterns of biomass (Blackard et al. 2008; Li et al. 2011).The use of machine learning, and notably random forests, has allowed for major advances in the capacity to make predictions of various forest attributes including biomass (Baccini et al. 2008; Li et al. 2011).

Our objectives here are to (1) develop a spatially explicit model depicting aboveground forest biomass for the Alaskan boreal forest using ground measured inventory plots, at a 1-km cell size, (2) evaluate model performance and compare the results to other region wide forest biomass estimates, (3) explore the contribution of the environmental predictors used to develop the biomass model and place them in ecological context, and lastly (4) use the resulting dataset to estimate and map forest biomass for the Alaskan boreal forest.

7.2 Methods and Materials

7.2.1 Biomass Data

Our dataset consisted of 694 permanent sample plots (PSPs) from the Cooperative Alaska Forest Inventory Database (CAFI; http://www.lter.uaf.edu/data_detail. cfm?datafile_pkey=452) (Malone et al. 2009) and the Fort Wainwright Forest Inventory Database (WAIN; http://www.usarak.army.mil/conservation) (Rees, personal communication). These databases consist of periodically re-measured PSPs located across interior and south-central Alaska north of 60°N (Fig. 7.1). The two databases were comparable because they had similar sampling designs and measurement protocols largely following the procedures outlined by (Curtis 1983). The plots are at a minimum of 300 m apart from one another. These data consist of periodically remeasured PSPs located across interior and southcentral Alaska north of 60° N (Fig. 7.1). These PSPs are located primarily on stocked forested lands. The CAFI plots are primarily located along the road system on Federal, State, Borough, and Native Corporation lands, while the WAIN plots are scattered across Military lands (Fig. 7.1).

The aboveground tree woody biomass, which includes biomass from the tree bole, stumps, branches and twigs, on each of the 694 PSPs was calculated for each tree greater than 2.54 cm in diameter at breast height (DBH) using the equations developed by Jenkins et al. (2003) for each of the seven possible tree species (*Picea glauca, P. marianana, Populus tremuloides, P. balsamifera, Betula neoalaska, B. kenaica,* and *Larix laricina*) present within a given PSP then aggregated to develop a megaton per hectare (Mg/ha) dry weight value. The Jenkins et al. (2003) calculations are used by the United States Forest Service Forest Inventory and Analysis Program (FIA) for their PSPs (Jenkins et al. 2004). We used the same calculations even though regional biomass calculations have been developed (see Yarie et al. 2007) so that the results from our model of aboveground tree woody biomass could be directly compared with other previously published results (Blackard et al. 2008).

7.2.2 Environmental Factors

Our original predictor dataset consisted of 39 variables, including the spatial structure (X and Y coordinates), climatic, topographic, vegetation, anthropogenic, and geophysical variables (Table 7.1). The climate variables were obtained from the Scenarios Network for Alaska Planning (SNAP; http://www.snap.uaf.edu/downloads/alaska-climate-datasets) which contains historical datasets derived from Climate Research Unit (CRU) data; shown to perform well in Alaska (Walsh et al. 2008). The spatial resolution of the monthly and annual temperature data were at 2km^2 grid size and were averaged over the years 1901–2009 while, the monthly and annual precipitation data are averaged from 1901–2006. The topographic variables were derived from 300 m digital elevation models (Alaska Geospatial Data Clearinghouse (AGDC; http://agdc.usgs.gov/agdc.html) using Spatial Analysis surface analysis tool and the TPI extension for ArcGis (Jenness 2006) within ArcGis 10.0 (ESRI 2011). Vegetation type was obtained from the 30 m National Land Cover Database (NLCD; http://seamless.usgs.gov/data_availability.php?serviceid=Dataset_13) for the year 2001 (Vogelmann et al. 2001). Forest type was obtained from the 250 m Forest Type Groups of Alaska map (Ruefenacht et al. 2008; http://fsgeodata.fs.fed.us/rastergateway/forest_type/) for the year 2008. The Normalized Difference Vegetation Index (NDVI) values were from the 14-day, band six Advanced Very High Resolution Radiometer (AVHRR) data from the NOAA polar-orbiting satellites covering the periods May through September 2011, which were obtained from the Geographic Information Network of Alaska database (GINA; http://docs.gina.alaska.edu/ndvi/how_to.html). The stand age was determined using data from the Alaska Interagency Coordination Center (http://afsmaps.blm.gov/imf_fire/imf.jsp?site=fire) fire perimeter data covering the years 1942–2011. Using this fire history data, a binary response variable was created for each location describing the forests as young (< 69 years) or mature (> 69 years) based on when the location last burned (Johnson et al. 2011). Extrapolated tree size basal area and tree species basal area diversity values derived from a combination of forest inventory and remotely sensed data within the study region were obtained from Young et al. (2016). We calculated the anthropogenic and the geophysical variables using data from AGDC and tools within ArcGIS 10.0 (ESRI 2011). The predictor variables with a spatial resolution greater than 1 km underwent either nearest neighbor resampling, if the data were categorical, or bilinear interpolation resampling, if the data were continuous. Those variables that had a native spatial resolution smaller than 1km^2 were rescaled. The predictor dataset was constructed by overlaying the individual datasets in ArcGIS 10.0, then at each PSP location the environmental variables where extracted, resulting in a table with aboveground tree woody biomass values as the response variable and the environmental variables as predictors (as per Ohse et al. 2009).

Table 7.1 Definition of variables used in the analysis

Variable	Description	Unit	Reference
Spatial structure			
X	Latitude (Alaska Albers)	10^5m	Magness et al. (2008)
Y	Longitude (Alaska Albers)	10^5m	Magness et al. (2008)
Climatic variables			
T_01	Mean temperature January	(°C + 100)	Ohse et al. (2009)
T_05	Mean temperature may	(°C + 100)	Ohse et al. (2009)
T_06	Mean temperature June	(°C + 100)	Ohse et al. (2009)
T_07	Mean temperature July	(°C + 100)	Ohse et al. (2009)
T_08	Mean temperature august	(°C + 100)	Ohse et al. (2009)
T_09	Mean temperature September	(°C + 100)	Ohse et al. (2009)
T_G	Mean temperature growing Season (may–September)	(°C + 100)	Ohse et al. (2009)
T_D	Mean temperature difference July–January	(°C + 100)	Ohse et al. (2009)
T_A	Mean annual temperature	(°C + 100)	Ohse et al. (2009)
P_05	Precipitation sum may	Mm	Ohse et al. (2009)
P_06	Precipitation sum June	Mm	Ohse et al. (2009)
P_07	Precipitation sum July	Mm	Ohse et al. (2009)
P_08	Precipitation sum august	Mm	Ohse et al. (2009)
P_09	Precipitation sum September	Mm	Ohse et al. (2009)
P_G	Precipitation sum growing season (may–September)	Mm	Ohse et al. (2009)
P_W	Precipitation sum winter (October–April)	Mm	Yarie (2008)
P_A	Precipitation sum annual	Mm	Ohse et al. (2009)
Topographic variables			
Solar	Potential maximum solar insolation	(kWH/m^2)	Fu and rich (1999)
Prod	Site productivity	Unitless	Stage and Salas (2007)
Sl	Slope	Percent	Stage and Salas (2007)
As	Transformed aspect	Unitless	Beers et al. (1966)
El	Elevation	m	Magness et al. (2008)
SPC	Slope position classification	Class	Murphy et al. (2010)
TPI	Topographic position index	Class	Murphy et al. (2010)
LC	Landform classification	Class	Johnson et al. (2011)
Vegetation variables			
Veg	Vegetation type	Class	Vogelmann et al. (2001)
FT	Forest type	Class	Ruefenacht et al. (2008)
Gmax	Maximum NDVI	NDVI	Magness et al. (2008)
Gmean	Mean NDVI growing season	NDVI	Magness et al. (2008)
Age	Stand age	Binary	Johnson et al. (2011)
H_s	Tree species diversity	Shannon's	Young et al. (in press)

(continued)

Table 7.1   (continued)

Variable	Description	Unit	Reference
H$_d$	Tree size class diversity	Shannon's	Young et al. (in press)
Anthropogenic variables			
Dtc	Distance to community	Km	Wurtz et al. (2006)
Dtr	Distance to roadway	Km	Wurtz et al. (2006)
Dtw	Distance to navigable waterway	Km	Wurtz et al. (2006)
Geophysical variables			
Perm	Permafrost	Class	Liang (2010)
Soil	Soil type	Class	Ohse et al. (2009)

7.2.3   The Calibration and Validation Datasets

For the 694 sites within our study region we had information about the aboveground forest woody biomass and the 39 environmental predictor variable and X and Y coordinates (Table 7.1). The dataset was randomly split into a calibration dataset (Cal, $n = 522$; 75% of the plots) and a validation dataset (Val, $n = 174$; 25% of the plots). The relationship between aboveground forest woody biomass and the environmental factors was modeled using the calibration dataset and the quality of the predictions was assessed using the validation dataset.

7.2.4   Statistical Methods

Our modeling approach involved determining the association between aboveground forest woody biomass and the 39 environmental factor predictors at each of the PSP locations. The environmental predictors were pre-selected using the Boruta package (Kursa and Rudnicki 2010) in R (Version 3.2.4). This algorithm assesses the relevance of each individual predictor by testing whether its importance, using P values at the 0.05 alpha level, is greater than a random permutation by running the random forests algorithm iteratively until all predictor variables are classified as "accepted" or "rejected" (Kursa and Rudnicki 2010). We computed the Boruta on these data with maxRuns = 1000 and ntree = 500 and the variables selected were determined to be significant based on their Z-score with an upper 95% confidence limit ($[Z] \leq 1.65$). The final set of predictors (Table 7.2) was then used in the development of the random forests models (RF; Breiman 2001) using the randomForest package (Liaw and Wiener 2002) implementation in R. RF is non-parametric and so it makes virtually no relevant assumptions about the distribution of input data (Drew et al. 2011). This allows for the capturing of non-linear relationships involving complex ecological and environmental interactions among variables and limits the problems associated with multicollinearity (De'ath and Fabricius 2000; Siroky 2009).

The RF as applied to these data used 2500 bootstrap samples each containing two-thirds of the Cal data. The remaining samples —out-of-bag (OOB)

Table 7.2 Variable importance (ranking) in determining aboveground forest woody biomass (biomass; Mg/ha dry weight) using percent increase in mean standard error (%IncMSE) for ranking purposes. The variables deemed important were used in the development of the final random forests model (RF)

Variables	Biomass %IncMSE	rank
H_d	90.62	1
Veg	36.40	2
P_06	31.86	3
Dtc	29.80	4
El	28.79	5
Y	27.52	6
Gmax	25.68	7
Prod	24.90	8
Sl	23.87	9
P_W	23.67	10
Gmean	23.05	11
Asp	22.04	12
P_07	21.65	13
T_A	21.00	14
X	20.94	15
TPI	20.74	16
P_A	19.37	17
FT	19.32	18
P_G	18.98	19
Perm	16.71	20
Solar	16.51	21
P_09	16.43	22
T_G	16.35	23
P_08	16.17	24
T_05	15.68	25
Age	14.45	26
Soil	14.20	27
T_07	13.96	28
T_09	13.33	29
P_05	13.30	30
T_08	12.61	31
T_06	12.41	32

observations— were used to assess model performance. For each bootstrap sample, an un-pruned regression tree was grown containing one-third of the predictor variables, which were randomly selected and used for binary partitioning. The average of all trees was then used to predict OOB observations, for cross validation (Breiman 2001) and to evaluate the overall error of the RF models. The importance values for each predictor was calculated by investigating the percent increase in mean squared error (MSE) when OOB data for each variable were permuted while all others were kept constant (Breiman 2001; Cutler et al. 2007).

External validations of the predictive capabilities of the RF models were conducted on the Val dataset. Because we used a continuous response, model performance was evaluated using Pearson's product-moment correlation coefficients (r), root-mean-square error (RMSE) and, mean absolute error (MAE) (Li et al. 2011). Partial dependence plots were constructed to visualize the marginal effects of the predictor variables in the RF estimates above ground woody biomass.

Considering that spatial autocorrelation affects statistical model predictions, due to a lack of independence (Legendre 1993), we tested for spatial autocorrelation as well as for large-extent spatial patterns within the residuals of the final RF models. We assumed that plots that were further apart would affect each other less than those that were closer (Cressie 1993) and therefore applied a spatial weight of inverse distance. Given the neighborhood structure, we then evaluated the residuals of the RF models using both Moran's I and Geary's C test statistics (Sokal and Oden 1978) within the spdep package in R (Bivand et al. 2007).

To further examine the relationships between aboveground forest biomass and the six most important environmental predictors for aboveground forest woody biomass determined through the RF, we employed regression tree analysis (RTA; De'ath and Fabricius 2000) in the 'party' package (Hothorn et al. 2006a) in R. RTA recursively partitions a dataset into subsets that are relatively homogeneous with regards to the response variable (De'ath and Fabricius 2000). The RTA as applied to these data used conditional inference trees (Hothorn et al. 2006b), which require a statistical significant P value (P < 0.01 for this analysis) determined through a Monte Carlo randomization procedure (9999 permutations used in this analysis). This technique reportedly minimizes bias and prevents over-fitting (Hothorn et al. 2006b). The results of the RTA are easily interpreted as a dendrogram containing a set of decision rules on the environmental predictor variables.

7.2.5   Predictive Maps

Once calibrated and validated, the RF model (Table 7.2) was then predicted to the 1km^2 resolution prediction grid developed for the boreal forest of Alaska and then converted to a raster using the Point to Raster tool in ArcGIS (10.0) in order to obtain an estimate of the aboveground forest biomass (Mg/ha dry weight) for the entire region.

7.3   Results

7.3.1   Variable Selection and Importance

The Boruta algorithm was used as the basis of our variable selection processes to predict aboveground forest biomass (biomass). The explanatory variables selected using Boruta algorithm that best predicted forest biomass improved overall model predictions. Applying the RF to the complete data set using the complete suite of prediction variables yielded a model with an RMSE = 19.73, MAE = 13.76, r = 0.96), compared to the Boruta variable selected model which yielded an RMSE = 19.02, MAE = 13.57 and, r = 0.97. Given the improvement in model performance for using the Boruta algorithm, we applied the final RF models to the reduced set of predictors as presented in Table 7.2.

Table 7.2 also shows the ranking of the predictor variables by their importance as determined by the percent increase in mean standard error (%IncMSE). The variable of tree size-class diversity (H_d) is by far the most important variable to predict biomass for the Alaskan boreal forest accounting for over 90%IncMSE. An additional vegetation variable, vegetation type (Veg), was the second most important variable accounting for 36.40%IncMSE. Of the climatic variables, June precipitation (P_06) was deemed the most important accounting for over 31%IncMSE. The anthropogenic variable of Distance to communities (DTC) was also found to be highly influential in predicting biomass contributing to nearly 30%IncMSE. Of the top ten variables to predict biomass three were vegetation variables (H_d, Veg, and Maximum NDVI (maxAV)), three are topographic variables (elevation (EL); the proxy of site productivity (Prod), which was derived from elevation, slope, and aspect (Stage and Salas 2007); and slope (Sl)), two climatic variables (P_06, and winter precipitation (P_W)), and one spatial structure variable (longitude (Y)), which is an indicator of continentality.

7.3.2 Random Forests Analysis Model Assessment

The RF model for aboveground forest biomass (as determined using the variables within Table 7.2) as applied to the calibration dataset (Cal) was able to explain 35.23% of the total variation in biomass (Fig. 7.2: RMSE = 19.02, MAE = 13.57, r = 0.97, P < 0.001). The RF model also provided predictions on the validation dataset (Fig. 7.2: $RMSE_{val}$ = 43.84, MAE_{val} = 32.14, r_{val} = 0.55, $P < 0.001$). Despite these highly significance levels, the RF model for biomass tended to slightly overestimate on the validation dataset for low biomass sites and underestimate in high sites. Additionally, we observed some variation in the magnitude of the errors; large errors are unlikely due to the relatively small difference between the $RMSE_{val}$ and MAE_{val} values. Our modeled average differences on the validation dataset between the predicted and the observed biomass value was 32.14 Mg/ha with a mean predicted value of 92.68 Mg/ha.

In comparison to our results, the predictions of Blackard et al. (2008) for aboveground forest woody biomass on the validation dataset (Val) were noticeably different (Fig. 7.3). Using the results from the model by Blackard et al. (2008) for biomass, we found a correlation coefficient of - 0.007 for the 176 PSPs. Additionally, the RMSE was 76.18 and the MAE was 59.00, suggesting a large magnitude of error and fairly low accuracy given that the average difference is found to be greater than the mean predicted value (44.46 Mg/ha) from the Blackard et al. (2008) model for aboveground forest biomass within this study region.

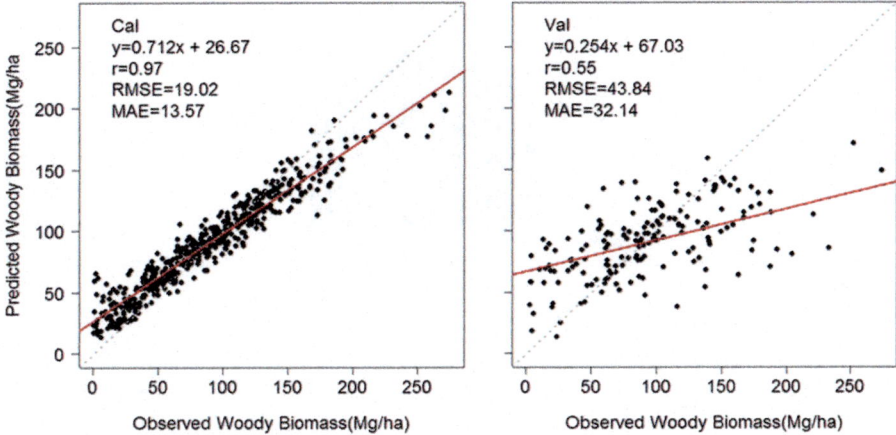

Fig. 7.2 Application of the random forests (RF) model for aboveground forest woody biomass (Mg/ha dry weight) to the calibration data set (Cal) and the validation data set (Val)

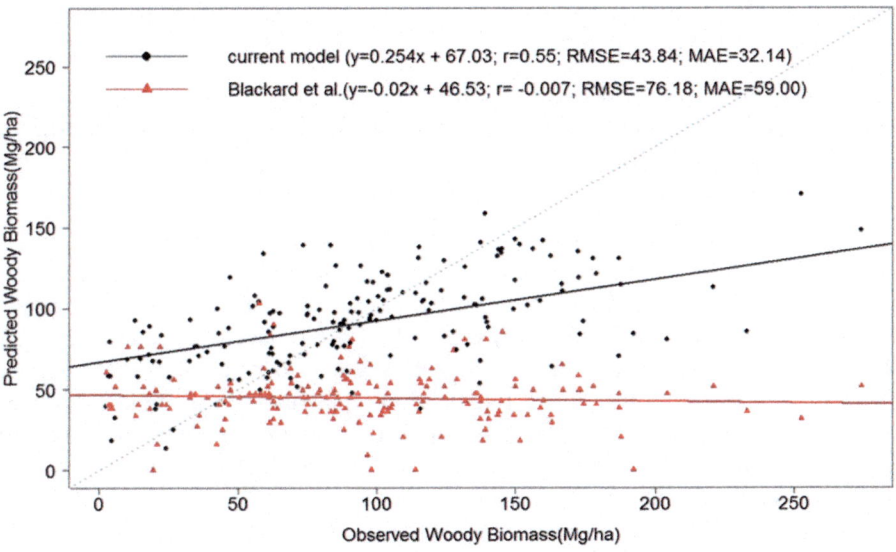

Fig. 7.3 Predicted and observed above ground full tree woody biomass (Mg/ha dry weight) for the 694 Sample Plots within the Alaskan boreal forest. The current model indicates the model presented in this paper while the other predictions are from Blackard et al. (2008)

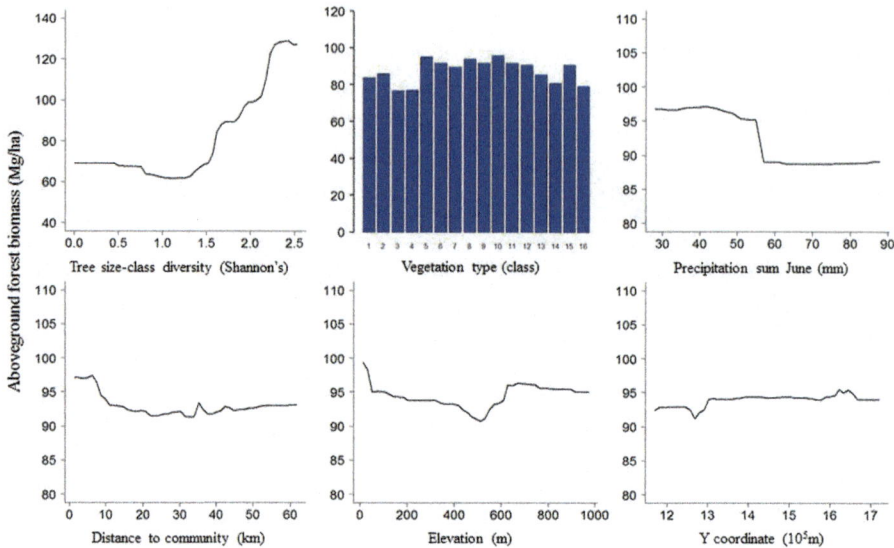

Fig. 7.4 Partial dependence plots representing the marginal effects of the six most important variables from the RF model on the estimates ofaboveground forest woody biomass while averaging out the effect of all the other variables

7.3.3 Influence of the Environmental Factors on Aboveground Forest Biomass

The partial dependency plots (Fig. 7.4) illustrate the relationships between aboveground forest biomass and the six most important environmental factors as determined by RF. The most important factor, tree size-class diversity (H_d) indicates a potential threshold at an H_d value of >1.5 biomass steeply increases but below this level, it is stable at ~ 70 Mg/ha. Vegetation type (Veg) is the second most influential with closed forest vegetation types having the greatest biomass values. June precipitation (P_06) also appears to express a threshold effect on biomass, with values less than 55 mm having higher biomass than those receiving more moisture. Distance to communities (Dtc) also displays a pronounced effect on biomass with decreasing biomass values with increasing distance. The effect of Elevation (El) on aboveground woody biomass suggests a depressed value at mid-elevations with increased amounts on either extreme. The effect of continentality, as measured by Longitude (Y), is also pronounced with a slightly positive effect on biomass.

7.3.4 Regression Tree Analysis Model Assessment

The RTA of aboveground forest biomass using the six most important variables determined from RF produced 8 terminal nodes (Fig. 7.5). The highest biomass values appear to occur when H_d is greater than 2.221 and on sites with diversity values between 1.903 and 2.221 that receive less than or equal to 54 mm of precipitation in June. The lowest biomass values occur on low diversity ($\leq$ 1.583) sites at elevations less than 417 m and greater than 30.7 km away from a community. Additionally, low biomass values are predicted for sites greater than 417 m in elevation which occupy micro sites dominated by open and or closed spruce forests or mixture of spruce shrub woodlands.

7.3.5 Spatial Dependency of Aboveground Forest Biomass

Forest biomass is strongly spatially autocorrelated (Table 7.3). The environmental predictors used in the RF model adequately accounted for the autocorrelation present in the forest biomass as evident by the lack of autocorrelation present in the residuals of the final models (Table 7.3). The capturing of the large-scale systematic spatial trend by the environmental predictors subsequently improves the accuracy of the predictions for the regions outside of the sample area.

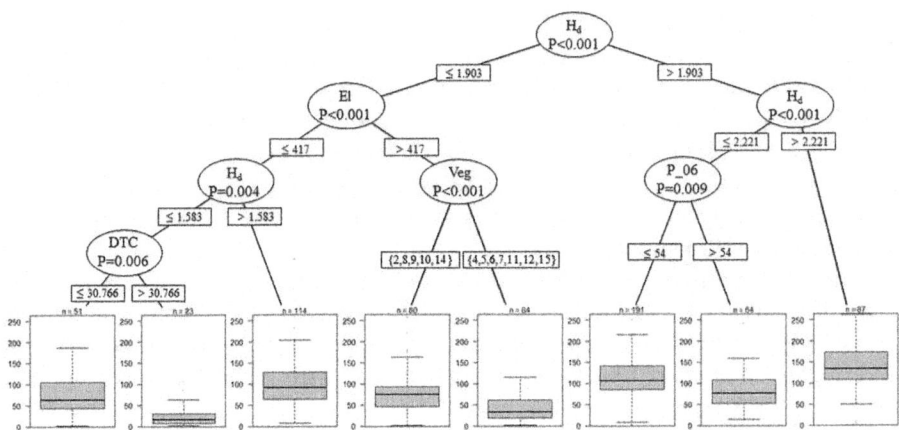

Fig. 7.5 Conditional Inference tree for aboveground forest woody biomass using the six most important predictor variables as determined by the random forests analysis (see Fig. 7.4, Table 7.2). The p-values at each of the nodes are from a Monte Carlo randomization test. In order for a split to occur the p-value must be <0.01. The box plots at the terminal nodes show the distribution of the data within that branch of the tree

Table 7.3 Spatial autocorrelation and its level of significance for aboveground forest woody biomass (biomass; Mg/ha dry weight) in the Alaskan boreal forest, and for the residuals of the random forests (RF) model used for predicting biomass from this dataset

	Moran's I	P-Value	Geary's C	P-Value
Biomass	0.3224	<0.001	0.6779	<0.001
Residuals of RF model for biomass	−0.0345	0.8906	1.0202	0.7535

7.3.6   Predicted Aboveground Forest Biomass Patterns

The aboveground forest woody biomass for the boreal forest of Alaska as predicted through random forests analysis using environmental predictors and forest inventory data ranged widely across the study region (Fig. 7.6). The predicted values ranged from 26.6–210.1 Mg/ha with a mean of 88.3 ± 16.4 Mg/ha. The highest values were primarily found within the central interior between Fairbanks and McGrath, around Glennallen, and north of Anchorage in the Matanuska-Susitna valleys.

7.4   Discussion

The data from the forest inventories (CAFI and WAIN datasets) were critical to our analysis in determining the abundance and spatial variation of aboveground biomass across the heterogeneous landscape of the Alaskan boreal forest. These datasets are getting more use (see Young et al. 2017), and we hope more can be done with those data in future investigations. So far, only the data contained within the CAFI and WAIN datasets comprise the necessary information for ground truthing large scale forest dynamics in the Alaskan boreal forest (Malone et al. 2009; Rees, personal communication). However; these data are not uniformly dispersed across the study region and a collection bias exist (see Fig. 7.1). To this end, random forests (RF; Breiman 2001) is ideally suited to studies such as this where other spatial prediction techniques (i.e. kriging) would potentially produce a single mean value for large portions of the study region (Cushman and Huettmann 2010; Drew et al. 2011).

The prediction for aboveground biomass was strongly influenced by tree size class diversity (H_d) This result was not surprising because the structure of a young forest is typically characterized by a single canopy layer, high stem density, few forest gaps, and trees of roughly the same size with generally lower biomass (Pretzsch 2005; Scherer-Lorenzen et al. 2005; Schulze et al. 2005) whereas older forests generally have a greater mixture of tree sizes in multiple canopy layers due primarily to niche differentiation (Harper et al. 2003; Kneeshaw and Gauthier 2003; McCarthy 2001), resulting in higher overall stand biomass (Kohyama 1993; Scherer-Lorenzen et al. 2005). Forest age was relatively unimportant (see Table 7.2) in this analysis, probably because it was represented in this dataset by a binary response, at

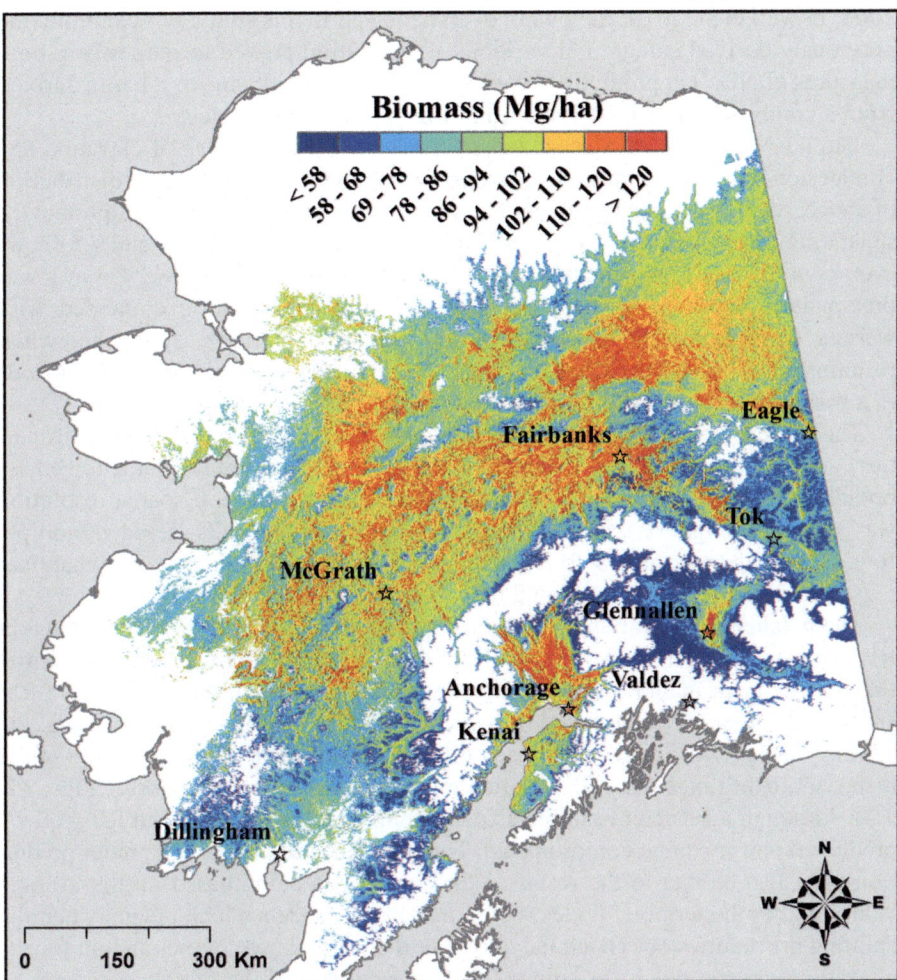

Fig. 7.6 A map of the aboveground forest woody biomass (Mg/ha dry weight) derived from the random forests analysis model using the CAFI and WAIN forest inventory plot biomass values as a function of environmental predictors (Table 7.2)

an age of 69 years, that may not reflect the age at which significant niche differentiation occurs within this forest type. The results suggest that forest managers could enhance biomass production by increasing tree size diversity in the boreal forest of Alaska if biomass production becomes a primary management goal.

Surprisingly, the AVHRR band 6 (NDVI) data (Gmax and Gmean variables) were not as important at predicting biomass within this study as some of the other vegetation, climatic, anthropogenic, and topographic variables (see Table 7.2). While reflectance variables have previously been shown to be good predictors of biomass (Baccini et al. 2004; Baccini et al. 2008; Blackard et al.

2008; Powell et al. 2010) they were overshadowed in this study. For example, the previously derived categorical variable of vegetation proved to outperform both measures of NDVI at predicting biomass, which was likely due to it being derived from a combination of NDVI and other topographical variables.

The Interior Forest of Alaska is characterized by a wide range of elevation and climate zones and these variables exert important controls on the spatial distribution of aboveground biomass. For example, average June precipitation was important for separating forests into larger and smaller timber volumes (Figs. 7.4 and 7.5) however, contrary to our expectations, higher biomass values were observed at lower precipitation amounts, perhaps because low June precipitation coincided with warmer temperatures in more continental regions or perhaps, spring snowmelt resulting in soil moisture recharge may have been sufficient to handle water demands in a warmer June with little precipitation present.

The presence of broadleaf trees mixed with conifers created particular difficulties, and the model tended to underestimate biomass in areas characterized by broadleaf and conifer mixtures. In this context, the use of 1 km^2 spatial resolution was a key challenge for this work because virtually all grid cells included multiple forest stands and mixtures of forests and shrubs. Future efforts using somewhat finer (e.g., 300 m) resolution data should help to resolve this problem.

The balance of the variable variance distribution also needs to be considered in RF modeling. The RF model in this study showed an overall trend of underestimation above the mean and overestimation below the mean because the diversity measures were either underestimated or overestimated. In RF regression modeling, the splitting and averaging algorithm used may have resulted in underestimation of the responses in the range where data points are scarce at the extreme ends (see Figs. 7.2, 7.3). Although a sufficient sample size can minimize this problem, an RF model's predictive power can be compromised. The phenomenon of over and under prediction may also be due to the result of the predictions being based on the average values within the terminal nodes within tree-based models which occurs when the splitting procedure stops (Breiman 2001), or the learning rate of the random forests algorithm was too quick (Friedman 2001). Regardless of the reason, a schema to increase the size of a training dataset with balanced predictor variance distributions is likely to help minimize this issue for future modeling efforts.

Assessment of the spatial variation of forest biomass across landscapes is challenging but vital in order to improve regional-scale assessments. However, due to the presence of autocorrelated environmental variables at multiple spatial scales, largely due to community processes (Legendre 1993); our ability to assess their spatial variation likely depends upon the spatial arrangement of our permanent sample plots (Lamsal et al. 2012). Across the heterogeneous landscapes of boreal Alaska, we had a dense but localized sampling effort that made it possible to detect localized patterns of biomass but our predictions for more distant locations is based strictly on correlation. Therefore, landscape- to regional-scale models of forest biomass distribution within the heterogeneous boreal forest of Alaska could benefit with a wider geographic array of permanent sample plots to assess this vital resource.

Predictive mapping is a powerful tool for landscape level planning and analysis (Franklin 1995). Consequently, landscape and regional scale predictive models often strike a balance between sample size and prediction performance. For example, RF models developed with small training sets are prone to low prediction performance (Breiman 2001). However, large sample sizes can be cost prohibitive particularly in remote locations such as Alaska. This modeling effort will lead to improved understanding of the current patterns of biomass and may assist land management agencies in their decision making processes in regards to sustainable forestry activities (Ogden and Innes 2009). Our proposed predictive mapping of above ground woody mapping at the 1-km^2 extent would best be used for broad landscape level planning, and it can be applied in an Adaptive Management framework. We believe that this model could be improved by adding more ground sampled data, open access data sources, a full emphasize on GIS data predictions, and a stronger collaboration with forest practitioners for policy improvements.

Acknowledgments We thank Douglas Hanson for his valuable comments and insights. The authors also thank Thomas Malone and Dan Rees and their field assistants for collecting and compiling all the forest inventory data. Support for this work was provided by the National Science Foundation, through its Integrative Graduate Education and Research Traineeship (IGERT, NSF 0114423) to the Resilience and Adaptation Program (RAP) at the University of Alaska Fairbanks; and the State of Alaska Department of Natural Resources Division of Forestry.

References

Alaska Energy Authority 2015 Alaska wood energy development task force project status; AWEDTG_Status_081315_8.511; AEA: Anchorage, AK, USA, 2015

Austin MP (2002) Spatial prediction of species distribution: an interface between ecological theory and statistical modelling. Ecol Model 157(2–3):101–118

Baccini A, Friedl MA, Woodcock CE, Warbington R (2004) Forest biomass estimation over regional scales using multisource data. Geophys Res Lett 31(10):L10501

Baccini A, Laporte N, Goetz SJ, Sun M, Dong H (2008) A first map of tropical Africa's aboveground biomass derived from satellite imagery. Environ Res Lett 3(4):045011

Beers TW, Dress PE, Wensel LC (1966) Aspect transformation in site productivity research. J For 64:691–692

Bivand RS, Anselin L, Berke O et al (2007) Spdep: spatial dependence: weighting schemes, statistics and models. R package version 0.4-9

Blackard J, Finco M, Helmer E et al (2008) Mapping U.S. forest biomass using nationwide forest inventory data and moderate resolution information. Remote Sens Environ 112(4):1658–1677

Bonnor G (1985) Inventory of forest biomass in Canada. Canadian Forestry Service, Environment Canada, Forestry Statistics and Systems Branch, Ontario

Botkin DB, Simpson LG (1990) Biomass of the North-American boreal forest - a step toward accurate global measures. Biogeochemistry 9(2):161–174

Breiman L (2001) Random forests. Mach Learn 45(1):5–32

Chapin FS III, Hollingsworth T, Murray DF, Viereck LA, Walker MD (2006) Floristic diversity and vegetation distribution in the Alaskan boreal forest. In: Chapin FS III, Oswood M, Van Cleve K, Viereck LA, Verbyla D (eds) Alaska's changing boreal forest. Oxford University Press, New York, pp 81–99

Cressie NAC (1993) Statistics for spatial data. John Wiley & Sons, New York

Curtis RO (1983) Procedures for establishing and maintaining permanent plots for silvicultural and yield research. p 56. Gen Tech Rep PNW-155. U.S. Department of Agriculture, Forest Service, Pacific Northwest Forest and Range Experiment Station, Portland

Cushman S, Huettmann F (eds) (2010) Spatial complexity, informatics, and wildlife conservation. Springer, Tokyo

Cushman SA, McKelvey KS (2009) Data on distribution and abundance: monitoring for research and management. In: Cushman SA, Huettmann F (eds) Spatial complexity, informatics, and wildlife conservation. Springer, Tokyo, pp 111–129

Cutler DR, Edwards TC Jr, Beard KH et al (2007) Random forests for classification in ecology. Ecology 88(11):2783–2792

De'ath G, Fabricius KE (2000) Classification and regression trees: a powerful yet simple technique for ecological data analysis. Ecology 81(11):3178–3192

Drew CA, Wiersma YF, Huettmann F (eds) (2011) Predictive species and habitat modeling in landscape ecology: concepts and applications. Springer Verlag, New York

ESRI (2011) Arc GIS desktop: release 10. Environmental Systems Research Institute, Redlands,CA

Fassnacht KS, Cohen WB, Spies TA (2006) Key issues in making and using satellite-based maps in ecology: a primer. For Ecol Manag 222(1–3):167–181

Ferrier S, Watson G, Pearce J, Drielsma M (2002) Extended statistical approaches to modelling spatial pattern in biodiversity in northeast New South Wales. I. Species-level modelling. Biodivers Conserv 11(12):2275–2307

Franklin J (1995) Predictive vegetation mapping: geographic modelling of biospatial patterns in relation to environmental gradients. Prog Phys Geogr 19(4):474–499

Fresco N (2006) Carbon sequestration in Alaska's boreal forest: planning for resilience in a changing landscape. PhD dissertation. University of Alaska Fairbanks

Friedman JH (2001) Greedy function approximation: a gradient boosting machine. Annals Stat 29(5):1189–1232

GAO (2005) Natural resources. Federal Agencies are Engaged in various efforts to promote the utilization of Woody biomass, but significant obstacles to its use remain. United States Government Accountability Office, Washington, DC

Grossmann E, Ohmann J, Kagan J, May H, Gregory M (2010) Mapping ecological systems with a random forest model: tradeoffs between errors and bias. USGS Gap Analysis Bulletin:16–22

Guisan A, Thuiller W (2005) Predicting species distribution: offering more than simple habitat models. Ecol Lett 8(9):993–1009

Harper K, Boudreault C, DeGrandpré L, Drapeau P, Gauthier S, Bergeron Y (2003) Structure, composition, and diversity of old-growth black spruce boreal forest of the Clay Belt region in Quebec and Ontario. Environ Rev 11(S1):S79–S98

Harrell PA, Bourgeau Chavez LL, Kasischke ES, French NHF, Christensen NL (1995) Sensitivity of ERS-1 and JERS-1 radar data to biomass and stand structure in Alaskan boreal forest. Remote Sens Environ 54(3):247–260

Hothorn T, Hornik K, Zeileis A (2006a) Party: a laboratory for recursive part (y)itioning. http://CRAN.R-project.org/

Hothorn T, Hornik K, Zeileis A (2006b) Unbiased recursive partitioning: a conditional inference framework. J Comput Graph Stat 15(3):651–674

Houghton RA (2005) Aboveground forest biomass and the global carbon balance. Glob Change Biol 11(6):945–958

Iverson LR, Prasad AM (2001) Potential changes in tree species richness and forest community types following climate change. Ecosystems 4(3):186–199

Jenkins JC, Chojnacky DC, Heath LS, Birdsey RA (2003) National-scale biomass estimators for United States tree species. For Sci 49(1):12–35

Jenkins JC, Chojnacky DC, Heath LS, Birdsey RA (2004) Comprehensive database of diameter-based biomass regressions for north American tree species. US Dept. of Agriculture, Forest Service, Northeastern Research Station, Delaware

Jenness J (2006) Topographic position index (tpi_jen.avx) extension for Arcview 3.x, v. 1.3a. http://www.jennessent.com/arcview/tpi.htm, Jenness Enterprises [EB/OL]

Johnson KD, Harden J, McGuire AD et al (2011) Soil carbon distribution in Alaska in relation to soil-forming factors. Geoderma 167-68(0):71–84

Kneeshaw D, Gauthier S (2003) Old growth in the boreal forest: a dynamic perspective at the stand and landscape level. Environ Rev 11(S1):S99–S114

Kohyama T (1993) Size-structured tree populations in gap-dynamic Forest - the Forest architecture hypothesis for the stable coexistence of species. J Ecol 81(1):131–143

Kursa MB, Rudnicki WR (2010) Feature selection with the boruta package. J Stat Softw 36(11):1–13

Lamsal S, Rizzo DM, Meentemeyer RK (2012) Spatial variation and prediction of forest biomass in a heterogeneous landscape. J For Res 23(1):13–22

Legendre P (1993) Spatial autocorrelation-trouble or new paradigm. Ecology 74(6):1659–1673

Li J, Heap AD, Potter A, Daniell JJ (2011) Application of machine learning methods to spatial interpolation of environmental variables. Environ Model Softw 26(12):1647–1659

Liang JJ (2010) Dynamics and management of Alaska boreal forest: an all-aged multi-species matrix growth model. For Ecol Manag 260(4):491–501

Liang JJ, Zhou M (2010) A geospatial model of forest dynamics with controlled trend surface. Ecol Model 221(19):2339–2352

Liaw A, Wiener M (2002) Classification and Regression by random forest. R News 2(3):18–22

Lloyd AH, Fastie CL (2002) Spatial and temporal variability in the growth and climate response of treeline trees in Alaska. Clim Chang 52(4):481–509

Loeffler D, Brandt J, Morgan T, Jones G (2010) Forestry based biomass economic and financial information and tools: an annotated bibliography. Gen Tech Rep RMRS-GTR-244WWW. Fort Collins, CO: U.S. Department of Agriculture, Forest Service, Rocky Mountain Research Station., p 52

Magness D, Huettmann F, Morton J (2008) Using random forests to provide predicted species distribution maps as a metric for ecological inventory & monitoring programs. In: Smolinski T, Milanova M, Hassanien AE (eds) Applications of computational intelligence in biology, studies in computational intelligence, vol 122. Springer, Berlin, pp 209–229

Major J (1951) A functional, factorial approach to plant ecology. For Ecol Manag 32:392–412

Malone T, Liang J, Packee EC (2009) Cooperative Alaska forest inventory. Gen Tech Rep PNW-GTR-785. Portland, OR: U.S. Department of Agriculture, Forest Service, Pacific Northwest Research Station

McCarthy J (2001) Gap dynamics of forest trees: a review with particular attention to boreal forests. Environ Rev 9(1):1–59

McRoberts RE, Winter S, Chirici G et al (2008) Large-scale spatial patterns of forest structural diversity. Can J For Res-Rev Can Rech For 38(3):429–438

Murphy MA, Evans JS, Storfer A (2010) Quantifying Bufo boreas connectivity in Yellowstone National Park with landscape genetics. Ecology 91(1):252–261

Niemelä J (1999) Management in Relation to disturbance in the boreal Forest. For Ecol Manag 115:127–134

O'Neill RV, Hunsaker CT, Jones KB et al (1997) Monitoring environmental quality at the landscape scale. Bioscience 47(8):513–519

Ogden AE, Innes JL (2009) Application of structured decision making to an assessment of climate change vulnerabilities and adaptation options for sustainable forest management. Ecol Soc 14(1):11

Ohse B, Huettmann F, Ickert-Bond SM, Juday GP (2009) Modeling the distribution of white spruce (Picea glauca) for Alaska with high accuracy: an open access role-model for predicting tree species in last remaining wilderness areas. Polar Biol 32(12):1717–1729

Parmentier I, Harrigan RJ, Buermann W et al (2011) Predicting alpha diversity of African rain forests: models based on climate and satellite-derived data do not perform better than a purely spatial model. J Biogeogr 38(6):1164–1176

Powell SL, Cohen WB, Healey SP et al (2010) Quantification of live aboveground forest biomass dynamics with Landsat time-series and field inventory data: a comparison of empirical modeling approaches. Remote Sens Environ 114(5):1053–1068

Prasad AM, Iverson LR, Liaw A (2006) Newer classification and regression tree techniques: bagging and random forests for ecological prediction. Ecosystems 9(2):181–199

Pretzsch H (2005) Diversity and productivity in forests: evidence from long-term experimental plots. In: Scherer-Lorenzen M, Körner C, Schulze E-D (eds) Forest diversity and function. Springer-Verlag, Berlin Heidelberg, pp 41–64

Ruefenacht B, Finco MV, Nelson MD et al (2008) Conterminous US and Alaska Forest type mapping using forest inventory and analysis data. Photogramm Eng Remote Sens 74(11):1379–1388

Scherer-Lorenzen M, Körner C, Schulze E-D (2005) Forest diversity and function: temperate and boreal systems. Springer Verlag, Berlin/Heidelberg

Schimel DS, House JI, Hibbard KA et al (2001) Recent patterns and mechanisms of carbon exchange by terrestrial ecosystems. Nature 414(6860):169–172

Schulze ED, Wirth C, Mollicone D, Ziegler W (2005) Succession after stand replacing disturbances by fire, wind throw, and insects in the dark taiga of Central Siberia. Oecologia 146(1):77–88

Siroky DS (2009) Navigating Random Forests and related advances in algorithmic modeling. Statistics Surveys 3:147–163

Sokal R, Oden N (1978) Spatial autocorrelation in biology 1. Methodology. Biol J Linn Soc 10(2):199–228

Stage AR, Salas C (2007) Interactions of elevation, aspect, and slope in models of forest species composition and productivity. For Sci 53(4):486–492

van Cleve K, Dyrness CT, Viereck LA, Fox J, Chapin FS, Oechel W (1983) Taiga ecosystems in interior Alaska. Bioscience 33(1):39–44

Viereck L, Little E (2007) Alaska trees and shrubs. University of Alaska Press, Fairbanks

Vogelmann JE, Helder D, Morfitt R, Choate MJ, Merchant JW, Bulley H (2001) Effects of landsat 5 thematic mapper and Landsat 7 enhanced thematic mapper plus radiometric and geometric calibrations and corrections on landscape characterization. Remote Sens Environ 78(1–2):55–70

Walsh JE, Chapman WL, Romanovsky V, Christensen JH, Stendel M (2008) Global climate model performance over Alaska and Greenland. J Clim 21(23):6156–6174

Wilmking M, Juday GP 2005 Longitudinal variation of radial growth at Alaska's northern treeline - recent changes and possible scenarios for the 21st century. In, 2005. Elsevier Science Bv, pp 282–300

Wurtz TL, Ott RA, Maisch JC (2006) Timber harvest in interior Alaska. In: Chapin FS III, Oswood M, Van Cleve K, Viereck L, Verbyla D (eds) Alaska's changing boreal Forest. Oxford University Press, New York, pp 302–308

Yarie J (2008) Effects of moisture limitation on tree growth in upland and floodplain forest ecosystems in interior Alaska. For Ecol Manag 256(5):1055–1063

Yarie J, Billings S (2002) Carbon balance of the taiga forest within Alaska: present and future. Can J For Res-Rev Can Rech For 32(5):757–767

Yarie J, Mead D (1982) Aboveground tree biomass on productive forest land in Alaska. Res Pap PNW-RP-298. Portland, OR: US Department of Agriculture, Forest Service, Pacific Northwest Research Station. 16 p

Yarie J, Kane E, Mack MC (2007) Aboveground biomass equations for trees of interior Alaska. Agricultural and Forestry Experiment Station Bulletin. Univesity of Alaska Fairbanks, Fairbanks

Young BD, Liang J, Chapin FS (2011) Effects of species and tree size diversity on recruitment in the Alaskan boreal forest: a geospatial approach. For Ecol Manag 262(8):1608–1617

Young BD, Yarie J, Verbyla D, Huettmann F, Herrick K, Chapin FS (2017) Modeling and mapping forest diversity in the boreal forest of interior Alaska. Landsc Ecol 32: 397–413

Part III
Data Exploration and Hypothesis Generation with Machine Learning

Artwork by Andrea Price and Catherine Humphries

"It is a capital mistake to theorize before one has data. Insensibly, one begins to twist the facts to suit theories, instead of theories to suit facts."
– Sherlock Holmes

Chapter 8
'Batteries' in Machine Learning: A First Experimental Assessment of Inference for Siberian Crane Breeding Grounds in the Russian High Arctic Based on 'Shaving' 74 Predictors

Falk Huettmann, Chunrong Mi, and Yumin Guo

8.1 Introduction

The Siberian crane (*Leucogeranus leucogeranus,* taxonomic serial number TSN *176185*) is an elusive species of global conservation concern listed as critically endangered since 2000 (BirdLife International 2001). It is known to breed in a dispersed fashion in the Russian high arctic and it is found in very low numbers in Asia during winter. There are appr. three populations in Asia. The Western/Central Flyway population is divided into Central Asian and Western Asian flocks. The Central Asian flock breeds on the basin of the Kunovat river, the north of West Siberia, Russia (Sorokin and Kotyukov 1987), and winters at the Keoladeo National Park, India. The Western Asian flock breeds in the basin of Konda and Alymka rivers, the centre of West Siberia, Russia (Sorokin and Markin 1996; Kanai et al. 2002), and winters in Fereydoonkenar in Iran. For more detail please see http://www.iucnredlist.org/details/22692053/0. The populations have distinct flyways (Kanai et al. 2002).

Overall population trends seem to be rather traumatic. The western population has experienced huge declines and is considered a remnant population, likely related to habitat loss and other disturbances including some political chaos in its flyway;

F. Huettmann (✉)
EWHALE Lab, Biology and Wildlife Department, Institute of Arctic Biology, University of Alaska-Fairbanks, Fairbanks, AK, USA
e-mail: fhuettmann@alaska.edu

C. Mi
Institute of Zoology, Chinese Academy of Sciences, Beijing, China

College of Nature Conservation, Beijing Forestry University, Beijing, China

Y. Guo
College of Nature Conservation, Beijing Forestry University, Beijing, China

© Springer Nature Switzerland AG 2018
G. R. W. Humphries et al. (eds.), *Machine Learning for Ecology and Sustainable Natural Resource Management*, https://doi.org/10.1007/978-3-319-96978-7_8

the eastern population is probably the biggest one right now but has been impacted and is directly confronted by hydro dam development, air pollution, hunting and a myriad of other impacts (e.g. Wu et al. 2009; Prentice 2010). Climate change impacts are an increasing concern with melting permafrost causing vastly changed breeding grounds including expansion of lakes and loss of islands and low-lying shorelines (Van Impe 2013). Conservation of major wetlands to protect habitat along migration routes and wintering habitat is reported to be critical to address population decline (Birdlife International 2001). Rivers and lake systems such as Poyang Lake, China, play a big role for this population (Wu et al. 2009) since the bird relies primarily on wetlands and this region is probably the biggest wintering area for this species with app. 3,750 individuals (Yu et al. 2008).

Due to its elegant appearance (Fig. 8.1) this species has reached a certain mystic celebrity status and it is considered sacred by many indigenous people (Matthiessen 2001). And since 2002, Crane Day Festivals have taken place across its range including West Siberia and Kazakhstan to promote appreciation and conservation (Moore and Ilyashenko 2009). The conservation of this species is covered in national laws in Russia, Iran, Mongolia and China due to the importance of its migration, and it is also featured in the Agreement on the Conservation of African-Eurasian Migratory Waterbirds (AEWA; www.unep-aewa.org/), as well as in the Bonn Convention (Convention of Migratory Species CMS; http://www.cms.int/).The Siberian crane also has a specific Memorandum of Understanding concerning 'Conservation Measures for the Siberian Crane' (https://www.ecolex.org/details/treaty/memorandum-of-understanding-concerning-conservation-measures-for-the-siberian-crane-tre-001318/). Indirectly this species is further covered by the RAMSAR convention (www.ramsar.org/). In reality though, and despite all those efforts, the numbers are widely declining and the conservation for this species is quite ineffective with funding assigned to this species being insufficient, conviction rates for harming the birds are difficult to track, and even exact population numbers and species distributions are not well known. In the Russian High Arctic for instance,

Fig. 8.1 Photo of the Siberian Crane, one of the most elegant birds in the world…

the Siberian Crane is rarely considered in coastal and associated marine conservation efforts and for protected areas, e.g. Klein and Magomedova 2003; Spiridonov et al. 2011 for overview). There is currently no good agreement for the nesting range of this species but predictive mapping using machine learning can provide great progress as the most suitable approach (see Ohse et al. 2009; Mi et al. 2017 and Han et al. 2017 for examples).

While machine learning cannot address management and legal questions directly, it can help to progress specific data analysis issues (Hastie et al. 2009; Mueller and Massaron 2016), and specifically habitat associations. Many examples for the beneficial use of such machine learning analysis exist already (e.g. Fielding 1999; Hochachka et al. 2007; Kandel et al. 2015; and specifically for cranes Cai et al. 2014; Jiao et al. 2014; Mi et al. 2017). With a generic failure of conservation (Mace et al. 2010 for United Nation assessment) and its underlying traditional analysis methods and parsimony (e.g. Guthery et al. 2005; Elith et al. 2006; Arnold 2010), here we investigated an advanced application and implementation of machine learning, 'batteries' (https://www.salford-systems.com/videos/tutorials/tips-and-tricks/data-mining-automation-with-spm-batteries) for conservation progress. 'Batteries' means in this context a collective term used for a set of experiments on how to look and mine-through predictors; here we use tree-based algorithms and it is implemented in an automated fashion in Salford System's Salford Predictive Modeler (SPM) to run by 'mouse-click'. There are over 28 different styles of batteries in SPM, and for this study we choose 'shaving' for our inquiry (https://www.salford-systems.com/videos/tutorials/tips-and-tricks/improve-model-quality-with-battery-shave). This battery iteratively 'shaves' the predictors by dropping the least relevant predictor and re-running the model, thereby allowing for an innovative way to assess unimportant and important predictors for inference. It follows a data mining concept but has its strength in the use of boosting as its underlying analysis engine. Here such type of analysis is presented for the first time.

8.2 Methods

8.2.1 Presence and Absence Points for Siberian Crane

We used 70 confirmed presence locations of Siberian crane in the nesting area. The data source was compiled by Chunrong Mi and was provided by a BirdLife International (2001) publication. For a spatially comprehensive characterization of environments in the study area (i.e., background sampling), we created random pseudo-absence locations (Barbet-Massin et al. 2012) using the random sampling tool in ArcGIS (ESRI Inc) resulting in 5046 data points (Fig. 8.2).

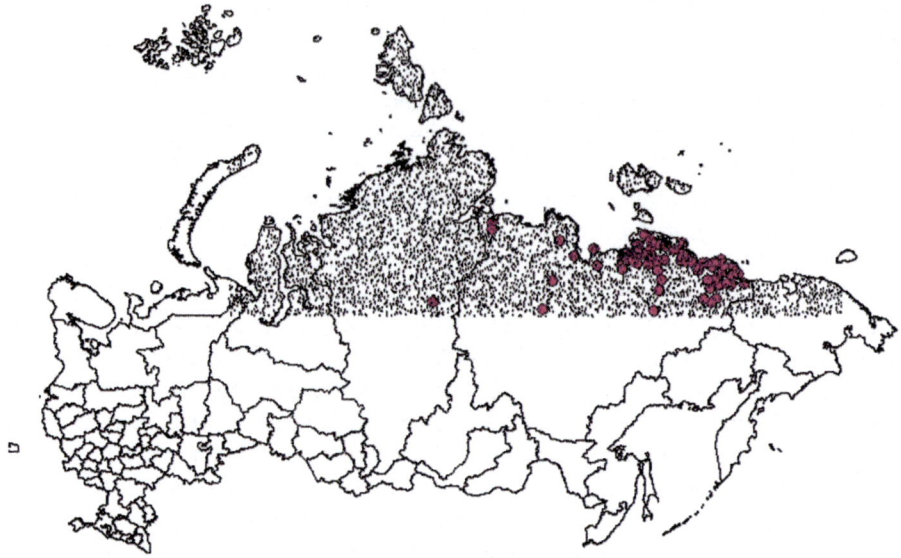

Fig. 8.2 Map of the wider study area and model area with presence (pink) and background samples (back dots)

8.2.2 GIS Predictors

To test for habitat associations for this species in summer, we used 74 environmental predictor layers including spatial patterns in climate (12 long term monthly averages as well as 12 minimum and 12 maximum predictors), 19 bioclim predictors, terrain topography, land cover, human infrastructure and proximity to critical habitat such as water bodies and coastlines. Latitude and longitude was used for GIS plotting purposes. This predictor set is meant to test for, and describe, the ecological niche of Siberian cranes in summer. Data sources are following Herrick et al. (2013) and were further extended by Chunrong Mi. Details about these data are presented and shown in Appendix 1.

8.2.3 'Battery' Runs in Salford System's Predictive Modeler (SPM7)

Using the assembled data cube (a set of cleaned columns and rows described above) we fitted a regular tree-based (Breiman et al. 1984) stochastic gradient boosting, TreeNet model in Salford Predictive Modeler (version 7; SPM7). TreeNet is a commercial and highly performing version of stochastic boosting, based on underlying classification and regression trees (Friedman 2001, 2002). It is freely available to the global public on a trial basis. We used the following model settings: 5 cross-validations (CV), balanced samples, logistic binary, 40 trees (because from pre-runs we found the optimum is approximately 30 trees). This resulted in a regular TreeNet output with an optimum found (details shown in Appendix 2).

We then used the battery setting and ran 'SHAVING' (bottom). This allows us from the 74 environmental predictors to drop the lowest predictor each time, and then rank predictors for impact and relevance on the model performance metric (ROC/AUC based on CV (Fielding and Bell 1997). We could describe models and predictors and rank them. In addition, we displayed them in ArcGIS and were able to visually assess them too. Table 8.1 shows the battery runs that were carried out and investigated.

The interactions were assessed as 2-way interactions in SPM7 using TreeNet. Table 8.2 shows the top 20 whole-variable interactions. From this table we decided to use the first three predictors and removed them from the model run (called 'Leaving out top 3 interacting predictors' in Table 8.1).

Table 8.1 Overview and justification of 10 battery runs for Siberian crane data using 74 environmental predictors

Model#	Model name	Battery details	Justification
1	'Kitchen Sink'	Use of all predictors	A usual approach in machine learning
2	TMax12	'Best' predictor	A classic parsimony univariate approach
3	BIO14	Second best predictor	Widely used in science
4	TMax12BIO14	Top and second predictor	A parsimonious model
5	Top5	First 5 predictors	Already a more holistic approach, more multivariate
6	Top10	First 10 predictors	Rarely applied but truly multivariate
7	Top29	First 29 predictors	Truly multivariate
8	Top35	First 35 predictors	Truly more multivariate
9	Bottom 44	44 lowest predictors (ignoring the top 35 predictors)	It tests one of 'the worst' models. If the predictor rankings are meaningful the lowest predictors should create the worse model outcome
10	Leaving out top 3 interacting predictors	'Kitchen sink' model but leaving out top 3 interacting predictors (Altitude, slope, distance to lake)	It's an aim of most traditional models and statisticians to have 'independent' data and predictors (parametric). Here we can test it

Table 8.2 Overview of the
top ten 2-way interactions for
the analysis in the dataset

Predictor	Interaction
Slope	50.3
Altitude	30.9
Distance to lake	21.8
Bio4	18.3
Bio2	15.8
Bio1	15.5
Bio14	14.0
Tmen2	13.3
Bio12	13.2
Tmin1	12.3

8.3 Results

8.3.1 Predictive Performance Metrics

The results indicate that the highest AUCs are achieved for models with many complex predictors (Table 8.3, Appendix 1). Except for univariate solutions, judged by a pure numerical aspect, most models perform really well regardless. This is a pretty interesting aspect because it makes the notion of model selection and predictor inference complex when using 'just' AUC. More insightful than just the 'pure numbers' is to look at the spatial predictions in the real world (Hilborn and Mangel 1997).

A summary of the Relative Index of Occurrence (RIOs) for each of the 10 models is shown in Fig. 8.3. It shows that the RIOs have a range of performances, and that most of the study area is probably driven by low RIOs, absence, predictions.

8.3.2 Visual Assessment of Prediction Maps

All of the maps are shown in Appendix 4 and summarized in Table 8.4. Due to lack of more data points freely available, this table lacks alternative/independent testing data for map accuracy assessment, but already when compared with known range maps our visual expert assessment shows, again, that parsimonious models perform more poorly than the least constrained approach. The maps are far more spatially explicit than the pure numeric interpretation of performance metrics which subsume much important information. This approach allows to simply identify the 'best' model as the 'kitchen sink model', all based on optimizations of the machine

Table 8.3 Predictive performance metrics for the individual battery models

Model name	AUC (from the testing data cube)	Interpretation
Kitchen sink model	96	The most multivariate non-parsimonious model predicts 'best'
TMax12	84	Single top predictor results into low AUC
BIO14	90	Single top predictor results into low AUC
TMax12BIO14	94	Top 2 predictors perform lower
Top5	95	Top 5 predictors perform high
Top10	95	Top 10 predictors perform high
Top29	95	Top 29 predictors perform high
Top35	96	Top 35 predictors also perform 'best'
Bottom 44	95	Bottom 44 predictors perform very high
Leaving out top 3 interacting predictors	96	Multivariate set of predictors performs very high

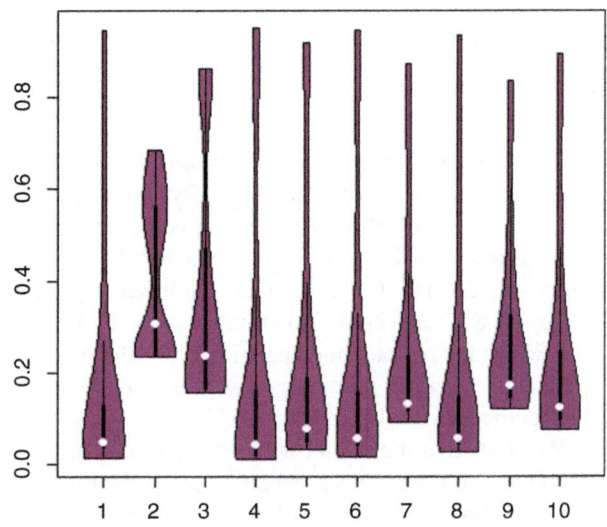

Fig. 8.3 Predictive performance of the 10 model runs (x axis) and tested, showing the distribution of the Relative Index of Occurrence (RIO) (y axis)

learning algorithm in SPM. Interactions do not really make a big difference nor do the drop of a few certain predictors. When based on powerful optimization algorithms, having a large data cube, overall, seems to be a guarantee for a decent outcome. Whereas, the actual composition of the data cube is less relevant (e.g. compare the closeness of model Top35 vs Bottom 44, and then compared to the Kitchen Sink Model overall).

Table 8.4 Visual assessment and rank of predictive performance metrics for the individual battery models

Model name	Rank	Justification and meaning
Kitchen sink model	1	All presences well fit
TMax12	9	Large overestimation of presences
BIO14	10	Large overestimation of presences, misses some presences
TMax12BIO14	8	Overpredicts and mis-predicts absences
Top5	7	Presences predicted very tight, some islands likely overpredicted
Top10	5	Presences predicted very tight
Top29	2	Decent presence regions
Top35	4	Some slight overpredictions
Bottom 44	6	Presence points well included, but some wider overprediction areas
Leaving out top 3 interacting predictors	3	A few overpredictions

8.4 Discussion

Our study set out to explain the ecological niche of Siberian crane in summer. It is based on a unique set of 74 environmental predictors which is rare to have available and to see employed in wildlife and ecological niche studies (Mi et al. 2017 for applications; see Herrick et al. 2013 for 40 layers, and Sriram and Huettmann unpublished using 104 GIS layers). Further, this is the first large-scale predictive model of Siberian crane distributions in the nesting grounds, as well as for the Siberian crane overall, and specifically for the high Arctic. The use of 'batteries' as a machine learning method is also a *novum* for this subject and for conservation management overall.

We show that the model with the top univariate predictor (Fig. 8.4) widely overestimates the range, beyond known presence locations and over-reaching pseudo-absences, when compared with the best model fit with the data and as expressed by the performance metric. Our results are in good agreement with studies that show parsimony as a failure for inference (e.g. Guthery et al. 2005; Elith et al. 2006; Arnold 2010; Mi et al. 2017).

The best possible map shows the birds to be mostly coastal, with a few locations along the rivers. The map shows a rather small available nesting area by now, and which is in support of indicating a conservation problem for this species; it is not a wide-spread species in the high arctic. Whether this is a new situation or a response to dramatic population declines needs more study.

We show that the so-called 'kitchen sink model' (all 74 predictors) in the TreeNet algorithm performs best for the Siberian crane when using a large number of environmental predictors. We believe that this presents a major result and progress for "model selection" as a scientific scheme (Chamberlin 1890; Akaike 1974) which is:

Fig. 8.4 Heatmap of the 'best possible' Siberian crane prediction, based on the 'kitchen sink model' of 74 environmental predictors using TreeNet. Red shows the highest predicted Relative Index of Occurrence (RIO), pink dots show compiled presence locations. For details see Methods

TreeNet (machine learning, boosting) performs in its default settings and with 'maximum numbers' of predictors as one of the best solutions for data mining and predictions (e.g. Elith et al. 2006). However, for learning about model structure and how individual predictors perform, data structure, batteries prove rather insightful and for learning about the data cube. That way we were able to locate additional predictor and predictor arrangements to the usual set and learn about their contributions. It made the modeling more informative and robust, and consequently the inference became better for conservation. From our findings one can easily assess and design protected area locations for this species in the breeding grounds.

One of the biggest surprises of this study probably was that 44 of the least relevant predictors perform almost as well as the top predictors. This is called predictor swapping and has major implications. This puts many questions towards traditional model selection (Burnham and Anderson 2002) as well as model selection, the supposed meaning of univariate predictor selections, predictor ranking, identification and use when using 'the best' predictor and such narrow interpretations. Instead of individually ranked predictors and parsimonious ones (as promoted by Burnham and Anderson 2002), we argue that a set of spatial predictors 'does the job' pretty well, if not equally well and when predictions and such inference are the goal (as per Breiman 2001). In other words, model fitting cannot achieve well (as stated by McArdle 1988 and others) and instead the multivariate perspective is much more powerful, informative and less biased than the univariate and groomed pre-made selection of predictors (as promoted in Manly et al. 2002 and Silvy 2012 for instance; but compare with McGarical et al. 2000).

When it comes to predictions, and inference from those settings (as per Breiman 2001), batteries show an improvement and increased insight for non-parsimonious solutions (see Fig. 8.4).

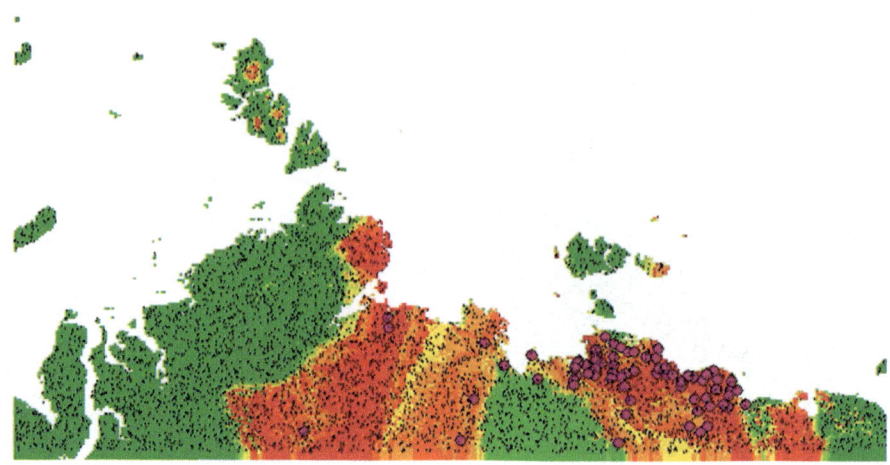

Fig. 8.5 Heatmap of the 'most parsimonious' Siberian crane prediction, based on the predictor 'TMAx12' using TreeNet. A comparison with Fig. 8.4 shows its shortfalls and overprediction. Red shows the highest predicted Relative Index of Occurrence (RIO), pink dots show compiled presence locations. For details see Methods

Despite their great promise and potential for ecological insight, battery applications in wildlife conservation are conspicuous by their absence. Here we just used 'shaving' from the wide list of batteries available (Fig. 8.5). Table 8.5 shows 28 types of batteries that can be run in SPM, for instance.

We think that batteries provide a powerful extension of traditional machine learning methods. They provide more insight into the data cube and model selected. As the reader will almost not be able to find relevant wildlife conservation publications on batteries (as listed in Table 8.4) they are currently almost unused for wildlife conservation and climate applications. More study is recommended using batteries and their variations (Table 8.4). Arguably, this will result in further assessment and decay of the AIC argument and of parsimony overall (Guthery et al. 2005; Arnold 2010), instead favoring models that address interactions and better predictions (all fully in line with Breiman 2001).

We suggest to use batteries as an informative and powerful exploratory tool to learn more about the underlying model structure and predictors for inference. Here, it has provided powerful inference on nesting Siberian cranes in the Russian high arctic. This species is widely overlooked for international conservation research and can benefit greatly from more large-scale studies for advanced conservation management.

Table 8.5 Overview of 28 batteries available in SPM; many of them await their testing for wildlife   conservation

Battery name	Explanation (taken from SPM7)	Comment
AddedVar	Treenet added Var battery	
Additive	Moves through the list of predictors, selecting one predictor at a time	Additive models based on machine learning
Bootstrap	Repeat with new learn sample (draw)	An additional bootstrap to boosting and bagging (which have versions of bootstrapping implemented)
CV	Number of folds in cross-validation	Traditional cross-validation
CVBIN	Creates a number of cross-validation, with binning defined by the several discrete variables	A more specific CV
CVR	Repeat CV with different random seeds	A further specific CV
Datashift	Roll learn and test samples	Rolls through data
Draw	Repeat with new learn sample draw (replacement)	Another version of bootstrapping
Flip	Reverse roles of learn and test samples	An innovative version of re-sampling
Keep	Select predictors at random, run and repeat, may include some required predictors	Random draw of predictors to include
Learnrate (for Treenet only)	Learnrate	Learnrate can be
LOVO	Drop one predictor and repeat for all in keep list	'Mills' through the entire dataset in a detailed fashion, one-by-one
MCT	Model overfitting test via Monte Carlo simulation	An often expressed concern for machine learning, but rarely a problem for bagging, for instance
Minchild	Size of smallest allowable terminal node	This is an essential and sensitive test for most tree-based models
Nodes	Maximum number of terminable nodes allowed	This allows to test for node depth, and indirectly, for interactions
Oneoff	One-predictor model for each predictor in the keep list	Specified univariate model predictor test run
Partition	Repeat with new learn, test and holdout samples drawn from the 'main'' data	An innovative approach to subsampling
Pboot	Parametric bootstrap models	Follows parametric assumptions in subsampling
Sample	Measure effect of learn sample size on error rate	Assesses subsampling effect
Seed	Randomforest seed	This could matter for bagging to find 'the best' model
Shaving	Drop least important predictor, re-run and repeat	A powerful approach to assess unimportant and important predictors in a multivariate approach
Stepwise	Builds model by forward-stepwise selection of predictors	Classic forward step-wise approach to modeling

(continued)

Table 8.5 (continued)

Battery name	Explanation (taken from SPM7)	Comment
Strata	Generates a series of models for each level of the strata variable and overall model	
Swapping	Replace one of the predictors in the model from the list of replacement predictors	An eye-opening approach to assess predictors
Target	Make each variable in keep list target	A very innovative approach to response and predictor assessment
Tnreg	TreeNet regression	
Tnsubsample	Varies the subsampling parameters	Allows for some sensitivity tests in boosting
Xony	Produces a series of models in which each predictor serves as the target, and the target serves as the sole predictor	Another innovative approach to response and predictor assessment

Acknowledgement We thank Dan Steinberg and Salford Systems Ltd. for a workshop with U.S. IALE at Snowbird, Utah, to introduce us to the power of batteries. FH acknowledges the kind and long collaboration with the Forestry University of Beijing, China, and the use of their data. U.S. IALE and S. Linke, C. Cambu, H. Hera, H. Berrios Alvarez and the -EWHALE lab- at UAF, are thanked for their support. This is EWHALE lab publication #185.

Appendix 1: Details of 74 GIS Environmental layers Used in the Model Prediction (+ 3 Additional Internal Columns)

#	Name and abbreviation of GIS layer	Source	Comment
1–12	Monthly mean temperature tmen_1–12	Worldclim.org	These are standard layers used for GIS modeling
13–24	Monthly minimum temperature tmin_1–12	Worldclim.org	(see above)
25–36	Monthly maximum temperature tmax 1–12	Worldclim.org	(see above)
37–48	Monthly precipitation prec_1–12	Worldclim.org	(see above)
49–67	Bioclim bio_1–19	worldclim.org/ bioclim	(see above)
68	Altitude	Worldclim.org	(see above)
69	Aspect	Worldclim.org	(see above)
70	Slope	Worldclim.org	(see above)

#	Name and abbreviation of GIS layer	Source	Comment
71	Landcover Landcv	Herrick et al. (2013)	Several of global landcover layers exist
72	Human infrastructure index Hii	Herrick et al. (2013)	Human footprint. Several human footprint layers
73	Distance to waterbody/lake Dislke	Mi unpublished	While essential for cranes, this layer is unlikely to be very accurate due to the huge and ephemeral wetlands worldwide
74	Distance to coastline Discsln	Mi unpublished	Relies on the coastline map resolution
75	x coordinate	ArcGIS	Not often used in most GIS model work but important for geo-referencing
76	y coordinate	ArcGIS	Not often used in most GIS model work but important for geo-referencing
77	Row index FID	ArcGIS	Not often used in most GIS model work but important for row identification

Appendix 2

List of Top 20 Predictors, as identified by TreeNet ranking

Predictor	Relative Importance
Bio12	100.0
Bio14	71.2
Bio17	44.2
TMEN9	40.1
Prec12	37.6
Distance to lake	35.1
TMAX12	29.8
Altitude	27.3
Slope	25.9
Tmin1	23.8
Bio1	23.0
Bio19	20.4
Tmen2	19.2
Bio3	18.9
Tmax3	17.9
Bio6	16.3
Tmen7	15.9
Prec6	14.3
Prec7	13.9
Tmin6	12.9

Appendix 3

Prediction Model Details for the Best Performing Model (the 'Kitchen sink model' with 74 predictors)

Siberian crane with a battery run on TreeNet (SPM7) balanced
The kitchensink model, all 74 environmental predictors

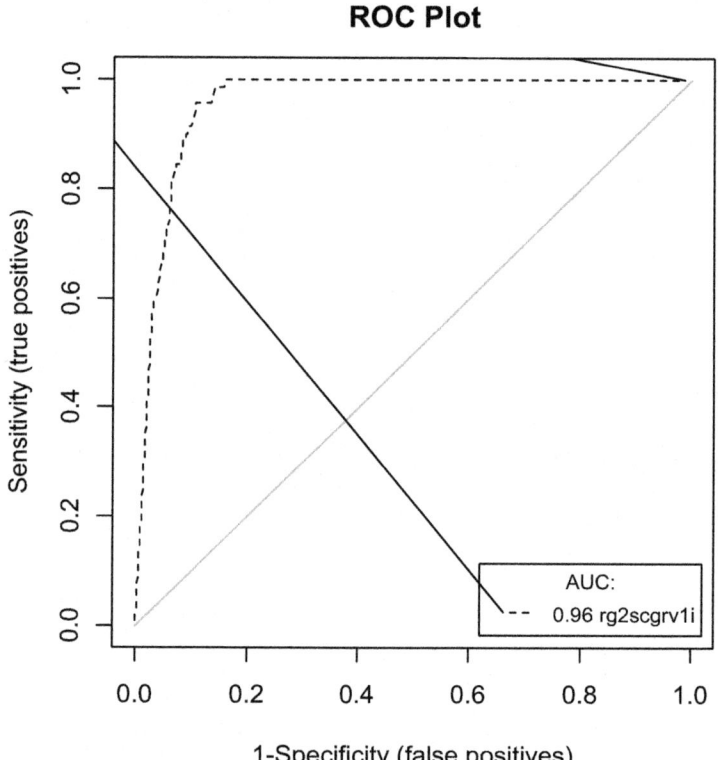

Frequency of Prediction Relative Index of Ocurrence (RIO 0-1) for known presence (1)

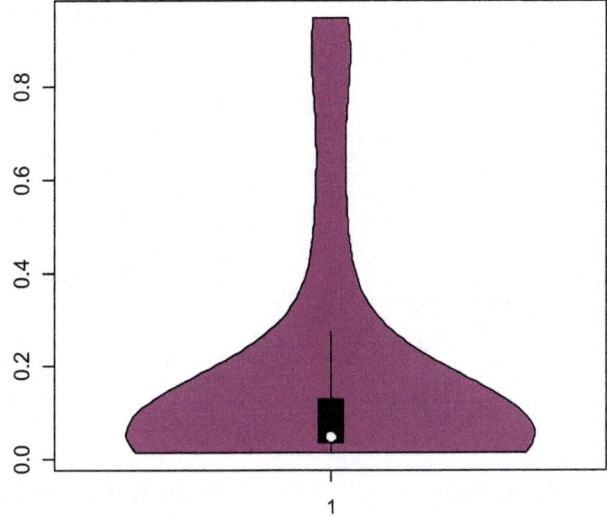

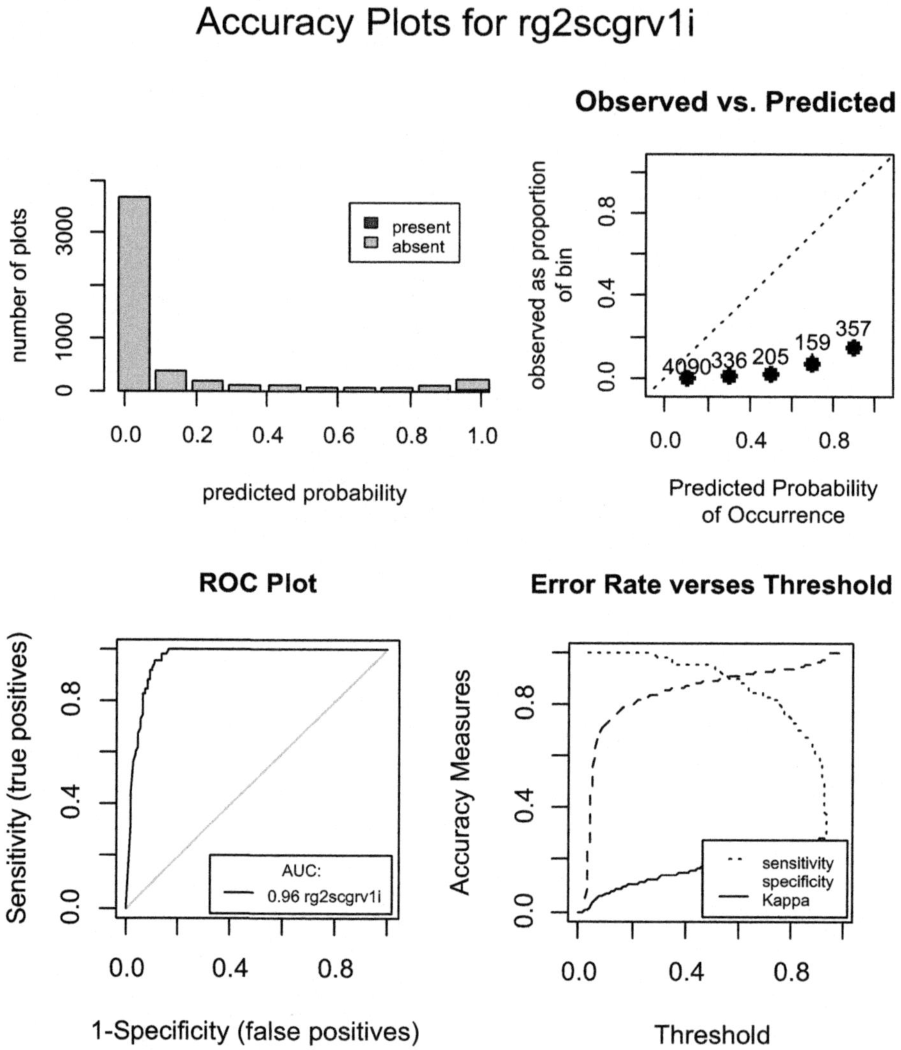

Accuracy Plots for rg2scgrv1i

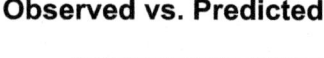

Observed vs. Predicted

Appendix 4

(For Prediction map 1 for the 'Kitchen sink model' see Fig. 8.4 in the text; for map legends please see this figure; same for all other appendix maps)

(For Prediction map 2 for the 'TMax12 model' see Fig. 8.5 in the text)

Prediction Map 3 for the 'BIO14 model'

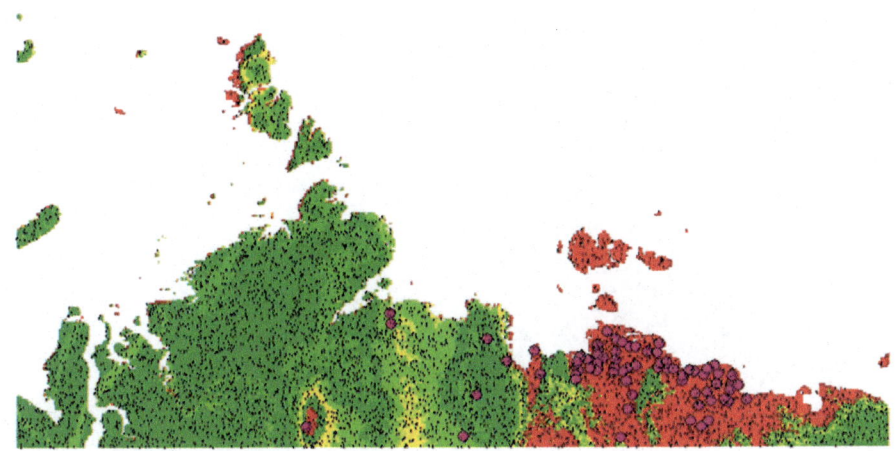

Prediction Map 4 for the 'TMax12BIO14 model'

Prediction Map 5 for the 'Top5 model'

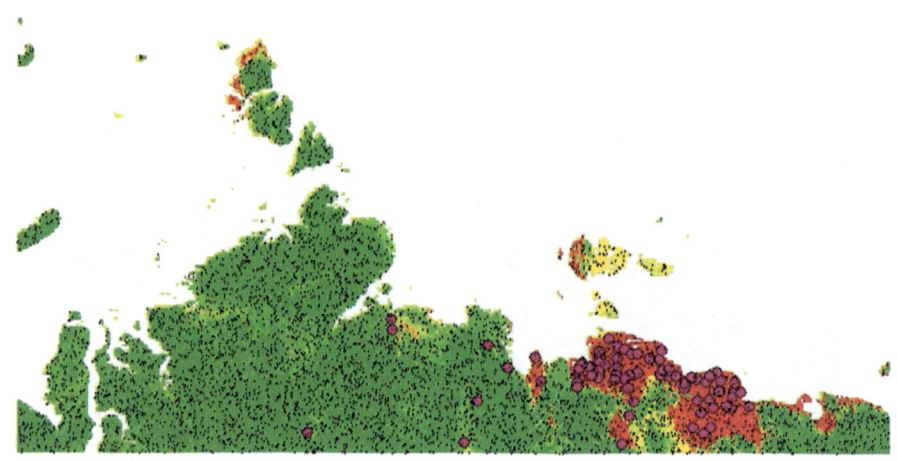

Prediction Map 6 for the 'Top10 model'

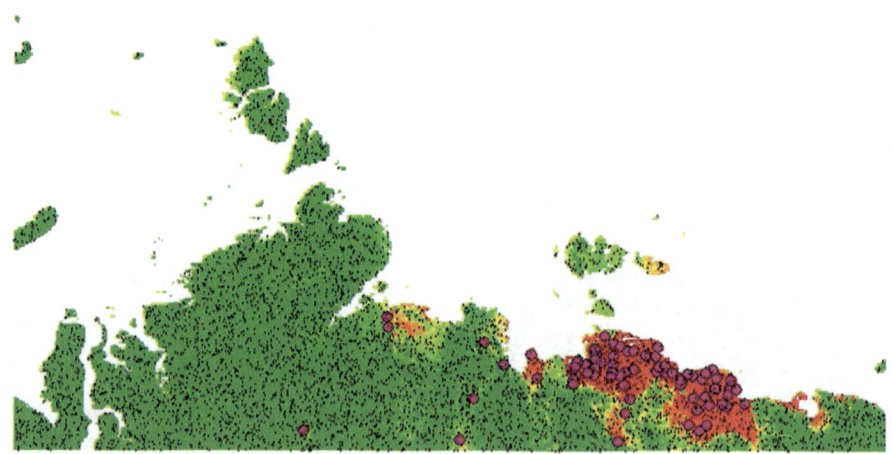

Prediction Map 7 for the 'Top29 model'

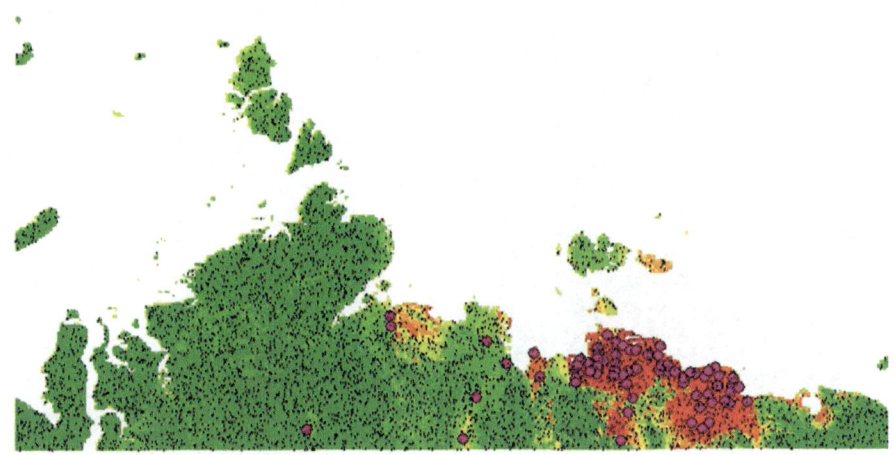

Prediction Map 8 for the 'Top35 model'

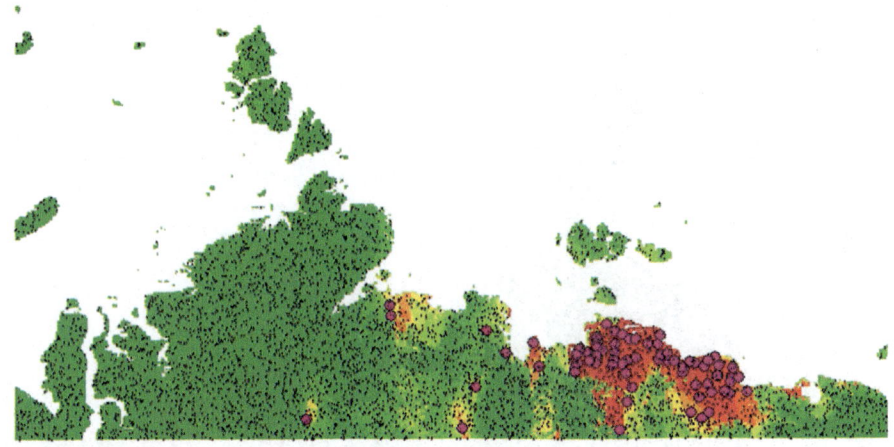

Prediction Map 9 for the 'Bottom 44 model'

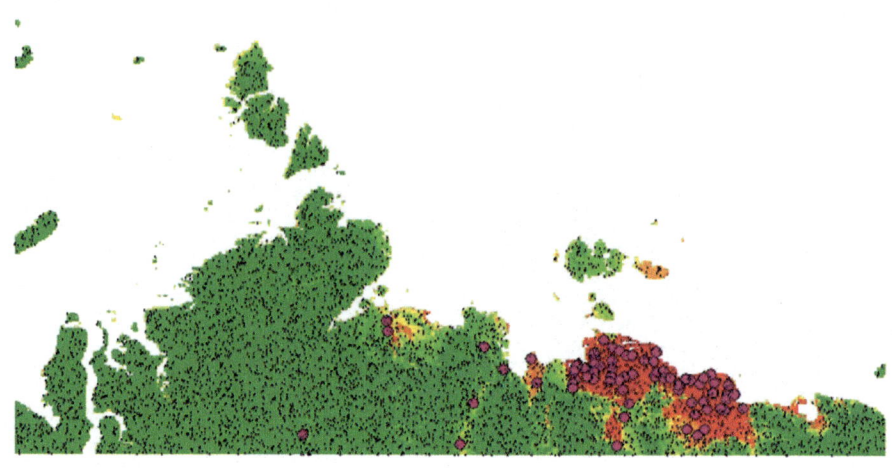

Prediction map 10 for the 'Leaving out top 3 interacting predictors model'

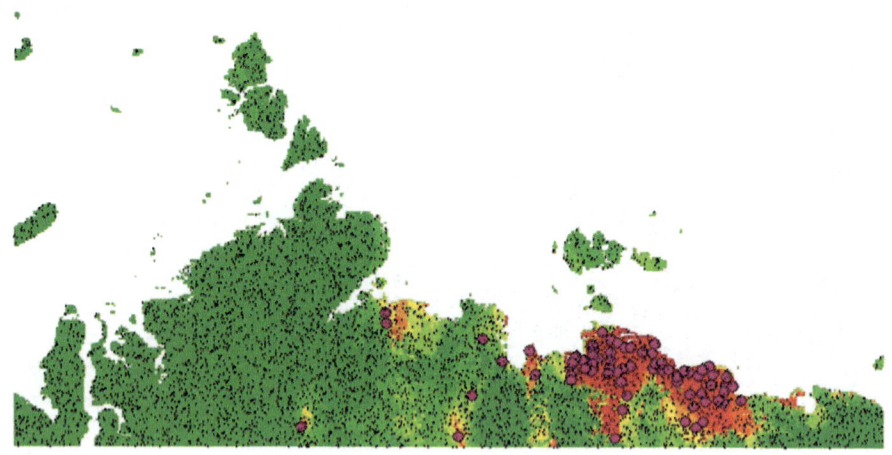

References

Akaike H (1974) A new look at the statistical model identification. IEEE Trans Automat Contr AC-19:716–23, Institute of Statistical Mathematics, Minato-ku, Tokyo, Japan

Arnold TW (2010) Uninformative parameters and model selection using Akaike's information criterion. J Wildl Manag 74:1175–1178

Barbet-Massin M, Jiguet F, Albert CH, Thuiller W (2012) Selecting pseudo-absences for species distribution models: how, where and how many? Methods Ecol Evol, 3:327–338. https://doi.org/10.1111/j.2041-210X.2011.00172.x

BirdLife International (2001) Threatened birds of Asia: the bird life international red data book, vol 1. Bird Life International Cambridge, Cambridge

Breiman L (2001) Statistical modeling: the two cultures (with comments and a rejoinder by the author). Stat Sci 16:199–231

Breiman L, Friedman J, Stone CJ, Olshen RA (1984) Classification and regression trees. CRC Press, Boca Raton

Burnham KP, Anderson DR (2002) Model selection and multimodel inference: a practical information-theoretic approach. Springer, New York

Cai T, Huettmann F, Guo Y (2014) Using stochastic gradient boosting to infer stopover habitat selection and distribution of hooded cranes *Grus monacha* during spring migration in lindian, Northeast China. PLoS ONE 9. https://doi.org/10.1371/journal.pone.0097372

Chamberlin TC (1890) The method of multiple working hypotheses. Science 15:92–96

Elith J, Graham CH, Anderson RP, Dudík M, Ferrier S, Guisan A, Hijmans RJ, Huettmann F, Leathwick JR, Lehmann A, Li J, Lohmann LG, Loiselle BA, Manion G, Moritz C, Nakamura M, Nakazawa Y, McC J, Overton M, Townsend Peterson A, Phillips SJ, Richardson K, Scachetti-Pereira R, Schapire RE, Soberón J, Williams S, Wisz MS, Zimmermann NE (2006) Novel methods improve prediction of species' distributions from occurrence data. Ecography 29:129–151

Fielding A (1999) Machine learning methods for ecological applications. Springer, Boston

Fielding A, Bell JF (1997) A review of methods for the assessment of prediction errors in conservation presence/absence models. Environ Conserv 24:38–49

Friedman JH (2001) Greedy function approximation: A gradient boosting machine. Ann Stat 29:1189–1232

Friedman JH (2002) Stochastic gradient boosting. Comp Stat Data Anal 38:367–378

Guthery FS, Brennan LA, Peterson MJ, Lusk LL (2005) Information theory in wildlife science: critique and viewpoint. J Wildl Manag 69:457–465

Han X, Guo Y, Mi C, Huettmann F, Wen L (2017) Machine learning model analysis of breeding habitats for the Blacknecked Crane in Central Asian Uplands under Anthropogenic pressures. Scientific Reports 7, Article number: 6114. https://doi.org/10.1038/s41598-017-06167-2. https://www.nature.com/articles/s41598-017-06167-2

Hastie T, Tibshirani R, Friedman J (2009) The elements of statistical learning: data mining, inference, and prediction, 2nd edn. Springer, New York

Herrick KA, Huettmann F, Lindgren MA (2013) A global model of avian influenza prediction in wild birds: The importance of northern regions. Vet Res. https://doi.org/10.1186/1297-9716-44-42

Hilborn R, Mangel M (1997) The ecological detective: Confronting models with data. Princeton University Press, Princeton

Hochachka W, Caruana R, Fink D, Munson A, Riedewald M, Sorokina D, Kelling S (2007) Data mining for discovery of pattern and process in ecological systems. J Wildl Manag 71:2427–2437

Jiao S, Guo Y, Huettmann F, Lei G (2014) Nest-Site selection analysis of hooded crane (*Grus monacha*) in northeastern china based on a multivariate ensemble model. Zool Sci 31:430–437

Kandel K, Huettmann F, Suwal MK, Regmi GR, Nijman V, Nekaris KAI, Lama ST, Thapa A, Sharma HP, Subedi TR (2015) Rapid multi-nation distribution assessment of a charismatic conservation species using open access ensemble model GIS predictions: red panda (*Ailurus fulgens*) in the Hindu-Kush Himalaya region. Biol Conserv 181:150–161

Kanai Y, Ueta M, Germogenov N, Nagendran M, Mita N, Higuchi H (2002) Migration routes and important resting areas of Siberian cranes (Grus leucogeranus) between northeastern Siberia and China as revealed by satellite tracking. Biol Conserv 106:339–346

Klein DR, Magomedova M (2003) Industrial development and wildlife in arctic ecosystems: Can learning from the past lead to a brighter future? In: Rasmussen RO, Koroleva NE (eds) Social and environmental impacts in the North. Kluwer Academic Publishers, The Netherlands, pp 35–56

Mace G, Cramer W, Diaz S, Faith DP, Larigauderie A, Le Prestre P, Palmer M, Perrings C, Scholes RJ, Walpole M, Walter BA, Watson JEM, Mooney HA (2010) Biodiversity targets after 2010. Env Sustain 2:3–8

Manly FJ, McDonald LL, Thomas DL, McDonald TL, Erickson WP (2002) Resource selection by animals: statistical design and analysis for field studies, Second edn. Kluwer Academic Publishers, Netherlands

Matthiessen P (2001) The birds of heaven. Travels with cranes. North Point Press, New York

McGarical K, Cushman S, Stafford S (2000) Multivariate statistics for wildlife and ecology research. Springer, New York

Mi C, Huettmann F, Guo Y, Han X, Wen L (2017) Why choose random forest to predict rare species distribution with few samples in large undersampled areas? Three Asian crane species models provide supporting evidence. PeerJ. https://doi.org/10.7717/peerj.2849

Moore GS, Ilyashenko E (2009) Regional flyway education programs: increasing public awareness of crane conservation along the crane flyways of Eurasia and North America. In: Prentice C (ed) Conservation of flyway wetlands in East and West/Central Asia. Proceedings of the project completion workshop of the UNEP/GEF Siberian Crane wetland project, 14–15 October 2009, Harbin, China. Baraboo (Wisconsin), USA: International Crane Foundation

Mueller JP, Massaron L (2016) Machine learning for dummies. For Dummies Publisher, 435 p

Ohse B, Huettmann F, Ickert-Bond S, Juday G (2009) Modeling the distribution of white spruce (*Picea glauca*) for Alaska with high accuracy: an open access role-model for predicting tree species in last remaining wilderness areas. Polar Biol 32:1717–1724

Prentice C (ed) (2010) Conservation of flyway wetlands in East and West/Central Asia. Proceedings of the project completion workshop of the UNEP/GEF Siberian Crane wetland project, 14–15 October 2009, Harbin, China. Baraboo (Wisconsin), USA: International Crane Foundation

Sorokin AG, Kotyukov YV (1987) Discovery of the nesting ground of the Ob River population of the Siberian Crane. In: Archibald GW, Pasquier RF (eds) Proceedings of the 1983 international crane workshop. International Crane Foundation, Baraboo, pp 209–212

Sorokin A, Markin Y (1996) New nesting site of Siberian Cranes. Newsletter of Russian Bird Conservation Union, Moscow

Spiridonov V, Gavrilo M, Krasnov MA, Nikolaeva N, Sergienko L, Popov A, Krasnova E (2011) Toward the new role of marine and coastal protected areas in the arctic: The russian case. In: Huettmann F (ed) Protection of the three poles. Springer, New York

Silvy NY (2012) The wildlife techniques manual: research and management, vol 2, 7th edn. John Hopkins University Press, Baltimore

Van Impe J (2013) Esquisse de l'avifaune de la Sibérie Occidentale: Une revue bibliographique. Alauda 81:269–296

Wu G, Leeuw J, Skidmore AK, Prins HHT, Best EPH, Liu Y (2009) Will the three gorges dam affect the underwater light climate of *Vallisneria* spiralis L. and food habitat of Siberian Crane in Poyang Lake. Hydrobiologia 623:213–222

Yu C, Yinghao W, Qing Y (2008) Ground survey of waterbirds in the Poyang Lake region in Winter 2007/2008. Siberian Crane Flyway News: 15

Chapter 9
Landscape Applications of Machine Learning: Comparing Random Forests and Logistic Regression in Multi-Scale Optimized Predictive Modeling of American Marten Occurrence in Northern Idaho, USA

Samuel A. Cushman and Tzeidle N. Wasserman

9.1 Introduction

The American marten (martes americana) is a species that is dependent on old conifer forest at middle to high elevations and is highly sensitive to habitat loss and
fragmentation in a scale dependent fashion (e.g., Hargis et al. 1999; Wasserman
et al. 2012a, b), and forest management is often influenced by considerations of how
management will affect extent and pattern of marten habitat. Due to their dependence on extensive, unfragmented forest landscapes and microhabitat structures
associated with late successional forest (Buskirk and Ruggiero 1994; Hargis et al.
1999), American marten are sensitive to fragmentation of late seral forest habitats,
such as that resulting from timber harvest and associated extraction routes and road
building (e.g., Cushman et al. 2011). Previous studies have consistently shown that
American marten habitat requirements include forests with high canopy cover
(Hargis and McCullough 1984; Wynne and Sherburne 1984), abundant near ground
structure (Chapin et al. 1998; Godbout and Ouellet 2008), high prey densities
(Fuller and Harrison 2005), and sufficient snow depth to provide subnivean spaces
during winter (Wilbert et al. 2000). These habitats are thought to provide opportunities for foraging, resting, denning, thermoregulation, and avoiding predation.
Perturbations, such as timber harvest, remove canopy cover, reduce coarse woody
debris, change mesic sites into xeric sites, remove riparian dispersal zones, and
change prey communities (Buskirk and Ruggiero 1994). American marten avoid

S. A. Cushman (✉)
U.S. Forest Service, Rocky Mountain Research Station, Flagstaff, AZ, USA
e-mail: scushman@fs.fed.us

T. N. Wasserman
School of Forestry, Northern Arizona University, Flagstaff, AZ, USA
e-mail: tnw23@nau.edu

© Springer Nature Switzerland AG 2018
G. R. W. Humphries et al. (eds.), *Machine Learning for Ecology and Sustainable Natural
Resource Management*, https://doi.org/10.1007/978-3-319-96978-7_9

areas with even relatively low levels of forest fragmentation and rarely use sites where more than 25% of forest cover has been removed (Hargis et al. 1999). Highly contrasting edge habitats, such as borders between late successional forest and harvested patches, and areas of open canopy are strongly avoided (Buskirk and Ruggiero 1994; Hargis et al. 1999; Cushman et al. 2011).

Recently, Wasserman et al. (2012a, b) predicted and mapped habitat suitability for American marten in northern Idaho, U.S.A. They used multiple scale habitat suitability modeling with logistic regression on a set of marten presence-absence locations collected non-invasively using genetic (hair) samples across a 3884 square kilometer region to quantify the relative importance of topographical, vegetation, and landscape metric variables in predicting marten occurrence. The Wasserman et al. (2012a, b) model identified strong and consistent relationships with various measures of landscape fragmentation: marten occurrence was positively associated with landscapes that contained high canopy closure, low density of all roads (including small forest roads), few past clear-cuts, and extensive late seral forest. Several of these variables had maximum influence on marten probability of occurrence at fairly broad spatial scales. At scales approximately the size of marten home ranges (500–1000 m radius; Tomson et al. 1999) within our study area, the Wasserman et al. (2012a, b) model showed that American marten select landscapes with high average canopy closure, low road density, and low forest fragmentation. Within these low-fragmentation landscapes, the model showed marten select foraging habitat at a fine scale (90 m) within middle-elevation, late-seral, mesic forests. This is consistent with the results of previous studies, which have shown high sensitivity to landscape fragmentation and perforation by non-stocked clear-cuts (Hargis et al. 1999; Cushman et al. 2011), and strong preference of American marten in northern Idaho for mesic riparian forest conditions in unfragmented watersheds (Tomson 1999; Shirk et al. 2014).

For a decade, logistic regression has been the dominant method in multi-scale habitat modeling (Hegel et al. 2010; McGarigal et al. 2016). Random forests (RF; Breiman 2001a, b) is increasingly used in a range of applications including digital soil mapping (Grimm et al. 2008), forest biomass mapping (Baccini et al. 2012), species distribution modeling (Evans and Cushman 2009), land cover change prediction (Cushman et al. 2017) and others given its often superior performance compared to other methods (Evans et al. 2011; Mi et al. 2017). However, there have been relatively few formal comparisons of the performance of multi-scale modeling between logistic regression and random forests. Recently, Cushman et al. (2017) compared the performance of logistic regression with random forests in a multi-scale optimized predictive modeling study of deforestation risk across Borneo. As found in virtually all of such investigations, the authors found that random forests substantially outperformed logistic regression. Our interest in this study is to conduct a similar comparison of logistic regression and random forests in multi-scale optimized predictive model of occurrence of a forest-dependent mammal species, the American marten (*Martes americana*) in northern Idaho USA.

The main purpose of this chapter is to compare the predictive power and the ecological interpretation of the Wasserman et al. (2012a, b) logistic regression model with a model produced on the same data using the same multi-scale optimization approach, but using random forests instead of logistic regression. Based on past work showing that random forests often outperforms other predictive modeling approaches (e.g. Evans et al. 2011; Cushman et al. 2017), we predicted that the random forests model would outperform the logistic regression model based on AUC (area under the receiver operator curve). Also, previous work has shown that marten habitat selection is highly scale dependent (e.g., Hargis et al. 1999; Wasserman et al. 2012a, b), and a recent review has demonstrated that multi-scale optimization is important for habitat modeling in general (McGarigal et al. 2016). Accordingly, an additional goal of this chapter is to see if the inferences about what variables are important and at what scales they are operative differ between models developed with random forests and GLM logistic regression.

9.2   Methods

9.2.1   Study Area

The study area is a 3884 km² section of the Selkirk, Purcell, and Cabinet Mountains, encompassing the Bonners Ferry and Priest River Ranger Districts of the Idaho Panhandle National Forest (2282 km²) and adjacent non National Forest System lands, including private land (986 km²), State (508 km²), tribal- and other federally managed land (Fig. 9.1). The topography is mountainous, with steep ridges, narrow

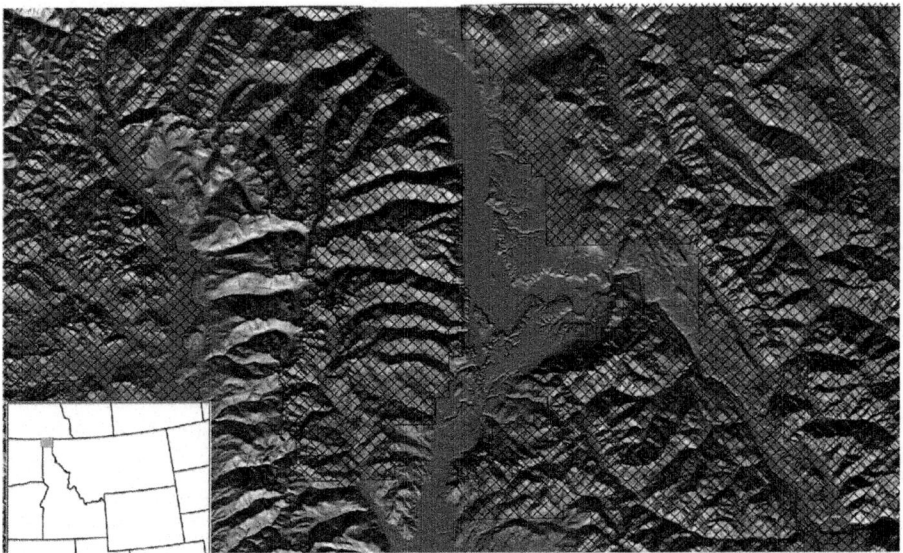

Fig. 9.1 Study area orientation map. Idaho Panhandle National Forest lands are cross-hatched

valleys, and many cliffs and cirques at the highest elevations. Elevation ranges from approximately 700 m to 2400 m above sea level. The climate is characterized by cold, moist winters and dry summers. The average daily maximum temperature at Bonners Ferry, the largest town in the study area, in the coldest month (January) is 0.2 °C, while that of the warmest month (July) is 27.8 °C. Average precipitation in the wettest month (December) amounts to 7.84 cm, while that of the driest month (July) is 2.33 cm, with an average annual total of 56.4 cm.

The area is heavily forested, with subalpine fir (*Abies lasiocarpa*) and Engelmann Spruce (*Picea engelmannii*) co-dominant above 1300 m, and a diverse mixed forest of Douglas-fir (*Pseudotsuga menziesii*), lodgepole pine (*Pinus contorta*), ponderosa pine (*Pinus ponderosa*), western white pine (*Pinus monticola*), grand fir (*Abies grandis*), western hemlock (*Tsuga heterophylla*), western red cedar (*Thjua plicata*), western larch (*Larix occidentalis*), paper birch (*Betula papyrifera*), quaking aspen (*Populus tremuloides*), and black cottonwood (*Populus trichocarpa*) dominating below 1300 m.

9.2.2 Occurrence Data and Logistic Regression Model

We decided to utilize a multi-scale habitat suitability model produced by (Wasserman et al. 2012a, b), who used multi-scale logistic regression modeling to predict habitat suitability from a presence/absence dataset collected non-invasively through hair snaring (e.g., Wasserman et al. 2010). To obtain data on American marten presence, Wasserman et al. (2010) deployed hair snare stations at 361 locations well distributed across a representative sample of topographical and ecological gradients over three winter seasons (2005, 2006, and 2007; 1 survey per site). Recently, Robinson et al. (2017) showed that this kind of non-invasive genetic sampling is consistent and has high success for species and individual identification across seasons and weather patterns. Genetic analysis confirmed the detection of American marten at 159 individual hair snare stations. (Wasserman et al. 2012a, b) selected variables *a priori* assumed to be related to American marten occurrence based on previous research (Buskirk and Ruggiero 1994; Hargis et al. 1999; Tomson 1999), including elevation, percent canopy closure, road density, patch density, percentage of the local landscape surrounding survey sites occupied by late seral forests, percentage of the landscape occupied by non-stocked clear-cuts, and probability of occurrence of each major tree species (western red cedar and six other species) in each cell across the landscape.

The first step undertaken by (Wasserman et al. 2012a, b) was to use bivariate scaling (Thompson and McGarigal 2002; Grand and others 2004) to identify the scale at which each of these independent variables was most strongly related to American marten occurrence. Given that environmental factors may be related to deforestation at a range of spatial scales (Wiens 1989), and given the critical importance of multi-scale optimization to correct inferences about habitat selection

(McGarigal et al. 2016), Wasserman et al. (2012a, b) calculated all predictor variables at 12 spatial extents including focal radii of 90 to 990 m at 90 m increments. This resulted in reduction of the model to seven variables significantly related to marten occurrence (Table 9.1). Wasserman et al. (2012a, b) then used logistic regression to test all combinations of these predictor variables, without interactions, and used model averaging, based on AIC weights, to produce parameter estimates for a final model predicting probability of marten occurrence. This model was then used to evaluate the impacts of past timber harvest and road building on the extent and quality of available marten habitat.

9.2.3 Predictor Variables for Analysis

A priori, we proposed several environmental and anthropogenic variables as predictors of marten occurrence. Following Wasserman et al. (2012a, b) we included road density and canopy closure, as well as a number of topographical and landscape composition and configuration metrics. Topographical variables included elevation and several terrain complexity measures produced using the Geomorphometry and Gradient Metrics Toolbox (ArcGIS 10.0; Evans et al. 2014). These included: topographical roughness, which measures the topographical complexity of the landscape

Table 9.1 Variables included in the Wasserman et al. 2012a, b habitat model used in the current analyses. There were seven variables in the habitat model, related to elevation, road density, canopy cover, patch density in the landscape mosaic, large saw timber, non-stocked clear cuts and western red cedar forest types. Each of these was included in the habitat model at a particular spatial scale (focal extent) at which it most strongly affected probability of occurrence. These scales ranged from 90 m in radius (western red cedar and large saw timber) to a maximum extent of influence of road density at a 1980 m radius. Each of these variables had different effects on marten probability of occurrence. Effect size in this table records the percent change in the probability of marten occurrence as the associated variable changes from the 10th to the 100th percentile value in the dataset, holding the other variables constant at their medians. Based on this measure of effect size, the most important predictors, in decreasing order of importance, are western red cedar forest type, percent canopy cover, road density, patch density, percent of the landscape in non-stocked clear cuts, elevation, and finally large saw timber

Predictor variable	Most significant scale (m)	Effect size
Elevation	1400	19.78
Road density	1980	−53.05
Percent canopy cover	990	61.05
Patch density	990	−46.26
Percentage of the focal landscape in large sawtimber	90	13.21
Percentage of the landscape in non-stocked conditions	990	−35.99
Western red cedar	90	77.21

within a defined focal extent (Blaszczynski 1997), relative slope position, which measures the relative position of the focal pixel within a defined extent on a gradient from valley bottom to ridge top (Evans et al. 2014), dissection index, which is the ratio between relative relief and to the absolute relief, curvature index, which measures the rate of change of local slope, heat load index, which predicts the total incident solar radiation as a function of latitude and topography, and compound topographical index, which models the cumulative aggregation of water flow through every cell in the landscape.

We also included FRAGSTATS metrics quantifying the extent and configuration of different land cover classes across a range of focal extents as predictor variables (McGarigal et al. 2012). The classes used in the analysis include: (1) large sawtimber (> 24 inches DBH), (2) small sawtimber (12–24 inches DBH), (3) pole timber (3–12 inches DBH), (4) sapling/seedling (< 3 inches DBH), (5) non-stocked forestland, and (6) non-forest (Wasserman et al. 2012a, b). For each of these classes we used FRAGSTATS 4.0 (McGarigal et al. 2012) to calculate five class-level (area-weighted mean patch size, Area_AM; edge density; ED, patch density, PD; percentage of the landscape, PLAND; area-weighed proximity index, PROXAM), and four landscape-level metrics (aggregation index, AI; contrast-weighted edge density, CWED; edge density, ED; patch density, PD). These metrics were chosen given that they measure several critical attributes of habitat extent and fragmentation that have been shown to have important influences on habitat selection (e.g., Chambers et al. 2016) and population connectivity (e.g., Cushman et al. 2013). Also, following Wasserman et al. (2012a, b) we calculated all variables within 12 focal scales ranging from 90 to 990 m radii around each sampling location to enable multi-scale model optimization.

9.2.4 Modeling Approaches

We used random forests machine learning and logistic regression to predict marten occurrence in the study landscape. We used the logistic regression model and results as published in Wasserman et al. (2012a, b). Random forests is a classification and regression tree (CART; De'ath and Fabricius 2000) - based bootstrap method that corrects many of the known issues in CART, such as over-fitting (Breiman 2001a, b; Cutler et al. 2007), multi-collinearity and variable interaction, and provides very well-supported predictions with large numbers of independent variables (Cutler et al. 2007). We used a modeling approach developed by Evans and Cushman (2009) to predict occurrence of marten using the random forests method (Breiman 2001a, b; Cutler et al. 2007) as implemented in the package 'randomForest' (Liaw and Wiener 2002) in R (R Development Core Team 2008).

We conducted the random forests in two steps, mirroring the approach Wasserman et al. (2012a, b) used in the original logistic regression model. We recognize that using random forests like a GLM does not unleash all its powers, but our purpose was to conduct a strict comparison keeping as many parameters as similar as possible to see how random forests and GLM differed in their predictions in this context. First, we ran univariate models across the multiple scales to identify the scale at which each variable had the strongest ability to predict marten occurrence, as suggested by McGarigal et al. (2016) as a robust approach for multi-scale model optimization, and as shown to work well for random forests by Cushman et al. (2017). To accomplish this, we ran a series of single random forests analyses for each variable across the 12 scales in each nation and used the Model Improvement Ratio (MIR; Murphy et al. 2010) to measure the relative predictive strength of each scale of the variable. The MIR calculates the permuted variable importance, represented by the mean decrease in out-of-bag error, standardized from zero to one. We compared the MIR scores for all scales for each variable, and retained the scale that had the highest MIR score for further multivariate modeling.

In the second step we used random forests to develop multivariate models predicting probability of marten occurrence as a function of landscape condition across the suite of scale-optimized variables. To identify the most parsimonious random forests model we applied the Model Improvement Ratio (MIR; Murphy et al. 2010). In model selection using MIR, the variables were subset using 0.10 increments of MIR value, with all variables above the threshold retained for each model. This subset was always performed on the original model's variable importance to avoid over-fitting (Svetnik et al. 2004). We compared each subset model and selected the model that exhibited the lowest total out-of-bag error and lowest maximum within-class error.

Model predictions for the random forests model were created by using a matrix of the ratio of majority votes to create a probability distribution. Random forests makes predictions based on the plurality of votes across all bootstrap trees and not on a single rule set. This votes-matrix can be scaled and treated as a probability given the error distribution of the model (Evans and Cushman 2009; Murphy et al. 2010). We used the function that (Evans and Cushman 2009) added to GridAsciiPredict (Crookston and Finley 2008) which uses the votes-probability function to write the probabilities to ASCII grids.

9.2.5   Model Assessment

There are a multitude of ways to assess the performance of predictions of the random forests and logistic regression models, and most previous studies have used the Kappa statistic (Cohen 1968) and similar measures of improvement of predicted classification compared to random assignment (e.g., based on the confusion matrix). However, following Ponitus and Milones (2011), we avoided the Kappa statistic given that it does not report a meaningful statistical measure of predictive success, even when corrected to address the two different aspects of prediction related to predicted amount and predicted location (Pontius and Si 2014). In addition, since the predictions we produced using random forests and logistic regression are in the form of predicted probabilities, it is more meaningful to assess the continuous pattern or predicted probability in comparison to the actual observed changes than to cross-tabulate observed vs. predicted change (Pontius and Si 2014) We chose this approach because transforming predicted probabilities into categorical responses requires using a threshold cut-point or probabilistic function, which loses information on the actual quality of the prediction (Pontius and Milones 2011; Pontius and Si 2014). We assessed the performance of the random forests and logistic regression predictions using area under the Total Operating Characteristic curve (Pontius 2014), as suggested by Pontius and Si (2014) and Pontius and Parmentier (2014). We also produced predicted probability of occurrence maps for both models and visually compared these to describe the differences in the pattern of predicted habitat suitability.

9.3   Results

9.3.1   Random Forests Univariate Scaling

The first step in the modeling approach was to identify the best scale for each individual variable out of the 12 scales considered (90–990 m, by 90 m increments), based on Model Improvement Ratio. For each variable we chose the scale with the largest Model Improvement Ratio, except in some cases we retained two scales if the second had an MIR value over 0.75 and differed substantially in scale from the scale with the highest MIR value (Table 9.2). There was a relatively broad range of scales selected across all variables (Fig. 9.2), with an apparent bimodal pattern where more variables were selected at either the broadest (greater than 630 m radius), or finest (less than 270 m radius) scales.

9.3.2   Random Forests Multivariate Model

The multivariate random forests model used the Model Improvement Ratio as a variable selection approach. The final model included 14 variables (Fig. 9.3). Five of these were selected at the broadest scale of 990 m, showing a stronger pattern of dominance by broad-scale relationships in the multivariate reduced model than in the univariate scaling.

We produced LOWESS splines of the pattern of presence vs. absence across the sampled range of each of the top eight variables. LOWESS (locally weighted scatterplot smoothing) is a non-parametric regression method that combine multiple regression models in a k-nearest-neighbor-based meta-model to produce non-linear

Table 9.2 Variables included in the random forests modeling and the scales retained in the univariate scaling step. Land – Landscape-level FRAGSTATS variable; Class – Class-level FRAGSTATS variable

Variable	Acronym	Top scale	Second scale retained
Agregation index (Land)	AI	630	180
Road density	AR	180	1440
Area-weighted mean patch size (Class)	areaam1	900	180
Area-weighted mean patch size (Class)	areaam2	720	
Area-weighted mean patch size (Class)	areaam3	540	
Area-weighted mean patch size (Class)	areaam4	990	
Area-weighted mean patch size (Class)	areaam5	450	
Area-weighted mean patch size (Class)	areaam6	990	
Mean canopy cover	canopy	180	630
Topographical curvature index	crv	810	630
Compound topographical index	cti	270	90
Contrast-weighted edge density (Land)	cwed	360	
Topographical dissection index	dis	90	
Edge density (Land)	ed	90	990
Edge density (Class)	ed1	630	
Edge density (Class)	ed2	90	
Edge density (Class)	ed3	180	
Edge density (Class)	ed4	990	
Edge density (Class)	ed5	450	
Edge density (Class)	ed6	900	
Elevation	elev90	720	
Heat load index	hil	720	270
Patch density (Land)	pd	990	630
Patch density (Class)	pd1	990	
Patch density (Class)	pd2	810	
Patch density (Class)	pd3	720	
Patch density (Class)	pd4	810	
Patch density (Class)	pd5	360	
Patch density (Class)	pd6	270	90
Percentage of the landscape (Class)	pland1	990	180
Percentage of the landscape (Class)	pland2	720	
Percentage of the landscape (Class)	pland3	990	
Percentage of the landscape (Class)	pland4	990	
Percentage of the landscape (Class)	pland5	360	
Percentage of the landscape (Class)	pland6	630	900
Proximity index (class)	proxam1	450	990
Proximity index (Class)	proxam2	450	
Proximity index (Class)	proxam3	540	
Proximity index (class)	proxam4	810	
Proximity index (Class)	proxam5	900	
Topographical roughness	r	90	900
Slope position	sp	810	

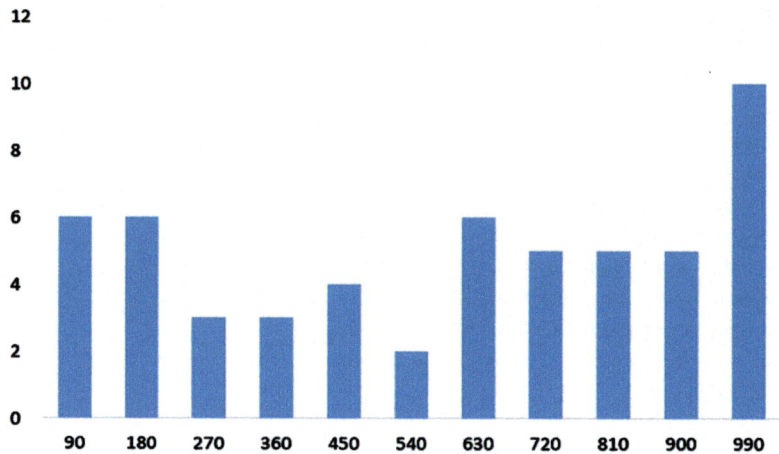

Fig. 9.2 Frequency of selected scales (in meters) across all variables for the random forests model

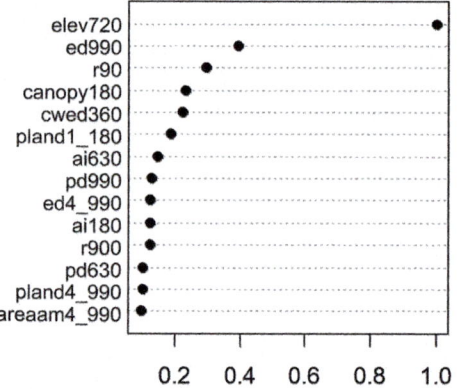

Fig. 9.3 Model improvement ratio plot for the selected variables. The most important variable is mean elevation within a 720 m focal radius (elev90720). The other variables are listed in order of their importance relative to elevation, with the x-axis indicating the relative additional model improvement when adding each successive variable

splines showing the response pattern in a bivariate scatter plot. The most important variable by far, based on the MIR, was mean elevation within a 720 m focal radius (Fig. 9.3). Marten occurrence has a strongly non-linear relationship with elevation; detections are very rare below 1000 m, rising rapidly to an apparent unimodal peak at approximately 1280 m, and then slowly declining at the highest elevations

(Fig. 9.4a). The second most important variable based on MIR was edge density within a 720 m focal radius. Marten occurrence has a nonlinear relationship with edge density as well, with the highest detection rates generally occurring at low edge densities (Fig. 9.4b). The third most important variable was topographical roughness at a 90 m focal radius, with marten occurrence increasing monotonically but nonlinearly with increasing topographical roughness (Fig. 9.4c). Mean canopy cover within a 180 m focal radius was the fourth most important variable in the random forests model, with marten detections increasing strongly, but again nonlinearly, at high levels of local canopy cover (Fig. 9.4d). The fifth most important variable based on MIR was contrast-weighted edge density, with marten occurrence declining with increasing density of high-contrast edges in the landscape mosaic

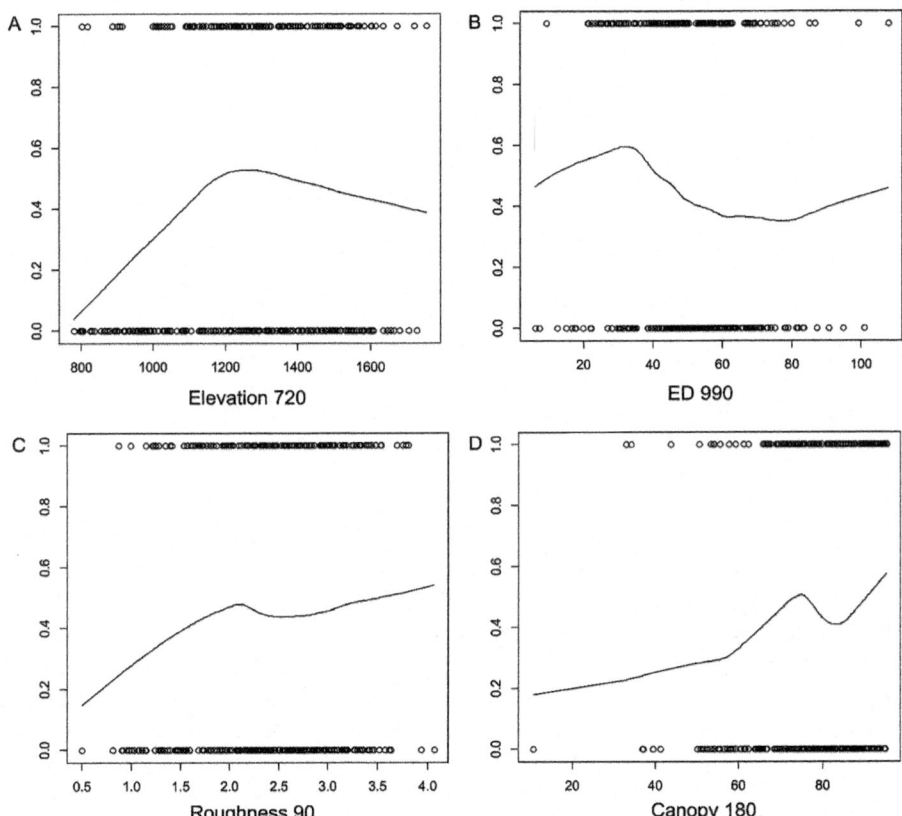

Fig. 9.4 (**a–d**) – part 1. Scatter plots of presence and absence and fitted LOWESS splines for the first four variables selected by the Model Improvement Ratio variable selection process for the multivariate random forests model

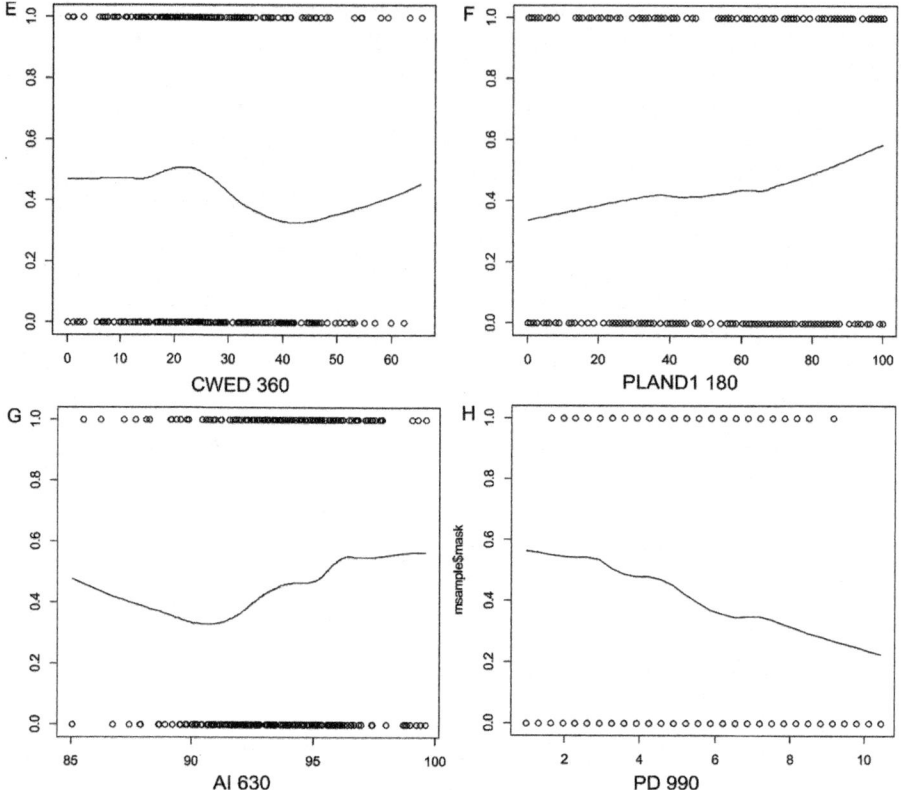

Fig. 9.4 (e–h) – part 2. Scatter plots of presence and absence and fitted LOWESS splines for the fifth through eighth variables selected by the Model Improvement Ratio variable selection process for the multivariate random forests model

(Fig. 9.4e). The percentage of a 180 m radius focal landscape occupied by large saw timber was the sixth most important variable, with occurrence frequency with occurrence frequency increasing monotonically with the amount of large, old forest in the local landscape (Fig. 9.4f). Landscape-level aggregation index was the seventh most important variable, with the frequency of marten occurrence increasing non-linearly but monotonically with increasing landscape aggregation within a 630 m focal radius (Fig. 9.4g). Landscape-level patch density within a 990 m focal landscape was the eighth most important variable, with monotonically decreasing frequency of marten as patch density increased (Fig. 9.4h).

9.3.3   Model Comparison

There was substantial similarity in the qualitative interpretation of the Wasserman et al. (2012a, b) logistic regression and the random forests model produced for this chapter. In both models occurrence was strongly predicted by a unimodal function

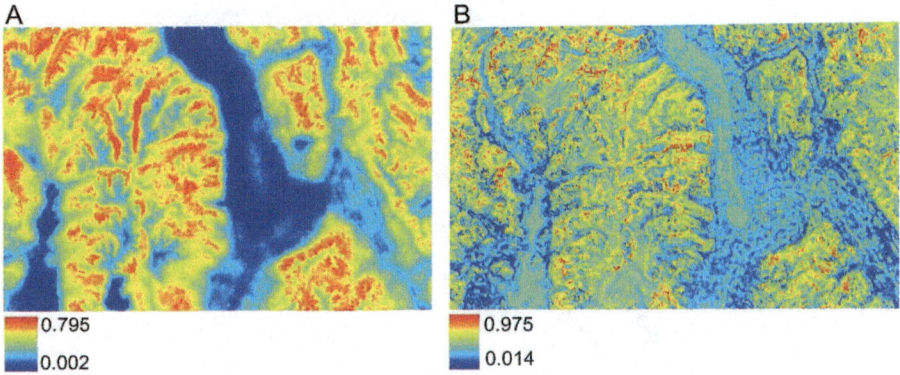

Fig. 9.5 Comparison of predicted probability of marten occurrence from the logistic regression (**a**) and random forests (**b**) models across the study area

of elevation, a non-linear function of canopy cover, a non-linear function of patch density, and the extent of the landscape in large conifer forest. However, there also were some important differences. First and foremost in performance (inference from predictions). Secondly, when looking at the predictors, road density and percentage of the landscape in non-stocked forestland were included in the final model-averaged logistic regression prediction, while these variables were not selected by the MIR in the random forests model.

In addition, a number of other variables were included in the random forests model that were not included in the logistic regression model, notably edge density, topographical roughness, contrast-weighted edge density and aggregation index. Together, these variables provide a substantially stronger "fragmentation signal" in the random forests model than the logistic regression model, with stronger identification of the negative effects of landscape heterogeneity than indicated by the logistic regression model.

In both models, extent of large sawtimber forest was a strong predictor at a fine spatial scale, while patch density was a strong predictor at the broadest spatial scale tested. This suggests that both models predict that optimal American marten habitat consists of patches of large, old forest within broad forested landscapes that have low levels of heterogeneity or fragmentation. However, the logistic regression model identified canopy cover as having the strongest effect at the broadest scale, while the random forests model identified a relatively fine scale effect of canopy cover.

A visual comparison of the predicted probability maps (Fig. 9.5) shows three main differences in the spatial prediction of marten habitat between the logistic regression and the random forests model. As also seen by Cushman et al. (2017), random forests produces predictions that are more discriminatory, with higher range of predicted probability and higher spatial heterogeneity than logistic regression. Logistic regression fits smooth linear functions of a linear combination of variables, which results in simple and smooth patterns of predicted occurrence. The logistic

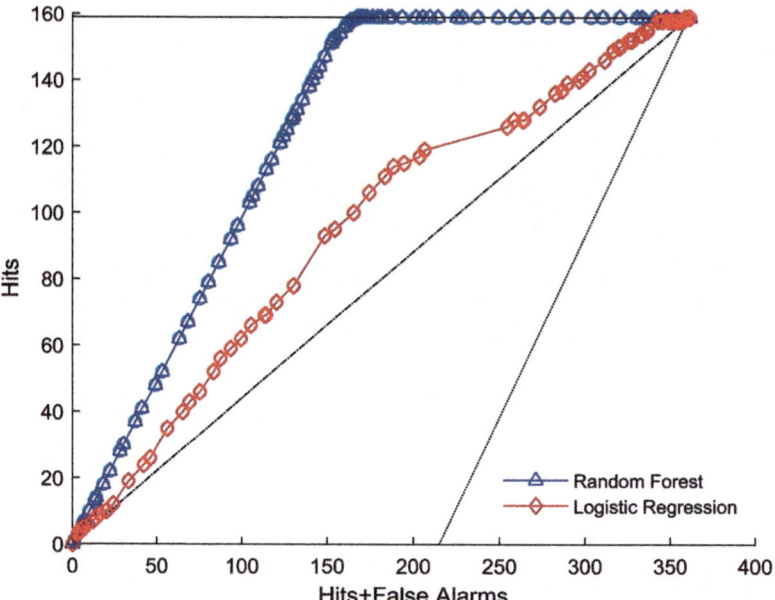

Fig. 9.6 TOC curves showing comparative model performance among the logistic regression, random forests, random forests without fragmentation variables, and the naïve model. Higher model performance is indicated by stronger convex curvature toward the upper left corner of the plot space. The AUC for the two models are 0.983 for random forests and 0.701 for logistic regression

regression map highlights areas of high canopy cover, high extent of old forest, and low fragmentation at middle to upper elevations. In contrast, the random forests model shows much higher heterogeneity of predictions, with steeper and stronger gradients of habitat quality across the landscape. Both models indicate that marten habitat quality is highest in middle to upper elevation areas with high canopy cover, low fragmentation, and high cover of old forest, but the random forests model shows that habitat quality varies more across space, with more areas of predicted very high occurrence probability, interspersed with areas of predicted lower quality, which are not seen in the logistic regression model predictions.

9.3.4 Model Performance

We assessed model performance based on the area under the TOC curve (Fig. 9.6). The logistic regression model has an AUC of 0.701, as previously reported by Wasserman et al. (2012a, b), indicating moderately good success in predicting presence vs. absence in the training dataset. By comparison, the random forests model had an AUC of 0.981, indicating very high predictive ability, and a much stronger ability to predict presences and absences in the training dataset than the logistic regression model. Expressed as a percentage, the random forests model had 28% higher performance, leading to much better prediction of habitat suitability, better inferences about habitat variables influencing marten occurrence, improved identification of scale dependency, and ultimately, therefore, better guidance to conservation and management (Breiman 2001b).

9.4 Discussion

Consistent with the results of other researchers who found that random forests outperforms other methods for prediction and classification (e.g., Cushman et al. 2010; Evans et al. 2011; Drew et al. 2010; Rodriguez-Galiano et al. 2012; Schneider 2012; Cushman et al. 2017), we expected that random forests would outperform logistic regression in predicting marten occurrences. Consistent with this expectation, random forests greatly outperformed logistic regression based on AUC measures of predictive success. This confirms the superiority of random forests as a modeling tool for habitat modeling, and we suggest that future studies use this powerful technique as a baseline.

Our analysis provided insight into patterns of scale-dependent habitat selection in American marten. The results were generally consistent with those found by Wasserman et al. (2012a, b). Specifically, the models show that American marten occurrence is highest in middle to upper elevation forested landscapes with high local canopy closure and high local cover of old-growth forest, and low levels of landscape heterogeneity and fragmentation at broader scales. In essence, our model reconfirms the description of Wasserman et al. (2012a, b) for optimal American marten habitat in northern Idaho: "... *at the scale of home ranges, marten select landscapes with high average canopy closure and low fragmentation. Within these low-fragmentation landscapes, marten select foraging habitat at a fine scale within late-seral, middle-elevation mesic forests. In northern Idaho, optimum American marten habitat, therefore, consists of landscapes with low road density and low density of non-forest patches with high canopy closure and large areas of middle-elevation, late successional mesic forest.*" Our analysis augments this interpretation with further emphasis on the importance of landscape heterogeneity at intermediate (CWED at 360 m) to broad scales (AI at 630 m, PD at 990 m), suggesting perhaps a larger importance of landscape fragmentation than suggested by the Wasserman et al. (2012a, b) analysis.

The random forests and logistic regression models were also quite different in their spatial predictions, with logistic regression producing smooth, monotonic patterns of predicted suitability, while random forests produced a map with higher heterogeneity and discrimination, showing stronger identification of areas of high suitability for marten. These differences are highly relevant if predictions from models are to be used effectively for management and conservation. Conservation prioritization based on habitat suitability would likely be quite different when based on either of these two maps, with the logistic regression producing coarse recommendations to protect middle elevation, unfragmented, old-growth forest in general, while the random forests would suggest the same general habitat niche but provide much stronger delineation of high priority areas.

Our analysis also provides an ability to assess patterns of scale-dependency in habitat relationships across a large number of predictor variables. This is an area of ongoing and increasing interest in landscape ecology (McGarigal et al. 2016). Relatively few studies have comprehensively evaluated patterns of scale dependence across pools of predictor variables. For example, Chambers et al. (2016) evaluated scale dependence of habitat associations and scaling patterns of landscape metrics in relation to bat occurrence or capture rate in forests of southwestern Nicaragua. They found that that edge density and patch density were the most important configuration variables across species, and percentage of the landscape was the most important class-level variable. In addition, they found that certain landscape and configuration metrics were most influential at fine (100 m) and/or broad (1000 m) spatial scales. Our results echo the importance of patch density and edge density as configuration predictor variables (the most important configuration variables in our analysis) and PLAND as a composition predictor variable (PLAND1 was the only composition variable in the random forests model).

One of the most important comparative differences between the logistic regression and the random forests models was their interpretation of scale dependence among the different predictor variables. In general both models found that landscape heterogeneity and forest fragmentation affected marten habitat suitability at broad scales, but the random forests analysis showed that fragmentation effects are active at both fine and broad scales, in contrast to the logistic regression which only identified these effects at broad scales. Also, the scales at which canopy cover and extent of old forest most strongly affected predictions were different between the models, indicating that the optimal scale of influence is highly sensitive to the method of modeling.

Random forests (Breiman 2001a, b) is a tree-based method based on "bagging" that is executed by bootstrapping (with replacement) 63% of the data and generating a weak learner based on a CART for each bootstrap replicate. Within the pre-set specification (e.g., node depth and number of samples per node) each CART is unconstrained (grown to fullest) and prediction is accomplished by tallying the 'majority votes' across all nodes in all random trees (Hegel et al. 2010). Independent variables are randomly selected at each node, with the number of variables selected at each node defined by m [sqrt(number of independent variables)]. These attributes provide several reasonable explanations for why random forests proved so much more powerful in predicting marten occurrence patterns in our northern Idaho dataset than did logistic regression. As seen in the LOWESS splines, there are strongly non-linear, often unimodal or multi-modal patterns of frequency of marten occurrence across the range of values of independent variables. Such complex non-linearity and non-monotonicity is a massive challenge to GLM modeling, such as logistic regression, even when, as in Wasserman et al. (2012a, b), nonlinear transformations are applied to the data. In contrast, the bootstrapping of CART within random forests provides the generation of a large number of trees which are combined across all nodes in all random trees. This enables immense flexibility to deal with non-linearity and multi-modality of response, resulting in random forests models predicting patterns of presence and absence in the training data much more tightly than is possible with GLM or similar functional relationship methods. This also enables random forests to accurately reflect complex multi-variate non-linear interactions among predictor variables, which are typically completely ignored in GLM modeling (see Chap. 10 by Baltensperger).

In our case the logistic regression model was fair at prediction (AUC = 0.7) while the random forests model was excellent (AUC = 0.98), even though both were applied to the same data, and included largely the same predictor variables. This suggests that the difference in prediction is primarily due to random forest's superior ability to reflect the complex non-linear relationships and multi-variate interactions in the American marten habitat relationships in northern Idaho.

9.5 Conclusion

Random forests is shown here to substantially outperform logistic regression in predicting patterns of marten occurrence. This suggests, consistent with other research, that random forests may generally be a superior approach when the goal is obtaining high predictive power. It should be by now the starting platform for any analysis of this sort. The random forests model produced an ecological understanding that was generally similar to that provided by the logistic regression model, but with some additional detail and clarity regarding variables and scales of influence. However, given the much higher predictive success, applications of the random forests model for mapping habitat quality and assessing the extent and pattern of habitat is likely to produce much more accurate and useful information.

References

Baccini A, Goetz SJ, Walker WS et al (2012) Estimated carbon dioxide emissions from tropical deforestation improved by carbon-density maps. Nat Clim Chang 2(3):182–185

Blaszczynski JS (1997) Landform characterization with geographic information systems. Photogramm Eng Remote Sens 63(2):183–191

Breiman L (2001a) Random Forests. Mach Learn 45(1):5–32

Breiman L (2001b) Statistical modeling: the two cultures (with comments and a rejoinder by the author). Stat Sci 16:199–231

Buskirk SW, Ruggiero LF (1994) The American marten. In: Ruggiero LF, Aubry KB, Buskirk SW, Lyon LJ, Zielinski WJ (eds) American marten, fisher, lynx, and wolverine in the western United States. Gen. Tech. Rep. RM-254. U.S. Department of Agriculture, Forest Service, Rocky Mountain Forest and Range Experiment Station, Fort Collins

Chambers CL, Cushman SA, Medina-Fitoria A, Martinez-Fonesca J (2016) Influences of scale on bat habitat relationships in a forested landscape in Nicaragua. Landsc Ecol 31:1299–1318

Chapin TG, Hamson DJ, Katnik DD (1998) In audience FH Is that the correct title? of landscape pattern on habitat use by American marten in an industrial forest. Conserv Biol 12:1327–1337

Cohen J (1968) Weighted kappa: Nominal scale agreement provision for scaled disagreement or partial credit. Psychol Bull 70(4):213–220

Crookston NL, Finley AO (2008) yaImpute: An r package for knn imputation. J Stat Softw 23(10):1–16

Cushman SA, Gutzwiller K, Evans JS, McGarigal K (2010) The gradient paradigm: a conceptual and analytical framework for landscape ecology. In: Cushman SA, Huettman F (eds) Spatial complexity, informatics, and wildlife conservation. Springer, Tokyo, pp 83–108

Cushman SA, Macdonald EA, Landguth EL, Halhi Y, Macdonald DW (2017) Multiple-scale prediction of forest-loss risk across Borneo. Landsc Ecol 32:1581–1598

Cushman SA, Raphael MG, Ruggiero LF, Shirk AJ, Wasserman TN, O'Doherty EC (2011) Limiting factors and landscape connectivity: The American marten in the rocky mountains. Landsc Ecol 26:1137–1149

Cushman SA, Shirk AJ, Landguth EL (2013) Landscape genetics and limiting factors. Conserv Genet 14:263–274

Cutler DR, Edwards TC, Beard KH et al (2007) Random forests for classification ecology. Ecology 88(11):2783–2792

De'ath G, Fabricius KE (2000) Classification and Regression Trees: A powerful yet simple technique for ecological data analysis. Ecology 81(11):3178–3192

Drew CA, Wiersma YF, Huettmann F (eds) (2010) Predictive species and habitat modeling in landscape ecology: concepts and applications. Springer Science & Business Media, New York

Evans JS, Cushman SA (2009) Gradient modeling of conifer species using random forests. Landsc Ecol 24(5):673–683

Evans JS, Murphy MA, Holden ZA, Cushman SA (2011) Modeling species distribution and change using random forest. In: Drew CA (ed) Predictive species and habitat modeling in landscape ecology: concepts and applications. Springer, New York

Evans JS, Oakleaf J (2012) Geomorphometry & Gradient Metrics Toolbox (ArcGIS 10.0)

Evans JS, Oakleaf J, Cushman SA, Theobald DM (2014) An ArcGIS toolbox for surface gradient and geomorphometric modeling, version 2.0-0. Accessed:2015 Dec 2nd. http://evansmurphy.wix.com/evansspatial

Fuller AK, Harrison DJ (2005) Influence of partial timber harvesting on American martens in north-central Maine. J Wildl Manag 69:710–722

Godbout G, Ouellet JP (2008) Habitat selection of American marten in a logged landscape at the southern fringe of the boreal forest. Ecoscience 15:332–342

Grand J, Buonaccorsi J, Cushman SA, Griffin CR, Neel MC (2004) A multiscale landscape approach to predicting bird and moth rarity hotspots in a threatened pitch pine–scrub oak community. Conserv Biol 18(4):1063–1077

Grimm R, Behrens T, Märker M, Elsenbeer H (2008) Soil organic carbon concentrations and stocks on Barro Colorado Island—digital soil mapping using random forests analysis. Geoderma 146(1):102–113

Hargis CD, Bissonette JA, Turner DL (1999) The influence of forest fragmentation and landscape pattern on American martens. J Appl Ecol 36:157–172

Hargis CD, McCullough DR (1984) Winter diet and habitat selection of marten in Yosemite National Park. J Wildl Manag 48:140–146

Hegel TM, Cushman SA, Evans J, Huettmann F (2010) Current state of the art for statistical modelling of species distributions. In: Cushman SA, Huettman F (eds) Spatial complexity, informatics and wildlife conservation. Springer, Tokyo, pp 273–312

Liaw A, Wiener M (2002) Classification and regression by random. Forest R news 2(3):18–22

McGarigal K, Cushman SA, Ene E (2012) FRAGSTATS v4: Spatial Pattern Analysis Program for Categorical and Continuous Maps. Computer software program produced by the authors at the University of Massachusetts, Amherst. Available at the following web site: http://www.umass.edu/landeco/research/fragstats/fragstats.html

McGarigal K, Wan HY, Zeller KA, Timm BC, Cushman SA (2016) Multi-scale habitat modeling: A review and outlook. Landsc Ecol 31:1161–1175

Mi C, Huettmann F, Guo Y, Han X, Wen L (2017) Why to choose Random Forest to predict rare species distribution with few samples in large undersampled areas? Three Asian crane species models provide supporting evidence. Peerj 5:e2849

Murphy MA, Evans JS, Storfer A (2010) Quantifying *Bufo boreas* connectivity in Yellowstone National Park with landscape genetics. Ecology 91(1):252–261

Pontius RG Jr, Milones M (2011) Death to Kappa: Birth of quality disagreement and allocation disagreement for accuracy assessment. Int J Remote Sens 32:4407–4429

Pontius RG Jr, Parmentier B (2014) Recommendations for using the relative operating characteristic (ROC). Landsc Ecol 29:367–382

Pontius RG Jr, Si K (2014) The total operating characteristic to measure diagnostic ability for multiple thresholds. Int J Geogr Inf Sci 28:570–583

Pontius RG Jr, Walker R, Yao-Kumah R, Arima E, Aldrich S, Caldas M, Vergara D (2014) Accuracy assessment for a simulation model of Amazonian deforestation. Ann Assoc Am Geogr 97:677–695

R Development Core Team (2008) R: a language and environment for statistical computing. R Foundation for Statistical Computing, Vienna

Robinson L, Cushman SA, Lucid M (2017) Winter bait stations as a multi-species survey tool. Ecol Evol 7:6826–6838

Rodriguez-Galiano VF, Ghimire B, Rogan J, Chica-Olmo M, Rigol-Sanchez JP (2012) An assessment of the effectiveness of a random forest classifier for land-cover classification. ISPRS J Photogramm Remote Sens 67:93–104

Samuel A. Cushman, Nicholas B. Elliot, Dominik Bauer, Kristina Kesch, Laila Bahaa-el-din, Helen Bothwell, Michael Flyman, Godfrey Mtare, David W. Macdonald, Andrew J. Loveridge (2018). Prioritizing core areas, corridors and conflict hotspots for lion conservation in southern Africa. July 5, https://doi.org/10.1371/journal.pone.0196213

Schneider A (2012) Monitoring land cover change in urban and peri-urban areas using dense time stacks of Landsat satellite data and a data mining approach. Remote Sens Environ 124:689–704

Shirk AS, Raphael MG, Cushman SA (2014) Spatiotemporal variation in resource selection: Insights from the American Marten (Martes americana). Ecol Appl 24:1434–1444

Svetnik V, Liaw A, Tong C, Wang T (2004) Application of breiman's random forest to modeling structure-activity relationships of pharmaceutical molecules. In: Roli F, Kittler J, Windeatt T (eds) Multiple classifier systems, lecture notes in computer science. Springer, Berlin/Heidelberg, pp 334–343

Thompson CM, McGarigal K (2002) The influence of research scale on bald eagle habitat selection along the lower Hudson River, New York. Landsc Ecol 17:569–586

Tomson SD (1999) Ecology and summer/fall habitat selection of American marten in northern Idaho. University of Montana. Thesis, Missoula, p 80

Wasserman TN, Cushman SA, Schwartz MK, Wallin DO (2010) Spatial scaling and multimodel inference in landscape genetics: *Martes americana* in northern Idaho. Landsc Ecol 25:1601–1612

Wasserman TN, Cushman SA, Wallin DO, Hayden J (2012a) Multi scale habitat relationships of *Martes americana* in northern Idaho, USA. Research Paper RMRSRP-94. USDA Forest Service, Rocky Mountain Forest and Range Experimental Station, Fort Collins

Wasserman TN, Cushman SA, Shirk AS, Landugth EL, Littell JS (2012b) Simulating the effects of climate change on population connectivity of American marten (Martes americana) in the northern Rocky Mountains, USA. Landsc Ecol. https://doi.org/10.1007/s10980-011-9653-8

Wiens JA (1989) Spatial scaling in ecology. Funct Ecol 3(4):385–397

Wilbert CJ, Buskirk SW, Gerow KG (2000) Effects of weather and snow on habitat selection by American martens (*Martes americana*). Can J Zool 78:1691–1696

Wynne KM, Sherburne JA (1984) Summer home range use by adult marten in northwestern Maine. Can J Zool 62:941–943

Chapter 10
Using Interactions among Species, Landscapes, and Climate to Inform Ecological Niche Models: A Case Study of American Marten (*Martes americana*) Distribution in Alaska

Andrew P. Baltensperger

10.1 Introduction

Machine learning algorithms are powerful analytical tools whose high predictive accuracy stems in part from their ability to predict non-linearly, handle missing data, and incorporate interactions among all predictors in a model (De'ath and Fabricius 2000; Breiman 2001a; Hastie et al. 2001; Cutler et al. 2007; Elith et al. 2008). These flexibilities provide algorithms such as boosted decision trees and stochastic gradient boosting analyses (e.g., TreeNet; Salford Systems Inc., San Diego, CA; www.salford-systems.com) with the ability to classify imperfect datasets and derive highly accurate predictions that can be applied across broad landscapes (Prasad et al. 2006; Cutler et al. 2007; Elith et al. 2008; Evans et al. 2011). Here, I use TreeNet to analyze the effects of interaction terms and variable combinations on the predictive performance of spatial models of American marten (*Martes americana*) distribution in Alaska.

10.1.1 Stochastic Gradient Boosting

There are a many types of machine learning algorithms which operate to iteratively develop non-linear predictive models from training data. One such algorithm, stochastic gradient boosting (implemented in TreeNet), acts by first developing a single decision tree that aims to estimate the main effect by accounting for the largest proportion of variance in a system (Hastie et al. 2001; Friedman 2002).

A. P. Baltensperger (✉)
National Park Service, Fairbanks, AK, USA
e-mail: abaltensperger@nps.gov

© Springer Nature Switzerland AG 2018   205
G. R. W. Humphries et al. (eds.), *Machine Learning for Ecology and Sustainable Natural Resource Management*, https://doi.org/10.1007/978-3-319-96978-7_10

In successive, iterative steps, additional trees are developed to explain residual error remaining left over from previous trees (Hastie et al. 2001; Elith et al. 2008). This ensemble-modeling process, which combines the error-reduction benefits of multiple tree-based models to reduce total error, is known as "boosting" and is fundamentally different from traditional frequentist statistical approaches that fit a single parsimonious model to data (Elith et al. 2008). Boosting is a method for improving model accuracy that relies on averaging the contribution from many satisfactory but imperfect models in a successive fashion rather than attempting to find a single model that best approximates the system in question (De'ath and Fabricius 2000; Hastie et al. 2001; Elith et al. 2008).

Whereas boosted decision tree analyses are not appropriate for quantifying system parameters among often collinear variables, they do serve as exhaustive exploratory tools for developing accurate and robust predictive models and identifying important contributing predictors. They allow the user to include many predictors, even when collinear, in models without penalty. Additionally, because machine learning analyses are not designed to test *a priori* hypotheses, datasets must not conform to assumptions of normality and so do not require prior data transformations. They are capable of quantifying complex, non-linear relationships simultaneously among categorical and continuous predictors, and are insensitive to outliers and missing data (Friedman 2002; Elith et al. 2008). Most importantly for the analyses here, stochastic gradient boosting algorithms automatically include and can quantify interactions among all predictors in a model (Breiman 2001b; Hastie et al. 2001; Elith et al. 2008).

10.1.2   Variable Interactions

Interactions occur when the effect of a variable on the response is magnified by the presence of a second variable, such that the effect synergistically alters or reverses the direction of either variable alone (Cox 1984; Friedman 2002; Ai and Norton 2003). More complex interactions among three or more predictors are also possible, though such high-order interactions are more difficult to conceptualize. In traditional frequentist statistical modeling approaches such as generalized linear regression (GLM), generalized additive models (GAM), or resource selection functions (RSF), the user decides *a priori* which interaction terms to include, where only a small subset of interactions can reasonably be incorporated in any model (Fielding 1999; Breiman 2001b; Hastie et al. 2001; Burnham and Anderson 2002; Johnson et al. 2004). Top parsimonious models identified using model-selection methods such as AIC (Akaike Information Criteria) are penalized for including an excessive number of variables or interaction terms because these do not represent parsimonious models and collinear variables make it difficult to accurately quantify model parameters (Burnham and Anderson 2002).

In contrast, because accurate predictions are based on explaining the highest proportion of system variance, machine-learning analyses often seek to incorporate as many contributions, however small, into the model, helping to increase prediction accuracy. Many machine learning models utilize dozens if not hundreds of predictors (e.g., Magness et al. 2008; Buchen and Wohlrabe 2010; Baltensperger and Huettmann 2015), translating into trillions of interactions. Because of the magnitude of interactions that occur in a multi-variate model, such complexity can only be handled using machine-learning techniques (Breiman 2001b; Hastie et al. 2001; Friedman 2002). For example, a linear model with n variables has 2^n terms including: a constant, each variable, and all of the interactions. From this, one can deduce the number of interaction terms to be: $2^n - n - 1$, because variables cannot interact with themselves (Friedman 2002). So for a relatively simple model with just 3 variables, there are 4 interaction terms. With 4 variables, this grows to 11 interaction terms, and in the study case here, 53 variables results in 9×10^{15} interaction terms! No linear model could incorporate this type of complexity (Hastie et al. 2001; Burnham and Anderson 2002).

The fact that such linear modeling approaches and our own comprehension breaks down with increasing interaction complexity is known as the "curse of dimensionality" (Bellman 1961). In omitting these higher order interaction terms from a model, some unknown portion of the explained variance is lost, and consequently the potential predictive accuracy of the model suffers (Hastie et al. 2001). However, it remains unknown the degree to which the overall predictive accuracies of machine-learning models can be improved by including all of the interactions among dozens of predictors. Here I examine the effects of different variable combinations and the strengths of interactions in determining the accuracies of stochastic gradient boosting models. Interaction indices and relative predictor importance values are also visualized in network graphs to demonstrate the complexities and relative strengths of multi-variate interactions in a modeled system.

10.1.3 Ecological Niche Models

From a landscape ecology perspective, analyses that can build interactions into species distribution models (SDMs) or ecological niche models (ENMs) are highly desirable, in order to make accurate predictions of niche space, not based exclusively on environmental predictors (Travis et al. 2005; Araujo and Guissan 2006). ENMs are, in fact, often criticized for failing to incorporate biotic interactions in their analyses (Austin 2002; Thuiller et al. 2005; Dormann 2007). It is theorized that in addition to environmental factors, competition from other species, symbiotic interactions among species, as well as the availability of prey and the prevalence of predators may also constrain where a species is able to live (Armstrong and McGehee 1980; Travis et al. 2005). It is true that most ENMs do not include variables that account for ecological interactions that may occur among species. This shortcoming, is in part due to a lack of spatial ecological data, but also to the inability of traditional analyses to incorporate numerous predictors and their interactions. Here I use boosted regression trees to address these modeling gaps by constructing a statewide ENM for American marten based on 53 interacting environmental and ecological predictors, including individual ENMs for 17 small mammal prey species as predictors. This example also provides the opportunity to examine the influence of prey on the distribution of a meso-carnivore predator and the effects of variable selection and interaction strength on the accuracy of ENM models. Additionally, I project a single best contemporary data model onto expected future environmental conditions in order to predict changes in marten distribution across Alaska by the year 2100. Such analyses complement existing research predicting changes in marten distribution near the northern limits of their range on the Kenai Peninsula, Alaska (Baltensperger et al. 2017) and near the southern limits in the Rocky (Wasserman et al. 2013, Chap. 9) and Appalachian Mountains (Carroll 2007).

10.2 Methods

10.2.1 Training Dataset

I downloaded 5990 occurrence records of marten from the Global Biodiversity Information Facility (GBIF; www.gbif.org). This collection of records is both free and publically available, but unfortunately represents only a portion of the known occurrences of marten in Alaska. While the Alaska Department of Fish & Game (2015) (ADFG) requires that harvested marten pelts be "sealed" (officially recorded), they do no not require trappers to report detailed locational information. Instead of identifying harvest locations with geographic coordinates, sealed marten are identified only to the minor drainage unit or uniform coding unit (UCU) from which they were harvested (e.g., Baltensperger et al. 2017). Because of the imprecise nature of these records, none of the state's thousands of marten sealing records were contributed to GBIF or exist in any public georeferenced database, and as such were unavailable for this study. These types of records should be made public in order to provide a more complete picture of marten distribution in the state. The size of the training dataset was further limited because it also contained numerous duplicate and imprecise records. These spurious records were filtered out, resulting in 774 unique locations with at least 100 m accuracy that could be used as training presence data in the models. Because this dataset was comprised of presence-only records, I also constructed a randomly-distributed set of 775 locations which served as a pseudo-absence dataset. The remaining presence and pseudo-absence data points were then attributed with 36 spatial environmental variables and the distributions of 17 potential small mammal prey species (Table 10.1; Baltensperger and Huettmann 2015), and served as the training dataset for the models.

10.2.2 Model Iterations

I used the stochastic gradient boosting algorithm, TreeNet, to develop 12 different sub-models using various combinations of predictors in order to assess the influence of predictors and their interactions in the performance of the overall model. I used

Table 10.1 Complete model predictor set, including each predictors type (continuous or categorical), whether it changes over time (dynamic or static) and if yes, is there future data to include to represent a predictor's dynamic nature

Variable	Type	Temporal	Future Data
Active layer thickness	Continuous	Dynamic	Y
Anadromous stream distance	Continuous	Static	
Aspect	Continuous	Static	
Cliome	Categorical	Dynamic	Y
Coast distance	Continuous	Static	
Dicrostonyx groenlandicus	Continuous	Dynamic	Y
Elevation	Continuous	Static	
Fall precipitation	Continuous	Dynamic	Y
Fall snow day fraction	Continuous	Dynamic	Y
Fall temperature	Continuous	Dynamic	Y
Fire history	Categorical	Dynamic	N
Freeze date	Continuous	Dynamic	Y
Geology	Categorical	Static	
Glacier distance	Continuous	Dynamic	N
Ground temperature	Continuous	Dynamic	Y
Growing season	Continuous	Dynamic	Y
Lake distance	Continuous	Static	
Lemmus trimucronatus	Continuous	Dynamic	Y
Max (March) sea ice extent distance	Continuous	Dynamic	Y
Microtus longicaudus	Continuous	Dynamic	Y
Microtus miurus	Continuous	Dynamic	Y
Microtus oeconomus	Continuous	Dynamic	Y
Microtus pennsylvanicus	Continuous	Dynamic	Y
Microtus xanthognathus	Continuous	Dynamic	Y
Min (September) sea ice extent distance	Continuous	Dynamic	Y
Myodes rutilus	Continuous	Dynamic	Y
NDVI	Continuous	Dynamic	N
NLCD Landcover	Categorical	Dynamic	N
Slope	Continuous	Static	
Soils	Categorical	Static	
Sorex borealis	Continuous	Dynamic	Y
Sorex cinereus	Continuous	Dynamic	Y
Sorex hoyi	Continuous	Dynamic	Y
Sorex minutisimus	Continuous	Dynamic	Y
Sorex monticolus	Continuous	Dynamic	Y
Sorex palustris	Continuous	Dynamic	Y
Sorex tundrensis	Continuous	Dynamic	Y
Sorex ugyunak	Continuous	Dynamic	Y
Spring precipitation	Continuous	Dynamic	Y
Spring snow day fraction	Continuous	Dynamic	Y
Spring temperature	Continuous	Dynamic	Y
Stream distance	Continuous	Static	

(continued)

Table 10.1 (continued)

Variable	Type	Temporal	Future Data
Summer precipitation	Continuous	Dynamic	Y
Summer snow day fraction	Continuous	Dynamic	
Summer temperature	Continuous	Dynamic	Y
Terrain	Continuous	Static	
Thaw date	Continuous	Dynamic	Y
Village distance	Continuous	Static	
Wetland distance	Continuous	Static	
Winter precipitation	Continuous	Dynamic	Y
Winter snow day fraction	Continuous	Dynamic	Y
Winter temperature	Continuous	Dynamic	Y
Zapus hudsonicus	Continuous	Dynamic	Y

TreeNet, despite RandomForests' generally superior predictive accuracy, because of its unique capability to quantify interactions and develop partial dependence plots for each variable (Hastie et al. 2001, www.salford-systems.com/products/treenet). For each sub-model, I grew 500 trees and systematically varied both the maximum number of nodes per tree and the number of minimum training cases per terminal node in order to obtain the most accurate models. Predictive accuracies for each sub-model were compared using the area under the receiver-operator curve (AUC ROC) and the Overall Balanced Misclassification Rate (OBMR), where the threshold differentiating presences from absences was set to the value which maximized the sum of sensitivity (correctly predicted absences) and specificity (correctly predicted presences; Manel et al. 2001; Jimenez-Valverde and Lobo 2007).

Before modeling, relationships among predictors were analyzed using a Spearman correlation varclus analysis (F. Harrell; https://github.com/harrelfe/Hmisc) in R 2.12, providing a means of visualizing clusters of predictors with similarly correlated spatial distributions. One should bear in mind that the collinear nature of many of these predictors is less of an issue in machine learning analyses, than it would be in a frequentist statistical analysis. Collinear predictors in machine learning models actually serve to reinforce tree splits, making models more robust. It is also important to remember that for a predictive model, unimportant variables are simply ignored, and so their inclusion does not harm overall model performance. The cluster analysis simply helped to identify predictors, whose contributions in the model were likely to be similar to one another.

10.2.3 Interaction Network Graphs

In order to quantify interaction strengths among predictors, TreeNet reported interaction indices for all predictors in the full model and the most accurate sub-model. This produced two metrics to evaluate predictor interactions for the two models: (1)

interaction strengths for each predictor with each of the other predictors, (2) interaction strengths for each variable aggregated across all other variables. Interaction strengths among variable pairs were graphed using Gephi (www.gephi.org), an open-source network analysis software, which helped to visualize connectivity, interaction strengths, and patterns among predictors in the models. Interaction indices between variable pairs (represented by line thickness), and variable importance (represented by node size) were input into network models for the full and continuous predictive models. I also created 2-variable joint partial-dependence plots for the most important interactions in the top sub-model to assess the directionality, trends, and thresholds in the interaction responses of predictors.

10.2.4   Landscape Predictions

After completing modeling iterations, I projected the most accurate versions of each of the 12 sub-models onto a lattice of 58,978 regularly-spaced (5 km) points that were attributed with the same set of variables. This resulted in several representations of marten distributions across Alaska for 2010 based on the different predictor sets. In order to select the best overall model, I then compared AUC and OBMR rates among sub-models and selected the one sub-model with the highest accuracy. This was used to predict the distribution of marten in the year 2100 using projected environmental conditions for that time frame. To do this, the most-accurate model was scored onto a 5-km lattice that was attributed with the future projections of dynamic predictors (static variables remained the same), namely downscaled values from the IPCC (International Panel on Climate Change) A2 climate projections.

Model predictions are represented by relative index of occurrence (RIO) values for each 5 km pixel across Alaska. These are not true probabilities of occurrence but do provide a metric to evaluate the relative prospect that marten would be found in one location versus another with a different RIO value. The top continuous models (2010 and 2100) were then converted to binary (presence/absence) models based on the balanced thresholds for each model. The 2010 model was then subtracted from the 2100 model using the ArcGIS 10.3 raster calculator in order to identify regions of species persistence, gain, and loss.

10.3   Results

10.3.1   Varclus Analysis

The varclus analysis using all 53 predictors resulted in the classification of 4 main groups of collinear variables ($\rho > 0.0$; Fig. 10.1). Most climatic variables (with the exception of Snow Day Fraction and Thaw Date) formed one cluster, topographic and physical features (except Coast Distance) comprised a second, small mammals and

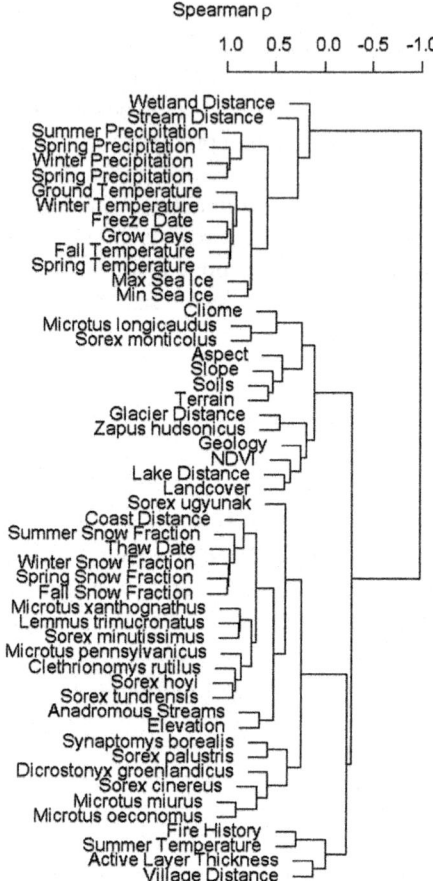

Fig. 10.1 Spearman correlation varclus analysis of full complement of variables. Groups of variables with root nodes greater than 0.0 are positively correlated, whereas those with root nodes less than 0.0 are negatively correlated

winter variables a third, and Fire History, Summer Temperature, Active Layer Thickness and Village Distance formed a small fourth cluster. The first cluster was highly negatively correlated ($\rho \cong -1$) with the other three, whereas each of the remaining three clusters were somewhat negatively correlated ($0 > \rho > -0.25$) with one another.

10.3.2   Full Model

Constructing a model with the full complement of 53 predictors (AUC = 94.7%; OBMR = 11.4%), I identified the top three variables (Soils, Geology, Landcover), whose variable importance scores exceeded 20.0 (Table 10.2). The order of these predictors remained largely consistent while model parameters (max nodes/tree and minimum cases/terminal node) were varied. Interaction strengths between each of

Table 10.2 Variable importance and interaction strength scores for the top 10 most important variables in the full model

Variable	Variable Importance		Interaction Strength	
	Continuous	Full	Continuous	Full
Soils		100.0		32.1
Geology		39.4		10.2
NLCD Landcover		30.3		7.7
Lemmus trimucronatus	100.0	11.1	16.0	2.2
Max (March) sea ice extent distance	72.3	11.5	19.5	2.5
Thaw date	41.1	12.1	3.8	2.6
Village distance	40.2	20.2	10.9	9.8
Min (September) sea ice extent distance	36.4	20.4	9.7	7.5
Cliome		17.3		2.9
Summer precipitation	35.9	16.4	5.4	4.3
Anadromous stream distance	31.5	15.7	7.4	3.6
Dicrostonyx groenlandicus	29.3	13.6	5.5	4.4
Spring snow day fraction	29.2	8.3	2.4	1.5
Sorex cinereus	29.2	10.1	4.6	1.7

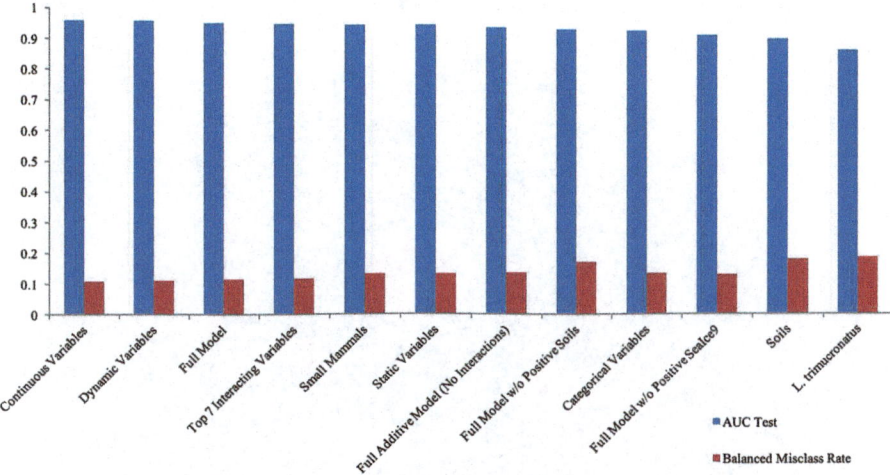

Fig. 10.2 Area under the receiver operator curve (AUC ROC) values and overall balanced misclassification/error rates (OBMR) for 12 analyzed sub-model variations

these and all other variables in the model were also related to their importance (Table 10.2). For example, the three most important predictors in the model also comprised three of the top four most interactive predictors in the model (Fig. 10.2).

The network graph for the full model provided a visualization of interaction complexity, strength, and variable importance among predictors (Fig. 10.3a). Soils was not only the most important predictor used in construction of the full model, but it also interacted most strongly with a small set of secondarily important variables including Village Distance (8.6), Distance to Minimum Sea Ice (7.4), Geology (6.0), and NLCD Landcover (5.3; Fig. 10.3a).

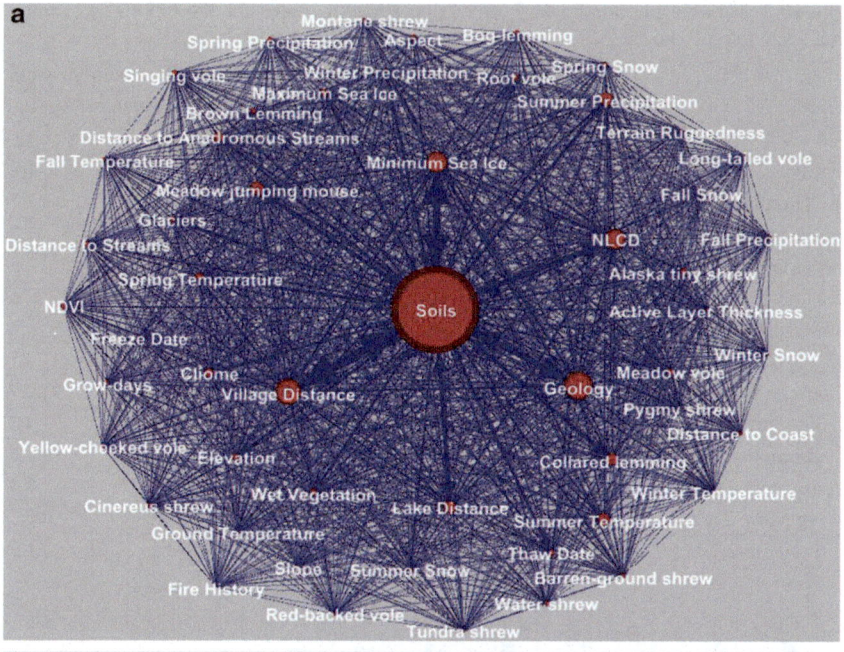

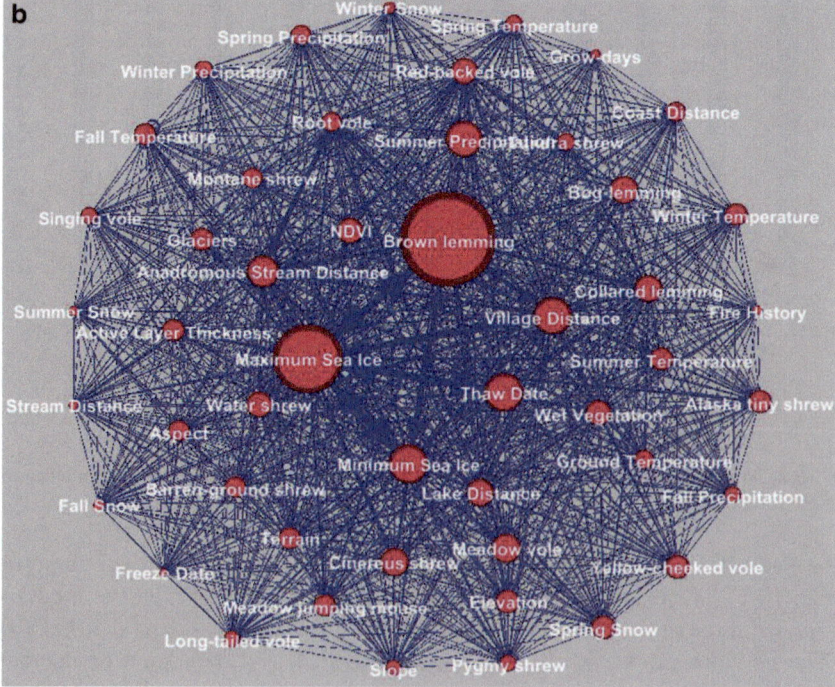

Fig. 10.3 Interaction networks among all interactions in the a) full model, and b) continuous submodel, where node size indicates relative variable importance and line thickness indicates relative interaction strengths

10.3.3 Top Sub–Models

Based on the rankings of predictors in the full model, I experimentally tested different combinations of predictors to determine whether a subset of predictors could
produce sub-models with similar or greater accuracy than the full model. I found
two predictor subsets that outperformed the full model using both AUC and OBMR
as evaluation metrics. The "Continuous" predictor sub-model, wherein I excluded
all categorical predictors (including the top three variables from the full model),
yielded the highest AUC value (95.7%) and an OBMR of 10.9% (Fig. 10.2). A second sub-model, the "Dynamic" predictor sub-model, wherein I excluded all static
variables (including many of the categorical variables as well), also outperformed
the full model, yielding an AUC of 95.5% and OBMR of 11.1% (Fig. 10.2).

After removing all categorical predictors, both interaction strengths and variable
importance scores were more evenly distributed across the Continuous sub-model
predictor set (Fig. 10.3b). In contrast to the full model, variable importance scores
for the top sub-model yielded 24 predictors with importance scores that exceeded
20.0, including the top predictor, *L. trimucronatus* (100.0), Maximum Sea Ice
(72.3), Thaw Date (41.1), and Distance to Village (40.2; Table 10.2). The relationship among interaction strength and predictor importance in the Continuous sub-
model was similar to the full model in that 3 of the 4 most important predictors were
also those with the largest interaction strengths (Table 10.2). The exception was the
third most important predictor, Thaw Date, who's whole variable interaction
strength (3.8) ranked 16th among the 46 predictors in the Continuous sub-model.
Among 2-way interactions, Distance to Maximum Sea Ice interacted most strongly
with *L. trimucronatus* (7.9), Minimum Sea Ice (7.9), and Village Distance, whereas
L. trimucronatus also interacted strongly Minimum Sea Ice (5.6) and Village
Distance (4.1; Table 10.2, Fig. 10.3b). Two-variable partial-dependence plots for the
top interacting variables in the Continuous sub-model illustrated the synergistic
relationships among predictors and their joint influence in the model (Fig. 10.4). For
example, martens were predicted to occur with increasing likelihood in regions far
from the Maximum and Minimum Sea Ice extents, especially in areas outside the
predicted distribution of *L. trimucronatus* and in close proximity to villages
(Fig. 10.4).

10.3.4 Other Sub–Models

Other sub-models were less accurate than the full model (Fig. 10.2). Even without
contributions from any environmental predictors, the Small Mammal sub-model,
using only the distributions of 17 small mammal prey species, outperformed 7 of the
10 other sub-models and was still only slightly less accurate (AUC = 94.0%,
OBMR = 13.0%) than the full model or the Continuous or Dynamic sub-models. To
determine the influence that predictor interactions have on model accuracy, I also

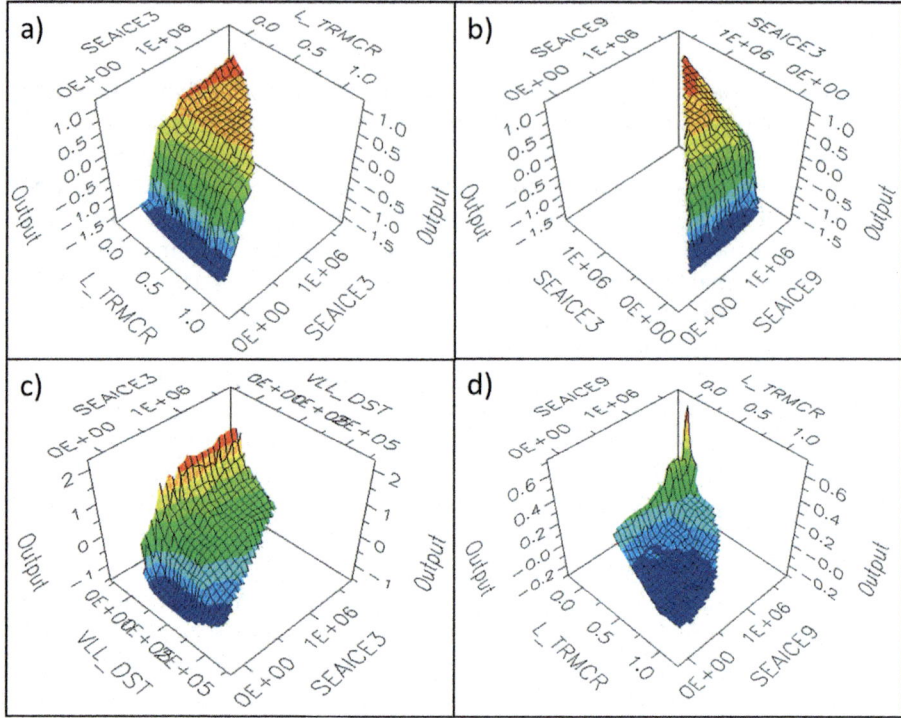

Fig. 10.4 Two-variable dependence plots for interactions between a) Maximum sea ice and the distribution of L. trimucronatus, and b) Maximum sea ice and distance to villages

forced TreeNet to exclude interactions in the modeling process. This resulted in an "Additive" sub-model (AUC = 93.0% and OBMR = 13.5%), which was the seventh best sub-model out of 12 (Figs. 10.2 and 10.5). Including interactions in the Continuous sub-model amounted to an increase in AUC of 1.7% and OBMR of 2.0% over the Additive sub-model. I also built two sub-models that removed those values of two predictors that were positively correlated with marten presence. This was done for both the top predicting categorical variable, Soils, and the top continuous variable, Maximum Sea Ice. In both cases, sub-models performed poorly compared to the full model (Fig. 10.2). For additional comparison and to test the predictive accuracy of sub-models based on single predictors, I built models using just the top predictor, Soils, and a second sub-model using only the top small mammal predictor, *L. trimucronatus*. Both of these sub-models performed markedly worse than the full model and the small mammal sub-model, respectively. The Soils Only sub-model scored 5.26% and 5.46% less than the respective AUC and OBMR of the full model. The *L. trimucronatus* sub-model also had an AUC score that was 8.36% lower and an OBMR that was 5.49% lower than the small mammal sub-model (Fig. 10.2).

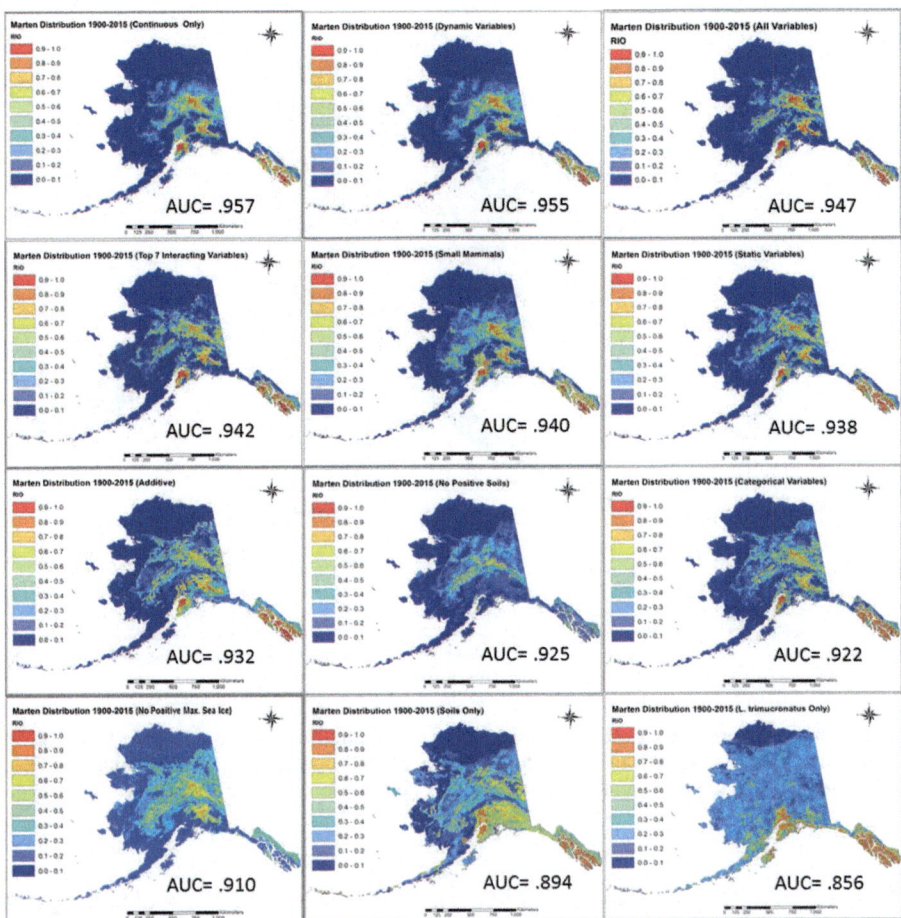

Fig. 10.5 Spatial representations of the 12 tested sub-models depicting predicted marten distributions across Alaska for 2010. AUC scores are also noted for each model

10.3.5   Spatial Models

Spatial depictions of the 12 sub-models demonstrated a range of predicted marten distributions, including many with similar range extents (Fig. 10.5). Most models were consistent in retaining predicted marten hot spots near occurrence clusters in the Yukon-Tanana Uplands, on the Kenai Peninsula, in the Copper River Basin, and in Southeast Alaska (Fig. 10.6). They also did not predict the presence of marten beyond the extent of latitudinal tree-line (Figs. 10.5 and 10.6). Whereas the top 7 models shared this pattern, the Soils Only, *L. trimucronatus* Only, No Positive Soils, and No Positive Maximum Sea Ice sub-models deviated noticeably. The No Positive Soils and No Positive Maximum Sea Ice maps did not predict marten to occur across the southern portions of the state. In contrast, the Soils Only and *L. trimucronatus*

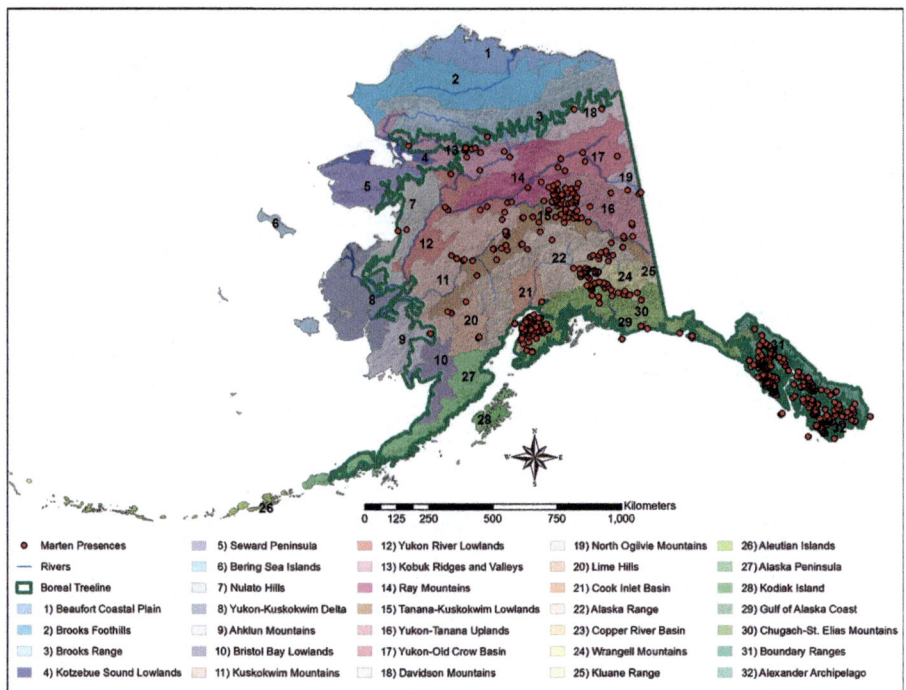

Fig. 10.6 Marten presences used as training data to build models. The green boundary depicts the boreal tree line (www.geobotany.uaf.edu), whereas shaded boundaries depict the major unified ecoregions of Alaska (http://agdc.usgs.gov/data/usgs/erosafo/ecoreg/index.html)

Only sub-models predicted marten to occur predominantly in the southern regions while failing to predict the occurrence of marten in the interior portion of the state, despite numerous training points indicating their detection throughout this region (Fig. 10.5).

Using the best performing sub-model (Continuous), I generated a map depicting the projected distribution of marten across Alaska in the years 2010 (Fig. 10.7a) and 2100 (Fig. 10.7b). Comparing the two models using a landscape change analysis, yielded predicted distribution gains and losses for marten in Alaska over the coming century (Fig. 10.7c). Much of the distribution did not undergo any change (129,063 km^2) during this time. These areas included the Kenai Peninsula, Southeast Alaska, much of the Copper River Basin, the Tanana-Kuskokwim Lowlands, and the Yukon-Tanana Uplands. However, losses to the distribution, amounting to 19,815 km^2, occurred around the periphery of the persistence regions in the Copper River and Tanana River basins. These losses were more than offset, however, by distribution gains of 187,459 km^2 that were projected to occur in the Cook Inlet Basin, Upper Kuskokwim Valley, Yukon-Old Crow Basin and other regions around Interior Alaska. Kodiak Island was also predicted to occur within the future ecologi-

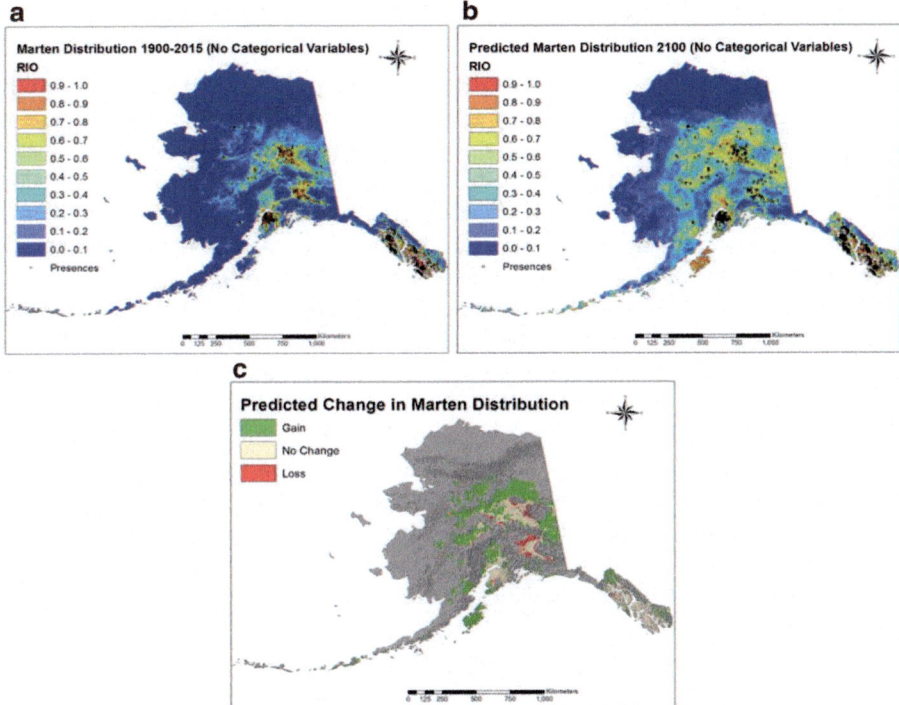

Fig. 10.7 Predicted current (**a**) and future (**b**) marten distributions using the best-performing sub-model (No Categorical Predictors) overlaid with marten occurrence records (open circles). The predicted distribution change map (**c**) depicts regions predicted to gain (green) or lose (red) marten, or where they are likely to persist (tan)

cal niche for marten, however the geographic isolation of the region makes this prediction less likely. Nevertheless, projected gains in marten distribution over the coming 85 years amounted to a net expansion of 167,644 km² across Alaska.

10.4   Discussion

From these modeling exercises, I demonstrated that (1) sub-models with less than the full complement of predictors are capable of outperforming models that include the full predictor set, (2) the top predictors and interaction strengths can be heavily influenced by certain "greedy" categorical predictors, and (3) marten distribution is predicted to expand across Alaska with climate change over the coming century. So what do these results indicate regarding the effects of interactions and optimal variable sets for most accurately predicting machine-learning models?

10.4.1 Parsimonious Versus Highly Interactive Models

In frequentist models, parsimonious variable sets are often desired since this avoids problems associated with collinearity among predictors. This approach allows for the quantification of individual variable effects and provides greater mechanistic inference (Burnham and Anderson 2002). However, smaller variable sets can also leave a large amount of unexplained variance in a model by failing to include the explanatory power of additional variables and the effects of n-dimensional interactions among many variables (Hastie et al. 2001; Friedman 2002).

In contrast, it is generally assumed that as the number of predictors in an optimized machine-learning model increases, the more accurate the model will be. This is a result of the increased predictive power of a larger variable set in explaining a higher proportion of the overall variance in a system. This should be true as long as each additional variable contributes even a small portion to the overall variance, and as such more variables should be included to improve the predictive accuracy of models (Hastie et al. 2001). However, the results here, as well as recommendations by Salford Systems, suggest that by removing the poorest-performing predictors and by treating high-level categorical (HLC) predictors (predictors with more than 20 categories) as continuous variables, the predictive accuracy of models can be improved (Steinberg 2011).

10.4.2 High Level Categorical (HLC) Predictors

In this case study, the accuracy of the marten distribution model improved slightly by removing categorical predictors and also by removing static variables. Using HLC predictors can be problematic for developing well-balanced and robust models. Because tree-based learning algorithms operate by splitting data into similar "nodes", predictors that are already organized into categories can be segregated using many more node combinations than would be possible using continuous variables (Steinberg 2011). The splitting power of categorical variables tends to disproportionately influence the model and only grows more problematic with more categories. A node can be split in as many as $2^{(K-1)} - 1$ ways, where K is the number of predictors in the model. For example, if K = 20 (maximum recommended), there are 524,287 ways to split the data at each node. However, when K = 253 (number of categories in the Soils predictor), a total of 7.24 x 10^{75} possible splits are possible! Not only does this slow down model processing, but the model is invariably driven disproportionally by a single predictor (Hastie et al. 2001; Friedman 2002; Steinberg 2011). When Soils and other HLCs such as Geology and NLCD are penalized, treated as continuous variables (TreeNet is adept at recognizing this condition), or removed from the model entirely, (such as was done here in the Continuous submode) the result is a higher-accuracy model that is more equally informed by splits using a diverse and robust range of predictors (Strobl et al. 2007; Steinberg 2011).

A similar improvement in model accuracy occurred when static variables were excluded from the "Dynamic" sub-model because many of the top static variables were also HLCs.

Additionally, because categories differ markedly from one another, interaction effects with categorical variables may be strong. A categorical variable's value may vary dramatically as raster cell neighborhoods cross spatial categorical boundaries. In contrast, the value of a continuous variable may vary only slightly among neighboring raster cells. The outsized change across categorical boundaries can lead to strong positive or negative interactive effects when combined with continuous predictors that vary more gradually across space (Strobl et al. 2007). Therefore, the categorical nature of Soils, and to a lesser extent of NLCD and Geology, resulted in strong interaction effects with other predictors.

10.4.3 Interaction Effects

Determining the contribution of interactions to model accuracy is somewhat more complicated. When building a model, the TreeNet algorithm first considers the "main effect," a subset of easily-categorized predictors (usually three or four predictors) at one time without incorporating their interactions (Hastie et al. 2001; Friedman 2002). A tree is constructed to describe the main effect. Next TreeNet considers the effects of 2-variable interactions, followed by higher order interactions (>2 variables) to boost the performance of the model by constructing trees that account for the remaining portions of unexplained variance (Friedman 2002). Thus the TreeNet algorithm is biased against strong, high-order interaction effects. The use of a subset of training records ("bagging") in the construction of each tree, allows for the creation of models that are similar in accuracy to the full model, even when the full predictor set has been manually reduced to a smaller subset (Fig. 10.2; Breiman 1996; Hastie et al. 2001; Prasad et al. 2006). In other words, as demonstrated here, a reasonably accurate model can be created using only lower-order interactions or only additive components (Hastie and Tibshirani 1990; Fig. 10.2). However, the accuracy is usually inferior to a full model that includes all of the interaction effects. This was demonstrated here by the lower performance of the additive model that was devoid of interactions (Fig. 10.2; Hastie et al. 2001; Friedman 2002). So while high-level interaction effects may contribute only marginally to the overall accuracy of models, they can improve accuracy enough to make their inclusion worthwhile. Additionally, by growing complex trees with more terminal nodes, more interactions may be allowed to occur, potentially increasing the explanatory power of the model.

Predictors without known biological relationships to the response can act to improve model performance via interactions with other predictors, illuminating potential new avenues of investigation. For example, the most important interactions in the final model highlighted the combined effects of sea ice extent with the distribution of *L. trimucronatus* in predicting where martens are likely to occur (Fig. 10.4a).

In this case, it was not the affiliation with either of these predictors, but rather inverse associations with both that were most influential in predicting marten distributions. Areas outside those heavily influenced by sea ice and where *L. trimucronatus* occurs, namely Arctic and coastal regions, were highly unlikely to contain martens (Fig. 10.4a). These results reaffirm the pervasive influence that sea ice has with coupled terrestrial systems (Deser et al. 2010) and demonstrates the magnifying effects that climate related changes can have on other ecological processes (Mantyka-Pringle et al. 2015). While sea ice loss occurs primarily in summer and fall, the largest effects are felt on land as warmer temperatures and increased precipitation in winter (Deser et al. 2010). While both effects would seem to favor the persistence of marten towards the northern extent of their range, the links between such correlated predictors begs further investigation. Associations with other small mammal distribution predictors were not highly interactive with other variables or especially predictive of marten occurrence, suggesting that marten are generalist predators that may be responding more to habitat conditions and prey abundance rather than associating themselves with any specific prey species. Conducting such an analysis using a more specialist predator and its potential prey items could yield stronger effects of prey on predator distributions.

10.4.4 Predicted Marten distribution in Alaska

The continuous spatial distribution sub-model for marten appropriately predicted marten to occur throughout the boreal forest biome of Alaska, and entirely within the latitudinal tree-line extent (Fig. 10.6). This is consistent with qualitative descriptions of marten as boreal forest generalists that require dense over-head cover, and adequate secondary forest structure in the form of coarse-woody debris (CWD) at ground level (Spencer 1987; Wiebe et al. 2014; ADF&G 2015). Other contemporary range and distribution estimates of marten using more general methods (MaxEnt, deductive habitat models, combined models, and minimum range polygons) show more expansive distributions that extend farther westward and northward (Gotthardt et al. 2013; IUCN 2016). The best model produced here exceeded the predictive accuracy of these other models using both AUC (+1.8%) and OBMR (+8.6–30.2%) for comparison, yet because of differences in test samples among the models, differences in accuracy may be obscured by statistical noise.

The contrasting extent of my models with other deductive methods of marten distribution and range in Alaska may stem from the lack of a comprehensive, public occurrence database which includes accurate locations of trapped and sealed marten in Alaska (Baltensperger et al. 2017). Because of such underreporting of marten occurrences and the limited accessibility of much of Alaska away from the road system to trappers, models were dominated by dense clusters of occurrence records in a handful of areas, without necessarily generalizing well to remote, unsampled regions, especially those far from human development. This is reflected by the high relative importance value of the Distance to Village predictor, which theoretically could indicate an affinity of marten for roads, but more realistically shows that a community/road related sampling bias exists (Table 10.2). Also, because I chose not to correct for autocorrelation among training samples, and because TreeNet used bagging to select testing samples from clustered occurrences, it is likely that the apparent accuracies of the models were somewhat inflated as compared to a truly independent testing set that would likely reflect lower actual accuracy.

Using this limited, but best-available training dataset, we can still make comparisons to future projections of marten distribution using the same methods. Predicted distributions were projected to expand considerably across Alaska during the coming decades (Fig. 10.7). This distribution expansion tracks the predicted growth in the Boreal Forest biome, and associated coastward and northward shifts in the boreal bioclimatic envelope (Murphy et al. 2010). While the direction of the predicted effect conforms to our knowledge of biome shifts in Alaska, the magnitude may be underrepresented because of the limited extent of training locations. However, the removal of HLC predictors, and the incorporation of interaction effects here provided improvements in producing the most accurate predictive model of marten distribution yet using publically available training data.

10.4.5 Conclusions and Suggested Practices

Having experimented with the predictive influence of a variety of predictor combinations and interactions in TreeNet, I can make four general recommendations for developing ENM models using TreeNet. (1) Include as many predictors as possible, as long as each can explain even a small portion of the overall system variance. Even collinear predictors or those without known biological associations to the response variable can increase model accuracy through interactive effects with other predictors. Relative importances of predictors can be surprising and may also ultimately prove to be highly informative, if only by highlighting correlative effects. (2) Treat HLC predictors as continuous, penalize them, or remove them entirely to avoid having a single greedy variable or limited variable set dominate a model. Predictive performance will likely improve, variable importance will be more informative, and models will be more robust by using a more diverse set of variables. (3) Incorporating the influence of other species into ENMs by including distributions of prey, predators, or competitors as predictors can be informative, but effects may be more dramatic when modeling specialist species. (4) When modeling complex ecological systems influenced by dozens of environmental predictors, machine learning approaches offer several advantages over frequentist approaches, namely that they can incorporate the interactive effects of all high level interactions among all (even collinear) predictors to achieve the most accurate predictive models.

References

Ai C, Norton EC (2003) Interaction terms in logit and probit models. Econ Lett 80:123–129

Alaska Department of Fish and Game. 2015. Alaska wildlife action plan. Juneau.

Armstrong RA, McGehee R (1980) Competitive-Exclusion. American Naturalist 115:151–170

Araujo MB, Guisan A (2006) Five (or so) challenges for species distribution modelling. J Biogeogr 33:1677–1688

Austin MP (2002) Spatial prediction of species distribution: an interface between ecological theory and statistical modelling. Ecol Model 157:101–118

Baltensperger AP, Huettmann F (2015) Predictive spatial niche and biodiversity hotspot models for small mammal communities in Alaska: Applying machine-learning to conservation planning. Landsc Ecol 30:681–697

Baltensperger AP, Morton J, Huettmann F (2017) Expansion of American marten (*Martes americana*) distribution in response to climate and landscape change on the Kenai Peninsula, Alaska. J Mammal 98:703–714

Bellman RE (1961) Adaptive control. Princeton University Press, Princeton

Breiman L (1996) Bagging predictors. Mach Learn 24:123–140

Breiman L (2001a) Random forests. Mach Learn 45:5–32

Breiman L (2001b) Statistical modeling: the two cultures. Stat Sci 16:199–231

Buchen T, Wohlrabe K (2010) Forecasting with many predictors - Is boosting a viable alternative. Munich Discussion Paper No. 2010–31, Ludwig-Maximilians-Universtät München, Munich, Germany, 1–7

Burnham K, Anderson D (2002) Model selection and multimodel inference: a practical information-theoretic approach. Springer, New York

Carroll C (2007) Interacting effects of climate change, landscape conversion, and harvest on carnivore populations at the range margin: Marten and Lynx in the northern Appalachians. Conserv Biol 21:1092–1104

Cutler DR, Edwards KH Jr, Cutler A., Hess K.T, Gibson J, Lawler J.J (2007) Random forests for classification in ecology. Ecology 88:2783–2792

Cox DR, Atkinson AC, Box GEP, Darroch JN, Spjotvoll E, Wahrendorf J (1984) Interaction. International Statistical Review 52:1–31

De'ath G, Fabricius KE (2000) Classification and regression trees: A powerful yet simple technique for ecological data analysis. Ecology 81:3178–3192

Deser C, Tomas R, Alexander M, Lawrence D (2010) The seasonal atmospheric response to projected Arctic sea ice loss in the late twenty-first century. J Clim 23:333–351

Dormann CF (2007) Promising the future? Global change projections of species distributions. Basic Appl Ecol 8:387–397

Elith J, Leathwick JR, Hastie T (2008) A working guide to boosted regression trees. J Anim Ecol 77:802–813

Evans JS, Murphy MA, Holden ZA, Cushman SA (2011) Modeling species distribution and change using Random Forest. In: Ashton Drew YFWC, Huettmann F (eds) Predictive species and habitat modeling in landscape ecology. Springer, Berlin, pp 139–160

Fielding AH (1999) An introduction to machine learning methods. In: Fielding AH (ed) Machine learning methods for ecological applications. Kluwer Academic Publishers, Norwell

Friedman JH (2002) Stochastic gradient boosting. Comput Stat Data Anal 38:367–378

Gotthardt T, et al (2013) Predicting the range and distribution of terrestrial vertebrate species in Alaska, The Alaska gap analysis project, University of Alaska

Hastie T, Tibshirani R (1990) Exploring the nature of covariate effects in the proportional hazards model. Biometrics 46:1005–1016

Hastie T, Tibshirani R, Friedman JH (2001) The elements of statistical learning: data mining, inference, and prediction. Springer, New York

IUCN (International Union for Conservation of Nature) 2016 *Martes americana* Version 2016–2. The IUCN Red List of Threatened Species

Jimenez-Valverde A, Lobo JM (2007) Threshold criteria for conversion of probability of species presence to either-or presence-absence. Acta Oecol Int J Ecol 31:361–369

Johnson CJ, Seip DR, Boyce MS (2004) A quantitative approach to conservation planning: using resource selection functions to map the distribution of mountain caribou at multiple spatial scales. J Appl Ecol 41:238–251

Magness DR, Huettmann F, Morton JM (2008) Using Random Forests to provide predicted species distribution maps as a metric for ecological inventory and monitoring. In: Smolinski TG, Milanova MG, Hassanien A-E (eds) Applications of computational intelligence in biology, Studies in computational intelligence, vol 122. Springer, Berlin/Heidelberg, pp 209–229

Manel S, Williams HC, Ormerod SJ (2001) Evaluating presence-absence models in ecology: the need to account for prevalence. J Appl Ecol 38:921–931

Mantyka-Pringle CS, Visconti P, Moreno DM, Martin TG, Rondinini C, Rhodes J (2015) Climate change modifies risk of global biodiversity loss due to land-cover change. Biol Conserv 187:103–111

Murphy K, Huettmann F, Fresco N, Morton J (2010) Connecting Alaska landscapes into the future: Results from an interagency climate modeling, land management and conservation project. Final Report, U.S. Fish and Wildlife Service, Anchorage, AK

Prasad AM, Iverson LR, Liaw A (2006) Newer classification and regression tree techniques: bagging and random forests for ecological prediction. Ecosystems 9:181–199

Spencer WD (1987) Seasonal rest-site preferences of pine martens in the northern Sierra Nevada. J Wildl Manag 51:616–621

Steinberg D (2011) Dan Steinberg's Blog [Internet]. San Diego: Salford Systems, Inc. Accessed 10 Oct 2016. Available from: https://www.salford-systems.com/blog/dan-steinberg/modeling-tricks-with-treenet-treating-categorical-variables-as-continuous

Strobl C, Boulesteix A-L, Zeileis A, Hothorn T (2007) Bias in random forest variable importance measures: illustrations, sources and a solution. BMC Bioinf 8:25

Thuiller W, Lavorel S, Araujo MB (2005) Niche properties and geographical extent as predictors of species sensitivity to climate change. Glob Ecol Biogeogr 14:347–357

Travis J, Brooker R, Dytham C (2005) The interplay of positive and negative species interactions across an environmental gradient: insights from an individual-based simulation model. Biol Lett 22:5–8

Wasserman TN, Cushman SA, Littell JS, Shirk AJ, Landguth EL (2013) Population connectivity and genetic diversity of American marten (*Martes americana*) in the United States northern Rocky Mountains in a climate change context. Conserv Genet 14:529–541

Wiebe PA, Thompson ID, McKague CI, Fryxell JM, Baker JA (2014) Fine-scale winter resource selection by American martens in boreal forests and the effect of snow depth on access to coarse woody debris. Ecoscience 21:123–132

Chapter 11
Advanced Data Mining (Cloning) of Predicted Climate-Scapes and Their Variances Assessed with Machine Learning: An Example from Southern Alaska Shows Topographical Biases and Strong Differences

Falk Huettmann

11.1 Introduction

Climate models are the method of choice for assessing the climate on a landscape scale (www.ipcc.ch/). As climate becomes an overarching role in the functioning of our ecosystems and economies (e.g. Stern 2006), these climate models are of utmost importance for human planning and natural resource management (e.g. Mi et al. 2017; Han et al. 2018). The information source used in such climate models tend to be based on extrapolations from other locations and their models because not all pixels of interest are, or can be, sampled across all spatial and temporal scales. As a matter of fact, most 'pixels' in landscapes - relatively small and individual areas of inference - lack good climate measurement data (see here for some sampling efforts https://databasin.org/datasets/15a31dec689b4c958ee491ff30fcce75). This increases the role and relevance of model predictions because predictions enable the extrapolation of local information to a larger spatial and temporal extent. This is possible due to the advanced statistical and mathematical methods using 'best available' data sources which alleviates some of the ineffectiveness of using other, less applicable types of mathematical methods. (e.g. Breiman 2001; Venables and Ripley 2002; Drew et al. 2011; Barri et al. 2014). Nowadays, predictive modeling methods are primarily computing-based algorithms, with machine learning being among the main platform for solutions (Fernandez-Delgado et al. 2014). Obtaining the best predictions means the use of, and competing for, the best algorithm that produces accurate generalizations from the data (Breiman 2001).

F. Huettmann (✉)
EWHALE Lab, Biology and Wildlife Department, Institute of Arctic Biology,
University of Alaska-Fairbanks, Fairbanks, AK, USA
e-mail: fhuettmann@alaska.edu

© Springer Nature Switzerland AG 2018
G. R. W. Humphries et al. (eds.), *Machine Learning for Ecology and Sustainable Natural Resource Management*, https://doi.org/10.1007/978-3-319-96978-7_11

Climate remains an overruling topic in ecology and for global conservation questions (e.g. Jamieson and Di Paola 2014). Consequently, the most relevant climate models are probably policy-oriented ones, which tend to be large spatial extents (Lawler et al. 2011; Huettmann et al. 2017; Mi et al. 2017; Han et al. 2018) to the global scale (e.g. CRU data used for instance in Chernetsov and Huettmann 2005). By now, the IPCC climate models (http://www.ipcc.ch/) have become a global narrative and have many implementations (for an overview and data, see here: www.ipcc-data.org/). These IPCC models are often updated and act on a relatively coarse global scale (c. 16-km^2 pixels), and often get further regionalized ("downscaled") for more localized decision-making on smaller pixels. In recent years, several locations started to run those local models, e.g. the EU, U.S., Canada, Japan and Australia (for the EU see for instance here https://sites.google.com/site/ rt3validation/europe). Alaska, in particular, is in a fortunate position to have such regional models (Walsh et al. 2008; see https://www.snap.uaf. edu/tools/ data-downloads for details) that are being applied to a variety of data needs, e.g. Alaska GAP project (http://akgap.uaa.alaska.edu/). Alaska is also part of continental climate data projects (e.g AdaptWest https://adaptwest.databasin.org/; PRISM http://prism.oregonstate.edu/), as well as specific Arctic and Polar data-driven partnerships (More details in Walsh et al. 2008 and with CAFF.org). These regional climate models ('climate-scapes') act on smaller scales and can be assessed on smaller landscape extents in space and time. These climate-scapes are ideal for local assessment, ground-truthing, and focused insights.

While climate models and such subjects have been assessed and criticized for their performance for decades (Sawitzki 1994a, 1994b; Refsgaard et al. 2005; Hayhoe 2010; Huettmann and Gottschalk 2011), by now, updated generations of climate models are improved and can be used for detailed pixel-based assessments and planning. Most flora and fauna relate closely to climate due to their metabolism (Fick and Hijmans 2017). In many cases, distribution and subsequent strategic planning and investment questions (Stern 2006) are directly related to climate models, e.g. expected temperature, sea level rise or wind questions for real estate planning, insurance costs and windpark site locations (see Moilanen et al. 2009 for generic applications).

There is virtually no doubt the Earth is experiencing a global warming trend (e.g. Giddens 2009 for an overview), but several other trends remain widely unclear and unresolved, (e.g. clouds, vapor, counter currents or a few increasing glaciers). Climate models, on a smaller scale, are known to still carry various problems, and those models are still not equal to weather forecasts. This chapter provides some examples for real-world problems (Table 11.1) and some additional technical climate model questions and issues of concern (Table 11.2).

Alaska sits at the forefront of climate change issues in the U.S., as well as in the Arctic and worldwide. Alaska's coastal zones are a specific area of interest (Hayward et al. 2017 for Chugach Forest). However, original data are relatively few (see Fig. 11.1 for Worldclim stations; worldclim.com; Fick and Hijmans 2017; see also http://akclimate.org/Climate). In this assessment, three climate models for Alaska were used and compared for a region in southern Alaska of high conservation

Table 11.1 Generic 'issues' with climate models that occur in virtually any complex landscape

Known weakness in climate models	Local example	Justification and explanation
Sampling efforts	Alaska	Lack of awareness and investment in sampling and climate stations.
Wind shadows	Tibetan plateau	The exact amount of wetness taken off the air mass is not easy to obtain.
Jetstream impacts	Himalaya	The functioning and local impacts of jetstreams are not well known, described and modeled, yet.
Ice thickness	Greenland glacier	While ice thickness is not part of climate models it is an inherent feature in the local climate.
Ocean coupling	Islands, El Nino, Arctic Sea ice, Monsoon	IPCC implemented ocean models just in recent global climate updates. These interactions are a big driver in many terrestrial and global weather patterns.
Steep slope terrain	Most steep and coastal mountain ranges	Pacific northwest in North America.
Rugged terrain	Most mountain ranges	Deviations from plain earth surfaces are difficult to map and to measure for climate (cold pools and warm peaks affected by aspect).
Valleys	Death Valley (U.S.), Rhine Valley (EU)	Valleys consist of steep slopes and lower elevations at the bottom which triggers cold air flow but leaves certain cliffs exposed to the sun.
Aspects	Virtually any mountain	Precise aspect maps are difficult to obtain, southern aspects tend to be much warmer than northern ones.
Large elevational peaks	Mt. Everest (Nepal, China), Denali (U.S.), Ometepe volcanoes (Nicaragua)	Large mountains are known to maintain a weather system of their own.
Weather forecast	Monsoon weather at a given pixel	Climate models do not relate directly to the weather but are instead statistical climate trends expressed on a pixel scale.
Specific day forecasts	A pixel on April 3rd 2067	It is virtually impossible to predict the global weather for an exact location in space and time for long-time ahead.
Forecasts at small scales/pixels	A pixel of 100 m	Climate models are not developed on that pixel size, yet.
Lack of funding for proper ground-truthing and quality assessment	Alaska	Ideally, every relevant pixel is to be visited, measured for its climate etc., all done long-term. Such a research and data luxury rarely exists though in the real world.

Table 11.2 Some methodological questions and problems with existing climate models (abbreviations and URL are explained in the text)

Known weakness in climate models	Local example	Reasoning
Lack of an appropriate research design for predictions and inference	IPCC, Arctic, Alaska	It needs a high-resolution and high-quality testing and independent assessment data set. While basic weather data station collecting was done for over a century - at a few sites - wider prediction views using modern methods were widely ignored.
Lack of machine learning	PRISM, SNAP-Alaska	Machine learning tends to outcompete any linear regression and differential equations
Lack of ISO compliant metadata	IPCC, SNAP-Alaska	Without a proper documentation the outcome is not transparent nor repeatable
Lack of underlying open access data made available	PRISM	Without showing the data used, the outcome is not transparent nor repeatable
Lack of open source code made available	PRISM	Without showing the code used, the outcome is not transparent nor repeatable
Lack of large-scale testing data	Most climate models	Without a proper assessment on all pixels the outcome is not reliable and does not carry a known error
Lack of re-runs	SNAP-Alaska	Without updating models they are not best-available science

management value (Chugach Forest and associated areas; Hayward et al. 2017 for details). The study area allows for a good test case because it includes landscape components listed in Table 11.1 and it has already achieved first climate assessments using regionalized climate data (Hayward et al. 2017). Following concepts by Hochachka et al. (2007) and Drew et al. (2011) climate and landscape data were mined for this climate-scape using machine learning and then investigated in a quantitative framework whether and what patterns and their predictors can be found in space and time.

11.2 Methods

11.2.1 GIS Data and Operations

Following Morton and Huettmann (2017) the Chugach forest outline was used as the study area based on an encompassing shapefile (Alaska NAD83 UTM Zone 6 in meters). Three commonly used climate models were tested for this region: Worldclim, Adaptwest and UAF SNAP. For each pixel the variance was computed among the three climate predictions for the monthly mean of January and July. See Fig. 11.1 for details.

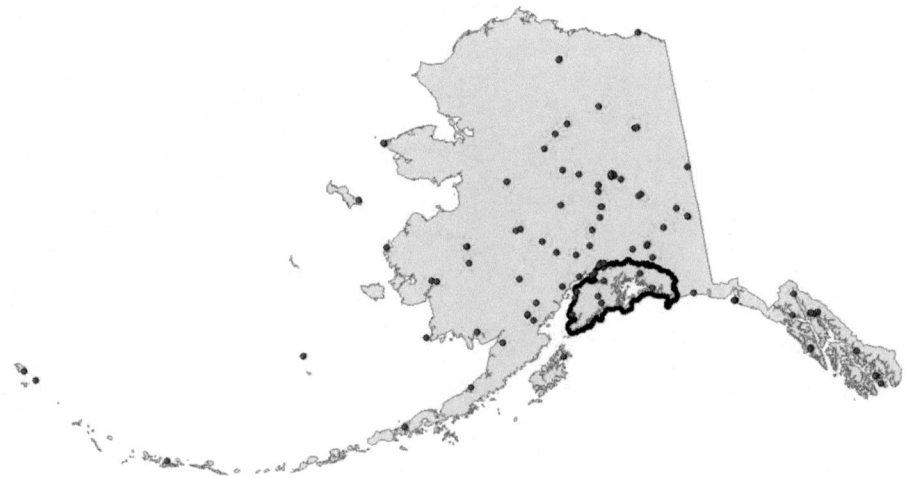

Fig. 11.1 Worldclim weather stations for Alaska and the study area (Chugach forest)

The predictors, elevation, slope, aspect, distances to coast, river and road, as well as vegetation landcover and island yes/no were overlaid. Next, overlays were done on a regular 5-km² lattice scale across the extent of the study area using ArcGIS and QGIS Open source GIS (Fig. 11.2). This resulted in 2130 terrestrial lattice points, which had the predictors attached as attributes to each point (see Fig. 11.3 for lattice map).

11.2.2 Data Mining

The resulting data cube for January (coldest month) and July (warmest month) was 'mined' using TreeNet (Salford Systems Predictive Modeling Suite SPM 8; https://www.salford-systems.com/). Each model was run with January and July data combined using the climate model variance as a response, and with the predictors mentioned above. Month became a flagged predictor in the data cube allowing for data cloning (i.e., repeated use of the data) to obtain better model fits and predictions (see Lele et al. 2007 and Jiao et al. 2016 for concepts and methods). The analysis to draw from three climate models might be perceived as a meta-analysis (Schmidt and Hunter 2014).

The resulting models, their outliers, and their variance trends were described and to obtain an improved inference and prediction. Overall, it was hoped to achieve improved insights into these climate models by using machine learning methods.

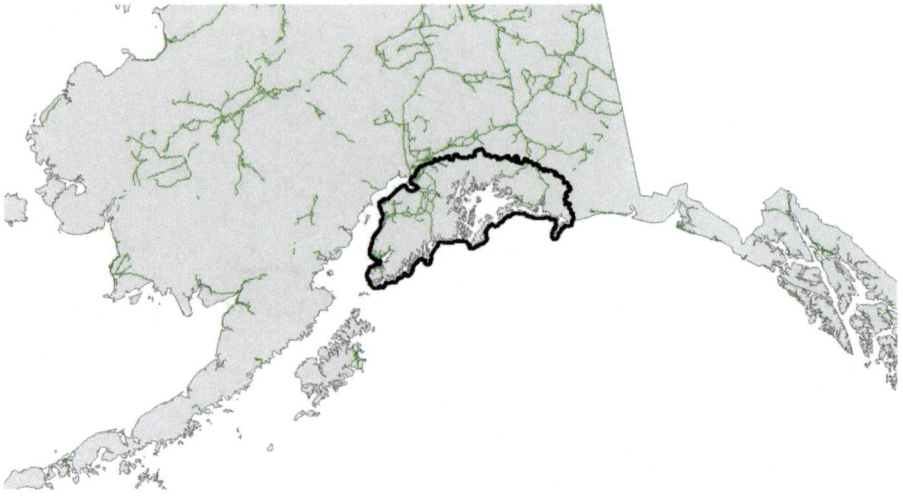

Fig. 11.2 Study area and location in Alaska

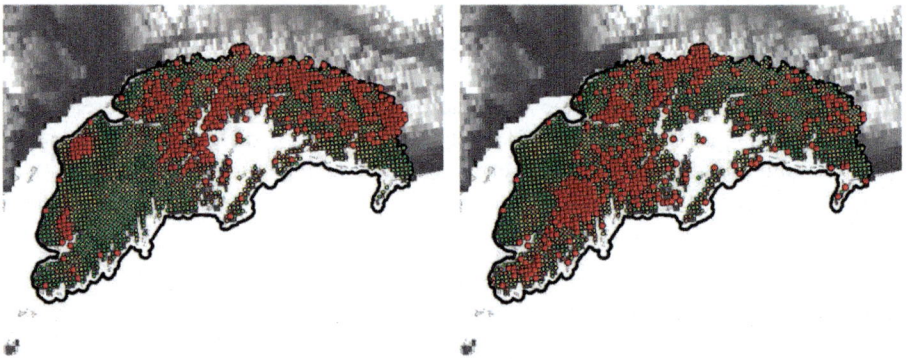

Fig. 11.3 Relative variance index of mean monthly temperatures among pixels from three climate models in January (**a**) and July (**b**). Red shows high variances, green low

11.3 Results

The climate model comparison in Table 11.3 generally shows that for the month of January the Worldclim model predicts cooler temperatures than ADAPTWEST and AK SNAP. However, for July AK SNAP predicts cooler temperatures than the other two models. Generally, results using descriptive statistics point towards a small difference of up to 2-degree C. between models. Map details are shown in Appendix A.

Despite just 'smaller' differences between models, the map of variances allows to show spatial discrepancies more clearly. The variance in January shows the biggest differences on islands and surrounding areas ('hinterland') (Fig. 11.3a), whereas the variance in July shows the biggest differences on glaciers and islands.

Table 11.3 Basic descriptive statistics for climate model comparisons

Model	Month	Minimum	Mean	Maximum
Worldclim	January	−22.00	−9.57	0.00
	July	0.00	10.43	14.70
AK SNAP	January	−21.24	−8.33	0.31
	July	−5.70	9.98	15.80
ADAPT-WEST	January	−20.80	−8.00	0.90
	July	0.00	11.19	15.40

Fig. 11.4 A selected 3-dimensional partial dependence plot showing the relevance of month (January and July) and elevation (AKDEM3) for the variance of three climate model pixels in the study area

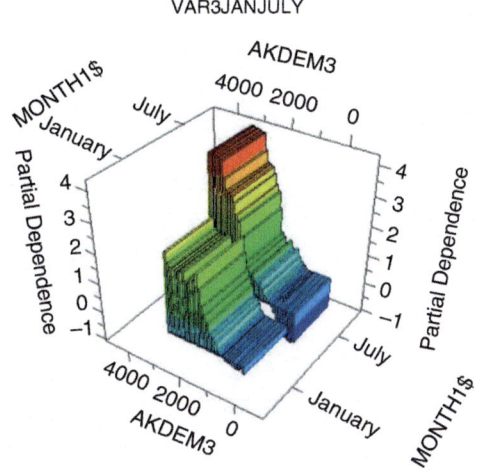

TreeNet obtained a 'good' model with a gains curve of c. 75% (Appendix for details) to explain variances for January and July. Using TreeNet allowed to extract nuanced but clear signals from the data to further indicate and explain differences among both climate models, as well as, describe patterns and identify relevant predictors. Across months, elevation is a major predictor explaining the discrepancies between both models (Fig. 11.3). Specifically, elevations ≥1100 m resulted in higher variance, with the highest being in July at c. 2500 meters. Additionally, the month of January had less variance than July (Fig. 11.4).

Further, the predictor 'slope' seems to drive high variances, specifically at higher elevations (Fig. 11.5). Fig. 11.6 shows also a distinct peak for the variance at pixels that are within 200 m of the coastline (a zone of generic uncertainty; that is presumably the tidal range). Figure 11.7 shows high variance for pixels that are on north-facing aspects (Figs. 11.8 and 11.9).

For landcover classes only class 12 (closed mixed forest) show high variances. Islands have a generally higher variance than mainland indicating poor estimates for islands. Further details of the TreeNet model are available in Appendix B. The overall ranking of all predictors is shown in Table 11.4.

Fig. 11.5 A selected
3-dimensional partial
dependence plot showing
the relevance of slope and
elevation for the variance
of three climate model
pixels in the study area

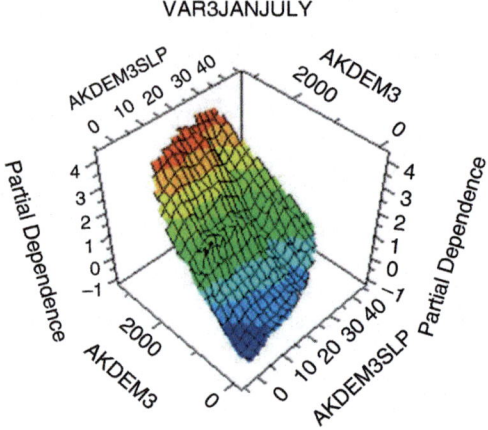

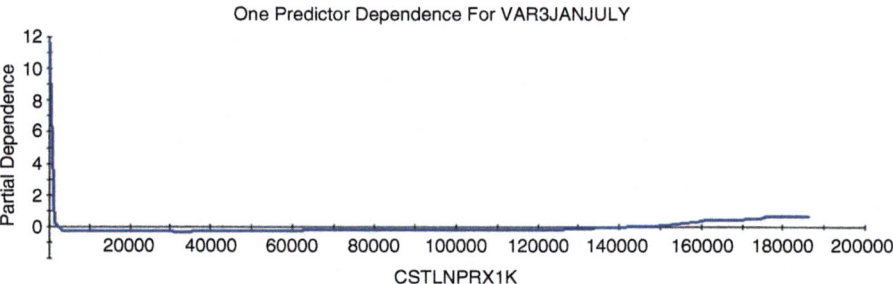

Fig. 11.6 A selected 2-dimensional partial dependence plot showing the relevance of proximity to coastline for the variance explained

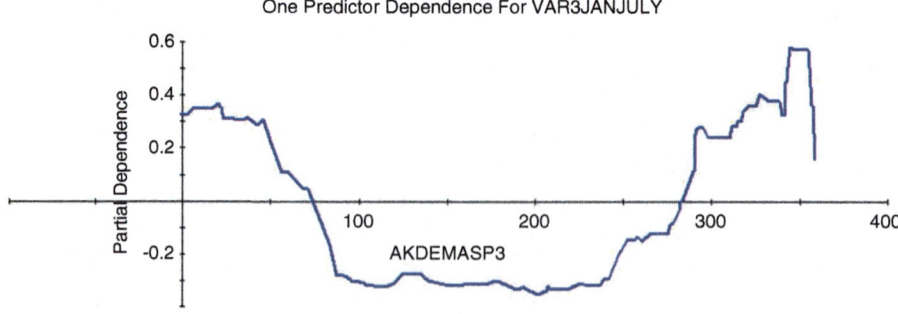

Fig. 11.7 A selected 2-dimensional partial dependence plot showing the relevance of north-facing aspect for the variance of three climate model pixels in the study area

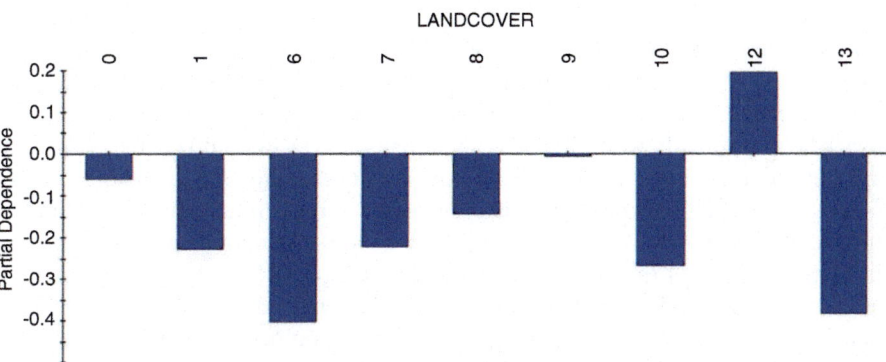

Fig. 11.8 A selected 2-dimensional partial dependence plot showing the relevance of landcover for the variance of three climate model pixels in the study area

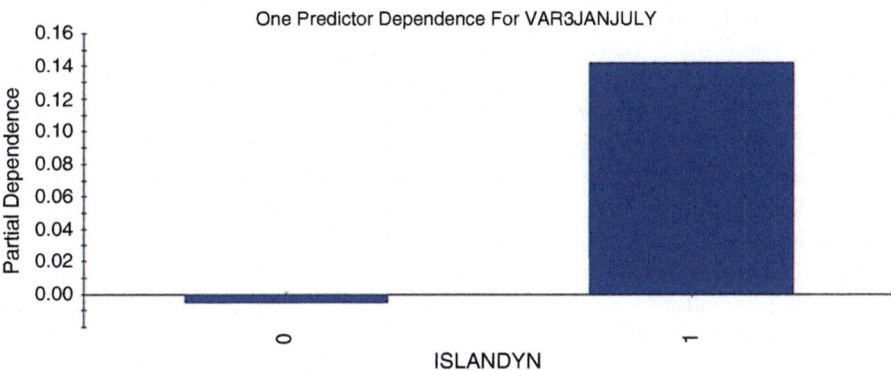

Fig. 11.9 A selected 2-dimensional partial dependence plot showing the relevance of island for the variance of three climate model pixels in the study area

Table 11.4 Ranking of predictors used in the TreeNet model

Rank	Predictor name	Continuous or Categorical	Importance Percent
1	Elevation	Continuous	100
2	Month	Categorical	66
3	Proximity to coastline	Continuous	57
4	Proximity to roads	Continuous	53
5	Aspect	Continuous	50
6	Slope	Continuous	49
7	Proximity to river	Continuous	45
8	Landcover	Categorical	40
9	Island yes/no	Categorical	10

11.4 Discussion

This is the first use of machine learning and advanced data mining on climate model assessments and comparison in the state of Alaska (Bieniek et al. 2015; see Huettmann and Morton 2017 for use of those methods for species model forecasting). Alaska is already well-known as an international posterchild for the observable and measureable effects of climate change on natural, cultural, and socio-economic systems. However, it is probably also known for some climate change denial, and the climate issue in Alaska remains a poorly addressed and a disputatious issue in the state, throughout the country, the circumpolar, and globally. Creating the most accurate climate models for Alaska reaches far beyond quantifiable evidence of climate change. Instead, it affects the 'global climate chamber', global processes, how foreign policy is implemented, and the laws that are to determine major industrial investments and their public justifications for overall regulation (or deregulation) and mitigation.

Alaska remains of strategic relevance, beyond 'just' natural resource management. But so far, Alaska remains of strategic relevance, beyond 'just' natural resource management. But so far, Alaska primarily employs mechanistic warming models, p-values (Gardner and Altman 1986), linear regressions or differential equations and the delta method to understand, model and forecast climate (Bieniek et al. 2015; https://www.snap.uaf.edu/methods/downscaling). Most models known to the author that are employed lack transparency of the decision-making process on their code development and the relevant GIS methods which makes for improper broad-based assessment, open peer-based scrutiny, and scientific repeatability. The notion of pixel sizes and geographic projections - or even autocorrelations, couplings and interactions - remain widely unresolved in Alaskan climate models and ignored resulting in unknown but widely documented errors.

Here a regular 5-km^2 lattice was used for model assessment. While this is relatively coarse still, it should favor coarser (=robust) trends in the climate models, is thus conservative in its findings and addresses well potential concerns of autocorrelation (Betts et al. 2009). Still, even on that scale, patterns of larger discrepancies in the data were already found and one can already point to relatively easily explainable patterns in the predictors, all based on data mining techniques for better insights.

To describe 'climate' correctly one needs a complex set of measurements from an appropriate research design. Here that discussion was started using only monthly mean temperature for summer (July) and winter (January). However, even with using those few metrics one can show bigger differences among models (all claiming high accuracy for Alaska and are used and applied), expand, what implications

Fig. 11.10 Coastal details and islands in the study area to exemplify abundance of islands and their exclusion from high-resolution climate-scape work

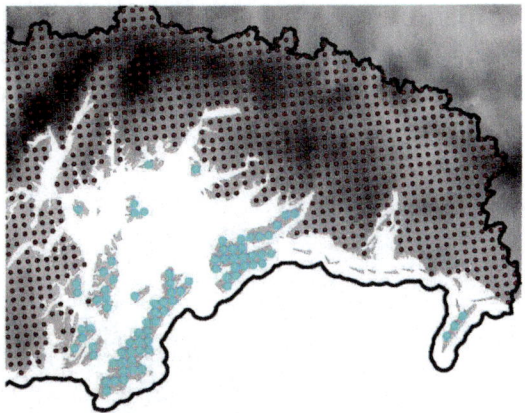

can result from this. One may assume that adding additional climate metrics would show even more differences and renders such climate layers even more unreliable overall without proper metadata and spatial accuracies.

Further, I would like to caution the interpretation of coastal, specifically island, pixels. Figure 11.10 shows a coastal zoom in, and many islands are either not captured in pixelated models (e.g. here 5-km^2 pixels by our lattice) or the existing island patterns differ from the described model even further (model details not shown here; available upon request from the authors). Climate models tend not to address islands specifically (e.g. due to lacking data and time required to assess each island; Alaska has easily thousands of islands) and thus they remain dubious in their validity (see Table 11.1 for details). This is in line with findings here that virtually all pixels, across months, have high variances when being located within 200 m of the coastline. One relevant topic is that islands are the prime areas of extinction and thus, good management should focus there but their climate data remain uncertain.

These findings here show that the use of machine learning brings new information and viewpoints to the table for climate-scape models. That is not only true for the data mining perspective, but also for understanding variances, outliers, for quantifications and later for correcting models through predictions and data for an improved consensus and meta-analysis among competing models. Considering the huge pressure on climate-related decision-making, globally, machine learning is likely a more reliable methodology for such research and conservation management as a 'best professional practice'. (Zuckerberg et al. 2011; see Silvy 2012 for lack of such entire concepts).

**Appendix A Example Map of Data Sets Used for this Machine
Learning Assessment of Climate Models: Adaptwest in July.
Raw Climate Surface and All GIS Maps are available
from the Authors on Request**

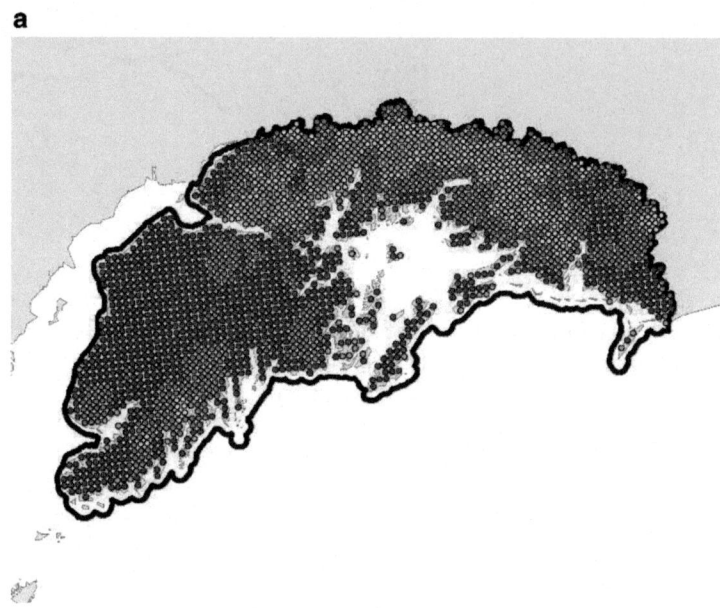

Appendix B Remaining Details of the TreeNet Model not shown in the text: (a) gains curve, and partial dependence plots for (b) proximity to road, (c) proximity to river

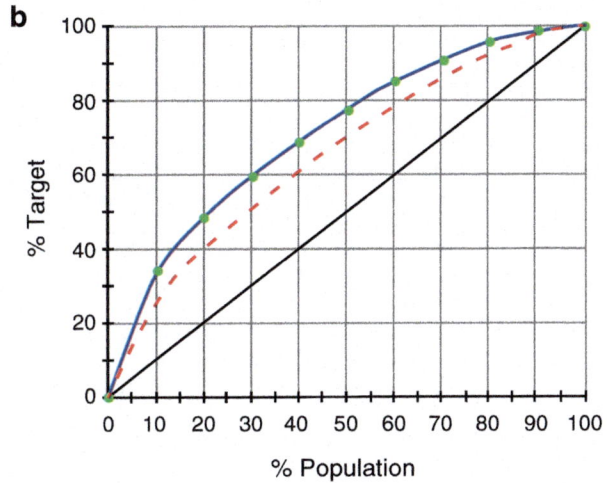

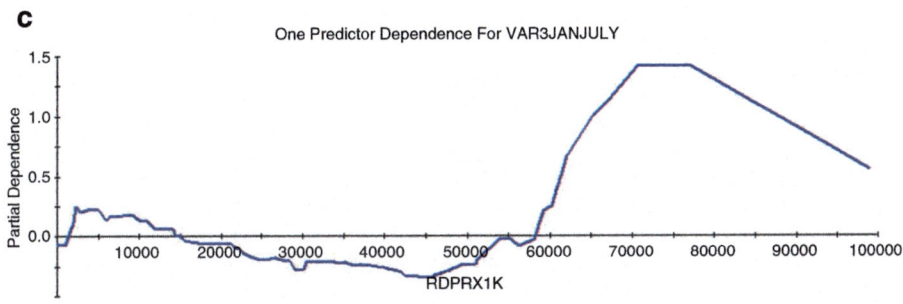

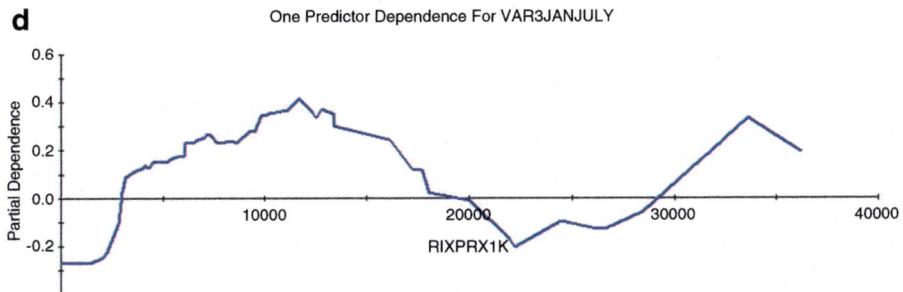

References

Bari A, Chaouchi M, Jung T (2014) Predictive analytics for dummies. Wiley, New York

Betts MG, Ganio L, Huso M, Som N, Huettmann F, Bowman J, Wintle WA (2009) Comment on "Methods to account for spatial autocorrelation in the analysis of species distributional data: a review". Ecography 32:374–378

Bieniek PA, Bhatt US, Walsh JE, Rupp TS, Zhang J, Krieger JR, Lader R (2015) Dynamical downscaling of ERA-interim temperature and precipitation for Alaska. JAMC https://doi.org/10.1175/JAMC-D-15-0153.1

Breiman L (2001) Statistical modeling: the two cultures (with comments and a rejoinder by the author). Stat Sci 16:199–231

Chernetsov N, Huettmann F (2005) Linking global climate grid surfaces with local long-term migration monitoring data: spatial computations for the pied flycatcher to assess climate-related population dynamics on a continental scale.Lecture notes in computer science (LNCS) 3482, International Conference on Computational Science and its Applications (ICCSA) Proceedings Part III: 133–142

Drew A, Wiersma Y, Huettmann F (eds) (2011) Predictive species and habitat modeling in landscape ecology. Springer, New York

Fernandez-Delgado M, Cernades E, Barro S, Amorim D (2014) Do we need hundreds of classifiers to solve real world classification problems? J Mach Learn Res 15:3133–3181

Fick SE, Hijmans RJ (2017) Worldclim 2: new 1-km spatial resolution climate surfaces for global land areas. Int J Climatol 37:4302–4313

Gardner MA, Altman DG (1986) Confidence intervals rather than P values: estimation rather than hypothesis testing. Br Med J 292:746–750

Giddens A (2009) The politics of climate change. Polity Press, New York

Han X, Huettmann F, Guo Y, Mi C, Wen L (2018) Conservation prioritization with machine learning predictions for the black-necked crane Grus nigricollis, a flagship species on the Tibetan plateau for 2070. Glob Environ Chang. https://doi.org/10.1007/s10113-018-1336-4

Hayhoe KA (2010) A standardized framework for evaluating the skill of regional climate downscaling techniques. Unpublished PhD thesis, University of Illinois

Hayward GD, Colt S, Mc Teague M, Hollingsworth T (eds) (2017) Climate change vulnerability assessment for the Chugach National Forest and the Kenai peninsula. General technical report PNW-GTR-000. USDA Forest Service. Pacific Northwest Research Station, Portland

Hochachka W, Caruana R, Fink D, Munson A, Riedewald M, Sorokina D, Kelling S (2007) Data mining for discovery of pattern and process in ecological systems. J Wildl Manag 71:2427–2437

Huettmann F, Gottschalk T (2011) Simplicity, complexity and uncertainty in spatial models applied across time. In: Drew CA, Wiersma Y, Huettmann F (eds) Predictive species and habitat modeling in landscape ecology. Springer, New York, pp 189–208

Huettmann F, Magnuson EE, Hueffer K (2017) Ecological niche modeling of rabies in the changing Arctic of Alaska. Acta Vet Scand 201759:18–31. https://doi.org/10.1186/s13028-017-0285-0

Jamieson DW, Di Paola M (2014) Climate change and global justice: new problem, old paradigm? Global Pol 5:105–111. https://doi.org/10.1111/1758-5899.12113

Jiao S, Huettmann F, Guo Y, Li X, Ouyang Y (2016) Advanced long-term bird banding and climate data mining in spring confirm passerine population declines for the northeast Chinese-Russian flyway. Glob Planet Chang. https://doi.org/10.1016/j.gloplacha.2016.06.015

Lawler JJ, Wiersma Y, Huettmann F (2011) Designing predictive models for increased utility: using species distribution models for conservation planning, forecasting, and risk assessment. In: Drew CA, Wiersma Y, Huettmann F (eds) Predictive species and habitat modeling in landscape ecology. Springer, New York, pp 271–290

Lele SR, Dennis B, Lutscher F (2007) Data cloning: easy maximum likelihood estimation for complex ecological models using Bayesian Markov chain Monte Carlo methods. Ecol Lett 10:551–563

Mi C, Zu Q, He L, Huettmann F, Jin N, Li J (2017) Climate change would enlarge suitable planting areas of sugarcanes in China. Int J Plant Prod 11:13–27

Moilanen A, Wilson KA, Possingham H (2009) Spatial conservation prioritization: quantitative methods and computational tools. Edited by Oxford, U.K. Oxford University Press, UK

Morton JM, Huettmann F (2017) Moose, caribou and Sitka black-tailed deer. In: Hayward GD, Colt S, McTeague M, Hollingsworth T (eds) Climate change vulnerability assessment for the Chugach National Forest and the Kenai Peninsula. General Technical Report PNW-GTR-000. USDA Forest Service, Pacific Northwest Research Station, Portland

Refsgaard JC, van der Sluijs JP, Brown J, van der Keura P (2005) A framework for dealing with uncertainty due to model structure error. Adv Water Resour 29:1586–1597. https://doi.org/10.1016/j.advwatres.2005.11.013

Sawitzki G (1994a) Testing numerical reliability of data analysis systems. Comput Stat Data Anal 18:269–286

Sawitzki G (1994b) Report on the reliability of data analysis systems. Comput Stat Data Anal (SSN) 18:289–301

Schmidt FL, Hunter JE (2014) Methods of meta-analysis: correcting error and bias in research findings, 3rd edn. Sage, Thousand Oaks

Silvy NY (2012) The wildlife techniques manual: research and management, vol 2, 7th edn, John Hopkins University Press, Baltimore

Stern N (2006) Review on the economics of climate change. Government of the United Kingdom, London

Venables WN, Ripley BD (2002) Modern applied statistical analysis, 4th edn. Springer, New York

Walsh JE, Chapman WL, Romanovsky V et al (2008) Global climate model performance over Alaska and Greenland. J Clim 21:6156–6174

Zuckerberg B, Huettmann F, Friar J (2011) Proper data management as a scientific foundation for reliable species distribution modeling. In: Drew CA, Wiersma Y, Huettmann F (eds) Predictive species and habitat modeling in landscape ecology. Springer, New York, pp 45–70

Chapter 12
Using TreeNet, a Machine Learning Approach to Better Understand Factors that Influence Elevated Blood Lead Levels in Wintering Golden Eagles in the Western United States

Erica H. Craig, Tim H. Craig, and Mark R. Fuller

Investigating the effects of environmental contaminants on individuals and, ultimately, on wildlife populations is challenging. This is partly because it is difficult to interpret results potentially influenced by many interacting variables. The complexity of such datasets can constrain the ability of researchers to obtain meaningful biological results using traditional statistical approaches. This can result in under-utilization of available information that is potentially useful to management decision makers (Craig and Huettmann 2009). Machine learning (ML) algorithms provide powerful tools that are increasingly seen as practical solutions for helping to address complex problems. They can be used as an end-product for prediction, to reveal patterns in data related to the incidence of contaminants in target species, and to guide future research efforts, or alternatively, to aid in hypothesis development or in conjunction with conventional statistical tools. ML has been used for decades in investigating the causes of disease in human populations (Collij et al. 2016; see e.g., Cooper et al. 1997; Cruz and Wishart 2006; Moradi et al. 2015; Shipp et al. 2002; Sriram et al. 2013) and the incidence of contaminants in the environment (see e.g., Hu and Cheng 2013; Wang et al. 2015), but remains under-utilized in ecology (Thessen 2016).

Pb from bullet fragments in hunter-killed game (Herring et al. 2016; Hunt et al. 2006; Knopper et al. 2006), and its potential effects on raptors and other avian species that scavenge carrion has received considerable attention in recent years (Cade 2007; see e.g., Church et al. 2006; Craighead and Bedrosian 2008; Cruz-Martinez

E. H. Craig (✉) · T. H. Craig
Aquila Environmental, Fairbanks, AK, USA
e-mail: goea.rs@gmail.com

M. R. Fuller
Boise State University, Raptor Research Center, Boise, ID, USA

© Springer Nature Switzerland AG 2018
G. R. W. Humphries et al. (eds.), *Machine Learning for Ecology and Sustainable Natural Resource Management*, https://doi.org/10.1007/978-3-319-96978-7_12

et al. 2012; Ecke et al. 2017; Herring et al. 2017, Herring et al. 2016). Golden Eagles (*Aquila chrysaetos*) are known to feed on carcasses and offal from hunter-killed game (referred to as game carcasses throughout the remainder of the text; Ecke et al. 2017). Although there may be other vectors, it is generally accepted that ingestion of Pb from game carcasses is the most widespread common source of eBLL in Golden Eagles (Bedrosian and Craighead 2009; Craig et al. 1990; Ecke et al. 2017; Fisher et al. 2006; Haig et al. 2014; Hunt 2012; Pattee et al. 1990; Stauber et al. 2010; Wayland et al. 2003; Wayland and Bollinger 1999). However, the issue of Pb contamination in Golden Eagle populations and how it affects the species is complex. Many factors influence the occurrence of eBLL in individual eagles and have the potential for subsequent population level consequences (see e.g., Finkelstein et al. 2012). Sublethal Pb contaminant loads may affect behavior (Ecke et al. 2017), and possibly even reproduction and survival; therefore it is important to identify members of a population at greatest risk and factors influencing exposure to Pb (Bedrosian and Craighead 2009; Ecke et al. 2017; Fisher et al. 2006; Herring et al. 2016). Recent US Fish and Wildlife Service (FWS) models indicate a downward trend in Golden Eagle populations in the western United States (US; USFWS 2016). In addition, long-term regional datasets and migration counts provide evidence for local declines (Hoffman and Smith 2003; Kochert and Steenhof 2002; Millsap et al. 2013). This trend is likely the result of a combination of different factors, including: changes occurring in shrub and grassland habitats critical to eagles (Kochert et al. 2002) that are due to urbanization and the rapid expansion of renewable energy development projects (Copeland et al. 2011), climate change and an associated increase in wildfires and drought (Abatzoglou and Kolden 2011; Dennison et al. 2014), and the subsequent changes in prey distribution and availability and eagle reproduction and productivity (Kochert et al. 1999). It is unknown if Pb contamination in eagles is a contributing factor to these declines.

In this paper we used TreeNet, a machine learning algorithm that utilizes stochastic gradient boosting (Friedman 2002; Friedman 2001), to 1) model the influence of six factors on eBLL in wintering Golden Eagles captured in Idaho, USA over a 9 year period, 2) compare winter blood lead levels (BLL) to levels in a small number of eagles sampled during late spring and summer, and 3) provide information relevant for guiding future research on Pb exposure in eagles.

12.1　Methods

We modeled factors influencing BLL using 317 blood samples (275 individuals) from Golden Eagles that were captured during winters 1989–90 through 1997–98; 42 birds were resampled up to 4 years later (see Craig and Craig 1998). We report the winter, rather than the year, in which an eagle was captured (e.g., eagles captured during winter 1989–90 are listed as being captured during winter 1990, those captured in 1990–91 as winter 1991, etc.). We assumed that the eagles we captured represented a random sample from the study area. The study area is located in a remote part of east central Idaho that is sparsely populated by humans. The

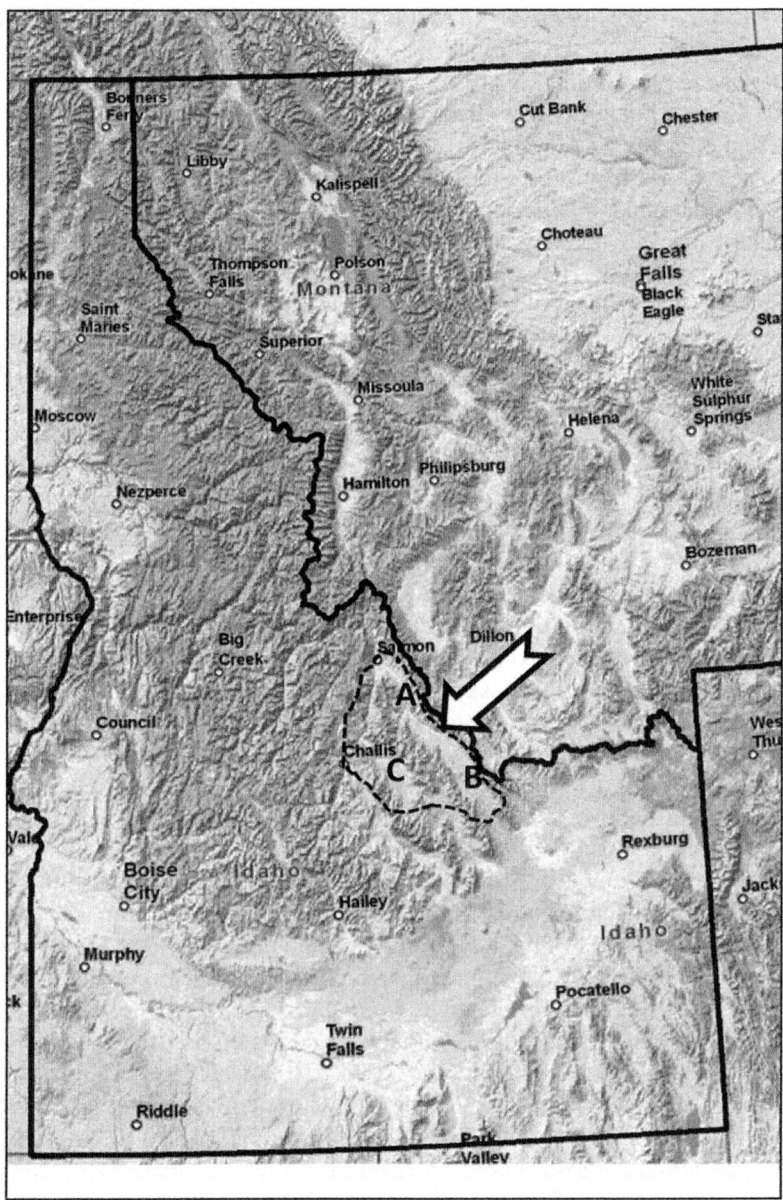

Fig. 12.1 Study area in Idaho where eagles were sampled for elevated blood Pb levels, winter 1989–90 through winter 1997–98. The letter **a** identifies the Lemhi Valley, **b**, the Birch Creek Valley, and **c,** the Pahsimeroi Valley

topography is varied with high mountain ranges that are dissected by rivers and streams; vegetation in the valleys is dominated by native sagebrush (*Artemisia spp.*) shrub-steppe habitat associations (Fig. 12.1). Most of our eagle trap-sites were located in the main valleys or along their larger tributaries. For modeling purposes, we grouped trap sites into three valleys in the study area. Two drain in a northerly direction and are part of the Salmon River drainage (Pahsimeroi and Lemhi Valleys); trap sites along the Salmon River were grouped with sites in the nearest of these two valleys. The third valley, Birch Creek, begins at the south end (head) of the Lemhi Valley and drains in a southeasterly direction to the edge of the Snake River Plain; we trapped eagles from this valley only during winters 1997 and 1998. Juvenile eagles were captured only during winters 1996, 1997 and 1998.

Golden Eagles in our study area reflect the partial migratory status of the species (Kochert et al. 2002). The eagles we trapped included year-round residents and regional or long-distance migrants that arrived from summer areas as much as ~5000 km away; the proportion of each group represented in the wintering popula-tion we sampled is unknown. We also report BLL from 12 nestlings sampled during summer just prior to fledging and five free ranging eagles captured or recovered dead outside the winter season; because these samples were not obtained during winter, they are not included in the models we developed for wintering eagles. We provide data on them for comparison with BLL of Golden Eagles sampled during winter.

All blood samples, except those collected during winter 1996, were analyzed for Pb levels with a Perkin-Elmer Zeeman 5011 PC Atomic Adsorption Spectrophotometer graphite furnace at a wavelength of 283.3 nm. During winter 1996, samples were analyzed for Pb using a VG Elemental PlasmaQuad through inductively coupled plasma-mass spectroscopy (ICP-MS; wavelength = 207.97 nm). An internal standard of indium was used at a wavelength of 114.9 nm. The lower limit of reportable Pb residues for samples, using both methods, was 0.01 ppm wet weight.

We used TreeNet to construct our models (Salford Systems, Inc.; Freidman 2002; Friedman et al. 2000; Friedman 2001) because it: 1) is known for its prediction accuracy (see e.g., https://info.salford-systems.com/predicting-customer-churn-with-gradient-boosting, accessed 25 October 2017), 2) provides graphs that visually depict the strength and form of relationship(s) and interactions among response and predictor variables, 3) is non-parametric, requiring no assumptions about the struc-ture of the data, 4) is not constrained by complex interactions in highly dimensional datasets, and 5) can handle datasets with missing or otherwise 'messy' data (Guisan et al. 2007; Hastie et al. 2009; Knudby et al. 2010). Machine learning algorithms such as this can also provide good results in spite of non-stationarity in data (Hastie et al. 2009).

We used a binary logistic approach for our models. Presence or absence of BLL above background (> 0.20 ppm: Kramer and Redig 1997; Cruz-Martinez et al. 2012) was the response variable. The six predictor variables that we used, repre-sented characteristics we measured for the wintering eagles sampled in our study area (eagle age class, gender, and winter, month, valley and time of day in which the bird was captured). We grouped the eagles we caught into three age classes based on plumage characteristics: adult, (≥ 4 years old), juvenile (first year birds still in their initial plumage), and subadult (eagles older than juveniles but not yet in adult plumage). We ran multiple models using different settings to achieve best model performance; our best model was developed using a subsample of 0.60 and the default settings (6 node trees, automatic learn rate, terminal node minimum training cases of 10). Although we specified a maximum number of 1000 trees be built in model development, the final number of trees for best model performance was determined by the software. This prevents overfitting the data by indicating the point at which the formation of additional trees explains no further variation in the data.

A realistic method for validating models is to evaluate performance when applied to an independent dataset (Elith et al. 2008) We randomly selected and withheld 10% of the samples in our dataset for independent testing and used the remaining 90% (called the learn data) for developing the model. We used prevalence (the default baseline; Liu et al. 2005) as our threshold when assessing models. We report the following five TreeNet metrics of model performance for the independent (test) data of our best model: 1) Precision: of eagles predicted to have eBLL, the proportion that were correct, 2) Sensitivity or True Positive: the proportion of eagles with eBLL that were correctly identified, 3) Specificity or True Negative: the proportion of eagles without eBLL that were correctly identified, 4) the F1 statistic for each model (values near 1 indicate the model is both Precise and has good Sensitivity), and 5) Receiver Operating Characteristic curve (Hastie et al. 2009; Powers 2011). To aid in interpretation of the models and the biological significance of the results, TreeNet produces an index ranking the relative importance of each predictor variable to the final model outcome. The top predictor with the greatest influence for predicting eBLL in Golden Eagles is given a score of 100; all other predictors are ranked in comparison with it. TreeNet also produces partial dependence plots (Friedman 2001; Hastie et al. 2009), which aid in graphically interpreting the functional relationship between the presence of eBLL in eagles (response) and the individual predictor variables. The graphs show the form and strength of the relationship of the variable(s) for predicting presence of eBLL after averaging out the influence of all the other variables. For comparison among variables, partial dependence values (PD) form the y axis for all graphs and represent an index of the likelihood of presence, i.e., $0.5 \, \text{logit}(p(X))$ (Cutler et al. 2007), where X is the presence of eBLL in wintering Golden Eagles in the study area. Univariate plots reveal the additive relationship of the individual variable with predicting eBLL in eagles while bivariate (interaction) plots indicate the shape and strength of 2-way interactions among predictors relative to eBLL. Figures for bivariate plots showing interactions between winter of capture and categorical variables are 'sliced' by category to aid in interpreting results.

12.2 Results

Overall accuracy of our best model was 76.7%, for predicting the presence or absence of eBLL in the independent sample of wintering eagles (also see Table 12.1). All six input variables contributed to the final model and were important predictors of eBLL in the eagles sampled (Table 12.2).

Table 12.1 TreeNet outputs measuring model performance for predicting presence or absence of elevated Pb levels in blood of 30 independent samples from wintering Golden Eagles in our study area

Model Performance Metrics	Percent
Specificity	80.00
Sensitivity	73.33
Precision	78.57
F1 statistic	75.86
ROC area	84.89

Table 12.2 Relative importance of variables for predicting the presence of eBLL in wintering Golden Eagles captured in Idaho. Variables are ranked relative to the top predictor

Predictor Variables	Relative Importance
Winter captured	100.00
Time caught	71.15
Month caught	47.34
Gender	43.45
Valley	41.58
Age class	40.60

The two most influential predictors were the year in which an eagle was captured, followed by the time of capture during the day; the other four predictors all contributed about equally to the final model.

The single variable dependence plots (Fig. 12.2) indicate that there is a positive association between eagles with eBLL during five of the nine winters of the study and a negative association with the other four winters (Fig. 12.2a). Birds captured mid-morning to mid-afternoon and eBLL were also positively associated; the strongest association was later in the afternoon from 1400–1500 hr. (Fig. 12.2b). Adults, females, birds captured during the months of December and January, and eagles captured in the Birch Creek Valley were also positively associated with the occurrence of eBLL, although the latter was a very weak association (Figs 12.2c–f).

The bivariate plots and their PD values indicated that the top predictor, winter in which an eagle was captured, interacted most strongly with gender, but also interacted strongly with all four other variables (Table 12.3; Figs 12.3a–f). These interaction effects were complex and had greater influence overall than individual variables in predicting the liklihood of eBLL in the wintering eagles sampled in the study area (Table 12.3). Capture month interactions and gender interactions exerted the next greatest influence on predicting eBLL. There was also a positive association with age class and valley of capture, although that interaction exerted the least influence on the final model (Table 12.3). All eagles with eBLL, but especially females, were more likely to be captured later in the afternoon during the months of December and January most years and during winters positively associated with eBLL (Figs 12.3a–f, 12.4a–f). Adult females, and to a lesser extent, subadult females that had eBLL were more likely to be captured later in the afternoon than adult

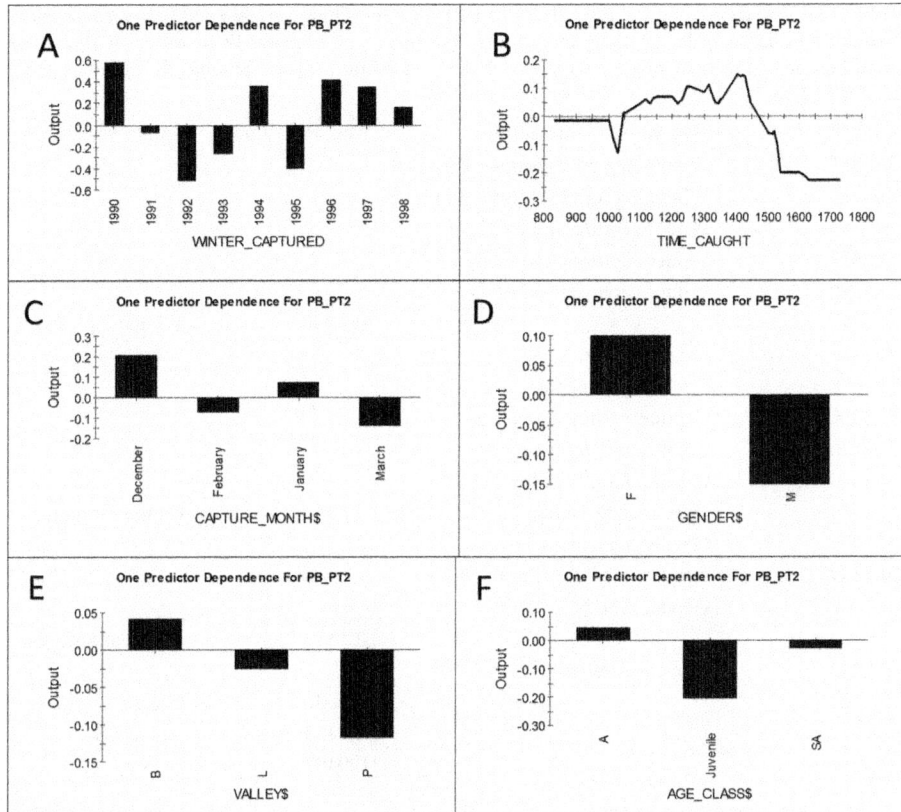

Fig. 12.2 Univariate partial dependence plots showing additive influence for each input variable on predicting the incidence of eBLL in wintering Golden Eagles sampled in the Idaho study area. Plots are listed in order of importance to the optimal model; Output on the y axis = 0.5 logit($p(X)$), where X is the presence of eBLL in wintering Golden Eagles in the study area. **a** Winter in which eagle was captured. **b** Time of day the eagle was captured. **c** Month of capture. **d** Age class of eagle (A = adult, Juvenile, and SA = subadult). **e** Valley in which the bird was captured (B = Birch Creek, L = Lemhi and P = Pahsimeroi. **f** Gender (F = female and M = male)

males and both were more likely to have eBLL than males. Subadult males and juveniles of either gender were least likely to have eBLL (Fig. 12.3e, 12.5a–c). Adults and subadults from the Birch Creek Valley and adults from the Lemhi Valley were more likely to have eBLL than juveniles or any age class from the Pahsimeroi Valley, but this association was relatively weak and varied by winter (Fig. 12.5d–f).

Samples taken from eaglets just prior to fledging revealed no or very low exposure to Pb during summer months. Six of the young eagles had BLL levels below the minimum detection level of 0.01 ppm; the remainder had BLL that are considered background (mean = 0.035 ppm, SE = 0.008 ppm; Kramer and Redig 1997; Cruz-Martinez et al. 2012). Two eagles were captured during the summer in the Pahsimeroi and Lemhi Valleys, an adult male (August 1991) and an eagle turned in

Table 12.3 Maximum partial dependence values (PD) for single variable and two-variable plots. Variables listed in order of their maximum PD value, not necessarily in order of their total contribution to the final model, which considers interaction effects among all variables when assigning rank

Max PD Value	Two Variable Interactions	Max PD Value	Single Variable
0.6511	Winter captured/gender		
0.6213	Winter captured /capture month		
0.5947	Winter captured/time caught		
0.5821	Winter captured/age class		
0.5790	Winter captured/valley		
		0.5761	Winter captured
0.2862	Capture month/gender		
0.2678	Time caught/gender		
0.2316	Capture month/time caught		
0.2126	Capture month/age class		
0.2080	Capture month/valley		
0.2064	Time caught/valley	0.2064	Capture month
0.1973	Gender/age class		
0.1889	Time caught/ age class		
0.1822	Gender /valley		
		0.1488	Time caught
0.1133	Age class/valley		
		0.1000	Gender
		0.0453	Age class
		0.0407	Valley

to the local Fish and Game Department for which age and gender were undetermined (August 1996); both had low BLL (0.06 and 0.01 ppm, respectively). A third eagle recovered dead in the early spring of 1993 in the Pahsimeroi Valley had a liver Pb level of 0.10 ppm wet wt; liver Pb levels of <2.0 ppm wet wt are considered background (Cruz-Martinez et al. 2012; Kramer and Redig 1997). Two adults, presumably a mated pair, were captured together just outside the study area during the fall (30 September 1991); their blood Pb levels were 0.95 ppm (male) and 0.45 ppm (female).

⟶

Fig. 12.3 (continued) captured interactions with time of day the eagle was captured. **b** Time of day the eagle was captured interactions with eagle gender. **c** Month of capture interactions with eagle gender. **d** Time of day the eagle was captured interactions with month of capture. **e** Age class of eagle (A = adult, SA = subadult and YOY = Juvenile) interactions with time of day the eagle was captured. **f** Valley in which the bird was captured (B = Birch Creek, L = Lemhi and P = Pahsimeroi. **g** Gender (F = female and M = male) interactions with age class of eagle

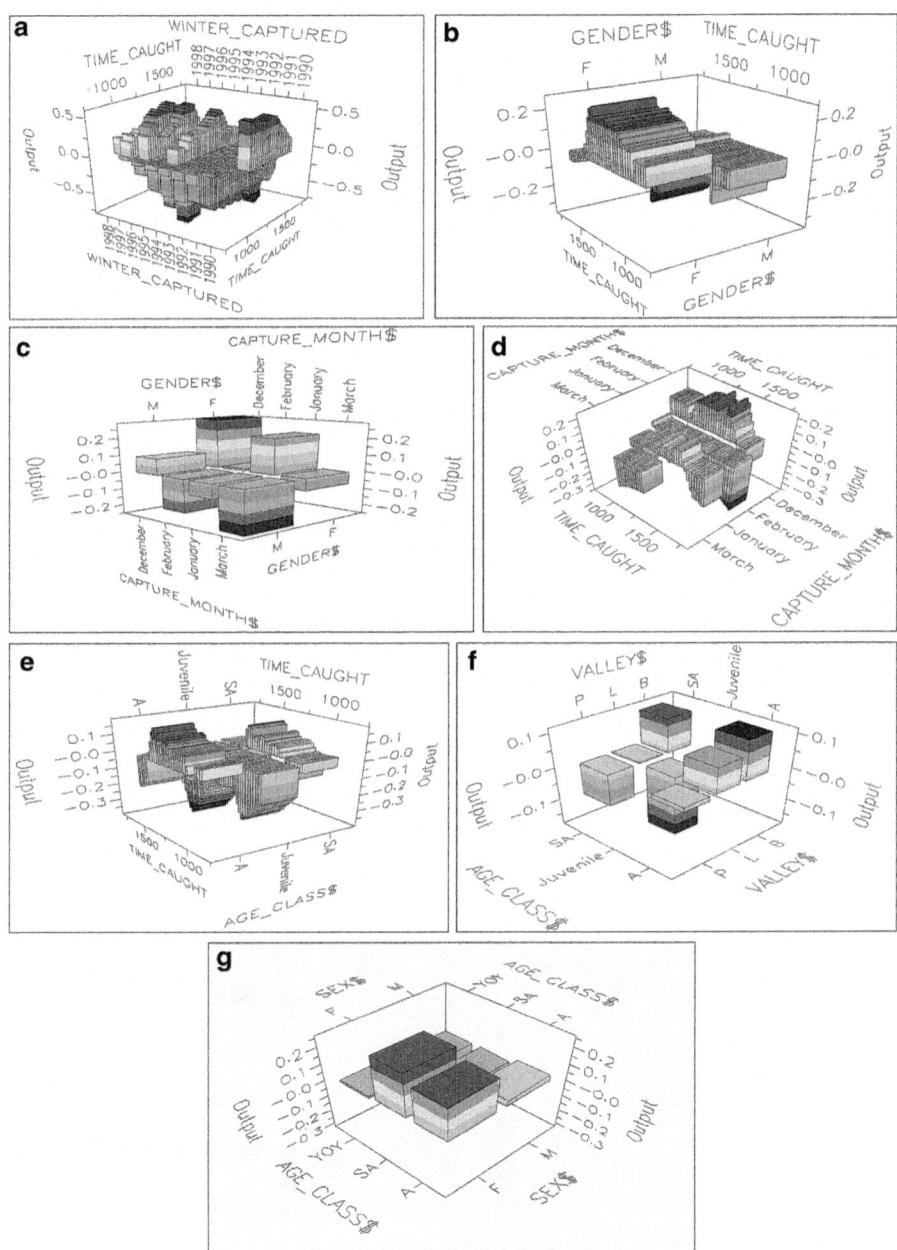

Fig. 12.3 Bivariate partial dependence plots showing two-way interactions among variables influencing the incidence of eBLL in wintering Golden Eagles sampled in the Idaho study area. Plots are listed in order of importance to the optimal model; Output on the y axis = 0.5 logit($p(X)$), where X is the presence of eBLL in wintering Golden Eagles in the study area. Note that the scale for the y axis may be different for each graph, so predicted association between eBLL and the height of the bars is not directly comparable across all graphs **a** Winter in which eagle was

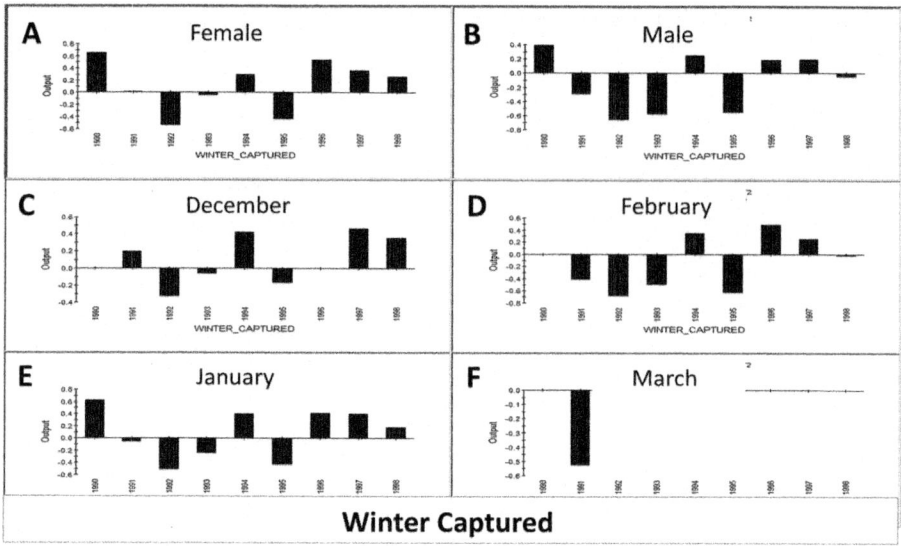

Fig. 12.4 Slices by categorical variable of bivariate partial dependence plots for predicting the likelihood of eBLL in wintering Golden Eagles in the Idaho study area. Slices show the relationship between eBLL and interaction effects between winter in which an eagle was captured and gender (**a** and **b**), or capture month (**c, d, e, f**), respectively. Output on the y axis = 0.5 logit($p(X)$), where X is the presence of eBLL in wintering Golden Eagles in the study area. Note that the scale for the y axis may be different for each graph, so predicted association between eBLL and the height of the bars is not directly comparable across all graphs

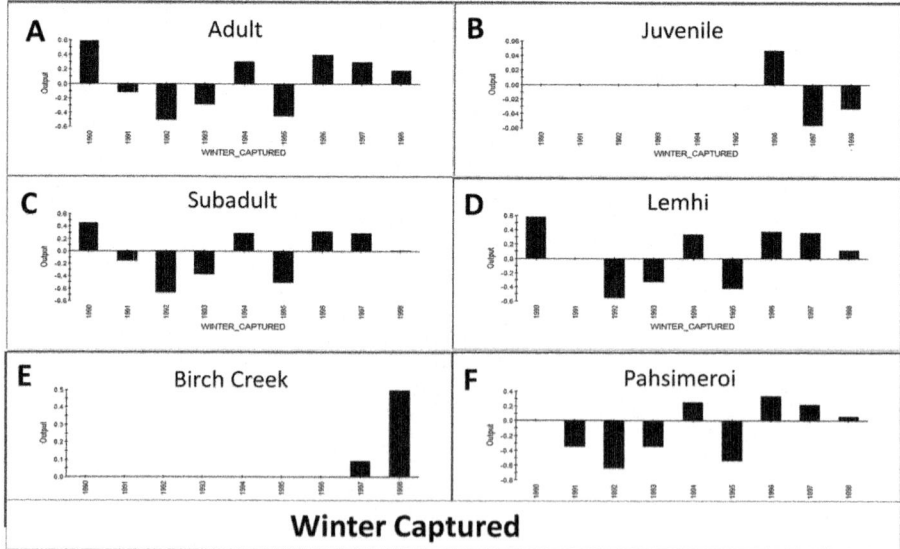

Fig. 12.5 Slices by categorical variable from bivariate partial dependence plots for predicting the likelihood of eBLL in wintering Golden Eagles in the Idaho study area. Slices show the relationship between eBLL and interaction effects between winter in which an eagle was captured and age class (**a, b, c**) or valley (**d, e, f**), respectively. Output on the y axis = 0.5 logit($p(X)$), where X is the presence of eBLL in wintering Golden Eagles in the study area. Note that the scale for the y axis may be different for each graph, so predicted association between elevated BLL and the height of the bar in each graph is not directly comparable. Eagles were only trapped in Birch Creek during winters 1997 and 1998. Juveniles were only captured during winters 1996, 1997 and 1998

12.3 Discussion

All six of the variables used in developing our best model interacted with one another and were important for predicting eagles with eBLL in the study area. The winter an eagle was captured was the top predictor but the likelihood of eagle exposure to contamination varied from year to year. This agreed with the results of our previous analyses using traditional statistics; we found a highly significant difference in Golden Eagle BLL among years (p = 0.000; Craig and Craig 1998). However, TreeNet results provided additional information that indicated the specific winters in which Golden Eagles were most likely to be exposed to Pb and how winter of capture interacted with the other input variables. Time of day an eagle was captured was the second-most influential predictor; it was not one of the variables included in our earlier, more parsimonious analyses (Craig and Craig 1998). There was a positive association between eagles captured from ~ 1000–1500 hr. and all years positively associated with eBLL. The strongest relationship was with eagles, particularly females, captured later in the day (early to mid-afternoon). Ecke et al. (2017) attributed changes in movement behavior of Golden Eagles in Sweden to sublethal Pb exposure. They reported negative effects on movement rates and flight heights and suggested that even low levels of Pb exposure affect eagle behaviors that could potentially reduce hunting success. It is possible that eagles with eBLL in our study area were more likely to be captured later in the day than other eagles because they had lower hunting success, and so had to hunt longer. Further research is needed into the behavioral effects of sublethal exposure to Pb in eagles to better understand the potential conservation implications.

Month in which an eagle was captured, gender, age class and valley of capture were also important contributors to the final model output. Identifying the influence of these variables individually was informative but identifying the purely additive effects of each was insufficient to adequately explain how they might have influenced the susceptibility of wintering eagles to contamination from Pb in east-central Idaho. The univariate and bivariate dependence plots helped in interpreting the form and strength of these relationships. Our model identified Golden Eagles captured during December and January, when game carcasses are often available, as positively associated with eBLL. Other studies, have also identified that eBLL in eagles are often associated with the timing of hunting seasons (Bedrosian et al. 2012; Ecke et al. 2017; Haig et al. 2014; Herring et al. 2017; see e.g., Pattee et al. 1990; Stauber et al. 2010). Big-game hunting seasons and sport hunting for jackrabbits (*Lepus spp.*) occur mainly during fall and winter in our study area; it is likely that these are also the primary sources of Pb exposure to eagles wintering there.

Our model also identified that adult females and eagles captured in the Lemhi River and Birch Creek Valleys were most likely to be associated with eBLL and indicated a positive but very weak interaction of these two valleys with both gender and age. Parts of both of those valleys have background Pb in soils and sites with historic Pb mining activity (Lemhi County History Committee 1992; Birch Creek and Lemhi; Umpleby 1913). Although, Pb contamination in some avian species has been linked to anthropogenic sources other than Pb bullets (see e.g., Henny 2003; Legagneux et al. 2014), bioaccumulation of Pb is generally thought to be low for raptors (Henny 2003). If background Pb from mining activity were available in portions of our study area it should be a year-round source of contamination. However, none of the nestlings we sampled, nor the free-ranging eagles we captured incidentally during the late spring and summer, had eBLL. Data are not available for Golden Eagles on the turnover rate of Pb in blood after exposure, but evidence from other avian species indicates that it is likely to be fairly rapid (see e.g., Fry et al. 2009; Pain et al. 2009; Herring et al. 2017). Pb in blood of birds generally has a half-life of ~2 weeks (Fry et al. 2009; Pain et al. 2009) and evidence from Bald Eagles (*Haliaeetus leucocephalis*) indicates that Pb effects from exposure may last up to 4 weeks (Hoffman et al. 1981). As a result, BLL samples taken from eagles during summer should reflect relatively recent exposure to Pb. We rarely observed Golden Eagles feeding on carrion during summer and game carcasses are not generally available then. All these factors indicate that background Pb from mining activity is not likely the source of contamination to wintering eagles in the study area. Further, most of the indications from our model point to game carcasses as the primary source of Pb to the sampled Golden Eagles. However, because potential for exposure may vary locally, we agree with Herring et al. (2017) that studies at smaller geographic scales, such as ours, are an important component of identifying the sources and assessing the effects of exposure to Pb on regional Golden Eagle populations.

Year-to-year and valley by valley differences in weather conditions and snow depth, hunting seasons, hunter success rates, and jackrabbit cycles are a few of the variables that may also influence the availability of Pb contaminated game carcasses to wintering eagles in the study area. Any of these factors could account for the yearly variation in risk to eagles that we observed. It is also possible that some of the migratory eagles with eBLL were exposed to Pb prior to arriving in the study area (Herring et al. 2017), or because of unidentified factors that we did not measure. Consequently, it is important to note that the six factors included in our model are not the only ones that may influence the occurrence of Pb in eagles. The inclusion of other variables and additional sampling in future research efforts could provide further insights toward understanding the conservation implications of the eagle-Pb contamination issue. A machine learning approach, with its ability to include all potential factors would be fundamental to these investigations.

Of the free-ranging eagles captured during winter, juveniles were identified by our model as the least likely to have eBLL. Similarly, more adult and subadult eagles in Canada had higher Pb levels than did immatures (Wayland et al. 2003; Wayland and Bollinger 1999). However, Wayland and Bollinger (1999) did not analyze BLL and results from all tissues (e.g., blood, liver) were combined by Wayland et al. (2003) in their analysis; it is known that Pb accumulates over time in liver, kidney and bone tissues in birds (see e.g., Herring et al. 2017; Pain et al. 2009). Because turnover rates for Pb in avian blood are generally high (Pain et al. 2009) and are not known to accumulate over long periods of time, it is doubtful that the differences we observed in BLL are due to age-related accumulation of Pb in blood.

It is possible that it may be a result of differences among age groups in feeding behavior at carcasses, habitat exploitation, or some other unknown factor.

Adult female Golden Eagles were identified by our model as the most likely demographic in the study area to have eBLL. Subadult females were less likely than adult females, but both subgroups were more likely to have eBLL than males of any age class. Our study is the first to report that female Golden Eagles, and often adults, were more susceptible to Pb contamination than males. In contrast, Pattee et al. (1990) found no significant difference by gender or age class in BLL of 162 Golden Eagles captured in California. However, adult female Bald Eagles have been reported to be the most sensitive age/gender group to Pb poisoning in the U.S. population (USFWS 1986). Other studies have reported that Marsh Harriers (*Circus aeruginosus*) and female Common Ravens (*Corvus corax*) were more likely to have eBLL than males during hunting seasons (Craighead and Bedrosian 2008; Pain et al. 1997). Pattee (1984) experimentally exposed American Kestrels (*Falco sparverius*) to Pb and reported higher levels of deposition in bones and liver of females than males. Craighead and Bedrosian (2008) suggested that the gender differences they observed in BLL of ravens may have resulted from different metabolic rates, behavioral differences in activity level between the sexes, or simply differences in ingestion rates of Pb. Any of these could explain the gender difference we observed in the eagles in our study area. Pain et al. (1997) suggested the gender differences they observed were related to partitioning of prey by marsh harriers; the larger females generally take bigger prey than males, which increased their susceptibility to ingestion of Pb fragments. It is unlikely that the size of prey is an important factor in explaining the gender differences identified by our model, since game carcasses are already dead and unlikely to influence selection by eagles that feed on them. However, behavioral differences may be a contributing factor. Golden Eagles, like most other raptors, exhibit reverse sexual dimorphism, with the female being larger than the male (Kochert et al. 2002). This size difference may give females an advantage when scavenging; females we observed wintering in our study area were often the first to feed on carcasses. Although it was not always the case, male Golden Eagles and other avian species (e.g., Bald Eagles, Common Ravens, Black–billed Magpies [*Pica hudsonia*]) often retreated when a larger female Golden Eagle flew onto a carcass. Similarly, Halley and Gjershaug (1998) observed conflicts by Golden Eagles over access to carcasses in Norway. Gender was not known for many of the conflicts they observed, but when gender could be determined, females were dominant at carcasses in five of seven encounters. This dominance could put females at greater risk to Pb contamination since the first bird on game carcasses probably starts feeding at the bullet entry or exit wound, where access is easiest and the site of highest Pb concentration (Hunt et al. 2006). Regardless of the reason that females, and especially adult females, may be more susceptible to Pb contamination than males, the demographic consequences of this potential difference merit further investigation.

The stability of populations has been linked to survival of adults that are capable of reproducing (Grier 1980; Newton 1989). Pb exposure may influence reproductive capabilities, and negatively affect flight behaviors and hunting success, making individuals more susceptible to mortality through factors like starvation or collisions (Ecke et al. 2017; Herring et al. 2017; Pain et al. 1997). Our initial results could be used to help inform population models using differential rates of exposure to Pb according to gender and age. Additional research is needed to determine how exposure to Pb is manifested in the survival and reproductive rates of Golden Eagles at local and regional scales. Recent evidence suggests that background blood Pb concentrations in Golden Eagles may be considerably lower than the 0.20 ppm currently used as a threshold in most studies, including ours (Craighead and Bedrosian 2008; Ecke et al. 2017). Consequently, the results we present should be considered a conservative representation of Pb exposure in the wintering Golden Eagle population in our study area.

Despite much research devoted to Pb contamination in avian scavengers, including eagles (see e.g., Haig et al. 2014; Herring et al. 2017), there is need for a better understanding of the sublethal effects of Pb exposure to Golden Eagles. How it affects gender and age subgroups and the conservation implications of potential subsequent population level consequences are important to determine. This is where TreeNet outputs are particularly valuable. Although it does not designate statistical significance with *p* values in the outputs we present, TreeNet accommodated our six input variables simultaneously and detected patterns in the data and factors that influenced eBLL in our sample population. Future research could include any number of additional variables in TreeNet models; TreeNet simply ignores those of no importance. Further, since there is no need for parsimony, it is not necessary to restrict, a priori, predictors that may be relevant and eliminate those with strong correlations. As a result, biases that might be introduced through data processing choices can be avoided. Our earlier, more parsimonious analyses using traditional statistics did not provide the more nuanced insights into the form and strength of interactions among variables provided by the TreeNet outputs. Our TreeNet model was further validated by the findings of others and our own earlier analyses using traditional statistical methods, but also identified additional information that was not previously reported. It revealed new information about timing of capture that may reflect behavioral effects of eBLL, the potential for gender related differences to Pb exposure within the wintering population and spatial distribution of eagles with eBLL that may warrant further attention. Our results provide direction for future investigation in other geographic areas and for environmental factors to include in future models.

Acknowledgments Many individuals have provided technical support for this project. We would particularly like to acknowledge the contributions of R. Craig, J. Craig, H. Craig McFarland and F. Huettmann. This project was partially funded by Aquila Environmental. Original funding for earlier aspects of this research was provided by the U.S. Bureau of Land Management, U.S. Geological Survey, University of Alaska Fairbanks, E-WHALE Lab, Western Ecological Studies Team, Idaho Department of Fish and Game, Golden Eagle Chapter of the Audubon Society, and The Idaho Wildlife Society. Trapping, banding and blood sample collection were authorized by the state of Idaho and the USGS Bird Banding Laboratory (Laurel, MD; permit no. 06714). Any use of trade, product, or firm names is for descriptive purposes only and does not imply endorsement by the U.S. Government.

References

Abatzoglou JT, Kolden CA (2011) Climate change in western US deserts: potential for increased wildfire and invasive annual grasses. Rangel Ecol Manag 64:471–478

Bedrosian B, Craighead D (2009) Blood lead levels of Bald and Golden Eagles sampled during and after hunting seasons in the greater yellowstone ecosystem. In: Watson RT, Fuller M, Pokras M, Hunt WG (eds) Ingestion of Lead from Spent Ammunition: Implications for Wildlife and Humans. The Peregrine Fund, Boise. ID, USA https://doi.org/10.4080/ILSA.2009.0209

Bedrosian B, Craighead D, Crandall R (2012) Lead exposure in bald eagles from big game hunting, the continental implications and successful mitigation efforts. PLoS One 7:e51978

Cade TJ (2007) Exposure of California condors to lead from spent ammunition. J Wildl Manag 71:2125–2133

Church ME, Gwiazda R, Risebrough RW, Sorenson K, Chamberlain CP, Farry S, Heinrich W, Rideout BA, Smith DR (2006) Ammunition is the principal source of lead accumulated by California condors re-introduced to the wild. Environ Sci Technol 40:6143–6150

Collij LE, Heeman F, Kuijer JP, Ossenkoppele R, Benedictus MR, Möller C, Verfaillie SC, Sanz-Arigita EJ, van Berckel BN, van der Flier WM, others (2016) Application of machine learning to arterial spin labeling in mild cognitive impairment and Alzheimer disease. Radiology 281:865–875

Cooper GF, Aliferis CF, Ambrosino R, Aronis J, Buchanan BG, Caruana R, Fine MJ, Glymour C, Gordon G, Hanusa BH, others (1997) An evaluation of machine-learning methods for predicting pneumonia mortality. Artif Intell Med 9:107–138

Copeland HE, Pocewicz A, Kiesecker JM (2011) Geography of energy development in western North America: potential impacts on terrestrial ecosystems, chapter 2. In: Naugle DE (ed) Energy development and wildlife conservation in western North America. Island Press, Washington, DC, USA, pp 7–22

Craig EH, Craig TH (1998) Lead and Mercury levels in golden and bald eagles and annual movements of golden eagles wintering in East Central Idaho, 1990–1997, Technical Bulletin No. 98–12. Idaho Bureau of Land Management, Boise, ID, USA

Craig E, Huettmann F (2009) Using "Blackbox" algorithms such as TreeNET and Random Forests for data-mining and for finding meaningful patterns, relationships and outliers in complex ecological data: an overview, an example using Golden Eagle satellite data and an outlook for a promising future. In: Wang H (ed) Intelligent data analysis: developing new methodologies through pattern discovery and recovery. IGI Global, Hershey, PA, USA pp 65–84

Craig TH, Connelly JW, Craig EH, Parker TL (1990) Lead concentrations in golden and bald eagles. Wilson Bull 102:130–133

Craighead D, Bedrosian B (2008) Blood lead levels of common ravens with access to big-game offal. J Wildl Manag 72:240–245

Cruz JA, Wishart DS (2006) Applications of machine learning in cancer prediction and prognosis. Cancer Informat 2:59–78

Cruz-Martinez L, Redig PT, Deen J (2012) Lead from spent ammunition: a source of exposure and poisoning in bald eagles. Human–Wildlife Interact 6:11 https://digitalcommons.usu.edu/hwi/vol6/iss1/11

Cutler DR, Edwards TC, Beard KH, Cutler A, Hess KT, Gibson J, Lawler JJ (2007) Random forests for classification in ecology. Ecology 88:2783–2792

Dennison PE, Brewer SC, Arnold JD, Moritz MA (2014) Large wildfire trends in the western United States, 1984–2011. Geophys Res Lett 41:2928–2933

Ecke F, Singh NJ, Arnemo JM, Bignert A, Helander B, Berglund AMM, Borg H, Bröjer C, Holm K, Lanzone M, others (2017) Sublethal lead exposure alters movement behavior in free-ranging golden eagles. Environ Sci Technol 51:5729–5736

Elith J, Leathwick JR, Hastie T (2008) A working guide to boosted regression trees. J Anim Ecol 77:802–813

Finkelstein ME, Doak DF, George D, Burnett J, Brandt J, Church M, Grantham J, Smith DR (2012) Lead poisoning and the deceptive recovery of the critically endangered California condor. Proc Natl Acad Sci 109:11449–11454

Fisher IJ, Pain DJ, Thomas VG (2006) A review of lead poisoning from ammunition sources in terrestrial birds. Biol Conserv 131:421–432

Friedman J (2002) Stochastic gradient boosting: non-linear methods and data mining. Comput Stat Data Anal 38:367–378

Friedman JH (2001) Greedy function approximation: a gradient boosting machine. Ann Stat 29:1189–1232

Friedman J, Hastie T, Tibshirani R, others (2000) Additive logistic regression: a statistical view of boosting (with discussion and a rejoinder by the authors). Ann Stat 28:337–407

Fry M, Sorenson K, Grantham J, Burnett J, Brandt J, Koenig M (2009) Lead intoxication kinetics in condors from California. In: Watson RT, Fuller M, Pokras M, Hunt WG (eds) Ingestion of Lead from Spent Ammunition: Implications for Wildlife and Humans. The Peregrine Fund, Boise. ID, USA, https://doi.org/10.4080/ilsa.2009.0301

Grier JW (1980) Modeling approaches to bald eagle population dynamics. Wildl Soc Bull 8:316–322

Guisan A, Zimmermann NE, Elith J, Graham CH, Phillips S, Peterson AT (2007) What matters for predicting the occurrences of trees: techniques, data or species' characteristics? Ecol Monogr 77:615–630

Haig SM, D'Elia J, Eagles-Smith C, Fair JM, Gervais J, Herring G, Rivers JW, Schulz JH (2014) The persistent problem of lead poisoning in birds from ammunition and fishing tackle. Condor 116:408–428. https://doi.org/10.1650/CONDOR-14-36.1

Halley DJ, Gjershaug JO (1998) Inter-and intra-specific dominance relationships and feeding behaviour of golden eagles *Aquila chrysaetos* and sea eagles *Haliatetus albicilla* at carcasses. Ibis 140:295–301

Hastie T, Tibshirani R, Friedman J (2009) The elements of statistical learning: data mining, inference and prediction, 2nd edn. Springer, New York, NY, USA

Henny CJ (2003) Effects of mining lead on birds: a case history at Coeur d'Alene Basin, Idaho. In: Hoffman DJ, Rattner BA, Burton GA Jr, Cairns J Jr (eds) Handbook of ecotoxicology, 2nd edn. Lewis Publishers, Boca Raton, FL, USA pp 755–766

Herring G, Eagles-Smith CA, Wagner MT (2016) Ground squirrel shooting and potential lead exposure in breeding avian scavengers. PLoS One 11:e0167926

Herring G, Eagles-Smith CA, Buck J (2017) Characterizing golden eagle risk to lead and anticoagulant rodenticide exposure: a review. J Raptor Res 51:273–292. https://doi.org/10.3356/JRR-16-19.1

Hoffman SW, Smith JP (2003) Population trends of migratory raptors in western North America, 1977–2001. Condor 105:397–419

Hoffman DJ, Pattee OH, Wiemeyer SN, Mulhern B (1981) Effects of lead shot ingestion on δ-aminolevulinic acid dehydratase activity, hemoglobin concentration, and serum chemistry in bald eagles. J Wildl Dis 17:423–431

Hu Y, Cheng H (2013) Application of stochastic models in identification and apportionment of heavy metal pollution sources in the surface soils of a large-scale region. Environ Sci Technol 47:3752–3760. https://doi.org/10.1021/es304310k

Hunt WG (2012) Implications of sublethal lead exposure in avian scavengers. J Raptor Res 46:389–393

Hunt WG, Burnham W, Parish CN, Burnham KK, Mutch B, Oaks JL (2006) Bullet fragments in deer remains: implications for lead exposure in avian scavengers. Wildl Soc Bull 34:167–170

Knopper LD, Mineau P, Scheuhammer AM, Bond DE, McKinnon DT (2006) Carcasses of shot Richardson's ground squirrels may pose lead hazards to scavenging hawks. J Wildl Manag 70:295–299

Knudby A, Brenning A, Ledrew E (2010) New approaches to modelling fish–habitat relationships. Ecol Model 221:503–511

Kochert MN, Steenhof K, Carpenter LB, Marzluff JM (1999) Effects of fire on golden eagle territory occupancy and reproductive success. J Wildl Manag 63:773–780

Kochert MN, Steenhof KM, McIntyre CL, Craig EH (2002) Golden Eagle (*Aquila chrysaetos*), version 2.0. In: Poole AF, Gill FB (eds) The Birds of North America. Cornell Lab of Ornithology, Ithaca, NY, USA. https://doi.org/10.2173/bna.684

Kramer JL, Redig PT (1997) Sixteen years of lead poisoning in eagles, 1980–95: an epizootiologic view. J Raptor Res 31:327–332

Legagneux P, Suffice P, Messier J-S, Lelievre F, Tremblay JA, Maisonneuve C, Saint-Louis R, Bêty J (2014) High risk of lead contamination for scavengers in an area with high moose hunting success. PLoS One 9:e111546

Lemhi County History Committee (1992) Centennial history of Lemhi County, Idaho, vol 1. Lemhi County History Committee, Salmon, ID, USA

Liu C, Berry PM, Dawson TP, Pearson RG (2005) Selecting thresholds of occurrence in the prediction of species distributions. Ecography 28:385–393

Millsap BA, Zimmerman GS, Sauer JR, Nielson RM, Otto M, Bjerre E, Murphy R (2013) Golden eagle population trends in the western United States: 1968–2010. J Wildl Manag 77:1436–1448. https://doi.org/10.1002/jwmg.588

Moradi E, Pepe A, Gaser C, Huttunen H, Tohka J, Initiative ADN, others (2015) Machine learning framework for early MRI-based Alzheimer's conversion prediction in MCI subjects. NeuroImage 104:398–412

Newton I (1989) Sparrowhawk. In: Newton I (ed) Lifetime reproduction in birds. Academic, London, England, pp 279–296

Pain DJ, Bavoux C, Burneleau G (1997) Seasonal blood lead concentrations in marsh harriers *Circus aeruginosus* from Charente-maritime, France: relationship with the hunting season. Biol Conserv 81:1–7

Pain DJ, Fisher IJ, Thomas VG (2009) A global update of lead poisoning in terrestrial birds from ammunition sources. In: Watson RT, Fuller M, Pokras M, Hunt WG (eds) Ingestion of Lead from Spent Ammunition: Implications for Wildlife and Humans. The Peregrine Fund, Boise, ID, USA. https://doi.org/10.4080/ilsa.2009.0108

Pattee OH (1984) Eggshell thickness and reproduction in American kestrels exposed to chronic dietary lead. Arch Environ Contam Toxicol 13:29–34

Pattee OH, Bloom PH, Scott JM, Smith MR (1990) Lead hazards within the range of the California condor. Condor 92:931–937

Powers DMW (2011) Evaluation: from precision, recall and F-measure to ROC, informedness, markedness and correlation. J Mach Learn Technol 2:37–63

Shipp MA, Ross KN, Tamayo P, Weng AP, Kutok JL, Aguiar RC, Gaasenbeek M, Angelo M, Reich M, Pinkus GS, others (2002) Diffuse large B-cell lymphoma outcome prediction by gene-expression profiling and supervised machine learning. Nat Med 8:68–74

Sriram TV, Rao MV, Narayana GS, Kaladhar DSVGK, VT (2013) Intelligent Parkinson disease prediction using machine learning algorithms. Int J Eng Innov Technol 3:212–215

Stauber E, Finch N, Talcott PA, Gay JM (2010) Lead poisoning of bald (*Haliaeetus leucocephalus*) and golden (*Aquila chrysaetos*) eagles in the US inland Pacific northwest region-an 18-year retrospective study: 1991–2008. J Avian Med Surg 24:279–287

Thessen AE (2016) Adoption of machine learning techniques in ecology and earth science. PeerJ PrePrints 4:e1720v1. https://doi.org/10.7287/peerj.preprints.1720v1

Umpleby JB (1913) Geology and ore deposits of Lemhi County, Idaho (No. Bulletin 528). Department of the Interior, United States Geological Survey, Government Printing Office, Washington, DC, USA

USFWS (1986) Final supplemental environmental impact statement on the use of lead shot for hunting migratory birds. Office of Migratory Bird Management, Washington, DC, USA

USFWS (2016) Bald and golden eagles: population demographics and estimation of sustainable take in the United States, 2016 update. USFWS, Division of Migratory Bird Management, Washington, DC, USA

Wang Q, Xie Z, Li F (2015) Using ensemble models to identify and apportion heavy metal pollution sources in agricultural soils on a local scale. Environ Pollut 206:227–235. https://doi.org/10.1016/j.envpol.2015.06.040

Wayland M, Bollinger T (1999) Lead exposure and poisoning in bald eagles and golden eagles in the Canadian prairie provinces. Environ Pollut 104:341–350

Wayland M, Wilson LK, Elliott JE, Miller MJR, Bollinger T, McAdie M, Langelier K, Keating J, Froese JMW (2003) Mortality, morbidity, and lead poisoning of eagles in western Canada, 1986–98. J Raptor Res 37:8–18

Part IV
Novel Applications of Machine Learning Beyond Species Distribution Models

Artwork by Catherine Humphries

"Artificial Intelligence, deep learning, machine learning—whatever you're doing if you don't understand it—learn it. Because otherwise you're going to be a dinosaur within 3 years."

– Mark Cuban

Chapter 13
Breaking Away from 'Traditional' Uses of Machine Learning: A Case Study Linking Sooty Shearwaters (*Ardenna griseus*) and Upcoming Changes in the Southern Oscillation Index

Grant R. W. Humphries

13.1 Introduction

The El Niño Southern Oscillation (ENSO) is a large-scale process which affects climate systems around the world (Rasmusson and Wallace 1983; Rasmusson et al. 1990; Latif et al. 1994; Timmermann et al. 1999; Turner 2004) and has also been shown to be correlated to population sizes of species (Stenseth et al. 2002; Velarde et al. 2004). Some of these correlations are straightforward: anchovy and sardine populations in the Humboldt Current fluctuate due to changes in upwelling that is directly related to the formation of an El Niño or La Niña (Schwartzlose et al. 1999; Ñiquen and Bouchon 2004; Gutiérrez et al. 2007). Other examples do not seem to have any immediately obvious mechanistic link: populations of owls in Chile have been linked to fluctuations in ENSO due to complex interactions between climate and abundance of prey items (Lima et al. 2002). The problem becomes further compounded when one begins to examine lagged effects. For example, a population that may be affected by ENSO may not begin to change until months after a climatic shift (e.g., Lima et al. 2002). This will occur in top predators because climate shifts that affect local conditions (e.g., total rainfall or wind speeds) will affect lower trophic level species (e.g., plants, insects, or even small mammals). Population changes in those affected lower trophic levels then propagate through the food web and are detected in top predators. However, oceanographic precursors to ENSO may also affect lower trophic levels, causing top predators to respond before the event. This would mean that changes in top predators (e.g., population) could be predictive of a shift towards an ENSO event. Studying the mechanisms of how and

G. R. W. Humphries (✉)
Black Bawks Data Science Ltd., Fort Augustus, Scotland
e-mail: grwhumphries@blackbawks.net

© Springer Nature Switzerland AG 2018   263
G. R. W. Humphries et al. (eds.), *Machine Learning for Ecology and Sustainable Natural Resource Management*, https://doi.org/10.1007/978-3-319-96978-7_13

why populations are predictive of El Niño and La Niña events will give insights into precursors of climate shifts. One species in New Zealand that has been linked to an early warning of ENSO is the sooty shearwater (*Puffinus griseus*).

Sooty shearwater chick harvesting from breeding colonies (annually in April through May around Stewart Island, New Zealand) is an important part of Rakiura (Stewart Island) Māori culture (Stevens 2006). Indices of chick quantity and size derived from harvest diaries were positively correlated with lagged values of the Southern Oscillation Index (SOI; Lyver et al. 1999). In years when there were fewer birds available to be harvested, negative values of SOI (i.e., El Niño conditions) tended to follow 4–14 months later. Similarly, when there were more birds available to be harvested, positive values of SOI (i.e., La Niña conditions) tended to follow 4–14 months later (Humphries and Möller 2017). This pattern was also reflected by the size of the chicks harvested where larger chicks indicated an oncoming positive shift in SOI, and smaller chicks indicated an oncoming negative shift in SOI (Humphries and Möller 2017). Obviously, birds do not cause the weather, therefore it is very likely that sooty shearwater adults are cuing into some oceanographic parameter(s) that acts as a precursor event to changes in SOI. Changes in oceanographic factors may lead to changes in prey quantity or quality, or in behavior which can alter how well birds are able to provide for their young (e.g., longer times at sea due to less favorable wind conditions).

Sooty shearwaters are flexible in the types of oceanographic habitats in which they forage during the breeding season. Therefore, it may be difficult to tease apart particular intricacies in the mechanisms that influence chick quantity and size. However, the importance of winds (speed and direction) for sooty shearwaters has been highlighted in the non-breeding (Adams and Flora 2009), and breeding (Raymond et al. 2010; Humphries 2014) seasons. Conditions which represent increased turbulence in the sub-Antarctic water region in March (i.e., significant wave height and wind speed in a region that sooty shearwaters use as a flyway enroute to foraging sites) are positively correlated with the size of chicks during the harvest by Rakiura Maori which occurs in mid-April to late-May (Humphries 2014). This is likely because increased wind speed increases the efficiency of adult flight, allowing birds to visit and return from foraging regions more quickly while using less energy. Signals from foraging efficiency in March would be detected in chick size during the period from which the population indices were derived (i.e., mid-April/late-May).

Due to the relationship between chick size and upcoming shifts in SOI, and the importance of March mean wind speeds in the sub-Antarctic water region in determining length of time of foraging trips and its relationship to chick size, I expect that wind speeds in the sub-Antarctic water region are also correlated with subsequent shifts in Southern Oscillation. Therefore, for this study I examined oceanographic variables associated with three important regions for sooty shearwaters (coastal NZ waters, offshore core foraging area, and sub-Antarctic water) against future lagged values of SOI. I do this in a machine learning context to demonstrate a novel use of generalized boosted regression models to search for oceanographic regions that explain variation in upcoming values of SOI.

13.2   Methods

13.2.1   Study Region

The study region was defined as the oceanographic extent around New Zealand where sooty shearwaters nesting on Whenua Hou (Codfish Island; 47° 47'S, 167° 38'E) were located during the 2004/2005 and 2005/2006 breeding seasons (defined as November first to mid-April of the following year) as shown from previously published light-sensing geolocator (GLS) data from Shaffer et al. (2006). The area extends from −16° to −66° latitude and 124° E to 124° W longitude (Fig. 13.1). Codfish Island is a small island, 3 km off the western coast of Stewart Island, and although not harvested by Maori, previous work has demonstrated that burrow densities are proportionally similar between this island and harvested islands (Lyver et al. 1999). I therefore make the assumption here that the population of shearwaters on Codfish Island is representative of birds from the nearby harvested islands.

13.2.2   Data

Monthly mean climatologies of physical oceanographic variables were downloaded from the European Centre for Medium-Range Weather Forecasts (ECMWF) Interim analysis project (https://apps.ecmwf.int) at a spatial resolution of 0.75 x 0.75 degrees for the years 1985 to the present. A list of the environmental data layers used in my analysis can be found in Table 13.1.

Environmental data were selected due to their potential relationships between bird behavior and physical oceanographic qualities which may affect atmospheric

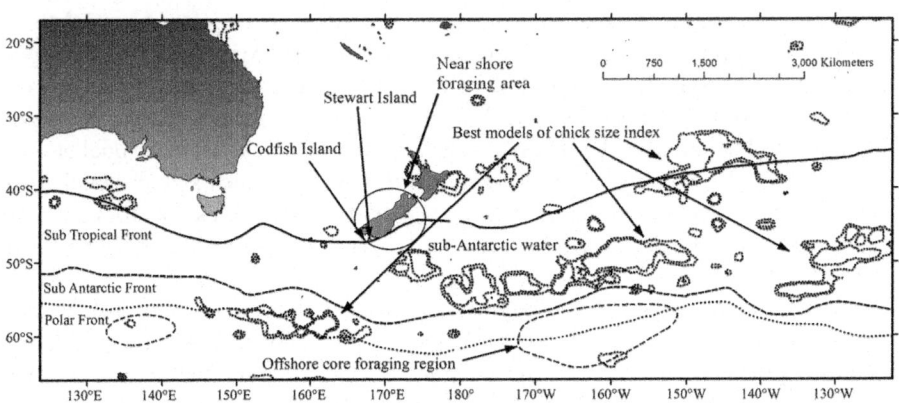

Fig. 13.1 Map of study area showing the regions more important for explaining variation in the chick size index, and offshore foraging areas in comparison to the Southern Ocean fronts. In this study, we focus on the eastern most offshore core foraging area

Table 13.1 European Center for Medium Range Weather Forecasting (ECMWF; https://apps. ecmwf.int/datasets) data downloaded for use in modeling exercises

Variable	Code	Units	Explanation
Charnock parameter	CHNK	–	Constant of atmospheric stress at ocean surface (Charnock 1955)
High cloud cover	HCC	%	Cloud cover at top level of ECMWF models
Low cloud cover	LCC	%	Cloud cover at lowest level of ECMWF models
Medium cloud cover	MCC	%	Cloud cover at mid-level of ECMWF models
Surface pressure	SP	Pa	Atmospheric pressure at surface of the ocean
Temperature at 2 meters depth	T2M	C	Ocean temperature at two meters depth
Total column water vapour	TCWV	kg*m^{-2}	Vertically integrated total mass of water vapour
Sea surface temperature	SST	C	Temperature at top microlayer of ocean
Significant wave height	SWH	m	Combined wind wave and swell height
Sea surface temperature gradient	SSTG	%	Percent change of sea surface temperature
Wind speed	WSPD	m/s	Wind speed from 0 to 10 m above surface of the ocean
Wind direction	WDIR	–	Classified compass bearing of wind direction (16 classes)
Wind differential	WDIF	Deg	Difference between direction of travel and wind bearing

circulation. For example, ocean temperature may affect the location where birds forage (due to prey types having specific oceanographic requirements) but may also influence evaporation and air temperature. Wind speeds directly impact bird flight but may also be indicative of atmospheric changes that could lead to shifts in the Southern Oscillation. Data were processed in ArcGIS 10.0 (www.esri.com), program R version 3.0.2 (R Core Team 2015), and NCL version 5.1.2 (NCAR Core Team 2013).

I also obtained chick size index data as described in Humphries and Möller (2017), which were representative of the months April and May due to the timing of the harvest. Mean chick size during the harvest (April and May) was available from 1985 to 2010 ($n = 25$). This was calculated from sooty shearwater harvest diaries by dividing the bucket size (10 liters) by the number of chicks in each bucket. The means were calculated by taking the average of the mean chick size per bucket for each season in each diary. I then calculated an overall mean chick size for each year by averaging the mean chick sizes across all diaries, thus creating a time series of mean chick sizes.

Sooty shearwaters begin breeding in November, with eggs hatching in January – early February at the latest. Adult birds feed chicks until mid-April at the latest, when the young have reached peak weights. It is at this time that chicks are harvested by local Maori. Humphries and Möller (2017) demonstrated that oceanographic parameters in March were important in determining the size of chicks in April – May (from harvest data) as this is the time of year when chicks are growing the fastest (Richdale 1945). Although it is possible there is a cumulative effect of oceanographic conditions on chick size over a season (i.e., variation in oceanographic conditions from hatching to April could impact overall growth rate and thus size of the chicks later in the season; Hedgren and Linnman 1979), conditions immediately prior to harvesting would remain the most important. For example, if conditions are good early in the season, chicks may be fed frequently or with high quality prey. However, in March when chicks are growing the fastest, if conditions change and adults are unable to forage for their young, the quality and size of birds can decrease dramatically. This was evidenced by personal field observations in February–March 2013 when chicks that weighed ~800 - 900 g would decrease to ~250 g and perish after 2 ½ – 3 weeks of not being fed. As such, I focused my analysis on environmental conditions in March.

I then used the near and offshore core foraging areas in March which was calculated using GLS tracking data from Shaffer et al. (2006) for March 2005 and 2006 from 14 birds tracked on Whenua Hou/Codfish Island. The 50% kernel densities were used to define the core foraging areas (Wood et al. 2000; Humphries 2014).

13.2.3 Spatial Models of SOI

I split the entire study area into 0.75 x 0.75-degree grid cells (matching the environmental data) and extracted the environmental data into each cell, including the temporal component. This created a time series for each of the environmental covariates (from 1985–2010, to match the shearwater data) in each grid cell. In each grid cell, I modeled the relationship between the environmental covariates and values of SOI from 1 to 24 months after March of the corresponding year (i.e., 24 time series models for each grid cell). March was used as the base month because it corresponds to when chicks reach their peak weight. I did this because I was interested in comparing results from these models on the same temporal scale that the chick size index was able to predict SOI values from Humphries and Möller (2017). For every 0.75 x 0.75-degree grid cell within the study area, I fit a generalized boosted regression model (gbm) model using the value of SOI as the response variable, and the oceanographic parameters as predictors.

Modeling was performed using the gbm algorithm in program R (Ridgeway 2007). The gbm algorithm is a non-parametric machine learning algorithm. I chose this particular algorithm to model as it is commonly used for predictive modeling and is not prone to over-fitting like more frequentist statistical techniques (Friedman 2002). It also has the ability to easily query relationships and interactions between explanatory and response variables and is not affected by inclusion of highly correlated predictors. This algorithm works by creating a series of regression trees (for continuous data), or decision trees (for categorical data) in an iterative fashion, minimizing the amount of error within the trees via cross validation (Friedman 2002). Cross validation tends to help avoid over-fitting of data and can lead to better overall predictive performance while taking into account complexities in datasets

(Breiman 2001). Yearly data of chick size were only available from 1985 to 2010 ($n = 25$), thus I was unable to perform a true predictive analysis on these data. This was because the sample size ($n = 25$) was too low to be able to perform statistically valid (i.e., independent) cross validation on the data. However, it was possible to assess the amount of variation captured within the data by comparing the root mean squared error (RMSE) or Spearman rho (r) values between models.

13.2.4 Comparing Ocean Variables to SOI

I calculated the spearman's r values for each month and oceanographic parameter for 24 months before after March to examine any potential overlap with correlation values from the chick size index. I did this to rule out the possibility that the chosen parameters were correlated to SOI prior to the period of interest. If strong correlations existed 24 months before March, it would mean that conditions in March were being affected by shifts in SOI which would further confound patterns in our data. I performed the analysis for three important oceanographic regions for breeding sooty shearwaters; the sub-Antarctic water region, the 50% kernel density core foraging area from GLS data, and the near shore foraging area. Only oceanographic parameters that had statistically significant correlations (as per basic linear regression techniques) with SOI after March were plotted in order to identify any of the potential mechanisms.

13.2.5 Models of SOI lags from Broad Oceanographic Regions

I next constructed four gbm models of SOI lags from 1 to 24 months after March using only oceanographic parameters as predictor variables: one for each of the three identified oceanographic regions and one model combining all three regions. This was done to demonstrate if oceanographic features alone from our identified areas could build a good model of upcoming shifts in SOI. The oceanographic parameters chosen (Table 13.1) were those that were identified to be significantly correlated to SOI values at the same time scale as correlations between SOI and chick size. Spearman r values for all four models were plotted, and the predictor variable ranks for the combined models from 4–14 months after March (i.e., the same time lag where significant correlations with chick size index were found) were calculated. I counted the numbers of times each variable occurred in the top three predictors to isolate which variables were most important to the model output. Finally, I created three-dimensional surface plots of the top three oceanographic predictor variables from the combined model with mean SOI values, and chick size values, from 4 to 14 months after peak chick size to demonstrate the overall relationship between chick size, SOI and the oceanographic features that link them.

13.3 Results

13.3.1 Spatial SOI Model Results

From one to three months after March, the oceanographic regions which best captured the variability in SOI (lowest RMSE values; 4.95 to 6.37) were to the Southeast of the study region, spanning an area from approximately 176° to 123° W, and from 50° to 65° S. This fell in a band which overlapped the southern parts of the sub-Antarctic water region, and both the sub-Antarctic and Polar fronts (Fig. 13.2). When comparing to Fig. 13.1, this region of low RMSE values overlapped with both the offshore core foraging area and the sub-Antarctic water region (Southeast of New Zealand) identified as important for sooty shearwaters in the 2004/2005 and 2005/2006 breeding seasons (Shaffer et al. 2006). This pattern remained until approximately 11 months after peak chick size (February of the following year), when variability was not well explained in any region. (Fig. 13.2). From 13 to 15 months after March (i.e., April, May and June 1 year after peak chick size), the region with the lowest RMSE values was a small patch to the South of New Zealand along the sub-Antarctic front. This dissipated at 16 to 17 months after peak chick size, but then from 18 to 23 months after March SOI values were best explained in the Tasman Sea, directly to the West of New Zealand. This pattern disappeared 24 months after peak chick size (Fig. 13.3).

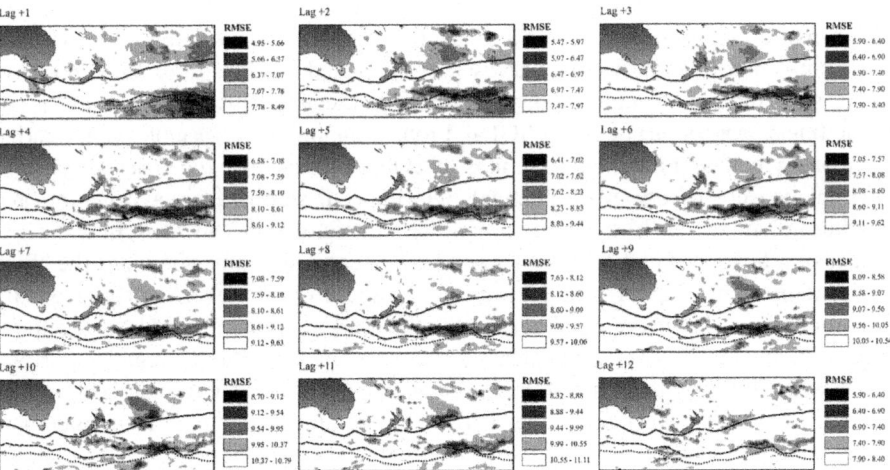

Fig. 13.2 Spatial generalized boosted regression models of Southern Oscillation Index (SOI) values for 1 to 12 months after the sooty shearwater peak chick feeding period (March). Maps represent the Root Mean Squared Error (RMSE) values for every grid cell in the study area to highlight oceanographic regions which best capture variability in SOI data using the entire suite of oceanographic variables. I did not keep the scale of RMSE the same between months because the target variables for each month differed and were thus not comparable in this way

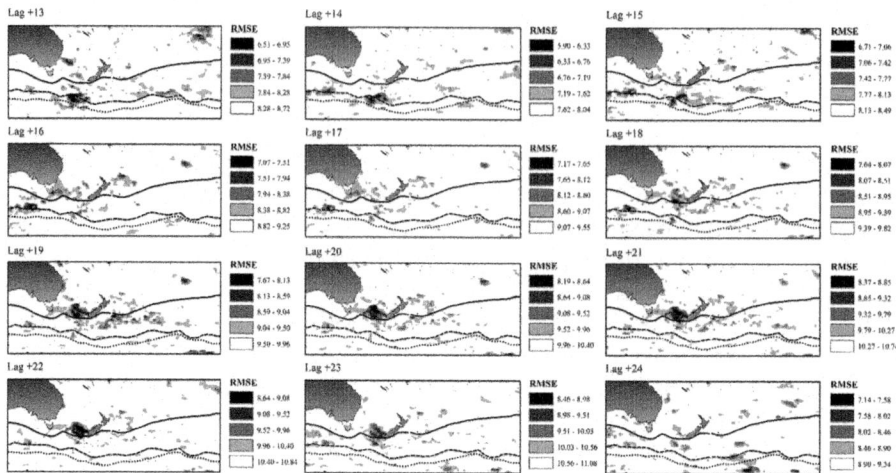

Fig. 13.3 Spatial generalized boosted regression models of Southern Oscillation Index (SOI) values for 13 to 24 months after the sooty shearwater peak chick feeding period (March). Maps represent the Root Mean Squared Error (RMSE) values for every grid cell in the study area to highlight oceanographic regions which best capture variability in SOI data using the entire suite of oceanographic variables. I did not keep the scale of RMSE the same between months because the target variables for each month differed and were thus not comparable in this way

13.3.2 Correlations between SOI and Ocean Parameters

Spearman correlations between the oceanographic parameters and values of SOI 24 months before and after the harvest were depicted in Figs. 13.4, 13.5 and 13.6. In the sub-Antarctic water region, there were significant positive correlations (as per basic linear regression) between SOI and four oceanographic variables that overlap temporally with significant correlations of SOI and chick size: the charnock parameter (a measure of ocean turbulence), wind speed, significant wave height, and medium cloud cover. In the case of wind speed, significant wave height and charnock, significant correlations overlapped with Spearman r values between the chick size index with a lag of 4 to 13 months. Medium cloud cover overlapped with significant correlations of chick size with a lag of 4 to 10 months. Sea surface temperature, low cloud cover, and surface pressure all had significant negative correlations with SOI after the harvest, however overlaps with the significant correlations from the chick size were fewer. Sea surface temperature had one overlapping value at lag month 4, while low cloud cover overlapped at lags 4, 5, 6, 12 and 13 months, and surface pressure overlaps with lags 8,9 and 10 months (Fig. 13.4).

In the core foraging region, correlation coefficients between SOI and sea surface temperature gradient remained weak and non-significant until 1 to 5 months after March (i.e., SOI for April, May, June, July and August). During this period, there was a negative significant correlation between sea surface temperature gradient and SOI (Fig. 13.5). Total column water vapor was found to be significantly, negatively

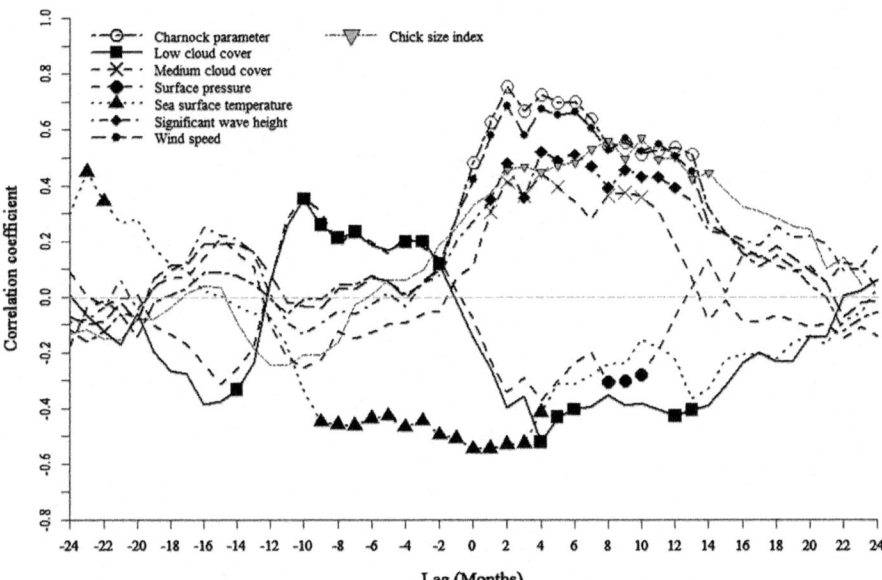

Fig. 13.4 Spearman's rho between March values of oceanographic variables from 1979 to 2010 in the sub-Antarctic water region Southeast of New Zealand and Southern Oscillation values from 24 months before and after peak chick size of sooty shearwater chicks. Variables plotted were only those that had significant correlations after the peak chick size period, and significant points are denoted by the corresponding marks on the lines as per the legend

correlated to SOI from 0 to 11 months after March. Surface pressure was also found to be significantly, negatively correlated to SOI from 0 to 10 months after peak chick size (March). Wind speed had a significant positive correlation from 0 to 14 months after March (i.e., chick size leads a change in wind speed). Low cloud cover was found to only have 3 points where there was significant correlations after peak chick size (4, 5 and 12 months), however, there was a trend of significant positive correlations with SOI from −10 to −1 months before March (Fig. 13.5).

In the near shore region (Fig. 13.6), only five variables were found to have significant correlations with SOI after March. Of the five variables, the only one that overlapped with the chick size correlations was sea surface temperature gradient from months 4 to 12, which has a significant positive correlation with SOI values.

In the sub-Antarctic water area and core foraging area, wind speed has a similar positive relationship to SOI. Wind speed also has a positive relationship to SOI after March in the near shore region however not at the time scale of interest. Medium cloud cover also shares a similar positive relationship to SOI in the sub-Antarctic water area and core foraging area, but not in the near shore area. Low cloud cover shares a similar negative relationship to SOI after March between all three regions. Surface pressure also shares a similar negative relationship between the sub-Antarctic water area and core foraging area, but not to the near shore region.

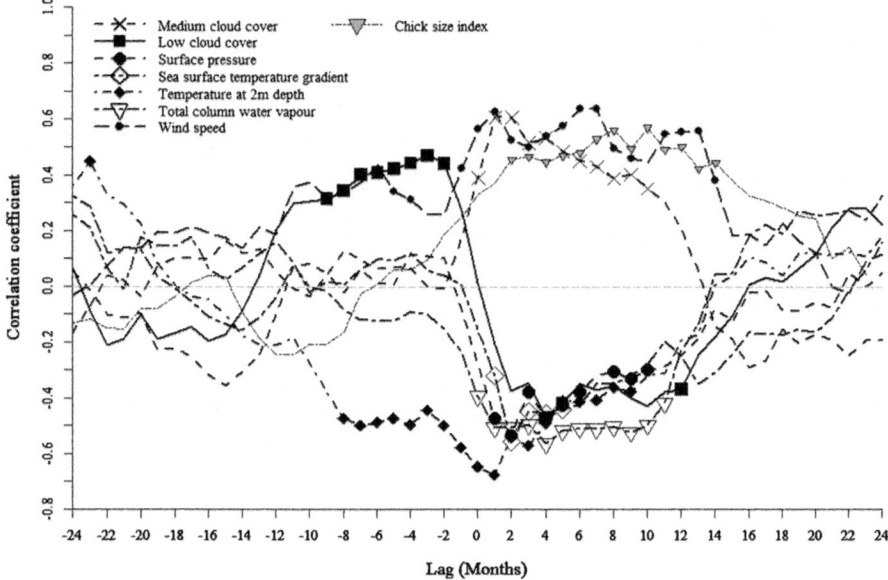

Fig. 13.5 Spearman's rho between March values of oceanographic variables from 1979 to 2010 in the offshore core foraging area Southeast of New Zealand and Southern Oscillation values from 24 months before and after peak chick size of sooty shearwater chicks. Variables plotted were only those that had significant correlations after the peak chick size period, and significant points are denoted by the corresponding marks on the lines as per the legend

13.3.3 Regional SOI Models

Gradient boosted regression models of SOI for the sub-Antarctic water region show that the best correlation coefficients of predicted versus observed values (r > 0.80) were from 1 to 14 months after March with the best predictions occurring at 2 months (r = 0.88). Assessment values then drop below r = 0.60 for 15 to 24 months after March. In the offshore core foraging area, the best model assessment values are from 1 to 7 months after March (r > 0.80). For the coastal NZ waters (i.e., near shore core foraging area), the best model assessments are from 16 to 21 months after March, coinciding with the timing of best predictions of SOI in the Tasman sea from Fig. 13.3 (Fig. 13.7).

When I ran a combined model of all the most important predictors (sea surface temperature, wind speed, medium cloud cover and significant wave height from the sub-Antarctic water region; sea surface temperature gradient from the near shore region; and wind speed, surface pressure, total column water vapor, temperature at 2 m depth and medium cloud cover from the core foraging area), model assessments were best from 1 to 13 months after March (r > 0.80) with the best prediction at month 9 (December; r = 0.86; Fig. 13.7). The best predictor variables for the combined model was wind speed in the sub-Antarctic water region, and in the offshore

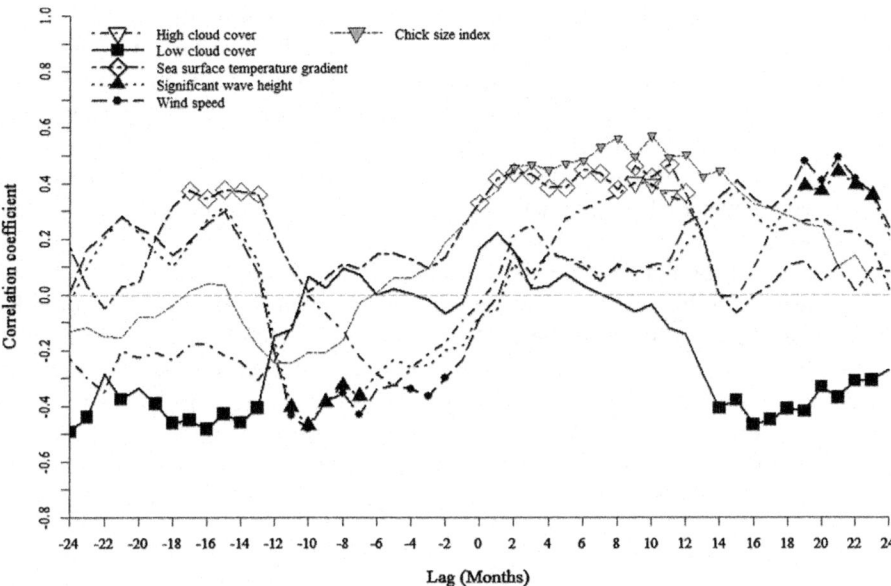

Fig. 13.6 Spearman's rho between March values of oceanographic variables from 1979 to 2010 in the near shore foraging area Southeast of New Zealand and Southern Oscillation values from 24 months before and after peak chick size of sooty shearwater chicks. Variables plotted were only those that had significant correlations after the peak chick size period, and significant points are denoted by the corresponding marks on the lines as per the legend

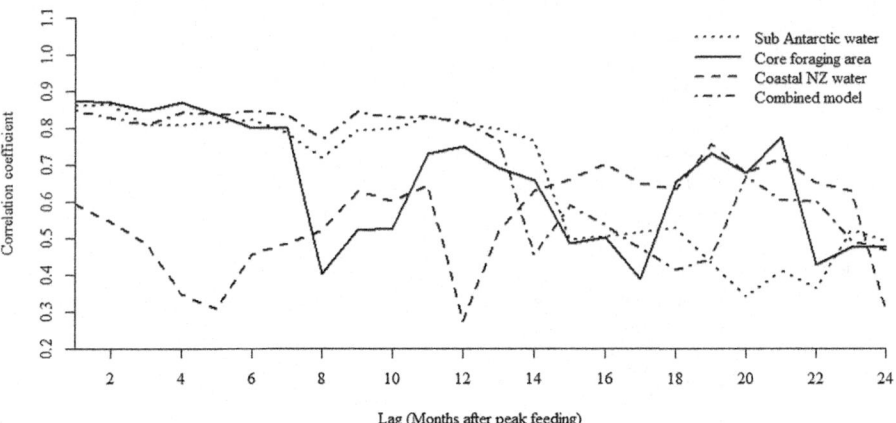

Fig. 13.7 Spearman's rho of predicted versus observed values for generalized boosted regression models of upcoming Southern Oscillation Index after March for the 1979–2010 time series using only oceanographic parameters from each region that were significantly correlated with SOI from 4–14 months after the sooty shearwater harvest. The core foraging area model is highlighted here with a solid line because I was interested in demonstrating how, although we have identified a potentially important area for sooty shearwater foraging, a better model is found in another area (Sub Antarctic water)

core foraging area, and the sea surface temperature gradient in the near shore core foraging area (Table 13.2). The highest values of mean SOI 4–14 months after peak chick size (generally corresponding to La Niña events) are noted when wind speed in both the sub-Antarctic water (r^2 = 0.52), and offshore core foraging areas (r^2 = 0.48) are higher, which also coincide with a higher chick size index (Figs. 13.8a and b). A similar pattern was found for sea surface temperature gradient in the near shore region, where higher chick size and gradient values are indicative of an upcoming La Niña (r^2 = 0.42; Fig. 13.8c).

Prior to a sustained El Niño 4–14 months after peak chick size, the chick size index during the harvest is on average 10% lower than prior to a sustained La Niña event (mean = 0.45 for an El Niño and mean = 0.50 for La Niña). I also found that mean wind speeds in the sub-Antarctic water region and offshore core foraging area were 9.29 m/s and 9.41 m/s respectively prior to El Niño conditions, which were 20% and 12% lower than wind speeds in both areas prior to a La Niña. In the near shore core foraging area, mean sea surface temperature gradient was 60.94% prior to El Niño conditions, and 72.85% prior to La Niña conditions (Table 13.3).

Table 13.2 Variable rankings from a generalized boosted regression model using a combination of the variables for each oceanographic region which had significant correlations with upcoming SOI values on the same time scale as the chick size index from (Humphries 2014)

Region	Variable	Number of times as top predictor	Number of times as second best predictor	Number of times as third best predictor	Total
Sub-Antarctic water	Wind speed	4	5	2	11
	Significant wave height	0	1	2	3
	Charnock parameter	0	0	0	0
	Medium cloud cover	0	0	0	0
Offshore core foraging area	Temperature at 2 m depth	0	1	1	2
	Wind speed	4	2	3	9
	Total column water vapour	0	0	1	1
	Medium cloud cover	0	0	1	1
Near shore core foraging area	Sea surface temperature gradient	3	2	0	5

Fig. 13.8 Three dimensional linear relationships between mean SOI from 4–14 months after peak chick feeding, the chick size index and the sub-Antarctic water (SAW) wind speed (**a**), offshore core foraging area (CFA) wind speed (**b**), and near shore sea surface temperature gradient (**c**)

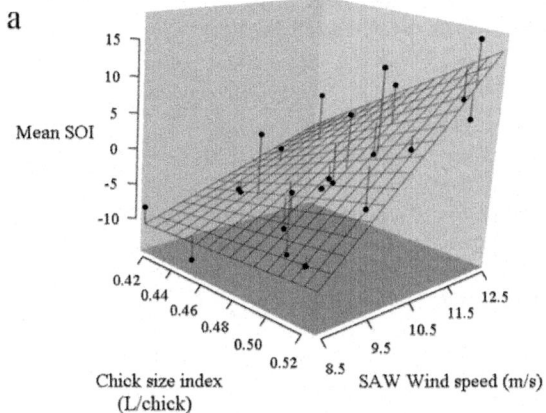

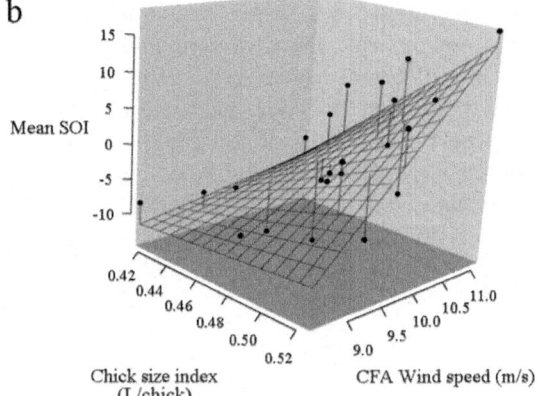

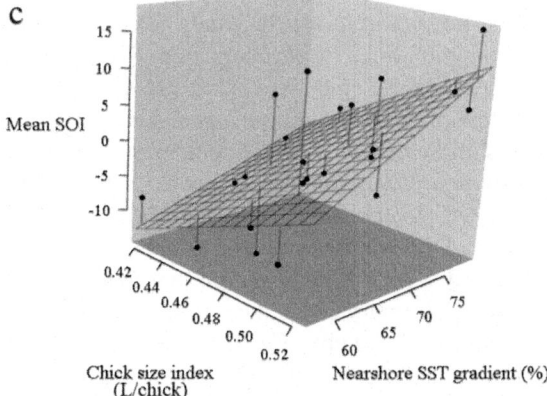

Table 13.3 Mean values of the chick size index and the top 3 predictor variables from the combined model prior to El Niño, normal, and La Niña conditions

Variable	Mean prior to El Niño conditions	Mean prior to normal conditions	Mean prior to La Niña conditions
Mean chick size (L / chick)	0.45	0.47	0.5
Sub-Antarctic water wind speed (m/s)	9.29	10.58	11.65
Offshore core foraging area wind speed (m/s)	9.41	10.12	10.67
Near shore sea surface temperature gradient (%)	60.94	66.84	72.85

13.4   Discussion

The importance of predicting upcoming shifts in climate cannot be understated due to the obvious role it has in agriculture (Hansen et al. 1998; Solow et al. 1998), economy (Tapia Granados et al. 2012), ecology (Veit et al. 1997; Stenseth et al. 2002; Stenseth et al. 2003), and society (Glantz et al. 1987). In this study my broad goals were to examine oceanographic factors in regions that were important for sooty shearwaters during the breeding season to determine if these factors could also explain variation in upcoming SOI shifts. My more specific goal of this study was to test if turbulent conditions (i.e., wind speed and significant wave height) in the sub-Antarctic water region (which affects size of chicks) were also predictive of upcoming shifts in SOI.

13.4.1   Considerations

All of the data used for this study were downloaded from ECMWF and represent model simulations of all parameters from 1985 to 2010. I was limited due to the time series of the chick size data however ECMWF data are available up to 2 months from the present, meaning the analysis could be extended if chick data were available. I chose to use ECMWF data because primary data obtained from remote sensing sources (e.g., satellites) do not extend back far in time (i.e., most data exist from mid-1990s to the present). Although the ECMWF data are generally calibrated using real time data from weather stations, they are not completely representative of measured values. It is also important to consider that I did not select biotic parameters (e.g., primary production or chlorophyll a or b) that have been correlated to the distributions of many seabirds (Hunt 1990; Suryan et al. 2006; Wells et al. 2008; Shaffer et al. 2009). Although there is mounting evidence that biology (particularly primary production) may be influencing climate on large scales (Charlson et al. 1987; Malin and Kirst 1997), I chose to only investigate physical ocean parameters which would be more likely to explain upcoming shifts in SOI. Sample size of the data is also a consideration when interpreting these results due to an n = 25 years for this dataset. When performing model assessment, the small sample size means that it is not feasible to perform independent cross-validation tests of model performance which would be indicative of the true predictive nature of the models. As such, gridded model assessment values are reported as the root mean squared deviation of the predicted versus observed within the data from 1985 to 2010.

13.4.2   Spatial Models of SOI Data

The spatial relationship between oceanographic conditions in March and upcoming values of SOI show a consistent pattern for 1 to 12 months after the sooty shearwater peak chick size period (March; Fig. 13.2). In general, the oceanographic region that best explains variation in the SOI data spans from 176° to 123° W and 50° to 65° S, with few other areas appearing in these maps (Fig. 13.2). For the first 3 months after March, the areas with the lowest RMSE overlap both the sub-Antarctic and Polar fronts. However, this situation changes by July (+4), when the Polar front seems to become a barrier to the areas of lowest RMSE, while overlap still exists with the sub-Antarctic front. The Polar front is the southern extent of the Antarctic Circumpolar Current (ACC), and as such acts as a barrier to biological and oceanographic processes (Convey et al. 2009). A pattern where the Polar front would act as a barrier to the low RMSE values is therefore not unexpected. When comparing the maps of core foraging areas for March 2005 and 2006, and the oceanographic area where the chick size index from Humphries and Möller (2017) is best explained (Fig. 13.1), there is relatively high overlap with both the offshore core foraging area and the sub-Antarctic water area and low RMSE values for upcoming SOI values. It is therefore likely that the link between SOI and chick size is driven by oceanographic conditions in the offshore core foraging area and sub-Antarctic water area.

Figure 13.3 depicts a region in the Tasman sea which has low RMSE values, suggesting that this region is important in predicting SOI from 16 to 24 months out. However, GLS data from Shaffer et al. (2006) and findings from Humphries and Möller (2017) do not support this region as a link between sooty shearwaters and SOI.

13.4.3   SOI and Oceanographic Factors

I examined spearman correlations between all the oceanographic factors for March and upcoming shifts in SOI in the sub-Antarctic water region, the offshore core foraging area, and the near shore region. In March in the sub-Antarctic water zone, I found that conditions that represent turbulence (i.e., significant wave height, wind speed and the charnock parameter) are highly correlated with upcoming SOI values Southern Oscillation Index (SOI):and oceanographic factors on the same time scale as significant correlations between chick size and upcoming SOI values. The relationship is positive, which indicates that when conditions are more turbulent in the sub-Antarctic region, there may be an upcoming La Niña event, and when conditions are less turbulent, there may be an upcoming El Niño event. A similar pattern is shown in wind speeds of the core foraging area. A combined model (using all variables that were correlated with SOI on the same time scale as the chick size index) confirms that the best predictors of upcoming shifts in SOI are wind speed variables in the offshore regions, and sea surface temperature gradient in the near shore region. The highest correlations between SOI values and oceanographic conditions in March in the sub-Antarctic water region occur at lags of two and three months, and then decrease, which suggests a gradual change in conditions prior to a shift in SOI.

The Southern Pacific sea-air flux is driven by two major latitudinally moving atmospheric circulations, the Hadley cell, which drives air from the equatorial Pacific towards the central South Pacific, and the Ferrell Cell, which drives air from the Antarctic region towards the central South Pacific (Schneider 2006). During a La Niña event, Hadley and Ferrel cells in the Pacific region are weakened, leading to increased storm frequency in the Southern Ocean between 170° to 120° W and 40° to 60° S regions due to a shift in the Southern jet stream. During El Niño events, the opposite is true, with strengthened Hadley and Ferrel cells causing generally more fine conditions in the same region due to a Northward shift of the jet stream (Rind et al. 2001). This means there are increased wind speeds during La Niña events, and decreased wind speeds during El Niño events, eventually leading to the formation of the Antarctic dipole (Yuan and Martinson 2000; Yuan and Martinson 2001; Liu et al. 2002). Because La Niña and El Niño events tend to take many months to form, it is likely that the pattern observed in Fig. 13.3 is due to the gradual southward shift of the Southern jet stream through the sub-Antarctic region, which would cause increased wind speeds and the presence of low pressure systems. A pattern of higher sea surface temperature gradients in the New Zealand coastal region prior to La Niña events could also be correlated to a southern shift in the southern jet stream due to wind speeds having a strong effect on the Southland current (Chiswell 1996). In this case, stronger wind speeds would lead to increased speed and strength of the Southland current, which would increase sea surface temperature gradient due to greater separation of colder offshore waters and warmer near shore waters.

13.4.4   Relating SOI, Chick Size and Oceanography

These findings show a 10% difference in the chick size index prior to La Niña events versus El Niño events. This corresponds to 20% difference in wind speed in the sub-Antarctic water region, a 12% difference in wind speed in the core foraging area, and a 16% difference in the near shore sea surface temperature gradient. Behavioral models from Humphries (2014) show strong relationships between wind speed and total time on long foraging trips. As suggested by these models, the likely explanation for the relationship between wind speed and the chick size index is due to foraging effort being altered, in that adults stay at sea for longer (or shorter) periods, thus affecting the size of chicks. However, the Polar front is on the edge of a major source of upwelling due primarily to wind speeds and direction in the ACC, and shifts in those speeds may, in some cases, cause changes in upwelling strength along the Polar front and therefore productivity (Convey et al. 2009). It could therefore be possible that there is a compounding effect of wind speed (via foraging effort) and general productivity in the foraging region, which would have consequences on chick size.

13.4.5 Future Steps

The reasons for shifts in SOI are complex and not well understood. Most models that currently exist are able to predict the onset of an El Niño or La Niña event by about 12 months. The results I have presented here rival current models and help resolve the mechanisms driving sooty shearwater dynamics during the breeding season. Further study is required to examine other aspects of these mechanisms including the effects of biological productivity shifts on the chick size index, and prey dynamics in foraging regions. More tracking studies over long-term periods would also help to determine if core foraging areas as determined from the Shaffer et al. (2006) tracking data remain stable over time, which would have effects on the breeding population. There are also unanswered questions as to any complex interactions between environmental covariates and sooty shearwater foraging activities with the formation of the Antarctic Dipole which drastically alters oceanographic conditions in the core foraging area (Yuan and Martinson 2001). The effect of the Antarctic Dipole on breeding sooty shearwaters needs to be assessed to answer these questions. Also, I have only examined oceanographic data from March in order to compare with the chick size index. There would be merit in examining oceanographic conditions across the entire breeding season (November–March) to determine if chick size (and perhaps SOI) can be predicted from cumulative effects over the course of the Southern hemisphere summer.

Finally, biological activity may play a major role in climate control on a global scale (Malin and Kirst 1997; Stefels et al. 2007; Le Clainche et al. 2010). Due to the large amount of biological activity in the Antarctic region (driven by upwelling from wind speed in the ACC region), it acts as a major sink for CO_2 in the atmosphere (Takahashi et al. 2002), at the same time potentially altering climate via cloud formation (Bates et al. 1987; Charlson et al. 1987). Therefore, it is also possible that the patterns I detected between wind speed, chick size index and SOI may be related to the fact that biological activity in the Antarctic, as determined by wind speed, has consequences in the upper atmosphere which leads to the relaxation of the trade winds and thus acting as a precursor to SOI shifts. However, this is speculation and no studies to this effect have yet been performed.

13.5 Conclusions

Understanding the dynamics between oceans, climate systems, and seabirds is important to understanding seabird population dynamics (Lloyd 1979; Croxall et al. 1999; Pinaud and Weimerskirch 2002; Ainley et al. 2005). It also highlights the importance of seabirds as sentinels of ecosystems (Piatt and Sydeman 2007; Parsons et al. 2008). In regions that were important for sooty shearwater chick size in the breeding season, oceanic factors that represented turbulence were also important in predicting shifts in the Southern Oscillation. Higher wind speeds prior to La Niña events in the sub-Antarctic water zone between the breeding colonies and the off-shore foraging regions allow birds to lower the amount of energy required for flight and thus spend less time foraging. These higher wind speeds are likely correlated with a transition from normal conditions to La Niña conditions as the Southern jet stream shifts south due to a relaxation of the Hadley and Ferrel cells in the Pacific Ocean. These relationships are only moderate and deeper studies of this phenomena could better resolve the connections between chick size, and ENSO. By studying this further it may be possible to monitor upcoming climate shifts by consulting groups like the Rakiura Māori about chick condition during the breeding season, which would have important implications in aspects of climate forecasting on a global scale.

Acknowledgements This project was funded by National Geographic, grant# WGS249-12 on behalf of the Waitt Foundation and the Department of Zoology at the University of Otago. Edits on early versions of the manuscript were given by D. Ainley, H. Möller, B. Raymond, and J.Overton. Thanks as well to the reviewers of this chapter. I would also like to thank the Rakiura Maori for the original sooty shearwater harvest data without which this study would not be possible. Finally, thank you to S. Shaffer for sharing GLS tracking data for Codfish Island. The data and the code used in this study are retrievable upon request from the author.

References

Adams J, Flora S (2009) Correlating seabird movements with ocean winds: linking satellite telemetry with ocean scatterometry. Mar Biol 157(4):915–929
Ainley D, Spear L, Tynan C, Barth J, Pierce S, Glennford R, Cowles T (2005) Physical and biological variables affecting seabird distributions during the upwelling season of the northern California current. Deep Sea Res II Top Stud Oceanogr 52:123–143
Bates T, Charlson R, Gammon R (1987) Evidence for the climatic role of marine biogenic sulphur. Nature 329:319–321
Breiman L (2001) Statistical modeling: the two cultures. Stat Sci 16(3):199–215
Charlson R, Lovelock J, Andreae M, Warren S (1987) Oceanic phytoplankton, atmospheric sulphur, cloud albedo and climate. Nature 326:665–661
Charnock H (1955) Wind stress on a water surface. Q J R Meteorol Soc 81(350):639–640
Chiswell S (1996) Variability in the Southland current New Zealand. N Z J Mar Freshw Res 30(1):1–17
Convey P, Bindschadler R, di Prisco G, Fahrbach E, Gutt J, Hodgson D, Mayewski P, Summerhayes C, Turner J (2009) Antarctic climate change and the environment. Antarct Sci 21:541–563

Croxall J, Reid K, Prince P (1999) Diet provisioning and productivity responses of marine predators to differences in availability of Antarctic krill. Mar Ecol Prog Ser 177:115–131

Friedman J (2002) Stochastic gradient boosting. Comput Stat Data Anal 38(4):367–378

Glantz M, Katz R, Krenz M (1987) The societal impacts associated with the 1982–83 worldwide climate anomalies. Report based on the workshop on the economic and societal impacts associated with the 1982–83 worldwide climate anomalies 11–13 November 1985 Lugano Switzerland

Gutiérrez M, Swartzman G, Bertrand A, Bertrand S (2007) Anchovy (Engraulis ringens) and sardine (Sardinops sagax) spatial dynamics and aggregation patterns in the Humboldt current ecosystem, Peru, from 1983–2003. Fish Oceanogr 16:155–168

Hansen J, Hodges A, Jones J (1998) ENSO influences on agriculture in the southeastern United States. J Clim 11:404–411

Hedgren S, Linnman Å (1979) Growth of guillemot Uria aalge chicks in relation to time of hatching. Ornis Scand 10:29–36

Humphries GRW (2014) Using long term harvest records of sooty shearwaters (Titi; Puffinus griseus) to predict shifts in the southern oscillation. Dissertation. University of Otago

Humphries GRW, Möller H (2017) Fortune telling seabirds: sooty shearwaters (Puffinus griseus) predict shifts in Pacific climate. Mar Biol 164(6):150

Hunt G (1990) The pelagic distribution of marine birds in a heterogeneous environment. Polar Res 8:43–54

Latif M, Barnett T, Cane M, Flügel M, Graham N, von Storch H, Xu J-S, Zebiak S (1994) A review of ENSO prediction studies. Clim Dyn 9:167–179

Le Clainche Y, Vézina A, Levasseur M, Cropp R, Gunson J, Vallina S, Vogt M, Lancelot C, Allen J, Archer S, Bopp L, Deal C, Elliott S, Jin M, Malin G, Schoemann V, Simó R, Six K, Stefels J (2010) A first appraisal of prognostic ocean DMS models and prospects for their use in climate models. Glob Biogeochem Cycles 24:1–13

Lima M, Stenseth NC, Jaksic FM (2002) Food web structure and climate effects on the dynamics of small mammals and owls in semi-arid Chile. Ecol Lett 5(2):273–284

Liu J, Yuan X, Rind D, Martinson D (2002) Mechanism study of the ENSO and southern high latitude climate teleconnections. Geophys Res Lett 29:8–11

Lloyd C (1979) Factors affecting breeding of razorbills Alca torda on Skokholm. Ibis 121(8):165–176

Lyver P, Möller H, Thompson C (1999) Changes in sooty shearwater Puffinus griseus chick production and harvest precede ENSO events. Mar Ecol Prog Ser 188:237–248

Malin G, Kirst G (1997) Algal production of dimethyl sulfide and its atmospheric role. J Phycol 33(6):889–896

NCAR Core Team (2013) NCAR command language (version 5.1.2) [software]. Boulder Colorado: UCAR/NCAR/CISL/VETS

Ñiquen M, Bouchon M (2004) Impact of El Niño events on pelagic fisheries in Peruvian waters. Deep Sea Res II Top Stud Oceanogr 51(6–9):563–574

Parsons M, Mitchell I, Butler A, Ratcliffe N, Frederiksen M, Foster S, Reid J (2008) Seabirds as indicators of the marine environment. ICES J Mar Sci 65:1520–1526

Piatt J, Sydeman W (2007) Seabirds as indicators of marine ecosystems. Mar Ecol Prog Ser 352:199–204

Pinaud D, Weimerskirch H (2002) Ultimate and proximate factors affecting the breeding performance of a marine top-predator. Oikos 99(1):141–150

R Core Team (2015) R: a language and environment for statistical computing (version 3.3.2) [software]. R foundation for statistical computing. Vienna, Austria

Rasmusson E, Wallace J (1983) Meteorological aspects of the El Nino/southern oscillation. Science 222(4629):1195–1202

Rasmusson E, Wang X, Ropelewski C (1990) The biennial component of ENSO variability. J Mar Syst 1:71–96

Raymond B, Shaffer S, Sokolov S, Woehler E, Costa D, Einoder L, Hindell M, Hosie G, Pinkerton M, Sagar P, Scott D, Smith A, Thompson D, Vertigan C, Weimerskirch H (2010)

Shearwater foraging in the Southern Ocean: the roles of prey availability and winds. PLoS One 5:e10960

Richdale L (1945) The nestling of the sooty shearwater. Condor 47(2):45–62

Ridgeway G (2007) Gbm: generalized boosted regression models. Documentation on the R package 'gbm' version 1.5

Rind D, Chandler M, Lerner J, Martinson D, Yuan X (2001) Climate response to basin-specific changes in latitudinal temperature gradients and implications for sea ice variability. J Geophys Res 106:20161–20173

Schneider T (2006) The general circulation of the atmosphere. Annu Rev Earth Planet Sci 34(1):655–688

Schwartzlose R, Alheit J, Bakun A, Baumgartner T, Cloete R, Crawford R, Fletcher W, Green-Ruiz Y, Hagen E, Kawasaki T, Lluch-Belda D, Lluch-Cota S, MacCall A, Matsuura Y, Nevárez-Martínez M, Parrish R, Roy C, Serra R, Shust K, Ward M, Zuzunaga J (1999) Worldwide large-scale fluctuations of sardine and anchovy populations. S Afr J Mar Sci 21:289–347

Shaffer S, Tremblay Y, Weimerskirch H, Scott D, Thompson D, Sagar P, Moller H, Taylor G, Foley D, Block B, Costa D (2006) Migratory shearwaters integrate oceanic resources across the Pacific Ocean in an endless summer. Proc Natl Acad Sci 103:12799–12802

Shaffer S, Weimerskirch H, Scott D, Pinaud D, Thompson D, Sagar P, Moller H, Taylor G, Foley D, Tremblay Y, Costa D (2009) Spatiotemporal habitat use by breeding sooty shearwaters Puffinus griseus. Mar Ecol Prog Ser 391:209–220

Solow A, Adams R, Bryant K, Legler D, O'Brien J, McCarl B, Nayda W, Weiher R (1998) The value of improved ENSO prediction to US agriculture. Clim Chang 39:47–60

Stefels J, Steinke M, Turner S, Malin G, Belviso S (2007) Environmental constraints on the production and removal of the climatically active gas dimethylsulphide (DMS) and implications for ecosystem modelling. Biogeochemistry 83:245–275

Stenseth N, Mysterud A, Ottersen G, Hurrell J, Chan K, Lima M (2002) Ecological effects of climate fluctuations. Science 297:1292–1296

Stenseth N, Ottersen G, Hurrell J, Mysterud A, Lima M, Chan K-S, Yoccoz N, Adlandsvik B (2003) Studying climate effects on ecology through the use of climate indices: the North Atlantic oscillation, El Nino southern oscillation and beyond. Proc R Soc B 270:2087–2096

Stevens M (2006) Kāi Tahu me te Hopu Tītī ki Rakiura: an exception to the "Colonial Rule"? J Pac Hist 41(3):273–291

Suryan R, Sato F, Balogh G, Davidhyrenbach K, Sievert P, Ozaki K (2006) Foraging destinations and marine habitat use of short-tailed albatrosses: a multi-scale approach using first-passage time analysis. Deep Sea Res II Top Stud Oceanogr 53:370–386. https://doi.org/10.1016/j.dsr2.2006.01.012

Takahashi T, Sutherland S, Sweeney C, Poisson A, Metzl N, Tilbrook B, Bates N, Wanninkhof R, Feely R, Sabine C, Olafsson J, Nojiri Y (2002) Global Sea–air CO2 flux based on climatological surface ocean pCO2, and seasonal biological and temperature effects. Deep Sea Res II Top Stud Oceanogr 49:1601–1622

Tapia Granados J, Ionides E, Carpintero Ó (2012) Climate change and the world economy: short-run determinants of atmospheric CO2. Environ Sci Pol 21:50–62

Timmermann A, Oberhuber J, Bacher A, Esch A, Latif M, Roeckner E (1999) Increased El Niño frequency in a climate model forced by future greenhouse warming. Nature 398:694–697

Turner J (2004) The El Niño–southern oscillation and Antarctica. Int J Climatol 24(1):1–31

Veit R, Mcgowan J, Ainley D, Wahl T, Pyle P (1997) Apex marine predator declines ninety percent in association with changing oceanic climate. Glob Chang Biol 3:23–28

Velarde E, Ezcurra E, Cisneros-Mata M, Lavin M (2004) Seabird ecology, El Nino anomalies, and prediction of sardine fisheries in the Gulf of California. Ecol Appl 14:607–615

Wells B, Field J, Thayer J, Grimes C, Bograd S, Sydeman W, Schwing F, Hewitt R (2008) Untangling the relationships among climate, prey and top predators in an ocean ecosystem. Mar Ecol Prog Ser 364:15–29

Wood A, Naef-Daenzer B, Prince P, Croxall J (2000) Quantifying habitat use in satellite-tracked pelagic seabirds: application of kernel estimation to albatross locations. J Avian Biol 31:278–286

Yuan X, Martinson D (2000) Antarctic Sea ice extent variability and its global connectivity. J Clim 13:1697–1717

Yuan X, Martinson D (2001) The Antarctic dipole and its predictability. Geophys Res Lett 28(18):3609–3612

Chapter 14
Image Recognition in Wildlife Applications

Dawn R. Magness

Digital camera equipment, data storage and image processing capacity have become cheaper and more accessible to ecologists. Camera trap stations, with the images delivered to our inboxes, are widely available (O'Connell et al. 2010). Ecologists and wildlife biologists are also deploying camera and videography equipment as a standard back up to traditional census techniques, such as observer counts of wildlife along transects. Drones can deliver high quality and detailed images of animals; for examples NOAA's recent release of Killer Whales collected images by drones (Fig. 14.1). The amount of collected images can quickly outpace our ability to analyze this data by hand. Can machine learning applications help ecologists and wildlife biologists leverage the information contained in these images?

In this chapter, I review some broad applications and uses of imagery for ecologists and wildlife biologists. Images can be used to (1) identify species for occurrence, (2) identify individuals for mark-recapture studies and other behavioral studies, and (3) count individual animals for population census. Machine learning can help us effectively process and extract information from images and in some cases; the methods are becoming more available to biologists without computer programming skills.

14.1 Identifying Species for Occurrence

Camera traps have been widely used in ecology and wildlife management across a spectrum of studies ranging from inventory to population estimates (O'Connell et al. 2010). Each camera trap station can produce large volumes of images; many

D. R. Magness (✉)
U.S. Fish and Wildlife Service, Kenai National Wildlife Refuge, Soldotna, AK, USA
e-mail: Dawn_Magness@fws.gov

© Springer Nature Switzerland AG 2018   285
G. R. W. Humphries et al. (eds.), *Machine Learning for Ecology and Sustainable Natural Resource Management*, https://doi.org/10.1007/978-3-319-96978-7_14

Fig 14.1 Drones now deliver high quality, detailed images of wildlife. NOAA has been capturing images of killer whale pods

without animals in the frame. Camera traps are an appealing method because they can be deployed in many places at the same time. However, annotating all the images is a challenge. Requiring an observer to go through every picture can be very time consuming. For example, researchers at Serengeti National Park, Tanzania deployed 225 camera traps over 3 years and collected over 1.2 million image sets (Swanson et al. 2015). The Snapshot Serengeti project used crowd sourcing to find volunteers to process the images and provide species identification. Snapshot Serengeti released this classified baseline dataset in 2015. Can computer scientists use machine learning to find new approaches to image processing and species identification for camera trap datasets?

Computer scientists are applying machine learning approaches to this problem. Image recognition has been an area of research developed in part to allow internet search engines to deliver images from the World Wide Web. Google recently released Tensorflow (https://www.tensorflow.org/), a scalable deep learning neural network useful for image recognition, as an open-source software library (Abadi et al. 2016). Machine learning can be used to identify images with animals in the frame and to classify the images by species. Machine learning uses an image dataset that has been annotated by an observer to train the model to categorize images. The model is then applied to a validation dataset to assess the predictive ability. Camera trap data can be difficult to classify with models because of unbalanced samples; for example, when there are many empty frames or the species of interest is rarely in the image when compared to a more common species. The model must also be robust enough to handle incomplete images and huge differences in the size, shape, and color of an animal that is due to where the animal is in relation to the camera and changing environmental conditions (Gomez and Salazar 2016). Often very large training datasets are needed to allow the model to train across the range of conditions or the image dataset must be filtered to a subset of high quality images.

Elias et al. (2016) used a very large dataset to train a Tensorflow computer vision model to classify images collected by a network of camera stations deployed to monitor wildlife on the University of California Santa Barbara Sedgwick Reserve. The large training dataset included empty images and augmented images that had bear, deer and coyote imaged cropped and overlaid on empty images. The augmented images used publically available images to develop the training dataset. The model was trained on the commercial cloud because it needed the computing power to handle the large datasets. When tested on a validation dataset of empty and images with animals, the model was able to accurately classify 87% bear, deer and coyote images (Elias et al. 2016). The classification ability of the trained model could be improved by including by including even more variation in image angles for deer, which had the majority of misclassified images (14% of images misclassified for deer; 1% for bear and coyotes).

Gomez and Salazar (2016) tested classification errors for 26 animals from the Snapshot Serengeti. They had classification accuracy of 35% when the dataset was unbalanced an included empty frames and frames with distant animals. Therefore, they filtered out the images without animals and trained the model on images with animals. When the dataset was preprocessed to remove empty frames and to only include images with optimal animal placement in the foreground, classification accuracy was 83%. The Snapshot Serengeti used human observers to annotate the entire dataset. Machine learning algorithms can also be used to pre-filter the training data. Earlier approaches filtered out empty frames by comparing pictures that are in a time series and use change between frames to indicate an animal (Figueroa et al. 2014). Figueroa et al. (2014) used change detection with some success for detecting ocelots, though understanding if the change was an animal entering or leaving the frame was problematic. Another preprocessing approach to filter images before classification is segmentation. Segmentation detects blobs or objects in the image. Figueroa et al. (2014) found that the SURF algorithm could be used to find ocelots in an image. Yu et al. (2013) used segmentation (SIFT) to crop the animal body out of the camera trap station images. These cropped images included 18 species and were classified with a multi-class SVM with 82% accuracy. Accuracy ranged from 58%–93%. The lowest accuracy was for Red brocket deer that were mostly misclassified as White-tailed deer.

14.2 Identifying Individuals for Mark Recapture

Beyond detection and occupancy studies, images can be used to identify individuals for population estimates via mark-recapture analyses. Tracking individuals can also give us insight into animal behavior, such as residency and movements (Marshall and Pierce 2012). Marine biologists have a long history of using images to study marine mammals. For example, researchers began photographing humpback whale flukes in the 1960s to use variation in fluke coloration to track individuals (S. Katona et al. 1979). Catalogs of humpback whale flukes are still published to allow researchers and volunteers, such as tour boat operators, to contribute sighting locations. These photographic datasets can be used to estimate population demographics via mark-recapture methods (Katona and Beard 1990; Smith et al. 1999). Photographic mark recapture studies have been used for a variety of marine species, such as whale sharks and some terrestrial species (Meekan et al. 2006; Van Tienhoven et al. 2007; Speed et al. 2007). Often a human observer is used to identify individuals, but this can be time intensive and limit the number of photographs that can be processed. Furthermore, matching errors increase as the size of the image catalog increases making observer matching for large populations less feasible.

Machine learning algorithms have been a focus of research groups interested in automating the identification of individuals. Image matching has been successfully deployed for a range of animals with coloration variability among individuals including fish, amphibians, and mammals (Bolger et al. 2012). Computer assisted identification can identify individuals through image matching faster than human observers and with similar accuracy (Dala-Corte et al. 2016). For image-based mark recapture, animal marking must be stable across time. In some species, natural marking and scar patterns can grow and change (Marshall and Pierce 2012). In general, the following steps are needed to match individuals in new images with databases of previously acquired images: (1) image preprocessing (2) creation and maintenance of image database, (3) use computer vision algorithm to extract information about animal markings from the image and (4) compare animal markings among images and rank potential matches (Bolger et al. 2012).

A variety of open-source software packages have been developed to aid in computer assisted image-based mark recapture studies (Table 14.1). These software packages reduce the programming skills needed by researchers interested in this approach. In most of these studies, animals are photographed to document marking that can be used to distinguish individuals, so software is useful for comparing these photographs, but will not help filter out images with no animals in the frame or with poor quality (see previous section).

In some cases, fully automated identification is possible. A colony of African penguins (*Spheniscus demersus*) on Robben Island, South Africa are being monitoring with a fully automated identification system (Sherley et al. 2010). Penguins are an ideal species for automated monitoring because they have contrasting plumage, stable and unique individual plumage, slow movements, and predictable travel routes. Images of the penguins are captured in the daytime by cameras placed along "penguin highways" which are the primary pathways to and from the sea (Sherley et al. 2010). Images were captured by an Ethernet camera networked to a computer for image processing. In order to identify individual penguins, machine learning algorithms were used to (1) locate each penguin in the images and identify the belly as the area of interest, (2) orient or pose each penguin to be front facing, and (3) identify individuals based on belly spot pattern (Burghardt 2008). Burghardt (2008)

Table 14.1 Open-source software packages developed for identifying individual animals in photographs.

Program	Pre-processing	Match algorithm	Final match	Citation	Website
Wild ID	Images need to be cropped to area of interest prior to importing into software	SIFT	RANSAC ranks matches and user inspection of best 20 matches	(Bolger et al. 2012)	http://dartmouth.edu/faculty-directory/douglas-thomas-bolger
Hotspotter	User identifies region of interest and orients image	RootSIFT	Local native Bayes nearest neighbor algorithm assigns identity	(Crall et al. 2013)	http://cs.rpi.edu/hotspotter/
SLOOP	Segmentation, illumination correction, and rectification tools available	Multi-scale PCA, scale-cascaded alignment, SIFT, SURF, ORB,	User inspection of best matches	(Duyck et al. 2015)	https://sloop.mit.edu/
FoTo Spottr	User outlines the area of interest	Spot segmentation and pixel by pixel comparison	User inspection of best 10 matches	(Schoen et al. 2015)	http://cnd.mcgill.ca/~aschoen/spottr/

developed the methods used to automate penguin counts in his well-documented dissertation. Sherley et al. (2010) found that 13% of the 1453 penguins that passed the camera could be processed to the individual with >90% accuracy. This sampling rate was similar to what could be achieved with flipper banding.

However, many software packages are semi-automated and not fully automated. In semi-automated software, animal images must be preprocessed by to crop and orient the animal. The user is also usually required to validate the final match by reviewing the top ranked candidates. A common algorithm for identifying units of animal markings is the Scale Invariant Feature Transform (SIFT) algorithm (Lowe 2004). SIFT is powerful because it can analyze images even when the scale (size of the animal in the frame) and rotation (angle of the animal in the frame) between two images are different. SIFT is too complex to be covered adequately here and it may not be important for most ecologists to understand exactly how it works (several websites include detailed descriptions such as, http://opencv-python-tutroals. readthedocs.io/en/latest/py_tutorials/py_feature2d/py_sift_intro/py_sift_intro. html). In simple language, SIFT identifies and describes features within an image (such as a distinct edge on a marking) in a way that is invariant to scale or orientation. SIFT then identifies candidate pairs of features present in pairs of images.

Image mark recapture studies have been successfully applied to wildlife studies. This technique was used to estimate demographic parameters for a giraffe population (Bolger et al. 2012). Success is largely dependent on the ability of computers to accurately identify individual animals in images. Misclassification can negatively bias survival estimates (Morrison et al. 2011). Error rates in most image recognition studies are low. For example, 95% of known matches were identified in a large database of marbled salamanders (*Ambystoma opacum*) (Gamble et al. 2008).

14.3   Counting Animals for Population Census

Many species require accurate population estimates for management applications. Aerial surveys can be an effective way to census animals (Jolly 1969). Wildlife managers and researchers can use imagery, from manned and unmanned aircraft, to aid in aerial census. Drones can collect data at low elevation and with high image resolution, which provides the opportunity to monitor smaller animals with aerial images. Small shorebirds are now possible to detect with low altitude drone missions (Chabot and Francis 2016). High resolution imagery collected in conjunction with observer-based aerial counting can be used to estimate observer errors or the images can be used as the primary census method. When images are used, the population census can be archived and repeat counting is possible. Observers have traditionally used photo editing software to hand mark and count animals in each image. Hand counting is time consuming, limits the area that can be sampled, and is subject to errors (Bajzak and Piatt 1990; Russell et al. 1996). Machine learning algorithms may be used to automate animal counts in order to make processing the images manageable.

There has been interest in using computers to count animals for nearly 30 years (Bajzak and Piatt 1990). Chabot and Francis (2016) provide an extensive review of studies that have used automated counting methods for birds. Of the 19 computer automated bird counts reviewed by Chabot and Francis (2016), 6 used spectral thresholding, 5 used spectral analysis and segmentation, and 5 used an object based classification. In other words, most studies to date have used approaches from the fields of remote sensing. Automating counting of animals in images must deal with the problem of many empty frames and for some spices, animals that occur in variable densities.

Spectral thresholding delineates the spectral signal of animals from their background and can provide adequate results for animals with high contrast. For example, snow geese could be counted with an accuracy of 2.8%, while caribou, which blend more with their background, had 10.2% count error (Laliberte and Ripple 2003). Software from the medical sector, such as ImageTool (http://compdent.uth-scsa.edu/dig/itdesc.html) and Image J (https://imagej.nih.gov/ij/), have tools that can delineate blobs and generate attributes that describe the shape and size of the blob. Blob size can be related to number of animals for animals in large flocks or that gather in groups (Pérez-García 2012; Laliberte and Ripple 2003). Blob characteristics, such as shape, can be used to filter out blobs that are not animals (Cunningham et al. 1996). Variable backgrounds between images, differences in lighting, and animal location can require that each image be preprocessed to sharpen the contrast of the animal of interest or to crop out problematic features in the picture (Chabot and Francis 2016).

Remote sensing software has also been utilized to segment and classify the spectral signatures of animals. For example, ArcGIS was used to conduct an unsupervised classification of a gull colony. The bird pixels were converted to polygons that were filtered by size to remove larger species. The resulting count was similar to visual counting methods (Grenzdörffer 2013). However, pixel based approaches did not work well to detect deer, an animal that occurs at low densities and blends with the background, because of high commission errors (Chrétien et al. 2016).

Object-based image analysis software, such as Ecognition, derives objects by grouping similar pixels for segmentation and classification (Blaschke 2010). Object-based image analyses use hierarchical rule-sets to segment imagery and classify objects. This flexibility is useful for separating animals from backgrounds. For example, seabird flocks can be identified and separated from variable sea conditions that cause counting errors prior to identifying individual birds (Chabot and Francis 2016). Object-based approaches have been used to count male and female scoters using aerial photographs (Groom et al. 2007). In the East African savannah, large mammals have been identified and counted using GeoEye-1 satellite images (2-m resolution pan sharpened to 0.5 m) with a 7% omission error and a 14% commission error; though species could not be distinguished due to the pixel resolution (Yang et al. 2014). Chrétien et al. (2016) had a detection probability of 0.5, a rate slightly lower than human observers, for deer using an object-based image analysis of thermal infrared and color images collected with an unmanned aircraft.

Other machine learning approaches have been applied to counting animals. Abd-Elrahman et al. (2005) used pattern recognition techniques from the field of computer vision to count wading birds. Both spectral signatures and image matching of photographed birds were used to recognize and count birds in the image. First, a normalized cross correlation, which provides a measure of the degree of similarity between a sample location and a known image of a bird, was used to identify potential birds in the images. Next, the center point of these potential birds is grown based on similarity to the spectral signatures of the image. The spectral characteristics within this region are them compared to the spectral signatures of photographed bird images. Finally, potential birds are filtered by size to remove areas that are too small. Mean omission error across images was 10% and mean commission error was 14%. Commission errors generally occurred when bright spots occurred due to combinations of sun, camera and vegetation angle.

Maire et al. (2015) used Convolutional Neural Networks (CNN) to detect Dugong (*Dugong dugon*), a marine mammal that can occur at very low densities. This approach stems from early work on face detection. A moving window feeds information to a classifier algorithm to detect the presence of an object of interest. The Pylearn2 framework in Python was used to implement the CNN and the code and dataset is archived by Maire et al. (2015). The best model correctly counted 41 Dugong, missed 10, and falsely identified 110. Maire et al. (2015) suggest that as more data is acquired, the model will become more accurate. Although these results would not be accurate enough for automated counting, the model does reduce the number of images that must be reviewed by a human counter. However, this approach cannot be applied when there are differences in image acquisition. The approach is scale dependent, so photos must be taken at a consistent altitude so that animals will not vary in apparent size.

Torney et al. (2016) use a rotation-invariant feature classification from the field of computer vision to perform a wildebeest population census. A 2009 survey of wildebeest collected 2017 images across the Serengeti National Park. Wildebeests in every image were counted twice, using 2 independent observers and compared to the computer automated count. Per image, the error rates are greater in the automated count. For the total census, the automated count is more accurate than either manual count alone because there is no systematic bias. However, when multiple human counts are averaged, they outperform the computer count. The computer correctly counted 1423 wildebeests, missed 235, and falsely counted 496. Torney et al. (2016) believe that the results are not accurate enough for fully automated counts, but the methods are promising.

References

Abadi M, Agarwal A, Barham P, Brevdo E, Chen Z, Citro C, Corrado GS, Davis A, Dean J, Devin M (2016) Tensorflow: large-scale machine learning on heterogeneous distributed systems. *arXiv Preprint arXiv*:1603.04467

Abd-Elrahman A, Pearlstine L, Percival F (2005) Development of pattern recognition algorithm for automatic bird detection from unmanned aerial vehicle imagery. *Surv Land Inf Sci* 65(1):37

Bajzak D, Piatt JF (1990) Computer-aided procedure for counting waterfowl on aerial photographs. *Wildl Soc Bull.* 1973–2006 18(2):125–129

Blaschke T (2010) Object based image analysis for remote sensing. *ISPRS J Photogramm Remote Sens* 65(1):2–16

Bolger DT, Morrison TA, Vance B, Lee D, Farid H (2012) A computer-assisted system for photographic mark–recapture analysis. *Methods Ecol Evol* 3(5):813–822

Burghardt T (2008) Visual animal biometrics: automatic detection and individual identification by coat pattern. University of Bristol

Chabot D, Francis CM (2016) Computer-automated bird detection and counts in high-resolution aerial images: A review. *J Field Ornithol* 87

Chrétien L-P, Théau J, Ménard P (2016) Visible and thermal infrared remote sensing for the detection of white-tailed deer using an unmanned aerial system. *Wildl Soc Bull 40*

Crall JP, Stewart CV, Berger-Wolf TY, Rubenstein DI, Sundaresan SR (2013) Hotspotter—patterned species instance recognition. In, 230–237. IEEE

Cunningham DJ, Anderson WH, Michael Anthony R (1996) An image-processing program for automated counting. *Wildl Soc Bull* 24(2):345–346

Dala-Corte RB, Moschetta JB, Becker FG (2016) Photo-identification as a technique for recognition of individual fish: A test with the freshwater armored catfish Rineloricaria Aequalicuspis Reis & Cardoso, 2001 (Siluriformes: Loricariidae). *Neotropical Ichthyology* 14(1)

Duyck J, Finn C, Hutcheon A, Vera P, Salas J, Ravela S (2015) Sloop: A pattern retrieval engine for individual animal identification. *Pattern Recogn* 48(4):1059–1073

Elias AR, Golubovis N, Krintz C, Wolski R (2016) Where's the bear?–automating wildlife image processing using iot and edge cloud systems. University of California, Santa Barbara, CA

Figueroa K, Camarena-Ibarrola A, García J, Villela HT (2014) Fast automatic detection of wildlife in images from trap cameras." In, 940–947. Springer

Gamble L, Ravela S, McGarigal K (2008) Multi-scale features for identifying individuals in large biological databases: An application of pattern recognition technology to the marbled salamander Ambystoma Opacum. *J Appl Ecol* 45(1):170–180

Gomez A, Salazar A (2016) Towards automatic wild animal monitoring: identification of animal species in camera-trap images using very deep convolutional neural networks." *arXiv* :1603.06169, no. Journal Article

Grenzdörffer GJ (2013) UAS-based automatic bird count of a common gull colony. In International archives of the photogrammetry, Remote sensing and spatial information sciences, Vol XL-1/W2, 2013 UAV-g2013. pp 169–174

Groom GB, Petersen IK, Fox T (2007) Sea bird distribution data with object based mapping of high spatial resolution image data. In *Challenges for earth observation-scientific, technical and commercial. Proceedings of the remote sensing and photogrammetry society annual conference*

Jolly GM (1969) Sampling methods for aerial censuses of wildlife populations. *East African Agricultural and Forestry Journal* 34(sup1):46–49

Katona SK, Beard JA (1990) Population size, migrations and feeding aggregations of the humpback whale (Megaptera Novaeangliae) in the western north atlantic ocean. Report of the international whaling commission (Special issue 12), no. Journal Article: 295–306

Katona S, Baxter B, Brazier O, Kraus S, Perkins J, Whitehead H (1979) Identification of humpback whales by fluke photographs. In: Behavior of marine animals. Springer, pp 33–44

Laliberte AS, Ripple WJ (2003) Automated wildlife counts from remotely sensed imagery. *Wildl Soc Bull* 31:362–371

Lowe DG (2004) Distinctive image features from scale-invariant keypoints. *Int J Comput Vis* 60(2):91–110

Maire F, Alvarez LM, Hodgson A (2015) Automating marine mammal detection in aerial images captured during wildlife surveys: a deep learning approach. In: Australasian joint conference on artificial intelligence. Springer, pp 379–385

Marshall AD, Pierce SJ (2012) The use and abuse of photographic identification in sharks and rays. *J Fish Biol* 80(5):1361–1379

Meekan M, Bradshaw C, Press M, McLean C, Richards A, Quasnichka S, Taylor J (2006) Population size and structure of whale sharks (Rhincodon Typus) at ningaloo reef Western Australia, no. Journal Article

Morrison TA, Yoshizaki J, Nichols JD, Bolger DT (2011) Estimating survival in photographic capture–recapture studies: overcoming misidentification error. *Methods Ecol Evol* 2(5):454–463

O'Connell AF, Nichols JD, Ullas Karanth K (2010) Camera traps in animal ecology: methods and analyses. Book, Whole. Springer Science & Business Media

Pérez-García JM (2012) The use of digital photography in censuses of large concentrations of passerines: The case of a winter starling roost-site. *Revista Catalana d'Ornitologia* 28:28–33

Russell J, Couturier S, Sopuck LG, Ovaska K (1996) Post-calving photo-census of the Rivière George caribou herd in July 1993. *Rangifer* 16(4):319–330

Schoen A, Boenke M, Green DM (2015) Tracking toads using photo identification and image-recognition software. *Herpetol Rev* 46(2):188–192

Sherley RB, Burghardt T, Barham PJ, Campbell N, Cuthill IC (2010) Spotting the difference: towards fully-automated population monitoring of african penguins Spheniscus Demersus. *Endanger Species Res* 11(2):101–111

Smith TD, Allen J, Clapham PJ, Hammond PS, Katona S, Larsen F, Lien J, Mattila D, Palsbøll PJ, Sigurjónsson J (1999) An ocean-basin-wide mark-recapture study of the North Atlantic Humpback Whale (Megaptera Novaeangliae). *Mar Mamm Sci* 15(1):1–32

Speed CW, Meekan MG, Bradshaw CJA (2007) Spot the match–wildlife photo-identification using information theory. *Front Zool* 4(1):1

Swanson A, Kosmala M, Lintott C, Simpson R, Smith A, Packer C (2015) Snapshot serengeti, high-frequency annotated camera trap images of 40 Mammalian Species in an African Savanna. *Sci Data* 2

Torney CJ, Dobson AP, Borner F, Lloyd-Jones DJ, Moyer D, Maliti HT, Mwita M, Fredrick H, Borner M, Hopcraft JGC (2016) Assessing rotation-invariant feature classification for automated wildebeest population counts. *PloS One* 11(5):e0156342

Van Tienhoven AM, Den Hartog JE, Reijns RA, Peddemors VM (2007) A computer-aided program for pattern-matching of natural marks on the spotted raggedtooth shark Carcharias Taurus. *J Appl Ecol* 44(2):273–280

Yang Z, Wang T, Skidmore AK, de Leeuw J, Said MY, Freer J (2014) Spotting East African mammals in open savannah from space. *PloS One* 9(12):e115989

Yu X, Wang J, Kays R, Jansen PA, Wang T, Huang T (2013) Automated identification of animal species in camera trap images. *EURASIP J Image and Video Process* 2013(1):1

Chapter 15
Machine Learning Techniques for Quantifying Geographic Variation in Leach's Storm-Petrel (*Hydrobates leucorhous*) Vocalizations

Grant R. W. Humphries, Rachel T. Buxton, and Ian L. Jones

15.1 Introduction

Many ecological studies have explored the numerous applications of machine learning (ML) algorithms. Primarily, these algorithms are used for species distribution models (Elith et al. 2006; Warren and Seifert 2011), where occurrence data (presences/absences) are modeled with environmental covariates to make predictions to larger areas. However, one can also apply these techniques to other interesting classification problems, including analysis of count, categorical, and continuous data (see also Chap. 1 in this book for an overview).

A potentially valuable application for ML techniques is classification of individuals into biologically relevant groups without any pre-conceived assumptions of what defines those groups. This is often done in genetics studies, where unsupervised techniques (e.g., clustering) are used to classify organisms based on their genetic makeup (Chen and Ishwaran 2012). For an ecologist, one might compare this to how you recognize an animal taxon in the field; the size, colors, vocalizations etc., that are all identification criteria for species. This gets more challenging when comparing sub-species, for example the split of the song sparrow (*Melospiza melodia*) into 23 different sub-species, all distinguished by subtle characteristics

G. R. W. Humphries (✉)
Black Bawks Data Science Ltd., Fort Augustus, Scotland
e-mail: grwhumphries@blackbawks.net

R. T. Buxton
Department of Fish, Wildlife and Conservation Biology, Colorado State University, Fort Collins, CO, USA
e-mail: Rachel.buxton@colostate.edu

I. L. Jones
Department of Biology, Memorial University, St. John's, NL, Canada
e-mail: iljones@mun.ca

© Springer Nature Switzerland AG 2018
G. R. W. Humphries et al. (eds.), *Machine Learning for Ecology and Sustainable Natural Resource Management*, https://doi.org/10.1007/978-3-319-96978-7_15

(morphology, genetics, and vocalizations; Arcese et al. 2002). Taxonomic splits like this usually result in heavy debate, conjecture and assumptions, and ultimately lead to questions about the definitions of "species" or "sub-species" (Winker 2010). We are now at a point, technologically, where we can quantify these traits in objective ways with useful ML based algorithms.

There are masses of data that could be collected from birds to help classify species, but perhaps the least intrusive are vocalization data (bird calls and songs). This is because we can set up automated recording systems to collect acoustic information that we can analyze at a later date. Most non-passerine bird groups have genetically determined vocalizations rather than imitative vocal learning, so vocal divergence is assumed to correlate with genetic divergence (Petkov and Jarvis 2012; Sosa-López et al. 2016). In colonial breeding birds like seabirds, individual recognition is a primary function of vocalizations, hence the extreme variability of call structure across individuals and low variability within individuals (e.g., Jones et al. 1987, Bretagnolle 1989, Bretagnolle et al. 1998). The study of avian vocalizations is becoming an increasingly important field, with focuses on sexual selection and geographic variation (Bretagnolle and Genevois 1997; Ferreira and Ferguson 2002; Alström and Ranft 2003; Hamao et al. 2016; Keighley et al. 2017; Lynch and Lynch 2017; Nelson 2017; Kriesell et al. 2018). Moreover, rapid advances in acoustic recording technology offers insights into avian song and diversity over extended spatiotemporal scales (see Textbox 15.1 for a brief overview of 'soundscapes'; Buxton et al. 2018).

Vocalizations are defined by subunits called notes and syllables (Kroodsma 2005) each of which contain information to be transmitted between the sender (vocalizer) and receiver (another animal hearing the vocalization). A note is the smallest of the subunits, where a syllable is a set of two or more notes repeated (Marler and Slabbekoorn 2004). Calls are sets of notes and syllables usually made by both sexes. You might hear these while strolling through the woods when you flush up a bird, which "chips" as it whizzes into the trees. Songs (learned in passerines) are generally complex (a variety of notes and syllables strung together) and can be used for warding off potential threats to the nest and attracting a mate (Marler and Slabbekoorn 2004; Kroodsma 2005). Each individual bird within a species can have unique note and syllable structures, allowing individual recognition (Lambrechts and Dhondt 1995).

Notes and syllables have characteristics that can be measured, such as amplitude and frequency. Amplitude refers to the energy or loudness of the soundwave, and frequency refers to the rapidity of the vibration or pitch (i.e. how frequently a soundwave oscillates). In many cases, notes are made up of a series of harmonics, characterized by a fundamental frequency, the lowest frequency harmonic in that note, and harmonic overtones that are multiples of the fundamental frequency (e.g., Fig. 15.1).

Vocalizations have been used to differentiate between populations in several species (Ainley 1980; Adkisson 1981; Bretagnolle and Genevois 1997). For example, the North American western grebe (*Aechmophorus occidentalis*) and morphologically similar Clark's grebe (*A. clarkii*) were distinguished from one another by way of vocalization, where western grebe were found to have one or two more syllables

Textbox 15.1 A brief introduction to soundscapes

Each landscape is characterized by a dynamic and unique suite of sounds, which allow organisms to gather information about their surroundings. Natural sounds drive essential behaviors, from predator deterrence to navigation, and thus many animals have evolved an acute ability to hear. Moreover, many animals have complex acoustic communication systems, with many species producing intricate song (e.g., humpback whales; Garland et al. 2011). If you, the reader, have been fortunate enough to visit the rainforest, you will have been witness to the cacophony of biological sounds ranging in pitch and volume. However, upon visiting more desolate landscapes like the Arctic, you might find the opposite; a quietness that seems to make the smallest sound seem louder.

The concept of the "soundscape" has been around since the 1960s, however it was used primarily in describing the sounds in urban environments (Southworth 1969). A soundscape is the assemblage of all anthropogenic, biological, and geological sounds emanating from a landscape (Schafer 1977). The acoustical characteristics of an area reflect natural processes and it is now evident that soundscapes reflect ecosystem health and biodiversity (Pijanowski et al. 2011; Mullet et al. 2016).

Because the soundscape is complex and related to the ecology of organisms and habitat, the evolution of species vocalizations could have implications for population biology. For example, sub-populations of species that are separated and being exposed to different soundscapes might eventually alter their vocalizations. But vocalization structures are complex, with potentially subtle changes in several different notes or syllables which can be difficult to detect. This chapter will examine how we can use ML techniques to detect differences in vocalizations.

than Clark's (Alström and Ranft 2003). This exercise can be more challenging in some less obvious cases, like that of the Leach's storm-petrel (*Hydrobates leucorhous*, formerly *Oceanodroma leucorhoa*), where the structure of vocalizations differs only subtlety in the frequency of notes.

The Leach's storm-petrel is one of the most widely distributed pelagic tubenose seabirds in the Northern Hemisphere (Order Procellariiformes, which includes storm-petrels, albatrosses, and shearwaters; Warham 1990). They are nocturnal birds that nest in earth burrows and dense grass on small islands in the North Pacific and North Atlantic and have an overall population estimated at eight million breeding pairs (Watanuki 1986; Huntington et al. 1996). The taxonomic classification of Leach's storm-petrel has fluctuated, divided into two to four subspecies of one to three species (Ainley 1980; Brooke 2004). We focus on *H. l. leucorhous*, which is wide-ranging across both the North-Atlantic and North Pacific. Bicknell et al. (2012) found striking genetic differences (mitochondrial and nuclear DNA) between Atlantic and Pacific populations of *H.l. leucorhous*, but with relative uniformity

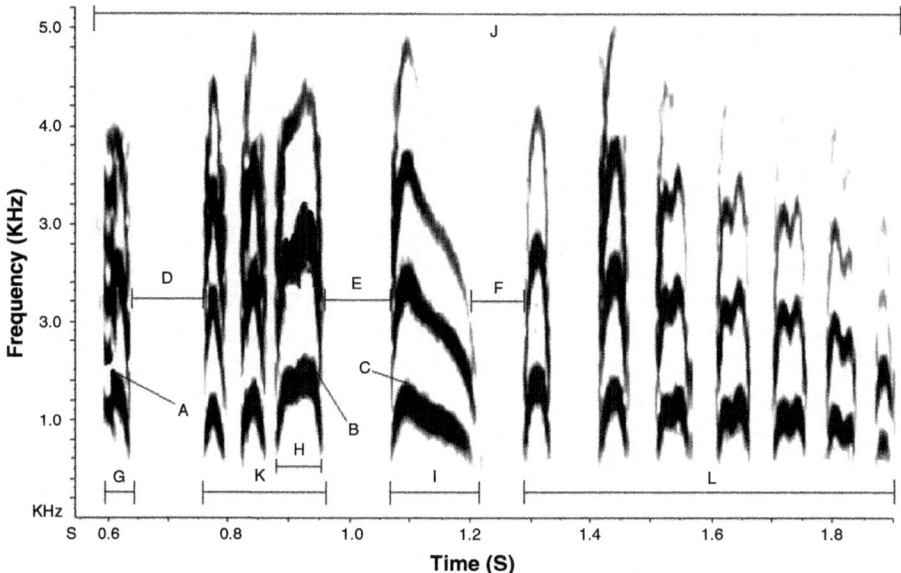

Fig. 15.1 The spectrographic display of a Leach's Storm-Petrel chatter call recorded on Buldir Island, Alaska and captured in Raven Lite software. The image shows all the parts of the call that were measured. Peak of Fundamental Frequency of: start note (**a**), accent note (**b**), and center note (**c**), duration between: start note and first note of first staccato section (**d**), accent note and center note (**e**), and center note and first note of second staccato section (**f**), duration of: start note (**g**), accent note (**h**), center note (**i**), and total call (**j**), number of notes in first staccato section (**k**) and second staccato section (**l**)

within the Atlantic and heterogeneity in the Pacific. Populations of *H. l. leucorhous* in the Aleutian Islands tend to have longer wings, bills, and tarsi than southern populations (i.e., Farallon Islands, Ainley 1980, Power and Ainley 1986).

Leach's Storm-petrel have a limited vocal repertoire consisting of a chatter call exhibited during flight around the colonies at night time, a purring call performed from the burrow by males, an alarm call, and a begging call (Ainley 1980; Bourne and Jehl Jr. 1982; Huntington et al. 1996). In Ainley's (1980) study, chatter calls were described as having a similar structure in the North Atlantic and the Pacific coast, but detailed analysis was not conducted. The chatter call consists of a series of approximately 10–12 notes in a characteristic rhythm with accentuations on 3 notes (Ainley 1980; Taoka et al. 1989). In the spectrogram below (Fig. 15.1), the accented notes are identified as the start note (A), the accent note (B) and the center note (C). Possible uses for the chatter call have not been thoroughly examined but may include sexual recognition and mate attraction, individual recognition, intraspecific competition, and nest defense (Huntington et al. 1996).

The Atlantic and Pacific populations of *H.l. leucorhous* are separated considerably geographically with limited evidence of any mixing (Bicknell et al. 2012). With their calls genetically determined like those of other non-passerine birds, we would expect chatter call structure to diverge due to either genetic drift or natural selection. Sound propagates differently through different habitats and this is considered to be an important factor in the evolution of bird vocalizations (Boncoraglio and Saino 2007). For example, birds that have calls with frequencies that carry better through trees are more likely to find their mates and pass on their genes in a forested habitat. Over time, birds with a particular frequency call may be more successful on forested islands, which differs from the frequency range on an unforested island. You might then ask, does that lead to speciation? If so, can we use vocalizations to differentiate sub-species? According to McCracken and Sheldon (1997), it might be possible to do this because the structure of calls (e.g., frequency) may reflect the physiology of the syrinx (an inherited quality).

Humphries (2007) found differences in chatter call structure between one Pacific (Buldir Island Alaska) and one Atlantic population (Gull Island, Newfoundland, Fig. 15.2).

Here we examined chatter call data in a ML framework of classification techniques. More specifically, we compared two implementations of generalized boosted regression models (Ridgeway 2007; H2O.ai 2016) and three implementations of random forests (Cutler et al. 2007; H2O.ai 2016) with respect to how well they categorized these geographically separate populations according to their chatter call properties. The goal was to assess performance and then provide some discussion on the efficiency of the algorithms to cluster different populations of birds based on their vocalization properties, as well as how they might play a role in future taxonomic efforts with Leach's storm-petrel.

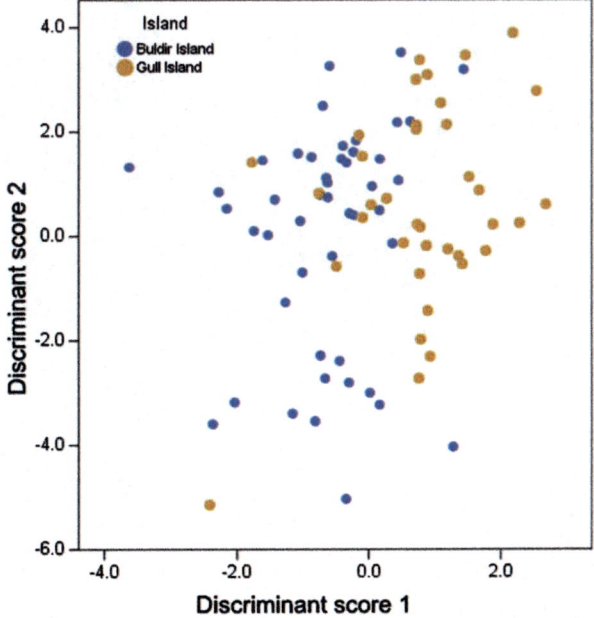

Fig. 15.2 Discriminant function analysis performed in Humphries (2007) showing a certain degree of differentiation between chatter calls of Leach's Storm-petrel recorded on Buldir island, Alaska and Gull island, Newfoundland

15.2 Recording and Measuring Vocalization Structure

15.2.1 Study Sites

Field recordings of Leach's storm-petrel chatter calls were taken at Buldir Island, Alaska (52.361°N, 175.920°E) and Gull Island, Newfoundland (47.261°N, −52.775°W) in the summer of 2006 (Humphries 2007), Amatignak Island (51.2622°N, 179.109°W) in 2008 (Buxton et al. 2013), and Grand Colombier Island, St. Pierre (46.783°N, 56.20°W) in 2009 (Roul 2010, Fig. 15.3).

15.2.2 Call Recording Technique

Chatter calls were recorded using dynamic microphones and digital audio recorders at night as birds were returning to colonies (Humphries 2007, Roul 2010, Fig. 15.4). Either an observer would stand with the microphone outstretched and record samples at random open areas around the colony (Shure Prologue microphone with Sony TCD 10 Pro II digital audio tape; Humphries 2007), or automated recorders (Song Meters, Model SM1, Wildlife Acoustics, Inc., Concord, MA) were deployed

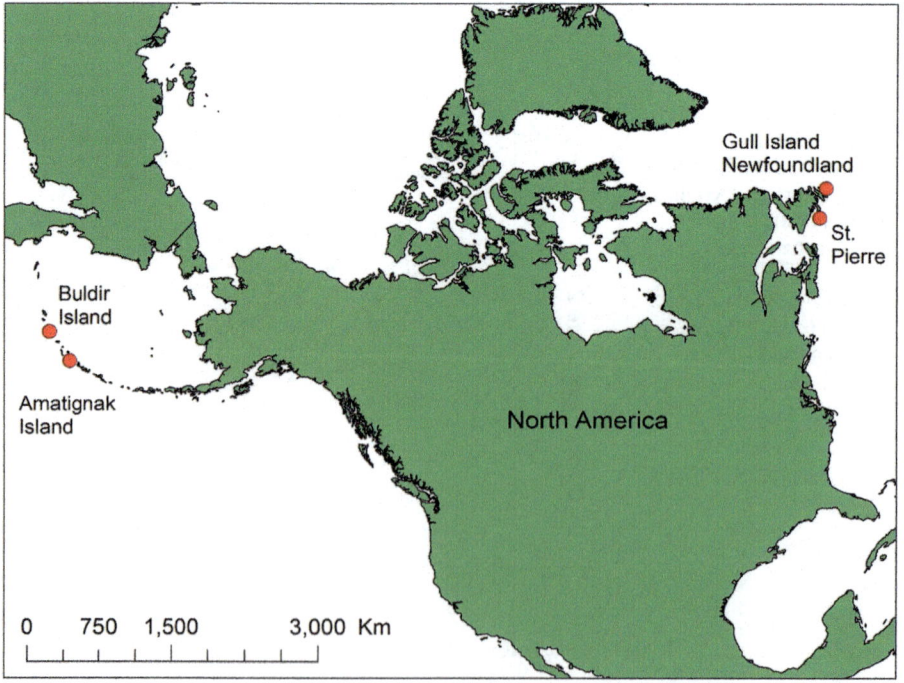

Fig. 15.3 Map of study sites

Fig. 15.4 Device used to record vocalizations of Leach's storm-petrel on Buldir Island (Alaska) and Gull Island (Newfoundland)

(Roul 2010; Buxton et al. 2013). We assumed, for our analysis, that there was little effect of device on comparability of our recordings.

15.2.3 Measuring Calls

The characteristics of Leach's storm-petrel chatter calls were measured as per methods in Humphries (2007). When identifying calls for measuring, we chose one call from each two-minute recording, which represented the clearest and most easily measurable call. This helped to minimize any issues around individual birds circling around the audio device. Raven Lite version 1.0 was used with screen scale set to 5.0 kHz x 1.2 seconds. We measured 14 parameters: peak frequency of the fundamental frequency of: start note, accent note, and center note; duration between: start note and first staccato element, accent note and center note, and center note and final staccato element; duration of: start note, accent note, center note, and total call; rate and number of elements of: first staccato element, and second staccato element (Fig. 15.1).

Each author measured the call elements manually in Raven Lite. Thus, the possibility existed that we might not be measuring vocalization characteristics the same way. To make sure that we were not adding more noise to the data we re-measured thirty of each other's measured calls, chosen randomly. A generalized extreme studentized deviate test was first run to test for outliers and none were detected ($p = 0.56$). Equal variances were also found between both datasets ($p = 0.32$) with a Levene's test. We compared the re-measurements using MANOVA tests to see if there were any significant differences. Because we found no difference between each author's measurements ($p = 0.23$), we included measurements from both observers in the same analysis.

15.2.4 Random Forests (Bagging techniques)

Many of the details of the random forests algorithm have been described elsewhere (Breiman 2001, Cutler et al. 2007, Chap. 1), and we therefore do not cover them in great detail here. In general, bagging techniques like random forests take the average (or uses a voting technique) across several models to determine the value given to predictions for data rows. Random forests does this by taking advantage of classification and regression trees (Breiman et al. 1984; De'ath and Fabricius 2000; Breiman 2001). Random forests will construct many trees, holding back a subset of data for each tree which is used to measure predictive accuracy. This algorithm has been shown to generally be more effective than boosting techniques for categorical data (Breiman 2001; Prasad et al. 2006; Cutler et al. 2007). Care must be taken by the user to ensure that cross-validation is properly used by ensuring the testing data are independent, otherwise there lies the risk of confounding inferences (Picard and Cook 1984).

15.2.5 Generalized Boosting (Boosting techniques)

As with the random forests algorithm, we do not detail the generalized boosted regression technique as it has been covered in other chapters of this book (see Chap. 1 in this book and elsewhere). Boosting techniques like those implemented in generalized boosted regression models create models in a step-wise fashion, aiming to minimize overall error in predictions. The generalized boosted regression model technique begins by building a regression tree, and then measuring the error associated with the model (e.g., predicted versus observed values). Using the information from the first tree, a second tree is constructed, and error is reduced. In other words, the algorithm learns and improves the estimates at every iteration until a minimum error is reached. The tree with the lowest amount of associated error is selected as the representative model for predictions (Friedman 2002; Ridgeway 2007; Elith et al. 2008).

15.2.6 Comparing Model Algorithms

When comparing modeling algorithms, there are a few aspects to consider. From a computing standpoint, we were interested in the time it takes to run models, as well as the ease of implementation, and memory requirements (i.e. if can it handle large datasets). We decided to use the two of the more commonly used programming languages: Python, and R, with libraries typically used for ML exercises. From a predictive standpoint, we were interested in how well the models predict to independent subsets of the data. Predictive performance was measured by the area under the receiver's operators curve (AUC; Bradley 1997). In our case, we are interested in determining which model best classified the ocean basin where the bird was recorded

(Atlantic versus Pacific). From a scientific standpoint, we were interested in understanding the mechanisms that account for differences between populations.

To perform the modeling in this study, we assigned the ocean (Pacific or Atlantic) as the target variable, with all variables measured in Fig. 15.1 as the explanatory (or predictor) variables.

Below, we compare these algorithms based on ease of implementation, memory requirements, predictive capability, and qualitatively discuss the scientific merits of each. For an added level of excitement, we'll also discuss and interpret our results in the context of Leach's storm-petrel biology.

15.3 Findings

15.3.1 Reproducibility

Before we delve into the results, let's report on the reproducibility. Using the thirty randomly sampled calls we performed a MANOVA on the results and found that there were no statistically significant differences between data measured by different observers ($p = 0.62$). This means that we could safely assume that there is little variance introduced due to observer error.

15.3.2 Computational Comparison between Algorithms

In the table below, we ranked the model algorithms based on speed, ease of setup and memory handling (Table 15.1).

For speed, the 'randomForest' package was by far the fastest implementation, followed by the H2o implementation of random forests, the 'gbm' package (generalized boosted regression models), the tensor forest implementation of random forests in Python, and finally the H2o implementation of generalized boosted regression. The randomForest package also ranked number 1 in ease of implementation, however this is very closely matched with the 'gbm' package. The h2o implementations of both algorithms come in at 3 and 4 but are very close to the first and second ranked implementations because of the similar set up. The difference with h2o was

Table 15.1 Ranking based on computational performance of 5 basic machine learning implementations

Implementation	Speed rank (mean time)	Ease	Memory
Random forests H2O	2 (5.64 sec)	4	2
Boosted trees H2O	5 (38.24 sec)	3	3
Random forests ('randomForest')	1 (0.16 sec)	1	5
Generalized boosted regression models ('gbm')	3 (8.78 sec)	2	4
Tensor forest	4 (14.23 sec)	5	1

mostly associated with the setup of the h2o package, which actually involves more steps than the regular "install.packages" used in R. Also, the h2o package requires that users set up specific data frame types before running. Finally, the tensor forest implementation in Python was the most difficult to set up. This is because it is implemented through the tensorflow package in Python and requires a few set up steps before being able to run (see Textbox 15.2).

For memory use, tensor forest was ranked the best because it was designed specifically for large datasets, and due to its coding in Python, does not struggle with memory blocks. The h2o implementations came in second and third for memory

---

Textbox 15.2 Sample code showing basic set up for each machine learning implementation

```
#### This assumes data are loaded into a data frame "X" that looks like this:

ISLAND        OCEAN  Center_Hz   Start_Hz    Accent_Hz   Total4
Amatignak_09  1      1031.2      1281.2      1406.2      0.180
Amatignak_09  1      1000.0      937.5       1187.5      0.105
Amatignak_09  1      1000.0      906.2       1156.2      0.076
Amatignak_09  1      1093.8      906.2       875.0       0.147
Amatignak_09  1      1281.2      875.0       1093.8      0.080
Amatignak_09  1      937.5       1031.2      1187.5      0.146

### After data have been split into training and testing data
### randomForest in R ###

randomForest(as.factor(OCEAN)~.,data=X,mtry=9,ntree=300,importance=T)
-----------------------------------------------------------------------------------------------
------------------
### gbm in R ###

gbm(as.factor(OCEAN)~., data = X, shrinkage = 0.001, cv.folds = 5,family =
'bernoulli', n.trees=2500)
-----------------------------------------------------------------------------------------------
------------------
### h2o in R ### (randomForest and gbm)

## h2o needs to be installed first, which is not done with the typical "install.packages" –
must be downloaded as a software package from http://h2o.ai/download
## data need to be converted to an h2o frame first

train.hex <- as.h2o(train)
test.hex <- as.h2o(h2otest)

h2o.gbm(training_frame = train.hex, validation_frame=test.hex, y=c(1),x =
c(2:17),learn_rate = 0.001, nfolds=5, distribution = 'bernoulli', ntrees=2500)

h2o.randomForest(training_frame = train.hex, validation_frame=test.hex,
mtries=9, y=c(1),x = c(2:17), sample_rate = 0.8, ntrees=300)
-----------------------------------------------------------------------------------------------
------------------
### Tensor forest in Python ###

X_train = np.array(X_train,dtype=np.float32)
Y_train = np.array(Y_train,dtype=np.float32)
X_test = np.array(X_test,dtype=np.float32)

hparams = tensor_forest.ForestHParams(
    num_trees=500,
    max_nodes=1000,
    num_classes=2,
    num_features=16,
    split_after_samples=20)
```

usage as they were also designed with large datasets in mind. Not only did H2o integrate with graphics processing units (GPUs), but could also be processed on multiple cores. However, we only let the h2o implementations run on one CPU. The worst two implementations for memory were 'gbm' and 'randomForest'. Many users of both will be very familiar with the memory error: "cannot allocate vector of size X Gb"; a frustrating experience. This means that neither algorithm is well equipped to handle large datasets unless you have a monstrously large CPU (usually only available on high performance super computers, or via Amazon web services).

15.3.3   Predictive Comparison

Interestingly, the top performing models came from the H2o and 'random forests' implementations, while Tensor forest landed in the fourth spot. Tensor forest is meant to be a more sophisticated approach (Colthurst et al. 2016) and is very flexible with regards to how the data are presented to the algorithm (Table 15.2).

15.3.4   Variable Importance

One commonality across all the algorithms was the variable ranking. In both random forests and gbm, variable importance was calculated by looking at which variables lowered overall predictive error the most across all trees (i.e., which variable contributed to the most number splits in the trees). However, we ran a few different models in this case (i.e. random forests and gbm from a few different packages), and we used our own implementation of a variable rank averaging (by way of a voting system) to make some broad inferences. The variable ranking was scored based on the most commonly selected variable across model runs in each rank. For example, if the variable "frequency of center note" was selected the most as the top predictor in all model runs, then this was selected as the top variable. In this case we only examined the top three variables. This is somewhat arbitrary, and we could dig further down into this if we so decided, but for the purposes of this chapter it should suffice.

The top variable selected was the total time of the accent note (variable H, Fig. 15.1). The second most important variable was the total time of the start note (variable G, Fig. 15.1). The third most important variable was the total time of the center note (variable I, Fig. 15.1). Interestingly, the top variables chosen across all the implementations were those which represented total length (in seconds) of notes (Fig. 15.5).

When the top three variables (total time of the accent, center and start notes) are examined, the accent and center notes tended to be longer in the Pacific population than the Atlantic population, and the start note tended to be shorter in the Pacific population (Fig. 15.6).

Table 15.2 AUC values from implementations of random forests and generalized boosted modeling for the storm-petrel vocalization data. Values above 0.8 are typically considered 'good' (Zhu et al. 2010)

Implementation	AUC
Random forests H2O	0.82
Boosted trees H2O	0.81
Random forests ('randomForest')	0.82
Generalized boosted regression models ('gbm')	0.78
Tensor forest	0.79

Fig. 15.5 Spectographic depiction of the first half of a Leach's storm-petrel chatter call with the 1st, second and third best predictors of population (Atlantic versus Pacific) highlighted. We show only the first half here because none of the top three predictors were in the second half of the call structure

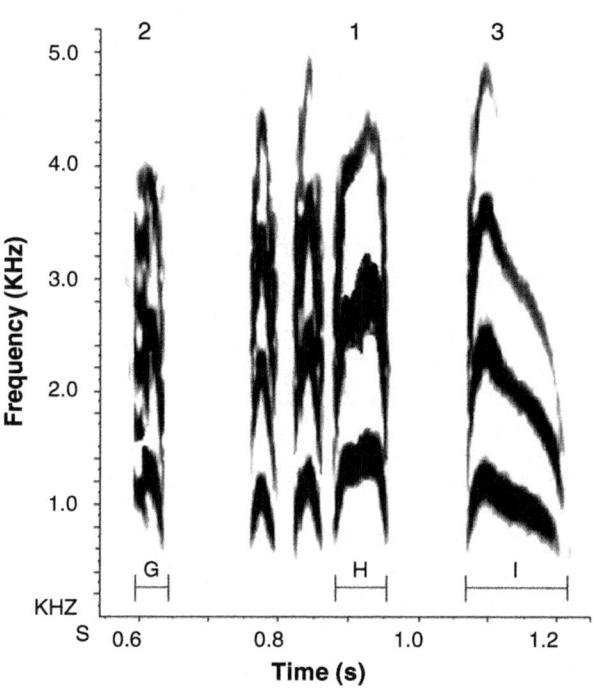

15.4 A Few Lessons Learned

This chapter briefly explored a few implementations of two popular ML algorithms widely used in ecology. Both the random forests and generalized boosted regression models offer insights into data structure, and most importantly, are non-parametric and able to handle large numbers of predictor variables. We demonstrate just one of many possible applications of ML algorithms to examine vocalization characteristics. In fact, these methods are much more automated (Acevedo et al. 2009; Huang et al. 2009) and integrated into more sophisticated techniques to identify species vocalizations in long-term acoustic recordings (e.g., convolutional neural networks). We conclude that these ML methods are easy to implement and powerful classifiers

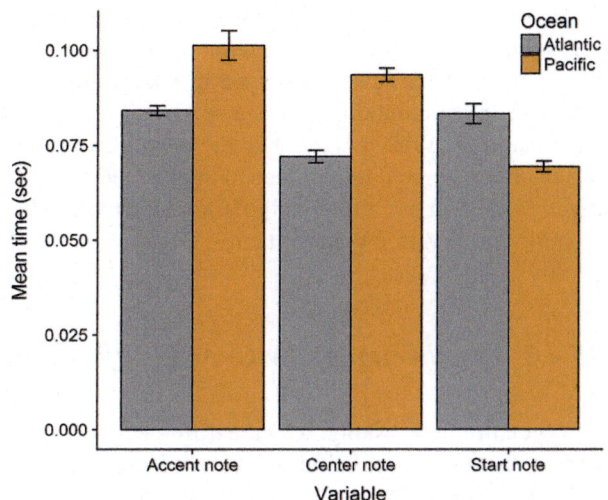

Fig. 15.6 Raw comparisons of mean time of Accent, Center and Start notes (top three predictors) between Atlantic and Pacific populations of Leach's storm-petrel

that offer interesting insights into the definition of seabird populations and sub-populations in the wild.

From the perspective of being able to identify species sub-populations, there are benefits to using algorithms that can efficiently tease apart complex relationships. Here, we compared Leach's Storm-petrel chatter call vocalization characteristics from four sites in two oceans. However, it is easy to imagine a plethora of other variables being included in such an analysis in future. For example, in some cases in addition to recording birds calls it may be possible to capture individual birds and collect morphometric measurements (i.e. bill length, mass, wing cord, etc.) and draw blood. At this point, we have genetic evidence, morphometric evidence, and vocal evidence that we can combine to make quantitative decisions on how populations might be different from each other. Excitingly, using ML algorithms we can examine sub-species categorization based on all of these data (for example, % of correctly classified individuals). The argument then moves from philosophical questions (e.g., what defines a species) to numerical questions (e.g., using data to define a sub-species).

15.4.1 Sexing Birds with Machine Learning

ML methods could also be adapted to determine the sex of individuals from acoustic recordings. Previous work found that individual Leach's storm-petrels calling from nest sites in Japan could be sexed by examining the structure of the chatter call (Taoka et al. 1989). They found that the highest peak of the fundamental frequency of the center note of the chatter call was higher in the male than that in the female, and that the female had more elements in the second staccato element of the chatter call.

However, it could be that call differentiation between sexes is more nuanced across their distribution; thus, more work would be required to develop an automated method for sexing birds using vocalization structure. Vocalizations play a vital role in differentiation among sexes in species that are sexually monomorphic and nocturnal (Taoka et al. 1989; Taoka and Okumura 1990). Therefore, ML methods could be of great importance in understanding certain aspects of Leach's storm-petrel (and other nocturnal seabirds') ecology and behavior, thus providing progress towards conservation management.

15.4.2 Environmental v Genetic Influences

Now you might be asking, why the difference in Leach's storm-petrel chatter vocalizations between populations as seen in Humphries (2007)? Bicknell et al.'s (2012) study indicated that Atlantic and Pacific populations are genetically distinct, have likely been separated for millennia, and have experienced minimal mixing of individuals between oceans. However, some evidence suggests that there is interchange of individuals and gene flow from Atlantic to Pacific populations (Bicknell et al. 2012). Differences in calls between Buldir Island and Gull Island (Humphries 2007) could thus be attributed to two factors, genetic drift or selection for different call structure. Environmental factors could have a role in determining Leach's Storm-petrel call structure based on optimal transmission of calls in the acoustic environment of the colony site (McCracken and Sheldon 1997). Lower frequency and shorter duration call elements are more effectively transmitted in dense vegetation, where trees and branches may absorb or echo sound, preventing backscatter. That is, sound with longer wavelengths travel better through dense vegetation than shorter wavelengths (Mccracken and Sheldon 1997). Characteristics of the acoustic environment of storm-petrel colony sites have not been measured, but is likely to vary between densely forested islands (Gull Island), grass covered islands (Buldir Island), and rocky barren (Grand Colombier Island). Because both Atlantic and Pacific colonies of Leach's storm-petrel occur on a variety of habitat types it is less likely that the differences we found between populations are due to environmental factors (Huntington et al. 1996). Alternatively, the differences we found may be due to genetic drift. Both factors could be studied in more detail with more recordings from other islands, quantitative characterization of their acoustic environments, and genetic data collected simultaneously.

15.4.3 The Machine Learning Algorithms

Why do some of the same implementations of algorithms work differently than others? In this case, the devil was in the details, where small code differences led to large changes in memory performance, and even predictive performance. First, it has to be noted that the random forests algorithm in R was developed by Leo Breiman and Adele Cutler (Breiman and Cutler 2007) and was originally licensed only to Salford Systems for use in their software suite. However, due to the rise of R, the implementation was adapted and uploaded as a package for open access use, which led to subtle changes in code. This original implementation has not been improved upon in R, save for some few package upgrades that made interpretation easier (latest upgrade was from 2015). As such, it is not really able to handle big datasets because the package was built prior to the big data revolution, hence the low score on the memory scale. For those who have used the randomForest package in R, you would be familiar with this when dealing with large datasets. H2o has been optimized to deal with huge datasets, and so the way data partitioning is done, and the loss algorithm used (i.e. mean squared error in most implementations) is calculated more efficiently. Similar to all the algorithms we tested, these slight differences could have led to differences in accuracy. In short, each algorithm was built to handle datasets differently, and thus differences would be expected. The drawback of the H2o implementations is that because it is meant to work on large datasets, it will take longer on small datasets compared to the other algorithms which have been optimized on small datasets. The Python implementation of tensor forest is an interesting case because it was transferred from the scikit-learn package with the intention of running it through tensorflow. Thus, there is a steep learning curve around its setup (for those not familiar with the technique). This algorithm was also optimized for different datatypes (big datasets, with hundreds of parameters), and as such, we could be seeing performance loss on smaller datasets due to this aspect.

We recommend, regardless of the algorithm, to try them all where possible and compare and contrast their predictive abilities. In the end, this can only lead to better understanding of the algorithms and can optimize ecological inference.

15.5 Conclusions

In our short study we have done two things: (1) compared the vocalizations of two sub-populations of Leach's storm-petrel which are of the same subspecies; (2) Compared the performance of a few different implementations of random forests and generalized boosted regression modeling. We show that there are differences between our two sub-populations as defined by our ML algorithms, mostly focused on the length of certain notes of the call. Finally, we show that there are differences in the performances of 5 different ML implementations, with random forests from the h2o and 'randomForest' packages performing best with regards to accuracy, 'randomForest' and 'gbm' performing best with regards to speed, and 'tensor forest' and 'h2o' implementations performing best with regards to memory. Consideration of these factors needs to be made when decided how to approach certain analyses.

Acknowledgments Thanks to S. Seneviratne for guidance with the original project in 2007 and to T. Miller for loan of the equipment used to record birds on Buldir and Gull islands. Thanks to A. Hedd and G. Robertson for field support, and for D. Fifield and S. Roul for providing some of the recordings used in our analysis.

References

Acevedo MA, Corrada-Bravo CJ, Corrada-Bravo H, Villanueva-Rivera LJ, Aide TM (2009) Automated classification of bird and amphibian calls using machine learning: a comparison of methods. Eco Inform 4:206–214

Adkisson CS (1981) Geographic variation in vocalizations and evolution of North American Pine Grosbeaks. Condor 83:277

Ainley DG (1980) Geographic variation in Leach's storm-petrel. Auk 97:837–853

Alström P, Ranft R (2003) The use of sounds in avian systematics and the importance of bird sound archives. Bull Br Ornithol Club 123A:114–135

Arcese P, Sogge MK, Marr AB, Patten MA (2002) Song sparrow (*Melospiza melodia*), version 2.0. In: Poole AF, Gill FB (eds) The birds of North America. Cornell Lab of Ornithology, Ithaca

Bicknell AWJ, Knight ME, Bilton D, Reid JB, Burke T, Votier SC (2012) Population genetic structure and long-distance dispersal among seabird populations: Implications for colony persistence. Mol Ecol 21:2863–2876

Boncoraglio G, Saino N (2007) Habitat structure and the evolution of bird song: a meta-analysis of the evidence for the acoustic adaptation hypothesis. Funct Ecol 21:134–142

Bourne WRP, Jehl JR Jr (1982) Variation and nomenclature of Leach's storm-petrels. Auk 99:793–797

Bradley AP (1997) The use of the area under the ROC curve in the evaluation of machine learning algorithms. Pattern Recogn 30(7):1145–1159

Breiman L (2001) Random forests. Mach Learn 45:5–32

Breiman L, Cutler A (2007) Random forests-classification description

Breiman L, Friedman J, Stone C, Olshen R (1984) Classification and regression trees. CRC Press, Boca Raton

Bretagnolle V (1989) Calls of Wilson's storm petrel: functions, individual and sexual recognitions, and geographic variation. Behaviour 111:98–112

Bretagnolle V, Genevois F (1997) Geographic variation in the call of the blue petrel: effects of sex and geographical scale. Condor 99(4):985–989

Bretagnolle V, Mougeot F, Genevois F (1998) Intra- and Intersexual Functions in the Call, of a Non-Passerine Bird. Behaviour 135(8):1161–1184

Brooke M (2004) Albatrosses and petrels across the world. Oxford University Press, Oxford, p 499

Buxton R, Major HL, Jones IL, Williams JC (2013) Examining patterns in nocturnal seabird activity and recovery across the western Aleutian Islands, Alaska, using automated acoustic recording. Auk 130(2):331–341

Buxton RT, McKenna MF, Clapp M, Meyer E, Angeloni L, Crooks K, and Wittemyer G (2018) Efficacy of extracting indices from large-scale acoustic recordings to monitor biodiversity. Conserv Biol

Chen X, Ishwaran H (2012) Random forests for genomic data analysis. Genomics 99(6):323–329

Colthurst T, Sculley D, Hendry G, Nado Z (2016) Tensorforest: scalable random forests on tensorflow, in: Machine learning systems workshop at NIPS

Cutler D, Edwards T, Beard K, Cutler A, Hess K, Gibson J, Lawler J (2007) Random forests for classification in ecology. Ecology 88:2783–2792

De'ath G, Fabricius K (2000) Classification and regression trees: a powerful yet simple technique for ecological data analysis. Ecology 81:3178–3192

Elith J, Graham C, Anderson R, Dudik M, Ferrier S, Guisan A, Hijmans R, Huettmann F, Leathwick J, Lehmann A, Li J, Lohmann L, Loiselle B, Manion G, Moritz C, Nakamura M, Nakazawa Y, Overton J, Townsend Peterson A, Phillips S, Richardson K, Scachetti-pereira R, Schapire R, Williams S, Wisz M, Zimmermann N, Overton JMC, Soberon J (2006) Novel methods improve prediction of species' distributions from occurrence data. Ecography 29:129–151

Elith J, Leathwick J, Hastie T (2008) A working guide to boosted regression trees. J Anim Ecol 77:802–813

Ferreira M, Ferguson JWH (2002) Geographic variation in the calling song of the field cricket Gryllus bimaculatus (Orthoptera: Gryllidae) and its relevance to mate recognition and mate choice. J Zool 257:163–170

Friedman J (2002) Stochastic gradient boosting. Comput Stat Data Anal 38:367–378

Garland EC, Goldizen AW, Rekdahl ML, Constantine R, Garrigue C, Hauser Nan D, Poole MM, Robbins J, Noad MJ (2011) Dynamic horizontal cultural transmission of humpback whale song at the ocean basin scale. Curr Biol 21:687–691

H2O.ai (2016) R Interface for H2O

Hamao S, Sugita N, Nishiumi I (2016) Geographic variation in bird songs: examination of the effects of sympatric related species on the acoustic structure of songs. Acta Ethologica 19:81–90

Huang C-J, Yang Y-J, Yang D-X, Chen Y-J (2009) Frog classification using machine learning techniques. Expert Syst Appl 36:3737–3743

Humphries GRW (2007) Leach's storm-petrel (Oceanodrama leucorhoa) vocalizations: a comparison of chatter calls between Buldir Island, Aleutian Islands, Alaska and Gull Island, Newfoundland. B.Sc. Thesis, Memorial University of Newfoundland. 80 pp

Huntington CE, Butler RG, Mauck RA (1996) Leach's Storm-Petrel (Oceanodroma leucorhoa). In: Poole A, Gill F (eds) The birds of north America. Academy of Natural Sciences, Philidelphia, PA and the American Ornithologists' Union, Washington DC

Jones IL, Falls JB, Gaston AJ (1987) Vocal recognition between parents and young of Ancient Murrelets (Synthliboramphus antiquus). Anim Behav 35:1405–1415

Keighley MV, Langmore NE, Zdenek CN, Heinsohn R (2017) Geographic variation in the vocalizations of Australian palm cockatoos (Probosciger aterrimus). Bioacoustics 26(1):91–108

Kriesell HJ, Aubin T, Planas-Bielsa V, Benoiste M, Bonadonna F, Gachot-Neveu H, Le Maho Y, Schull Q, Vallas B, Zahn S, Le Bohec C (2018) Sex identification in King Penguins Aptenodytes patagonicus through morphological and acoustic cues. Ibis. https://doi.org/10.1111/ibi.12577

Kroodsma D (2005) The singing life of birds: the art and science of listening to birdsong. Houghton Mifflin Company, New York

Lambrechts MM, Dhondt AA (1995) Individual voice discrimination in birds. In: Power DM (ed) Curr Ornithol, vol 12. Springer, Boston

Lynch MA, Lynch HJ (2017) Variation in the ecstatic display call of the Gentoo Penguin (Pygoscelis papua) across regional geographic scales. Auk 134(4):894–902

Marler P, Slabbekoorn HW (2004) Nature's music : the science of birdsong. Elsevier Academic, San Diego

McCracken KG, Sheldon FH (1997) Avian vocalizations and phylogenetic signal. Proc Natl Acad Sci 94:3833–3836

Mullet TC, Gage SH, Morton JM, Huettmann F (2016) Temporal and spatial variation of a winter soundscape in south-central Alaska. Landsc Ecol 31:1117–1137

Nelson DA (2017) Geographical variation in song phrases differs with their function in white-crowned sparrow song. Anim Behav 124:263–271

Petkov CI, Jarvis ED (2012) Birds, primates, and spoken language origins: behavioral phenotypes and neurobiological substrates. Front Evol Neurosci 4:12

Picard RR, Cook RD (1984) Cross-validation of regression models. J Am Stat Assoc 79(387):575–583

Pijanowski BC, Villanueva-Rivera LJ, Dumyahn SL, Farina A, Krause BL, Napoletano BM, Gage SH, Pieretti N (2011) Soundscape ecology: the science of sound in the landscape. BioScience 61:203–216

Power DM, Ainley DG (1986) Seabird geographic variation: Similarity among populations of Leach's storm-petrel. Auk 103:575–585

Prasad A, Iverson L, Liaw A (2006) Newer classification and regression tree techniques: bagging and random forests for ecological prediction. Ecosystems 9:181–199

Ridgeway G (2007) gbm: generalized boosted regression models. Documentation on the R Package "gbm", version 1.5

Roul S (2010) Distribution and status of the Manx Shearwater *Puffinus puffinus* on islands near the Burin Peninsula, Newfoundland in 2009, as inferred from automated acoustic monitoring devices. B.Sc Honours thesis, Memorial University of Newfoundland

Schafer RM (1977) Tuning of the World. Knopf, New York

Sosa-López JR, Martínez Gómez JE, Mennill DJ (2016) Divergence in mating signals correlates with genetic distance and behavioural responses to playback. J Evol Biol 29(2):306–318

Southworth (1969) The sonic environment of the cities Environment and behavior I, 49±70

Taoka M, Okumura H (1990) Sexual differences in flight calls and the cue for vocal sex recognition of Swinhoe's storm-petrels. Condor 92:571

Taoka M, Sato T, Kamada T, Okumura H (1989) Sexual dimorphism of chatter-calls and vocal sex recognition in Leach's storm-petrels (*Oceanodroma leucorhoa*). Auk 106:498–501

Warham J (1990) The petrels – their ecology and breeding systems. Academic Press, London, p 440

Warren D, Seifert S (2011) Ecological niche modeling in Maxent: the importance of model complexity and the performance of model selection criteria. Ecol Appl 21:335–342

Watanuki Y (1986) Moonlight avoidance behavior in Leach's storm-petrels as a defense against slaty-backed gulls. AUK 103:14–22

Winker K (2010) Subspecies represent geographically partitioned variation, a gold mine of evolutionary biology, and a challenge for conservation. Ornithol Monogr 67(1):6

Zhu W, Zeng N, Wang N (2010) Sensitivity, specificity, accuracy, associated confidence interval and ROC analysis with practical SAS implementations. NESUG proceedings: health care and life sciences, Baltimore, Maryland, 19, p 67

Part V
Implementing Machine Learning for Resource Management

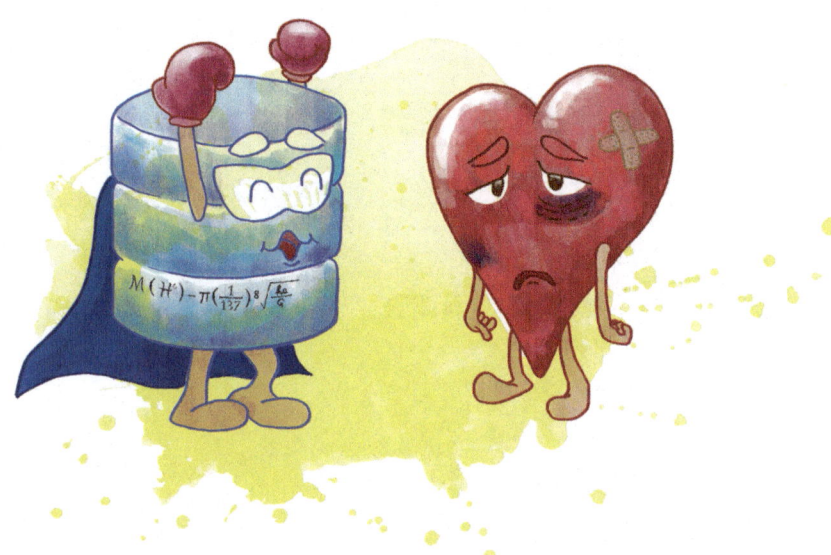

Artwork by Grant Humphries and Catherine Humphries

"Data beats emotions."
– Sean Rad

Chapter 16
Machine Learning for 'Strategic Conservation and Planning': Patterns, Applications, Thoughts and Urgently Needed Global Progress for Sustainability

Falk Huettmann

*"National Parks are just a rather poor transitional protection
scheme by the currently dominating western world trying to put
the more serious destruction on hold before better systems are
found to truly achieve global sustainability"*

*Source unknown, Writing Center, University of Alaska
Fairbanks campus, 2014*

Arguably, the modern western life-style and culture has dramatically marginalized the state of the global environment and its natural resources (Daly and Farley 2011; see Czech et al. 2000 for impacts and Walther et al. 2016 for wider implications). It comes then to no surprise that - as a global pattern - the environment is by now in a state of global crisis (Mace et al. 2010 and Cockburn 2013 for a generic assessment). While humans have used the earth for millennia and made a certain footprint (see Groube et al. 1986 for an example of 40,000 years of documented human occupancy and Flannery 2002 for its benign impacts), the specific anthropogenic footprint and impact of the last four decades remain unprecedented in terms of extinction and global climate change (Rockström et al. 2009, Baltensperger and Huettmann 2015). It is noteworthy that the predominant governance paradigm during this period is globalization, based on Americanization (Czech 2000; Stiglitz 2003). The last few decades are arguably the worst managed in human history (Paehlke 2004; Alexander 2013). Actually, the history of the earth and universe as we know it has not produced such a destruction of life (by non-cosmic events) ever before (Cushman and Huettmann 2010); consider the outlook of what will easily be 10 billion people

F. Huettmann (✉)
EWHALE Lab, Biology and Wildlife Department, Institute of Arctic Biology,
University of Alaska-Fairbanks, Fairbanks, AK, USA
e-mail: fhuettmann@alaska.edu

© Springer Nature Switzerland AG 2018   315
G. R. W. Humphries et al. (eds.), *Machine Learning for Ecology and Sustainable Natural
Resource Management*, https://doi.org/10.1007/978-3-319-96978-7_16

in the next 100 years while global temperatures are also on the rise and most natural resources are already used up (Rockström et al. 2009)!

While I leave it here to others to assign that blame and to document impacts in great detail (Rich 1994; Czech et al. 2000; Huettmann 2011; Cockburn 2013), a way out of this crisis - moving forward in the best-possible fashion - is likely to be pro-active (e.g. to avoid problems before they occur) and to be pre-cautionary (e.g. to identify problems ahead of time and act carefully; Huettmann 2007a, b). The notion of "the need to know impacts ahead of time before they occur" are not new concepts in conservation though (Silva 2012), and they are by now mandatory by the U.N. and part of 'best professional practices' (http://unesdoc.unesco.org/images/0013/001395/139578e.pdf). This is a key ingredient for good governance and for a trusted leadership. Machine learning plays a central role in this approach, and for achieving the best-possible predictions, to be obtained by the best-available science (Huettmann 2007a, b) before impacts occur, e.g. for Alaska see Murphy et al. (2010), Baltensperger et al. (2015), Huettmann et al. (2017).

However, it is easy to show that, thus far, Strategic Conservation Planning does not use much machine learning (see Ardron et al. 2008 and Moilanen et al. 2009 for 'best professional practices' and textbook). Instead, the Strategic Conservation Planning tool has been almost entirely a stand-alone approach not connecting with machine learning. Mostly it relies just on optimization algorithms to find the best solution, such as for instance the 'simulated annealing' algorithm (as employed by MARXAN http://www.marxan.org/; Martin et al. 2008). Further, most MARXAN applications "do not look much into the future" , e.g. by using future scenarios obtained from machine learning and optimizing those ones (see Nielsen et al. 2008 and Murphy et al. 2010 for an application). Instead, latest developments actually deal with optimizing in 'zones' - subunits- (http://marxan.net/index.php/ marzone; Watts et al. 2009). I find this to be a problem on three accounts: (i) The latest science, machine learning, and related potential got ignored. (ii) Breaking down a spatial optimization problem into small separate, parsimonious, zones loses the overall optimization power, and (iii) relevant progress got somewhat stifled by forc-ing creative minds and their solutions back into existing administrative boundaries and units and thus just re-confirming existing problems and patterns. In this chapter, I outline how machine learning has been used and how it could play a larger role in Strategic Conservation Planning projects towards true progress beyond circular rea-soning and traditional mind traps. I am adding relevant perspectives on carrying capacity, limits of the earth and global governance to achieve global sustainability.

16.1 How Machine Learning Predictions Feed into Strategic Conservation Planning: A Common Sense Workflow Still Widely Underused for its Conservation Potential

One of the largest Strategic Conservation Planning projects was designed to make recommendations for the marine protected area (MPA) networks. Similar to the ter-restrial national parks of the world, it is meant to protect the world's oceans and assure that a fixed percentage (~10%) set aside (Aichi targets for 2020 by the U.N.: https://www.cbd.int/sp/targets/)! Strategic Conservation Planning has also been applied to terrestrial systems, although many of these earlier approaches were just *ad hoc* and not optimized much for relevant conservation gains (see Table 16.1). Global applications that involve wider and holistic concepts such as the atmosphere and its protection are still lacking (but see Huettmann 2012 how to include those aspects). Arguably though, one cannot achieve sustainable solutions on a small scale because all is highly connected and influenced globally.

In the marine world, the Great Barrier Reef, Australia, has one of the longest histories of Strategic Conservation Planning (UNESCO Great Barrier Reef Marine Park Authority 1981. Fernandes et al. 2005). The task was here to identify the most relevant reef locations for human enjoyment in perpetuity (see end of the chapter for its outcome, thus far) and much can be learned from that 'experiment' and exercise.

Widely used software for such Strategic Conservation Planning are MARXAN, but also ZONATION (Lehtomäki and Moilanen 2013), and C-Plan (http://www.edg.org.au/edg-free-resources/cplan.html) as well as many derivatives across operating systems and software implementations (e.g. https://www.aproposinfosystems.com/solutions/qmarxan-toolbox/). Such software helps users by finding 'the optimal' solution through an approximation for an assumed truth. The underlying theoretical differences between software packages is not described here but can be found in their respective manuals and URLs, as well as elsewhere in the conservation literature.

For being successful and without relevant errors, Strategic Conservation Planning projects usually require substantial input of information (data). Ideally, maps of species ranges and conservation features for the study area and its planning units are to be available. The study area consists of planning units (PUs, which are often 'bins' e.g. hexagons or pixels). This highly detailed information is widely missing for larger study areas, however, conservation decisions must still be made while destruction is ongoing (Alidina et al. 2008). That is specifically true for a global scale and on a macro-ecology perspective (see Forman 1995 for an effective balance using 'a regional scale'). How can we overcome the problem of data gaps fast enough and with reliable information so that we can make informed conservation decisions in the best way possible for large areas of the globe?

In the past, so-called *ad hoc* decisions were made with political convenience and opportunism driving the agenda (see for instance Huettmann and Hazlett 2010 for Alaska). The protected area network for most of the circumpolar Arctic, in Russia (Spiridonov et al. 2012 for the Russian arctic) or all of North America reflects just that ('*protection of rock and ice*' Scott et al. 2001). Experts got used to identify and fill data gaps in Strategic Conservation Planning projects. In addition, scoping meetings are often held with commercial stakeholders. This is widely referred to as a delphi process (a non-scientific process that simply banks on agreements and compromises made during a session). The use of experts is known though to be biased and has been widely criticized for years and was assessed accordingly (e.g. Perera et al. 2012; see Gonzales et al. 2003 for a real-world example). Those planning efforts may not be representative; and often just the most effluent, vocal or

Table 16.1 Overview of a global selection of conservation priority areas and conservation networks

Project name and citation	Location	Relevance	Computational tools involved
Path of the panther ('Jaguar Trail'; http://www.ecoreserve.org/tag/path-of-the-panther/)	Central America	Probably the biggest initiative for connecting North and South America in ecological terms	None
YtoY (Yellowstone to Yukon; https://y2y.net/)	North America, Rocky Mountains	Tries to protect and connect areas in the Rocky Mountains, also relevant for major watersheds	None
10% protection of polar temperature and associated animals (Huettmann 2012)	Arctic, Antarctic and Hindu Kush Himalaya	The global climate chamber, endemic species as a world heritage	Marxan
Africa's traditional protected areas (e.g. http://www.critical improv.com/index.php/surg/article/view/1987/2670)	Africa	African wildlife and national parks as a world heritage	None
RAMSAR convention (http://www.ramsar.org/)	Global	Global wetland conservation policy	None
Important bird areas (IBAs; http://www.birdlife.org/worldwide/programmes/sites-habitats-ibas)	Global	Global waterbird conservation	None
Great barrier reef (e.g. Fernandes et al. 2005)	NW Australia	One of the finest coral reefs in the world	Marxan
California coastal protected zones (http://www.californiamsf.org/pages/about/strategicplan.html)	California	California as a land and coastal area of global relevance	Marxan, CALZONE etc.
Circumpolar Arctic	Polar	A global climate chamber, global endemism	None (Spiridonov et al. 2012; but see Huettmann 2012, Spiridonov et al. 2017 and Spiridonov et al. 2017 for Marxan)
Global MPA network (http://www.mpatlas.org/)	World-wide	The globe's ocean protection	Marxan

wealthy stakeholders drive the process. The delphi method can easily compromise the entire objectivity, fairness, quality and transparency of Strategic Conservation Planning and of its science, process and trustworthiness overall.

It is here where machine learning offers help. It can provide a solution to the problem of filling data gaps with the delphi method. As shown in Drew et al. (2011)

SDMs can be produced with transparent, repeatable methods using open access data (see GBIF.org for species data, and worldclim.org for environmental data) with open source models (e.g. Openmodeler http://openmodeller.sourceforge.net/om_desktop.html or commercial models (SPM https://www.salford-systems.com/). The SDMs with the highest accuracy tend to use machine learning; see Elith et al. 2006 and Fernandez-Delgado et al. 2014 for a review of the highly performing algorithms available.

However, despite the easy access, very few SDMs are actually run for, or used by, Strategic Conservation Planning projects (Moilanen et al. 2009; but see Beiring 2013 for an example how SDMs are employed for Strategic Conservation Planning). There are several reasons why that problem still exists, as outlined in Table 16.2.

In my experience, available SDM applications are never really sufficient and not truly complete for Strategic Conservation Planning exercises (see Table 16.3 for shared experiences). In addition, SDM accuracy is another topic to watch for. That's because SDM accuracies are usually inconsistent across species in a study area, and which does not allow then for consistent inference or application. Sometimes SDM performances can be too low to be useful or to be employed to larger areas (Aycrigg et al. 2015). Overall, I often find that limited and sparse raw data cannot provide consistent and year-round spatial estimates of important demographic and ecological parameters such as fecundity, winter survival and migration risk for instance. SDMs rarely are applied yet to provide demographically relevant spatial estimates such as mating places, productivity hotspots and mortality landscapes (often summarized as 'sources and sinks'; Pulliam 1988). On the one hand, the detail that would be ideal and needed is rarely possible to achieve with demographic tools

Table 16.2 Some reasons for SDMs not being used in strategic conservation planning projects

Reason	Implication	Fix	Comment
SDM does not make predictions widely available	SDM results are just 'shiny' and not used	Request all SDM model files to be fully open access with ISO metadata	This is a common problem in SDM projects, e.g. see Guisan and Zimmermann (2000); Guisan and Thuiller (2005); Franklin and Miller (2009) but see Drew et al. (2011)
Strategic conservation planning project runs out of time	Not all species considered and some information under-utilized	Realistic time window needed for planning	Many agencies do not have adequate resources for all planning projects
Strategic conservation planning project runs out of money	Not all species considered and some information under-utilized	Realistic budget planning; cost-effective methods needed	Many agencies do not have adequate resources or technical capacity for all planning projects
Strategic conservation planning project ignores ecological complexities involving 'all' species	A reductionist and simplistic approach gets applied	Admission of incompleteness; focus on multivariate approaches	A so-called pragmatic [a]approach is frequently applied

[a]Being pragmatic does not solve the initial problem and creates problems of its own

Table 16.3 Shared experiences for SDM approaches in strategic conservation planning type projects

Project	SDM application	Strength	Weakness
Conservation assessment for Asian passerine migrants (Beiring 2013)	Provide species range estimates as input into MARXAN	First models and migratory bird estimates produced	Lack of good data. Lack of stakeholder support. SDM accuracy rel. low
Alaska corridors (Murphy et al. 2010)	Provide species range estimates as input into MARXAN	First models and estimate for 4 species produced	Legal constraints not allowing to address land ownership issues and buy-outs or such discussions and planning. Virtually left unused by stakeholders.
Arctic protection (Huettmann and Hazlett 2010, Spiridonov et al. 2017, Solovyev et al. 2017)	Suggested to use SDMs to start strategic conservation planning	GIS data and model discussion starts the conservation gap and management work	Unless designed specifically with local knowledge and citizen science, it can be too much driven by GIS and disconnected from implementation networks
St. Lucia island, Caribbean (Evans et al. 2015)	No SDM directly applied, but employs concepts of risk	Allows for simulations and to test concepts and assumptions	No direct species occurrences and abundances used
Bears in US/CAN (Proctor et al. 2004, Singleton et al. 2004)	None (Habitat Suitability Analysis HSI, Resource Selection Functions (RSF))	None	No quantitative progress; potential left unused; ambiguous results
Alaska (Semmler 2010)	Species distributions for major predators and their food chains	Overcomes existing data gaps	Model assessments for each pixel. Not used by agencies and deciders

(Amstrup et al. 2006). On the other hand, conservation decisions are urgent and in times of a global conservation crisis (Rockström et al. 2009, Mace et al. 2010). And so even basic SDMs, such as species occurrence, help to provide information that is useful for the overall Strategic Conservation Planning process. By now, virtually all what is not parsimonious (e.g. using Generalized Linear Models (GLMs) and Akaike's Information Criteria (AIC); Arnold 2010) can be achieved as progress (Breiman 2001) considering the global environmental crisis mankind is facing.

Overall, the use of machine learning for SDM approaches has greatly improved Strategic Conservation Planning processes (Table 16.3). These projects are often the first of their kind allowing for these methods and approaches to be introduced to Strategic Conservation Planning in the region. While data must be available to run SDMs (but often are not openly shared in SDM publications, and hardly in webportals like Movebank https://www.movebank.org/) and model outputs could easily be improved, these projects then allow for a subsequent machine learning 'culture' of conservation to be set up, also based on stakeholders in a public forum. It's rather transparent that way; the contribution of GBIF in that context remain unchallenged (see for instance Beiring 2013). Beyond data and model accuracies, the actual introduction of such a new conservation culture may be the biggest contribution. The need for data models, transparency and stakeholder buy-in is essential for implementing management of natural resources (Clark 2002). Having that awareness and accept such a need of wider community buy-in might well present the main contribution in SDMs and Strategic Conservation Planning projects for a global sustainable society.

16.2 How Machine Learning Predictions can Also Be Used Directly for Strategic Conservation Planning: How it's 'ought to be' and towards 'Better' Solutions

Strategic Conservation Planning is usually employed with an optimization framework; rarely, it is used with forecast scenarios directly. The actual optimization is usually based on methods like 'solvers', namely the simulated annealing algorithm (Ardron et al. 2008), whereas machine learning just provides the input GIS layers for describing generic patterns in the landscape. The use of future scenarios is possible and has been increasingly applied though (see Murphy et al. 2010 for an example). SDMs built with explanatory variables that have future forecasts, such as downscale global climate models and for 2100 let's say, can be used to forecast future conditions for species. While Population Viability Analysis (PVA) lack much of the spatially explicit aspects (e.g. Proctor et al. 2004), a spatial population viability analysis (sPVA) offers interesting and relevant possibilities for forecasting future conditions for Strategic Conservation Planning (Nielsen et al. 2008 for an example). These techniques link demographic PVA approaches with GIS habitat data and future scenarios, all based on optimizations from Strategic Conservation Planning. It tends to represent the best science available!

Often, sPVAs themselves fall short on some of the principles of Strategic Conservation Planning, or leave them unaddressed (Table 16.4), such as lack of optimization and not comparing several scenarios in parallel but just favoring singularity and reductionism. However, the strengths of sPVAs linked with scenario planning (Peterson et al. 2003) are that they can be much better and directly applied and tested for specific management questions, including demographic sensitivities and outlooks. The use of well-thought out scenarios provide policy alternatives (e.g. Gonzales et al. 2003; Nielsen et al. 2008) as compared to the narrow, singular and

Table 16.4 Some principles of conservation planning and protected area design (as per Martin et al. 2008, Moilanen et al. 2009)

Principle	Meaning	Relevance
Efficiency	The process and protected area includes no 'waste' of effort	Conservation is time critical
Compactness and/or connectedness	The trade-off in the spatial arrangement is clear and correctly implemented	Species dispersal and gene flow
Flexibility	Alternative options exist to achieve the goal still	Reaching the goal regardless of obstacles
Complementarity	The process and solution matches the context	Taking into account ongoing and other efforts
Selection frequency vs irreplaceability	Unique sites are considered appropriately	Endemic hotspots vs. generalists
Representativeness	The protected area represents the wider landscape and all of its components	The solution is complete and meaningful, unbiased
Adequacy	The process and protected area is sufficient to achieve meaningful goals	Adequate effort and outcome
Optimizations based on decision-theory and mathematical programming	The best solution is found using best-available science	Best solution that humans can achieve, an ethical mandate
Best-available data used	The process and result is based on best-available information	An ethical mandate to find the best-available solution

traditionally used Strategic Conservation Planning solution (which often consist of just 'one' solution wiping all other thoughts off the agenda). Computationally, and for all work that includes machine learning, this can easily be achieved. However, the sPVA option and when applied with scenarios is not yet widely employed in wildlife management and it is not required by law (see Huettmann et al. 2005; Nielsen et al. 2008) but it can come rather close to the ideal of adaptive wildlife management (Walters 1986; Huettmann 2007a; b). In the meantime, scenario-planning starts to become more common (Peterson et al. 2003). e.g. for climate change outlooks (IPCC; http://www.ipcc.ch/). The use of scenarios is widely known in the social sciences but so far less common in the traditional North American Model of wildlife management (Organ et al. 2012; Silva 2012), or in most other natural resource management schemes in the world. Clearly, such work is computational intense, and the role of coding and linking tools and codes across operating platforms becomes a power tool to achieve such conservation solutions! It's all part of machine learning either way!

IMPUTATION: More Ways with Machine Learning to Fill Data Gaps and Smooth–Out Predictors with Information Gain for an Effective Science–Based Conservation Management

Most statistical analysis default on data gaps. It's a '*no go*' zone for modern science and it tends to result in statements of convenience such as '*no evidence exist.*' So what to do when your data set is gappy, missing records and gets labeled '*no go*' zone? This seemingly old question remains a 'hot button' item, and becomes now a key bottleneck to overcome for progress. This problem actually made it to the forefront of modern data analysis (e.g. Graham 2009). Many reasons can be envisioned why data gaps occur. But if that can be resolved, then new insights can be won, and a subject can be moved forward (see Kenward 2013 for health applications). Thus, finding methods to substitute and fill data gaps make for a classic but very relevant and modern problem that data miners and machine learners have to deal with. It's a typical case in conservation that data are gappy while decisions are to be made though.

Reason for data gap	Detail of the data gap	Comment
Predictor has a data gap	Data column is incomplete	This is a common problem.
Response has a data gap	Data column is incomplete	In times of open access data this probably the biggest problem.
Missing data *per se*	Entire absence of information	Probably the biggest issue for holding us back for making inference
Lacking data due to research design problems	Entire lack of a specific information	The 'best' research design is usually not known ahead of the study. At minimum, research design can always be improved in hindsight.
Wrong entries	Entry is not correct	Depending on the term 'wrong' and its definitions some datasets are said to feature 50% of such 'errors' and 'wrong' entries.
Entries got erased by accident	Hole in the data set	This has been observed quite a lot and when data are modified, re-formatted and sorted in spread sheets for instance, or when they get operated by inexperienced people.
Entries had to be removed due to uncertainties	Hole in the data set	This is a common problem, e.g. when metadata are missing and with older data. Doubts about geo-referencing, taxonomy and collection times are typical examples. Arguably, it's better to keep weak data than having none.

Reason for data gap	Detail of the data gap	Comment
Entries had to be removed due to copyrights	Hole in the data set	This is also a common problem when data are not shared.
Error simulations	Based on certain assumed backgrounds, some data might be acceptable whereas under other assumptions might make this data set viable	For advanced inference, data error simulation can provide new insights. While this can present questions for high-end data mining, it can be a relevant feature for high-end information gain and inference.

One can easily envision a situation where these 'real world' data gaps occur, and then, where they 'magically' could be overcome! So why not simply imputing them? [1] Conceptually, imputation means to replace data holes and to fill (substitute) them (Enders 2010). In reality that means they are to be modeled! It usually makes use of existing, neighboring, associated and surrounding attributes and data points. Based on those relationships in the data, data gaps can get filled. With the help of advanced computing and data science, it is less a question 'whether' that can be done (no problem really to do so). Instead, the issue is more 'how good' is the accuracy obtained for a certain purpose (many applications are happy to have a 75% modeled accuracy when compared to no data at all)? In a way, imputation is a specific method to model-predict data gaps. And this can be done, all with a certain estimation accuracy. A 100% prediction accuracy can probably not be achieved, but often it is not needed even. Instead, one can fill errors in a decent way, the models do not default, and which helps to move the overall process and progress forward for an analysis topic.

By now, imputation, as a statistical discipline, is evolving fast and many methods exist (see also for updates at the Wikipedia site https://en.wikipedia.org/wiki/Imputation_(statistics). Major techniques are for instance single imputations such as hot-deck, cold-deck, mean-substitution, regression. Multiple imputations are other powerful techniques. Often, these methods are

---

[1] It should be emphasized here that many other methods exist to overcome data gaps, including data cloning to stabilize models on poor data or to extend the data (Lele et al. 2007; Jiao et al. 2016 for machine learning application). The other, and equally exciting approach for overcoming data gaps is to explain why data gaps actually occur (forensics), often based on 'common sense'. It tends to work nicely because most data gaps have a reason for their existence! For instance, some field research data gaps are due to bad weather (rain), or inaccessibility of steep slope elevations. Those factors, data gaps, can then serve as explanations for the absence of certain events in the data. Tree-based models, and specifically the work from Friedman (2002), make use of such approaches with good success.

linked with sampling and re-sampling approaches. Of particular relevance are the geo-imputation methods as they allow interpolations in geographic information systems (GIS), linked with the discipline and tools from Spatial Statistics. Those are popular in forestry, canopy, remote sensing and image analysis too. Entire textbooks and journals are devoted to that topic, e.g. https://www.journals.elsevier.com/spatial-statistics.

There are several software approaches possible that allow to run imputations; see for instance Enders (2010). Similarly, Salford Systems Ltd. offers in TreeNet (stochastic gradient boosting) options to run analysis with 'gappy data' (Friedman 1999, Salford Systems Ltd. https://www.salford-systems.com/products/treenet)). In the R language, YAImpute is one of those packages that can 'impute' data based on using nearest neighbor observations (kNN; Crookston and Finley (2008; https://cran.r-project.org/web/ packages/ yaImpute/ index.html). See also applications of such R code by J. Evans (http:// evansmurphy.wixsite.com/evansspatial).

While this is all pretty exciting, developing and moving forward fast, the sad news is that in wildlife conservation management, and for many natural resource applications, imputation convinces in the pure absence. The mainstream literature is extremely poor on making use of those methods and for advancing fields like conservation remote sensing, geo-locators, telemetry, wildlife surveys, disease outbreaks and citizen science. Two notable exceptions can be found though, namely climate as well as some forest work (Eskelson et al. 2009).

Literature

Crookston NL, Finley AO (2008) yaImpute: an R Package for kNN Imputation. J Stat Softw 23: 1–16

Enders CK (2010) Applied missing data analysis. Guilford Press, New York

Eskelson BN, Temesgen H, Lemay V, Barrett TM, Crookston NL, Hudak AK (2009) The roles of nearest neighbor methods in imputing missing data in forest inventory and monitoring databases. Scand J For Res 24: 235–246

Friedman JH (2002) Stochastic gradient boosting. Comp Stat Data Anal 38: 367–378

Graham JW (2009) Missing data analysis: making it work in the real world. Annu Rev Psychol 60: 549–576

Jiao S, Huettmann F, Guo Y, Li,Ouyang Y (2016) Advanced long-term bird banding and climate data mining in spring confirm passerine population declines for the Northeast Chinese-Russian flyway. Glob Planet Chang https://doi.org/10.1016/j.gloplacha.2016.06.015

Kenward MG (2013) The handling of missing data in clinical trials. Clin Investig 3: 241–250

Lele SR, Dennis B, Lutscher F (2007) Data cloning: easy maximum likelihood estimation for complex ecological models using Bayesian Markov chain Monte Carlo methods. Ecol Lett 10: 551–563

16.3 A Wider Perspective against the Local Techno-Fix: Good Ethics and Ecological Governance Foundations to Achieve a Conservation Break–Through with Open Access Machine Learning and Strategic Conservation Planning Carrying a Global Perspective for Mankind

An honest assessment of the environment, and the strategy needed for its conservation will expose nothing but a crisis, which is footed on a failed management and leadership in an overall destructive global framework (Ostrom 2015). We are running out of space while most relevant species and habitats are not protected, or at least not protected well or optimal! (Table 16.5).

Table 16.5 Selected examples for lack of achievement with National Park concepts for conserving species effectively

Species	Status	Habitat area protection through National Parks
Snow leopard	Vulnerable (or endangered, as per recent debate)	Found widely outside of protected areas
Songbirds, virtually worldwide	Large declines	Found widely outside of protected areas, e.g. Beiring (2013), Jiao et al. (2016)
Shorebirds	Large declines, e.g. for arctic species	Found widely outside of protected areas
Tree Kangaroo	Large declines (several species; Australia and Papua New Guinea)	Found widely outside of protected areas (FH unpublished)
Langures	Large declines	Found widely outside of protected areas (FH unpublished
Red panda	Globally threatened	Found widely outside of protected areas, e.g. Kandel et al. (2015)
Great panda	Vulnerable (1,000 individuals in the wild)	Not well protected within the protected area (Xu et al. 2014)
Black-necked cranes	Vulnerable	Found widely outside of protected areas (Xuesong et al. 2017)
Grizzly Bear (Canada)	Species at risk, partly extinct	Not well protected within protected area (Gallus 2010)
Atlantic Puffin	Vulnerable	Almost no marine protected areas (MPA), e.g. Huettmann et al. (2016)
Sharks	Widely declining	Almost no marine protected areas (MPA)
Commercial food fish species worldwide	Widely declining	Almost no marine protected areas (MPA; no take-zones)
Gharial	Critically endangered	Not well protected within protected area and outside

The New World Order, and starting with Bretton Woods in 1944, The World Bank and its subsequent institutions such as IUCN and UNEP show us nothing but that (Rich 1994). All too often we then just get presented a techno fix used to present us with progress when there actually is none (Czech et al. 2000; Cockburn 2013); even basic principles of strategic conservation are consistently violated (Table 16.4). By now, the list of those technical 'innovations' and fixes are very long, almost comical if it were not that tragic. It is easy to see that machine learning, or optimization algorithms from strategic conservation planning could fall into that category. The challenge now remains to show that it is not and to apply them in a good ecological setting. On the one hand machine learning cannot really break out of the techno-trail. It's a technological high-end application. It also requires energy and resources as input, eventually. Some of the stakeholder workshops also leave a big financial and carbon footprint, for instance. However, on the good side, the outcome is more than its individual pieces. That is nowhere so true than in machine learning: just consider what the phrase *'many weak learners make for a strong learner'* (Schapire 1990) means in real life.

So, while we are easily trapped in our institutions and minds with certain techno-arguments and its neoliberal world, there can be a decent output for the better, and hopefully with a good life- balance to be found eventually. Tables 16.4 and 16.6 show some of the core components of Strategic Conservation Planning projects to be successful, but which are widely missing in real world applications still (Huettmann 2007a, b, 2008a, b for projects and related data and publications). Table 16.7 shows known failures and a mis-use of Strategic Conservation Planning.

Table 16.6 Components to further improve Strategic Conservation Planning with a 'good' machine learning component

Wider topic	Justification	Example and citation	Status in strategic conservation planning projects
Best predictions	Best predictions	SPM8 (https://www.salford-systems.com/)	Not fully employed, yet
Use of best data	Best information assures best inference	Open access, e.g. Freedom of information act (https://www.foia.gov/)	Not used to the full potential yet
Make final project data available	Repeatable and transparent conclusions	Kandel et al. (2015)	Almost never achieved
Ethics	Avoid mis-use and destructive science	Weaver (1996), Bandura (2007), Daly and Farley (2011) Steady state economy mother earth	Virtually ignored

(continued)

Table 16.6 (continued)

Wider topic	Justification	Example and citation	Status in strategic conservation planning projects
Dynamic re-runs	MPA conditions change and need to be constantly adjusted	Dynamic MPAs at sea that reflect ocean currents and climate change (Murphy et al. 2010)	Discussed for the marine environment but virtually left without implementation, e.g. for terrestrial applications
Enforcement and policy link	Laws are only as good as their enforcement culture and related budget	Legal theory	Virtually ignored (see for missing links in wildlife management, conservation biology and strategic conservation planning textbooks, e.g. Bolen and Robinson (2002), Silva (2012), Primack (2016), Ardron et al. (2008)
Ecological economics	The only known economy to achieve a 'steady state'	Czech (2000), Daly and Farley (2011)	Rarely considered, yet.

Table 16.7 Known failures of strategic conservation planning projects and their suggested fixes

Strategic conservation planning project	Known failure in the goal	Effective fix	Outlook
Great barrier reef	Overruled by political and global forces, e.g. Indian mining concessions in the wider watershed	Implement a better global governance	It is currently not realistic to assume a global governance that puts harmony and environmental balance at its core serving the wider public good
Terrestrial national parks	10% protection level too small for being effective or meaningful, Marxan not well used	30% protection levels, implement a better global governance	Unlikely to happen any time soon in ecological relevant terms
Global MPA ocean network	Most of the MPAs are usually not 'No Take' zones, lacking enforcement either way, greatly improve coverage	30% protection levels, implement a better and realistic global governance and enforcement	Unlikely to happen any time soon in ecological relevant terms
OSPAR (https://www.ospar.org/work-areas/bdc/marine-protected-areas)	Use of a MPA far offshore to mitigate climate change. Put MPAs where the diversity actually sits	Reduce industrial consumption as the underlying cause of climate change, improve ocean protection and management	Unlikely to happen any time soon in ecological relevant terms
Russian Arctic MPAs (Spiridonov et al. 2017, Solovyev et al. 2017)	Climate change futures not included	Re-run with climate change future outlook	Unlikely to happen any time soon in ecological relevant terms

In a way, I hope, machine learning, e.g. for Strategic Conservation Planning, can at least help to strike that balance better reaching a steady-state (Daly and Farley 2011). We can actually afford to spend some energy, as long it is sustainable and not excessive, on machine learning with Strategic Conservation Planning and for good decision-making reaching sustainability on a global level. As a matter of fact, if we have any energy, or effort for that matter handy, it should be invested into great decision-making, achieved with machine learning-aided Strategic Conservation Planning making good use of those techniques available to mankind. This can eventually lead to a global society respecting 'mother earth'. For global sustainability to become real, wider questions come to play, including universe ones, belief systems, spirituality, governance structures, distribution of wealth and the balance of life (Weaver 1996; Stiglitz 2003). But one way or another, machine learning is available and involved by now, and all one can ask for then is to make good and best-suitable use of this method; all done with good ethics and outcome for the wider public, global good. Anything that is not destructive science would be progress in that regard. Now, who would not agree?

References

Alexander JC (2013) The dark side of modernity. Polity Press

Alidina HM, Fisher D, Stienback C, Ferdana Z, Lombana A, Huettmann F (2008) Assessing and managing data. In: Ardron J, Possingham H, Klein C (eds) Marxan good practices handbook, Vancouver, pp 14–20 http://www.pacmara.org/

Amstrup SC, McDonald TL, Manly B (2006) Handbook of capture-recapture analysis. Princeton University Press

Ardron J, Possingham H, Klein C (eds) (2008) Marxan good practices handbook, Vancouver http://www.pacmara.org

Arnold TW (2010) Uninformative parameters and model selection using Akaike's information criterion. J Wildl Manag 74:1175–1178

Aycrigg J, Beauvais G, Gotthardt T, Huettmann F, Pyare S, Andersen M, Keinath D, Lonneker J, Spathelf M, Walton K (2015) Novel approaches to modeling and mapping terrestrial vertebrate occurrence in the northwest and Alaska: an evaluation. Northwest Sci 89:355–381. https://doi.org/10.3955/046.089.0405

Baltensperger AP, Huettmann F (2015) Predicted shifts in small mammal distributions and biodiversity in the altered future environment of Alaska: an open access data and machine learning. PLoS One. https://doi.org/10.1371/journal.pone.0132054

Bandura A (2007) Impeding ecological sustainability through selective moral disengagement. Int J Innovat Sustain Dev 2:8–35

Beiring M (2013) Determination of valuable areas for migratory songbirds along the East-Asian Australasian Flyway (EEAF), and an approach for strategic conservation planning. Unpublished Master Thesis. University of Vienna, Austria

Bolen EG, Robinson W (2002) Wildlife ecology and management, Fifth edn. Pearson Publisher

Breiman L (2001) Statistical modeling: the two cultures (with comments and a rejoinder by the author). Stat Sci 16:199–231

Clark TW (2002) The policy process: a practical guide for natural resource professionals. Yale University Press

Cockburn A (2013) A colossal wreck. Verso Publishers, New York

Cushman S, Huettmann F (2010) Spatial complexity, informatics and wildlife conservation. Springer Tokyo, Japan, p 448

Czech B (2000) Shoveling fuel for a runaway train: errant economists, shameful spenders, and a plan to stop them all. University of California Press, Berkeley

Czech B, Krausman PR, Devers PK (2000) Economic associations among causes of species endangerment in the United States. Bioscience 50:593–601

Daly HE, Farley J (2011) Ecological economics: principles and applications. Island press, Washington, DC

Drew CA, Wiersma Y, Huettmann F (eds) (2011) Predictive species and habitat modeling in landscape ecology. Springer, New York

Elith J, Graham CH, Anderson RP, Dudík M, Ferrier S, Guisan A, Hijmans RJ, Huettmann F, Leathwick JR, Lehmann A, Li J, Lohmann LG, Loiselle BA, Manion G, Moritz C, Nakamura M, Nakazawa Y, McC J, Overton M, Townsend Peterson A, Sj P, Richardson K, Scachetti-Pereira, Schapire RE, Soberón J, Williams S, Wisz MS, Zimmermann NA (2006) Novel methods improve prediction of species' distributions from occurrence data. Ecography 29:129–151

Evans JS, Schill S, Raber GT (2015) Chapter 26: A systematic framework for spatial conservation planning and ecological priority design: an example from St. Lucia, Eastern Caribbean. In: Huettmann F (ed) Conservation of Central American biodiversity. Springer, New York, pp 603–623

Fernandes L, Day J, Lewis A, Slegers S, Kerrigan B, Breen D, Cameron D, Jago B et al (2005) Establishing representative no-take areas in the great barrier reef: large-scale implementation of theory on marine protected areas. Conserv Biol 19:1733–1744

Fernandez-Delgado M, Cernadas E, Barrow S, Amorim D (2014) Do we need hundreds of classifiers to solve real world classification problems? J Mach Learn Res 15:3133–3181

Flannery T (2002) The future eaters: an ecological history of the Australasian lands and its people. Grove Press, New York

Forman RTT (1995) Land mosaics: the ecology of landscapes and regions. Cambridge University Press

Franklin J, Miller JA (2009) Mapping species distributions: spatial inference and predictions. Cambridge University Press, USA

Gallus J (2010) Grizzly Bear manifesto. Rocky Mountain Books, Toronto

Gonzales EK, Arcese P, Schulz R, Bunnell FL (2003) Strategic reserve design in the central coast of British Columbia: integrating ecological and industrial goals. Can J For Res 33:2129–2140 https://doi.org/10.1139/x03-133

Groube L, Chappell J, Muke J, Price D (1986) A 40,000 year-old human occupation site at Huon Peninsula, Papua New Guinea. Nature 324:453–455. https://doi.org/10.1038/324453a0

Guisan A, Thuiller W (2005) Predicting species distribution: offering more than simple habitat models. Ecology Letters 8:993–1009. https://doi.org/10.1111/j.1461-0248.2005.00792.x

Guisan A, Zimmermann NE (2000) Predictive habitat distribution models in ecology. Ecol Model 135:147–186

Huettmann F (2007a) Modern adaptive management: adding digital opportunities towards a sustainable world with new values. Forum Public Policy: Climate Change Sustain Dev 3:337–342

Huettmann F (2007b) The digital teaching legacy of the International Polar Year (IPY): details of a present to the global village for achieving sustainability. In: Tjoa M, Wagner RR (eds) Proceedings 18th international workshop on Database and Expert Systems Applications (DEXA) 3–7 September 2007, Regensburg, Germany. IEEE Computer Society, Los Alamitos, pp 673–677

Huettmann F (2008a) Appendix II: literature & references. In: Ardron J, Possingham H, Klein C (eds) Marxan good practices handbook. Vancouver, Canada, pp 139–153 http://www.pacmara.org/

Huettmann F (2008b) Marine conservation and sustainability of the sea of Okhotsk in the Russian far east: an overview of cumulative impacts, compiled public data, and a proposal for a UNESCO World Heritage Site. In: Nijhoff M (ed) Ocean year book, vol 22. Halifax, Canada, pp 353–374

Huettmann F (2011) From Europe to North America into the world and atmosphere: a short review of global footprints and their impacts and predictions. The Environmentalist. https://doi.org/10.1007/s10669-011-9338-5

Huettmann F (2012) Protection of the three poles. Springer, New York

Huettmann F, Hazlett S (2010) Changing the arctic: adding immediate protection to the equation. Alaska Park Science, Fairbanks, pp 118–121

Huettmann F, Franklin SE, Stenhouse GB (2005) Predictive spatial modeling of landscape change in the Foothills Model Forest. For Chron 81:1–13

Huettmann F, Riehl T, Meissner K (2016) Paradise lost already? A naturalist interpretation of the pelagic avian and marine mammal detection database of the IceAGE cruise off Iceland and Faroe Islands in fall 2011. Environ Syst Decis. https://doi.org/10.1007/s10669-015-9583-0

Huettmann F, Magnuson EE, Hueffer K (2017) Ecological niche modeling of rabies in the changing Arctic of Alaska. Acta Vet Scand 201(759):18–31. https://doi.org/10.1186/s13028-017-0285-0

Jiao S, Huettmann F, Guoc Y, Li X, Ouyang Y (2016) Advanced long-term bird banding and climate data mining in spring confirm passerine population declines for the Northeast Chinese-Russian flyway. Glob Planet Chang. https://doi.org/10.1016/j.gloplacha.2016.06.015

Kandel K, Huettmann F, Suwal MK, Regmi GR, Nijman V, Nekaris KAI, Lama ST, Thapa A, Sharma HP, Subedi TR (2015) Rapid multi-nation distribution assessment of a charismatic conservation species using open access ensemble model GIS predictions: Red panda (*Ailurus fulgens*) in the Hindu-Kush Himalaya region. Biol Conserv 181:150–161

Lehtomäki J, Moilanen A (2013) Methods and workflow for spatial conservation prioritization using Zonation. Environ Model Softw 47:128–137

Mace G, Cramer W, Diaz S, Faith DP, Larigauderie A, Le Prestre P, Palmer M, Perrings C, Scholes RJ, Walpole M, Walter BA, Watson JEM, Mooney HA (2010) Biodiversity targets after 2010. Environ Sustain 2:3–8

Martin T, Smith J, Royle K, Huettmann F (2008) Is Marxan the right tool? In: Ardron J, Possingham H, Klein C (eds) Marxan good practices handbook. PACMARA, Vancouver, pp 58–74 http://www.pacmara.org/

Moilanen A, Wilson KA, Possingham H (2009) Spatial conservation prioritization: quantitative methods and computational tools, First edn. Oxford University Press

Murphy K, Huettmann F, Fresco N, Morton J (2010). *Connecting Alaska landscapes into the future.* U.S. Fish and Wildlife Service, and the University of Alaska.Fairbanks, Alaska. http://www.snap.uaf.edu/downloads/connecting-alaska-landscapes-future

Nielsen SJ, Stenhouse GB, Beyer HL, Huettmann F, Boyce MS (2008) Can natural disturbance-based forestry rescue a declining population of grizzly bears? Biol Conserv 141:2193–2207

Organ, JF, Geist V, Mahoney SP, Williams S, Krausman PR, Batcheller GR, Decker TA, Carmichael R, Nanjappa P, Regan, R, Medellin RA, Cantu R, McCabe RE, Craven S, Vecellio GM, Decker DJ (2012) The North American model of wildlife conservation. The Wildlife Society technical review 12-04. The Wildlife Society, Bethesda, Maryland, USA

Ostrom E (2015) Governing the commons: the evolution of institutions for collective action (Political Economy of Institutions and Decisions). Cambridge University Press, New York

Paehlke RC (2004) Democracy's dilemma: environment, social equity, and the global economy. MIT Press, Cambridge

Perera AH, Drew AC, Johnson C (eds) (2012) Expert knowledge and its application in landscape ecology. Springer, New York

Peterson GD, Cumming GS, Carpenter SR (2003) Scenario planning: a tool for conservation in an uncertain world. Conserv Biol 17:358–366

Primack R (2016) Essentials of conservation biology, Sixth edn. Sinauer Press, Boston

Proctor MFC, Servheen M, Miller SD, Kasworm WF, Wakkinen WL (2004) A comparative analysis of management options for grizzly bear conservation in the U.S.–Canada trans-border area. Ursus 15:145–160

Pulliam HR (1988) Sources, sinks, and population regulation. Am Nat 132:652–661. https://doi.org/10.1086/284880

Rich B (1994) Mortgaging the earth: the world bank, environmental impoverishment and the crisis of Development. Island Press

Rockström J, Steffen W, Noone K, Persson Å, Chapin FS III, Lambin EF, Lenton TM, Scheffer M, Folke C, Schellnhuber HJ, Nykvist B, de Wit CA, Hughes T, van der Leeuw S et al (2009) A safe operating space for humanity. Nature 461:472–475

Schapire RE (1990) The strength of weak learnability. In: Machine learning, vol 5. Kluwer Academic Publishers, Boston, pp 197–227. https://doi.org/10.1007/bf00116037

Scott M, Davis FW, McGhie RG, Wright RG, Groves C, Estes J (2001) Nature reserves: do they capture the full range of America's biological diversity? Ecol Appl 11:999–1007

Semmler M (2010) Spatial and temporal prediction models of Alaska's 11 species mega-predator community: towards a first state-wide ecological habitat, impact, and climate assessment. Unpublished M.Sc. thesis. Georg-August University Goettingen, Germany

Silva NJ (2012) The wildlife techniques manual: research & management, vol 2, Seventh edn. The Johns Hopkins University Press

Singleton PH, Gaines WL, Lehmkuhl JL (2004) Landscape permeability for grizzly bear movements in Washington and southwestern British Columbia. Ursus 15:90–103

Solovyev B, Spiridonov V, Onufrenya L, Belikov S, Chernova N, Dobrynin D, Gavrilo M, Glazov D, Yu K, Mukharamova S, Pantyulin A, Platonov N, Saveliev A, Stishov M, Tertitsk G (2017) Identifying a network of priority areas for conservation in the Arctic seas: practical lessons from Russia. Aquat Conserv Mar Freshwat Ecosyst 27(S1):30–51

Spiridonov V, Gavrilo M, Yu Krasnov, A. Makarov, N. Nikolaeva, L. Sergienko, A. Popov, and E. Krasnova (2012). Toward the new role of marine and coastal protected areas in the Arctic: the Russian case. in: Protection of the three poles, by F. Huettmann (Ed.). Springer, New York. pp 171–202

Spiridonov V, Solovyev B, Chuprina E, Pantyulin A, Sazonov A, Nedospasov A, Stepanova S, Belikov S, Chernova N, Gavrilo M, Glazov D, Yu K, Tertitsky G (2017) Importance of oceanographical background for a conservation priority areas network planned using MARXAN decision support tool in the Russian Arctic seas. Aquat Conserv Mar Freshwat Ecosyst 27(S1):52–64

Stiglitz JE (2003) Globalization and its discontent. Norton Paperback, New York

UNESCO Great Barrier Reef Marine Park Authority (1981) Nomination of the Great barrier reef by the Commonwealth of Australia for inclusion in the world heritage list: United Nations educational scientific and cultural organization Great barrier reef marine park authority, Townsville, Australia

Walters C (1986) Adaptive management of renewable resources. Blackburn Press

Walther BA, Boëte C, Binot A, By Y, Cappelle J, Carrique-Mas J, Chou M, Furey N, Kim S, Lajaunie C, Lek S, Méral P, Neang M, Tan B-H, Walton C, Morand S (2016) Biodiversity and health: lessons and recommendations from an interdisciplinary conference to advise Southeast Asian research, society and policy. Infect Genet Evol 40:29–46

Watts MI, Ball I, Stewart RS, Kleina CJ, Wilson K, Steinback C, Lourival R, Kircher L, Possingham HP (2009) Marxan with zones: software for optimal conservation based land- and sea-use zoning. Environ Model Softw 24:1513–1521

Weaver J (1996) Defending mother earth: native American perspectives on environmental justice. Orbis Books

Xu W, Viña A, Zengxiang Q, Zhiyun O, Liu J, Hui W (2014) Evaluating conservation effectiveness of nature reserves established for surrogate species: case of a Giant Panda nature reserve in Qinling Mountains, China. Chin Geogr Sci 24:60–70

Xuesong H, Guo Yu, Mi C, Huettmann F, Wen L (2017) Machine learning model analysis of breeding habitats for the Blacknecked Crane in Central Asian uplands under anthropogenic pressures. Scientific Reports 7, Article number: 6114 doi:10.1038/ s41598–017–06167-2. https://www.nature.com/articles/s41598-017-06167-2

Chapter 17
How the Internet Can Know What You Want Before You Do: Web-Based Machine Learning Applications for Wildlife Management

Grant R. W. Humphries

17.1 Importance of Web–Based ML Applications

In the twenty-first century, we are inundated with terabytes of data on a daily basis. Furthermore, there is a global increase in access to internet services, with nearly 4 billion users as of mid-2016 (http://www.internetworldstats.com). Combined with this is the threat of over-population (Toth and Szigeti 2016), climate change (Bestion et al. 2015), and resource depletion (Schneider et al. 2015). I therefore pose a question to the reader: What is the fastest and cheapest way to get information from source to manager, while engaging the public? The answer is, of course, the internet.

The growth in global infrastructure now means information is at our fingertips, (i.e., we are now able to learn and adapt at the click of a button). For example, you are driving from New York to Washington, D.C. and you are using a web-based mapping application (e.g. Google Maps) which analyzes traffic patterns while giving directions. Suddenly, there is an accident on the route which you were taking (we'll call it a fender bender to appease the sensibilities of the reader). Your mapping application, which uses machine learning and data-mining tools to analyze traffic conditions, then detects the accident and re-routes you, saving money, time, and sparing you the look of disappointment in your partner's eyes because you were late for supper. In other words, you used a web-based application to make direct and efficient decisions, which helped to increase your quality of life. Wildlife management could take this form if scientists, managers and developers worked together to build web-based tools and applications that actively engaged the public in an open access manner. One use for such an application might be in the enforcement of illegal fishing, where machine learning algorithms could be used to estimate risk of a

G. R. W. Humphries (✉)
Black Bawks Data Science Ltd., Fort Augustus, Scotland
e-mail: grwhumphries@blackbawks.net

© Springer Nature Switzerland AG 2018
G. R. W. Humphries et al. (eds.), *Machine Learning for Ecology and Sustainable Natural Resource Management*, https://doi.org/10.1007/978-3-319-96978-7_17

vessel engaging in illegal activities. If the risk is calculated to be high enough, then enforcement officers would have to board the vessel and inspect the goods.

Web tools are ideal delivery methods for scientific data and for management because they: a) create a storage space for data and information that can be accessed from anywhere in the world (as opposed to dusty file cabinets in a retired professor's basement), b) act as a near real-time, peer-review system for scientific findings, c) engage the public in an open access fashion (and seeing as they pay for most of our science, why not allow them to poke the bear a bit?), d) allow for full transparency of decision making processes as things can be easily re-created and usually the code is housed in repositories like GitHub, e) mean that users do not have to install any additional software or have technical knowledge of the code to use them, and f) allow for easier collaborations. Because web-based technologies are more efficient, cheaper, and more transparent, with better feed-back mechanisms, I believe we are moving towards a new medium of scholarly achievement. In my humble opinion, the "publish or perish" attitude is likely to change drastically, where more emphasis will be placed on the use or creation of these sorts of tools ad products. Not to say that publishing will end completely as scholarly articles are still important and will remain so. However, getting that job after 10 years or so of University may soon be easier with a track record of building tools for conservation rather than just studying the academic questions.

17.2   Open Access

The free access of data has become a vastly important issue in science for a variety of reasons (Engleward and Roberts 2007). Most importantly is so that all stakeholders (e.g., scientists, managers, and the public) have free and easy access to the best available science and information. Arguably, this should be the policy for any publicly funded science. Funding organizations such as the National Science Foundation, or Landscape Conservation Cooperatives (LCC) require all studies to make their data and finding open access after publication and reporting periods have passed, however this is generally un-enforced (however, with some LCCs, support can be pulled for non-compliance of making data available). Existing web tools, or those under development, can alleviate issues with open access data, and offer amazing platforms to apply machine learning techniques for decision support. That is to say, web applications like eBird (Sullivan et al. 2009), GBIF (Flemons et al. 2007), or the Birdlife International Seabird Tracking database (http://www.seabirdtracking. org) are designed to ingest data from the public and scientists in a way that minimizes the amount of pre-processing. Once on the server, or in the appropriate databases, machine learning algorithms can mine those data and make forecasts or predictions that may help inform the pre-cautionary principle or other management guidelines. We could be nearing a point where computers can tell us what we need to do before we know what we need (and don't worry, we aren't describing HAL

9000 from "2001: A Space Odyssey"). These web-tools also give us access to pertinent information about the data we are accessing.

17.3   Metadata

The 'data about the data' (A.K.A. the metadata) is a vital piece of information that should come along with any data download (Fegraus et al. 2005). Many style formats exist however the International Organization for Standardization (ISO) have several internationally recognized standards for metadata structure. Geographic data (e.g., ESRI shapefiles, rasters, etc....) for example, is dictated by ISO standard 19115, and states that metadata must contain information about spatial extent, resolution, projection (and datum), as well as the organizations or individuals who created the files and how to contact them. Programming techniques and machine learning algorithms can automatically generate metadata, and append them to data being downloaded. Web-tools are again by far the most effective way of disseminating metadata records, and therefore opens the door for major citizen science initiatives.

17.4   Citizen Science

One of the real advantages of web-based applications is the ability to integrate science and public knowledge (Dickinson et al. 2012). For example, over 75% of people in the United States have access to the internet (Zickuhr 2013). If, of those approximately 229 million people who have internet access, only 5% are wildlife enthusiasts, that leaves us with ~ 11.5 million people. Further, we will assume that only 2% of those people are avid bird watchers who would report their sightings via a web application (~229,000 individuals). We then assume that they report one time every month (maybe a conservative estimate for some of the truly avid birders out there). That means that over the course of a single year, you could have access to nearly 2.8 million sightings of various species. From an ecological standpoint, this is quite substantial, and is the very model that eBird uses to get their data. Those data are then processed and used for various modeling exercises (though most of the results are yet to become publicly accessible).

Another example of citizens accessing and providing data via web-tools is on the website "Penguin Watch" (http://www.penguinwatch.org). Scientists from the United Kingdom have placed a number of cameras around the Antarctic peninsula, which are taking an array of images over the course of a year at colonies. These images are uploaded to the web, where citizens can go online and count any of the penguins they see in those images. This project is still relatively new as of when we wrote this piece, but plans are underway to make population models and other data open access. This is an important aspect of any scientific work to ensure public engagement and because we owe these data to the public as they are often the funding source (i.e., in the case of government funding).

17.5   Public Access

Public engagement is vital to the success of science (Stilgoe et al. 2014). Engagement ensures transparency and fills an obligation that scientists have towards their funders. As I have stated in previous sections, a large proportion of individuals world-wide have access to the internet. Web-tools are the fastest way of getting information from scientists to the public. I think it goes without saying that public access / engagement is the best way of making sure that science continues to get funding into the future.

However, enough of my prattling on about the importance of these tools (I would imagine as a reader of this book, you may already be aware of the issues). Let us move on to the juicy bits. As a budding machine learning user or data scientist you want to know HOW to implement these sorts of systems. In the next few sections, I will identify some common platforms, and give examples and a "how-to" with some (hopefully) easy to follow code.

17.6   How to Develop your Own Application

Prior to any software or application development, it's important to clearly identify your GOALS, OBJECTIVES and AUDIENCE. This allows you to build an application around their needs. Once this has been done, the following steps can be implemented:

1) Select a database framework
2) Select a programming framework
3) Choose your modeling algorithm (this might depend on your question, or your application)
4) Build your database (a plethora of useful frameworks exist for this; SQLite, mySQL, PostgreSQL)
5) Construct your model code
6) Build your static web content
7) Integrate your model with the static web content
8) Select a web host
9) Test and push Beta version to the web
10) Get feedback
11) Re-test and push live version to the web

12) Make results from modeling and data downloadable with ISO standard metadata
13) Track website usage with analytics programs (e.g., Google)

17.7   Databases and Choosing the Right One for you

In another chapter, some of my colleagues will go over databases in more detail. However, to build a good web application, database selection is a key feature, so I will briefly touch on it here.

There are several key things to consider when choosing a database structure:

1) Usability: Is it easy to use for you? How much training will you need?
2) Security: How much security do you require for your data to prevent database corruption due to web-bots?
3) Functionality: Does the framework do everything you need it to do with respect to storing and extracting data on demand?
4) Integration: How easily can you integrate with your web application?
5) Scalability: Can you continually grow the database?
6) Hosting: Can your chosen database framework be hosted easily on the web?
7) Long-term maintenance?

Once you have examined your needs, you can wade through an ocean of database management systems. They range from completely free to wildly expensive, and scalability and security are often directly linked to the cost. Oracle RDBMS, for example, can be upwards of tens of thousands of dollars per annum (or more), depending on the side of the database. Oracle can be programmed, and has a very intuitive database construction tool, as well as unlimited support for security, updates and troubleshooting. MySQL is a completely open source database system, and does not have the same logistical support, nor a straightforward tool for database construction. Typically, MySQL databases are programmed, and there can be a steep learning curve associated with it. However, MySQL is useful for most generic databases. PostgreSQL is a powerful open source database management system that not only has a relatively intuitive interface (PGADMIN III), but also links to a variety of web app frameworks, and can be enabled with PostGIS so you can store spatial data quite easily. Not only can you interface with PostgreSQL through PGADMIN III, but using packages in R or in Python, you can easily access and program the database. A more commonly used program for those of you who don't program databases is MS Access, which is a pretty simple graphical user interface system for building SQL enabled relational databases. However, it suffers from several drawbacks including a finite amount of space available for a database, meaning it's only good for small programs, and it has very poor security overall (to name a couple) (Textbox 17.1).

Textbox 17.1 Sample code to construct database table in R. The database itself (in our example, "censusdata"), must be created ahead of time, and then R is used to build tables, load data, etc... In my below example, I've created a database using PGADMIN III with a username "me", and a password "password". From there, I assign an SQL command as a string of text to a variable and the use the dbGetQuery command to run the SQL sequence on the table

Access database via R package 'RPostgreSQL'

```
pw <- "password"
drv <- dbDriver("PostgreSQL")
con <- dbConnect(drv, dbname = "censusdata", host = "localhost", port = 5432, user = "me", password = pw)
```

Create SQL query that will create the table in the database
```
sql <- "
DROP TABLE IF EXISTS preprocessed;
CREATE TABLE preprocessed (
  preprocessed_id SERIAL PRIMARY KEY,
  common_name TEXT,
  day SMALLINT CHECK (DAY >= 1 AND DAY <= 31),
  month SMALLINT CHECK (MONTH >= 1 AND MONTH <= 12),
  season SMALLINT NOT NULL CHECK (SEASON > 1000),
  count INTEGER CHECK (COUNT >= 0),
  accuracy SMALLINT CHECK (ACCURACY >= 1 AND ACCURACY <= 5)"
```

```
dbGetQuery(con, sql)
```

Create data table that will be used in the database, then load into PostgreSQL database.

```
X <- ## Data table that will be sent to database
```

```
dbWriteTable(con=con, name="censusdata", value=X, row.names=FALSE, append=TRUE, match.cols=TRUE)
```

17.8 Selecting a (relatively) Painless Programming Framework

Only in recent years have wildlife ecologists begun to truly delve into the world of computer science. Particularly since open access statistical languages (e.g., R) have become commonplace for analysis. This arguably means that we (as wildlife ecologists) are still disconnected from the fundamentals of computer science, and are somewhat forced to stick with easy to understand scripting and programming languages. I could drone on about various languages like PERL, RUBY, C, C++, F92 etc...., but instead I will focus on the two most commonly used languages in our field (i.e., R and Python), and the advantages and disadvantages of both (briefly, as this topic alone could be a thesis). As a note before proceeding, languages like C, and C++ are much more powerful than R or Python, yet are much more complex and carry the risk of 'up-ending' an operating system if you are not careful. There is a good infographic available at: https://www.datacamp.com/community/

Table 17.1 Comparison of key points for R versus Python for web applications and data mining/ machine learning based on author's experience and conversations with colleagues. Points are ranked as either Yes or No, or Poor, Moderate and Good

Feature	Python	R
Free	Yes	Yes
User contributed packages	Yes	Yes
Memory management	Good	Poor
Web deployment	Good	Poor
Data visualization	Good	Good
Ease of statistical analysis	Moderate	Good
Spatial data handling	Moderate	Good
Learning curve	Moderate	Poor
Processing speed	Good	Moderate
GPU access	Good	Moderate
Operating systems	All – Works best with Linux	All
GitHub integration	Good	Good
Multi/General-purpose	Good	Poor
Script management	Good	Good
Online community / Help	Good	Good

tutorials/r-or-python-for-data-analysis#gs.jrKw75w. Both R and Python are comparable and your choice of language depends wholly on your application (see Table 17.1 for a 'subjective' comparison).

17.9 The R Statistical Programming Language

The language R (R Core Development Team 2015) was developed as a statistical language, and due to its relatively easy interface, has become one of the most widely used languages in data science and wildlife ecology (this is based on extrapolation from conversations with colleagues, and the known availability of statistics training in ecology programs). Many of the readers of this book are likely already R users, or are at least thinking about it. Because it has a wide user base, and most code are open-access, it makes for a great platform for database and machine learning implementation. However, with respect to deployment for web applications, R still has a way to go and is somewhat limited. The "RShiny" add on for R and RStudio is the most used method for building web applications in R and I will use this in the below examples.

17.10 Python

Python, named for the Monty Python acting troupe; (Python Software Foundation 2015) is a powerful language that is used very heavily by data scientists around the world, directly "competing" with R (though they can be used in conjunction with each other). It has a huge online community and a number of user developed packages that make scientific programming easy to do. Furthermore, it has several powerful spatial packages, and interfaces directly with ArcGIS (if you are lucky enough to have a license for it). In a general sense, Python has better spatial packages and can integrate with web development much better than R, however has fewer statistical packages and can be more complex to learn than R. Django is a library that is commonly used for web application development in Python (though others exist), and I will use this package to illustrate my examples below.

17.11   Choose your Weapon: Algorithms for your Modeling Pleasure

Throughout this book, you have been introduced to a variety of machine learning algorithms which can be applied to address different problems. Your choice of algorithm when you build a web application will be dependent on your questions or ultimate goal. For example, image recognition applications might require the use of convolutional neural networks (Krizhevsky et al. 2012). Time series models might benefit from the use of long short term memory neural networks (Hochreiter and Schmidhuber 1997), and spatial modeling efforts could benefit from the use of random forests (Breiman 2001; Cutler et al. 2007). No matter which algorithm you choose, there are some excellent libraries in Python and R that can be easily integrated into web applications. In Table 17.2 I present some of the more commonly used packages in R and Python. There are other options on the "cloud" as well for machine learning applications on the web, and although there are some issues to be resolved (Dillon et al. 2010), they offer powerful alternatives to straight models in R or Python alone. Both Google (https://cloud.google.com), and Amazon (https://aws.amazon.com/machine-learning) offer machine learning solutions on the cloud, which allow users to deploy web applications efficiently.

17.12   Building your Code: Database Construction, Mining and Modeling

No matter which language or algorithm chosen, there is a basic structure that should be adhered to when developing an application that uses machine learning for decision support. This process typically starts with development of the question and data cleaning or pre-processing. Once the database has been created, it can be queried and modeled and then relevant output reported (Fig. 17.1).

Table 17.2 A few useful packages and libraries in Python and R for data mining, machine learning and web deployment. Further libraries can be found at http://www.kdnuggets.com/2015/06/top-20-python-machine-learning-open-source-projects.html for Python and https://cran.r-project.org/web/views/MachineLearning.html for R

Language	Package	Purpose
Python	*Django*	*Allows for the "model-view-template" structure to develop web apps in Python*
	Numpy	*Includes many options for array style maths and data*
	Scipy	*Includes functions for linear algebra, interpolation and other scientific processing*
	Scikit-learn	*Contains an array machine learning algorithms*
	Dendropy	*Can be used for a variety of basic statistical analyses*
	Bottleneck	*Increased functionality for statistics in Python*
	Pandas	*Useful for handling reading and writing of csv files*
	psycopg2	*Used for accessing PostgreSQL databases*
R	*RShiny*	*This creates a server / ui structure to develop web apps in R*
	Dplyr	*Used for cleaning and organizing data in R*
	Tidyr	*More functions for data cleaning and organization*
	randomForest	*Implements Breiman's randomForest algorithm in R*
	Dismo	*Has a suite of machine learning algorithms and functions for species distribution models*
	ggplot2	*Package for creating professional plots in R*

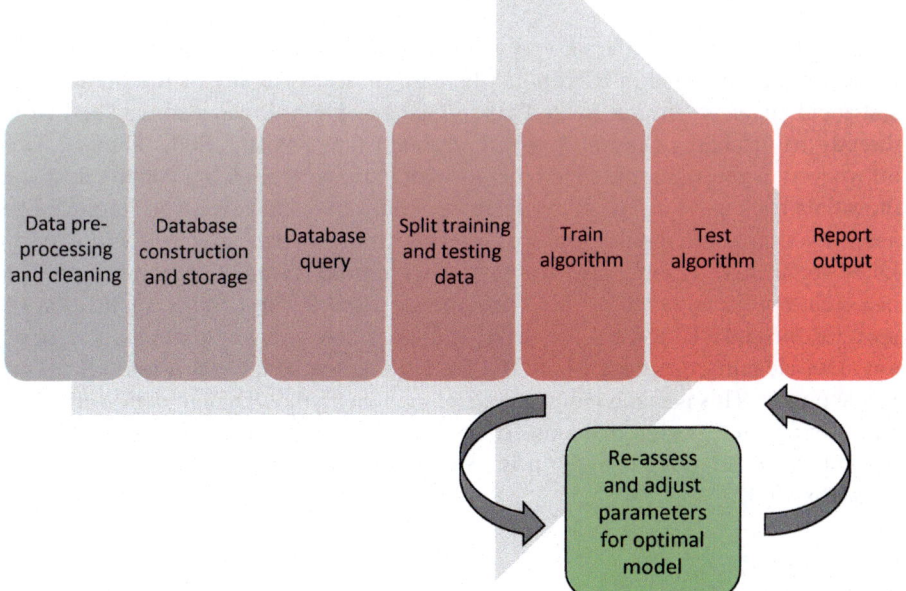

Fig. 17.1 General steps for modeling within a web application

17.13   Static Content and Integrating your Models

Unfortunately for you budding data scientists, the work has yet to finish. Ideally, you will be part of a team which not only consists of scientists and programmers, but also, web designers. Web design skills are a must when constructing new open access tools, and that means understanding core concepts of HTML, CSS and Javascript. Despite this hurdle, there are many online resources to learn these languages, and many good style guidelines that exist on the web through some simple searches. Furthermore, those who are already talented programmers might find the jump to these languages relatively easy, and may already have a flare for good web designs.

17.13.1   In the Python/Django Framework

Applications are developed around .html, .css and .js files which contain all pertinent code for processing static content. The HTML will create objects like paragraphs, headers, images, etc...., while CSS is used for styling those objects (fonts, locations on page, size, etc....). Javascript code is used for handling data objects or HTML objects (e.g., animating elements, changing elements dynamically). In the Django framework, Python is used to first access the database and manipulate data (e.g., modeling). From Python, data objects are sent as a HTTP response that can be accessed via HTML code, which can further be manipulated with Javascript, and styled with CSS. Textbox 17.2 shows a minimized example of how this works in Django.

The library RShiny offers a very simple option for fast deployment of applications, and because it is in R, you get the ease of access to any of R's libraries for statistical analysis. Similar to the Python/Django option, R can also read and write directly to databases (like postgreSQL through the 'RPostgreSQL' library). This allows you to get past some of R's memory limitations by allowing you to select and draw data back and forth from the database. In this case, however, you would not be aided by simple "Pythonic" ways of filtering from the database; you will need to know some SQL. RShiny also works best within the RStudio environment as you can create projects which already have the structure for an RShiny application (A server.R file and a Ui.R file). The server.R file is where you would run models, or do any data cleaning or setup, where the Ui.R file is the script you use to set up the visual output. This is much more simplistic than the Python/Django option and does not offer as much design freedom. However, it is faster to set up, and is particularly helpful for those trained solely in R (which is commonplace in wildlife ecology) (Textbox 17.3).

Textbox 17.2 Sample code snippit from Django/Python framework showing some basic code used in a web application. The first glance of this will seem complicated to those who have not used Django. A wide variety of resources exist on the web, including a great tutorial by the Microsoft team. This code snippit is meant to simply demonstrate the different approach taken by Django versions R (see Textbox 17.1)

```
Database object (models.py):  ## This creates a table for the database
class Census(models.Model):
    preprocessed_id = models.AutoField(primary_key=True)
    site = models.ForeignKey('Site', blank=True, null=True)
    common_name = models.ForeignKey('Species', db_column='common_name', blank=True, null=True)
    day = models.SmallIntegerField(blank=True, null=True)
    month = models.SmallIntegerField(blank=True, null=True)
    season = models.SmallIntegerField()
    count = models.IntegerField(blank=True, null=True)
    accuracy = models.SmallIntegerField(blank=True, null=True)


Data rendering (views.py):  ## This accesses the data from the database and renders it for HTML
def Model(request):
    AllData = Census.objects.all() # Accesses all data in the database – but you can filter these as well using .filter() instead
    ### ACCESS DATA AND PERFORM FUNCTIONS. e.g., modeling, etc... ###
    Output = ## Some sort of model output, usually a LIST type, or NUMPY ARRAY type, STR types as well.
    Context = RequestContext(request, {'output',Output}) ## Sends variable "Output" to the HTML variable 'output'
    Template = loader.get_template('styles/Layout.html')
    return HttpResponse(Template.render(Context))


Static content (Layout.html):  ## This outputs the data and can be styled using CSS and
JAVASCRIPT
<head>
    // Some content here
</head>
<body>
    // Use some HTML code in here:
    <p> There is some model output found here: </p>
    {{ output }}  // This is from DJANGO, and Using JAVASCRIPT, this could be converted into a chart or other
</body>
```

Textbox 17.3 Sample code showing very basically how RShiny works with the server.R and ui.R setup. This is vastly different from the Django/Python setup and is much faster overall in going from nothing to a functioning app. However, RShiny is much less flexible overall than Django. I would recommend that RShiny is great for test or Beta level applications, while Django (or similar frameworks) are better for full web application launches

```
Database object (server.R):  ## This accesses data and creates output

### First we access the PostGRESQL database
library("RPostgeSQL")
pw <- "password"
drv <- dbDriver("PostgreSQL")
con <- dbConnect(drv, dbname = "censusdata", host = "localhost", port = 5432, user = "me", password = pw)
rm(pw)

### Then we query some data from it (assuming "census" is a table in the database, "censusdata")
dat <- dbGetQuery(con, "SELECT * from census")

### Next we run our analysis like we would …
### Model model model model
### Then we render the output to a table, or figure

output$results <- renderPlot({
    ##Do your plots as you normally would here
})

Data rendering (Ui.R):  ## This gives us the visual output

ui <- fluidPage(
        titlePanel("The title")
                sidebarLayout( ### sidebar layout code in here )
        mainPanel(plotOutput("results"),
                p("There is data in the above graph")) // This is equivalent to <p> in HTML
                )
```

17.14 Pushing to the Web: Selecting a Host, Test Test Test, and Reassess

Once you have created a functioning version of your web application on your local machine (a.k.a. local server), the next step is to select a good web host which can support your database and static content. There are many good web hosts that can be used with Django, some of which have options for good database storage. Most web hosts are paid services, particularly those that function with Django, and for a good application (i.e., that which has high security and reliability), it is recommended you find one you can afford. Some web hosts connect directly to Amazon cloud services, and even go so far as to create Git repositories so that code and data can be easily shared.

The R/RShiny option is somewhat more limited in that there are few web hosts that can currently work with them. RShiny does offer a web server (RShiny.io), which you can pay for, or use the free service (which requires you to use a given URL as opposed to a custom one). In order to deploy an RShiny app to a custom URL with the paid service, there is some requirement to set up your own server (which is, in essence, a computer that is constantly running). This can be done via Amazon Cloud services (or similar).

Once a host is selected, the next steps are to test and re-test, then test again. This is done by accessing your application on a variety of platforms, operating systems and devices. Once you are satisfied your application works, you can go live with a Beta version, which can be tested by "naïve" users to determine the usefulness or accessibility of your tool. Surveys are a great aid during this stage to help fine tune any glitches in the program.

After the survey phase, you can deploy your web app and "go live". However, your web application will require maintenance from time to time. You should also ensure that you have code to track website usage (e.g., Google analytics), which is a typical part of reporting to funding agencies. Finally, your website will need to be "refreshed" on a frequent basis to ensure engagement. Part of this is a smart use of data-mining and machine learning techniques which give convincing and scientifically valid results.

17.15 Case Examples

In wildlife ecology, there is still a lack of web applications as I have described them in this chapter (i.e., using machine learning specifically to make forecasts or models). In industry however, there are number of examples ranging from traffic mapping, fraud detection and search engine optimization. The National Aeronautics and Space Administration (NASA) has funded a series of biodiversity forecasting projects; some of which have yet to go live. A list of these projects is found here: http://appliedsciences.nasa.gov/programs/ecological-forecasting-program. Below I highlight several live examples that nearly meet the criteria as I have specified in this chapter, and offer a few simple suggestions that may improve their usefulness.

17.15.1 FLIRT

The FLIRT program (https://flirt.eha.io/) was designed to simulate the spread of Zika virus using flight pathway analysis from a database of airports and flights (Huff et al. 2016). Their web application is a simple mapping application that allows users to select a major airport with certain parameters such as number of passengers per day and how full the average aircraft is. Using Generalized Linear Models (Nelder and Wedderburn 1972), the application makes predictions on outbreaks of

Zika on a global scale. The application is very clean, open access, and easy to use. Furthermore, the maps and data are relatively simple to access. This application could be improved by applying a more sophisticated algorithm (e.g., random forests, or a type of network analysis).

17.15.2 Movebank

Movebank is a web application that allows users to upload GPS tracking data, then have those data temporally and spatially annotated with climatic or environmental data (e.g., sea surface temperature, etc.…). This application was designed more specifically with marine species in mind, but could be applied to terrestrial species in some circumstances. Movebank has a relatively simple interface, but does require a user log-in to function. Once logged in, users have access to all relevant data, and can download any GPS tracks uploaded to their database. However, the application does not include any analysis methods or tools where a user could request behavioural predictions or a home-range analysis (for example). If Movebank were to employ machine learning in this specific case to make custom predictions and analyses for users, the functionality would be greatly enhanced.

17.15.3 BirdVis

The BirdVis program (http://www.birdvis.org/) is an application that was developed using eBird data and climate variables to create predictive species maps for the United States (Ferreira et al. 2011). This is an open access program that can be downloaded and run on any system, and presents temporal and spatial data based on semi-parametric machine learning models (Fink et al. 2010). This is a useful tool, and although it uses open access data from the web (eBird), it requires installation on a local machine. The program was written using C++ and in order to run, must be compiled, which can be challenging for those without the proper skills. If the program were developed into an open access web application, it would be a valuable tool for bird conservation.

17.15.4 MAPPPD

The Mapping Application for Penguin Populations and Projected Dynamics (MAPPPD; http://www.penguinmap.com) is a recently developed application which delivers population data for four species of Antarctic penguins (Humphries et al. 2017). The application is completely open access, and all data are downloadable. Behind the website lies a Bayesian model which incorporates meta population dynamics as well as environmental data to create population projections up to the present year. The website is map-based and can be queried a number of ways. However, the program could be improved by including machine learning population forecasts and metadata for users to download.

17.16   Web Based Programs for Landscape–Scale Conservation

Despite the variety of legislation, parks, and attempts to conserve biodiversity, wildlife management schemes to date are failing with respect to conservation goals (Butchart et al. 2010, 2015). Whole ecosystem management techniques (where all aspects of an ecosystem are accounted for; (Slocombe 1998, Visconti and Joppa 2014) must therefore be considered as viable alternatives to replace current, failing schemes. Web-based tools have the ability to assist and drive a shift towards this type of management system as they can be open access, transparent, and integrate the latest (and greatest) in statistical techniques and visualization.

As you may have learned by reading along this book, machine learning algorithms can integrate large number of predictor variables. These variables can be drawn from other web sources (if needed), or stored on your database. The variables can represent various factors from human economy, to climate, habitat, and predator/prey responses. Using these data to train a model would encapsulate the very essence of whole ecosystem based management. In a web framework, an interactive application could be developed where users could alter future climatic, economic or ecological factors to determine how populations may change. A good machine learning algorithm would be able to actively predict these shifts in either time or space, and managers would be able to plan for worst-case scenarios in a more meaningful capacity.

Because these data and applications could be open access, the general public could easily engage with the website as well and view various scenarios as they needed. It would fall to the individuals (e.g., you) to ensure that all relevant model assessment data (e.g., cross-validation success values, etc....) are available to download, or expressed in the static content. Furthermore, any workshops or meetings to determine management objectives would be enhanced as all individuals would have access to the same information with the best available science.

17.17   Concluding Remarks

Web-based tools for decision support have been around for some time, but in wildlife ecology/management, are still only a decade or so old. Furthermore, the use of machine learning in these tools is very minimal. This is changing however, as easy-to-use data mining and machine learning libraries become available in commonly used scripting/programming languages (e.g., R and Python). Further, more and more wildlife ecologists are being trained in these languages and techniques. In time, it is conceivable that whole-ecosystem management could be implemented globally using web-based tools due to their ability to be open access and transparent. However, this means training a new generation of ecologists in proper machine learning, data management and data-mining techniques, as well as web-design and programming skills, while retaining their knowledge of ecological systems. Furthermore, we require a significant coordination between federal agencies regarding data access and portals. This means that data management skills need to be improved in wildlife conservation. The merging of ecological knowledge, data-mining/machine learning, computational science, and web design is a growing field, and the basics of web applications as outlined in this chapter are important. A team based approach here would be best, where individuals with specific skill sets can be combined to build these applications; only then will natural resource managers be able to better conserve wildlife with the latest and greatest in web technologies.

Acknowledgements I would like to thank P. McDowall for initial bits of code which kick-started my interest in web based applications. I would also like to thank my co-editors and the reviewers of this chapter.

References

Bestion E, Teyssier A, Richard M, Clobert J, Cote J (2015) Live fast, die young: experimental evidence of population extinction risk due to climate change. PLoS Biol 13:e1002281

Breiman L (2001) Random forests. Mach Learn 45:5–32

Butchart SHM, Walpole M, Collen B, van Strien A, Scharlemann JPW, Almond REA, Baillie JEM, Bomhard B, Brown C, Bruno J, Carpenter KE, Carr GM, Chanson J, Chenery AM, Csirke J, Davidson NC, Dentener F, Foster M, Galli A, Galloway JN, Genovesi P, Gregory RD, Hockings M, Kapos V, Lamarque J-F, Leverington F, Loh J, McGeoch MA, McRae L, Minasyan A, Morcillo MH, Oldfield TEE, Pauly D, Quader S, Revenga C, Sauer JR, Skolnik B, Spear D, Stanwell-Smith D, Stuart SN, Symes A, Tierney M, Tyrrell TD, Vie J-C, Watson R (2010) Global biodiversity: indicators of recent declines. Science 328:1164–1168

Butchart S, Clarke M, Smith R, Sykes R, Scharlemann J, Harfoot M, Buchanan G, Angulo A, Balmford A, Bertzky B, Brooks T, Carpenter K, Comeros-Raynal M, Cornell J, Ficetola G, Fishpool L, Fuller R, Geldmann J, Harwell H, Hilton-Taylor C, Hoffmann M, Joolia A, Joppa L, Kingston N, May I, Milam A, Polidoro B, Ralph G, Richman N, Rondinini C, Segan D, Skolnik B, Spalding M, Stuart S, Symes A, Taylor J, Visconti P, Watson J, Wood L, Burgess N (2015) Shortfalls and solutions for meeting national and global conservation area targets. Conserv Lett 8:329–337

Cutler DR, Edwards TTC, Beard KKH, Cutler A, Hess KKT, Gibson J, Lawler JJJ (2007) Random forests for classification in ecology. Ecology 88:2783–2792

Dickinson J, Shirk J, Bonter D, Bonney R, Crain R, Martin J, Phillips T, Purcell K (2012) The current state of citizen science as a tool for ecological research and public engagement. Front Ecol Environ 10:291–297

Dillon T, Wu C, Chang E (2010) Cloud computing: issues and challenges. IEEE International conference on advanced information networking and applications 24:27–33

Engleward B, Roberts R (2007) Open access to research is in the public interest. PLoS Biol 5:e48

Fegraus E, Andelman S, Jones M, Schildhauer M (2005) Maximizing the value of ecological data with structured metadata: an introduction to ecological metadata language (EML) and principles for metadata creation. Bull Ecol Soc Am 86:158–168

Ferreira N, Lins L, Kelling S, Wood C, Freire J, Silva C (2011) Birdvis: visualizing and understanding bird populations. IEEE Trans Vis Comput Graph 17:2374–2383

Fink D, Hochachka W, Zuckerberg B, Winkler D, Shaby B, Munson M, Hooker G, Riedewald M, Sheldon D, Kelling S (2010) Spatiotemporal exploratory models for broad-scale survey data. Ecol Appl 20:2131–2147

Flemons P, Guralnick R, Krieger J, Ranipeta A, Neufeld D (2007) A web-based GIS tool for exploring the world's biodiversity: the global biodiversity information facility mapping and analysis portal application (GBIF-MAPA). Eco Inform 2:49–60

Foundation PS (2015) Python Language

Hochreiter S, Schmidhuber J (1997) Long short-term memory. Neural Comput 9:1735–1780

Huff A, Allen T, Whiting K, Breit N, Arnold B (2016) FLIRT-ing with zika: a web application to predict the movement of infected travelers validated against the current zika virus epidemic PLoS currents outbreaks 1

Humphries G, Che-Castaldo C, Naveen R, Schwaller M, McDowall P, Schrimpf M, Lynch H (2017) Mapping application for penguin populations and projected dynamics (MAPPPD): data and tools for dynamic management and decision support. Polar Record

Krizhevsky A, Sutskever I, Hinton G (2012) ImageNet classification with deep convolutional neural networks. Adv Neural Inf Proces Syst 2:1097–1105

Nelder AJ a, Wedderburn RWM (1972) Generalized linear models. J R Stat Soc Ser A (General) 135:370–384

Schneider L, Berger M, Finkbeiner M (2015) Abiotic resource depletion in LCA—background and update of the anthropogenic stock extended abiotic depletion potential (AADP) model. Int J Life Cycle Assess 20:709–721

Slocombe D (1998) Defining goals and criteria for ecosystem-based management. Environ Manag 22:483–493

Stilgoe J, Lock S, Wilsdon J (2014) Why should we promote public engagement with science? Public Underst Sci 23:4–15

Sullivan B, Wood C, Iliff M, Bonney R, Fink D, Kelling S (2009) eBird: a citizen-based bird observation network in the biological sciences. Biol Conserv 142:2282–2292

Team RC (2015) R: a language and environment for statistical computing. R Foundation for Statistical Computing, Vienna

Toth G, Szigeti C (2016) The historical ecological footprint: from over-population to over-consumption. Ecol Indic 60:283–291

Visconti P, Joppa L (2014) Building robust conservation plans. Conserv Biol 29:503–512

Zickuhr K (2013) Who's not online and why. Pew research Center's internet and American life project

Chapter 18
Machine Learning and 'The Cloud' for Natural Resource Applications: Autonomous Online Robots Driving Sustainable Conservation Management Worldwide?

Grant R. W. Humphries and Falk Huettmann

18.1 Introduction

18.1.1 What is 'the cloud'?

Throughout human history people have struggled with the concept of clouds (i.e. those big puffy things in the sky); they are usually somewhat fuzzy and difficult to describe and measure. In the technological era, the same seems to apply to the modern cloud computational infrastructure. The actual technical definition of the cloud is debated in the global community and in the computing world, and as such it can be described in several ways (SYS-CON Media Inc. 2008; Youseff et al. 2008; Armbrust et al. 2009). What is generally accepted is that 'the cloud' (in its modern technical definition) is essentially an infrastructure that exists on the world-wide web (www) and that is accessible by most people with an internet connection (and/ or appropriate logins for specific cloud services) for a variety of purposes (n.b., some people argue that "the cloud" is just another way of saying "the internet"). Generally speaking, it is a series of data servers that are located at data centers accessible to those with an internet connection (see Youseff et al. 2008 and Armbrust et al. 2009 for deeper reviews on the cloud). It has been used heavily for purposes like database construction and hosting, problem optimization, banking scenarios

G. R. W. Humphries (✉)
Black Bawks Data Science Ltd., Fort Augustus, Scotland
e-mail: grwhumphries@blackbawks.net

F. Huettmann
EWHALE Lab, Biology and Wildlife Department, Institute of Arctic Biology,
University of Alaska-Fairbanks, Fairbanks, AK, USA
e-mail: fhuettmann@alaska.edu

© Springer Nature Switzerland AG 2018
G. R. W. Humphries et al. (eds.), *Machine Learning for Ecology and Sustainable Natural Resource Management*, https://doi.org/10.1007/978-3-319-96978-7_18

and gaming solutions. This makes the cloud a central item in information-creating efforts and tools, e.g. in data mining efforts using machine learning on big data. Or put in other terms: the cloud is a new strategic sword to tackle progress and control issues of global well-being and performance (Fig. 18.1).

Throughout this chapter, when we refer to the cloud, we are referring primarily to cloud-computing (i.e. the ability to handle large, complex tasks remotely), and cloud storage (i.e., data storage through secure or open access servers). These can be paid or free services, and certain cloud computing organizations (e.g., Amazon Web Services, Google Cloud Services we describe further below) offer both options depending on the computing power required by users. Furthermore, the cloud is platform-independent, meaning that anyone on any computer with an internet connection can take advantage of any services that exist there. This makes the cloud generic and universal for usage globally, making its powerful potential readily understood.

In the following, we will provide an overview of 'the cloud', particularly with a focus on how it links with machine learning for natural resource / environmental management. Such focus is rarely done, yet very important because it is widely ignored in discussions about 'the cloud' (e.g. see citations we provided above), secondly, wildlife-habitat applications are widely lacking (e.g. see Silva 2012), and the potential and responsibilities for use are 'huge' (Hochachka et al. 2007; Huettmann 2015a).

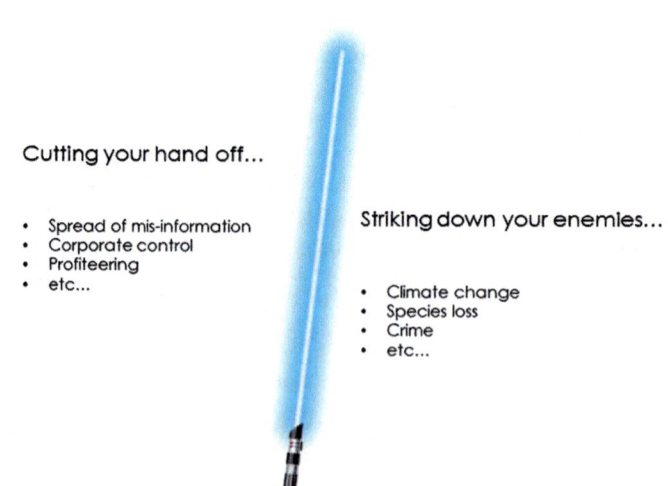

Cutting your hand off...

- Spread of mis-information
- Corporate control
- Profiteering
- etc...

Striking down your enemies...

- Climate change
- Species loss
- Crime
- etc...

Fig. 18.1 A comic representation of the digital sword. A powerful weapon (that looks really cool) when used carefully and mindfully, allowing people to strike down the evil forces of climate change, and biodiversity loss. However, when mis-used, is just as likely to "cut your own hand off" via spreading of mis-information, allowing for corporate influence on our daily lives and freedoms, and profiteering (amongst others)

18.1.2 The Role of WWW and 'the cloud' in Daily Life

Popular use and access to the cloud has led to some major players driving much of the framework and guidelines which are adhered to on the www (e.g., Microsoft, Google, W3C, Mozilla, Apple, hackers and blockchain currencies). These organizations, actors and their products are run primarily by cloud computing technologies and are also primary service providers to most people with the internet. The cloud is global and plays an active role in the every-day life of people with access to the internet (and to a degree, those without). The subsequent links with democracy, governance and its institutions have quickly become obvious thanks to how the cloud is used for information transfer and analysis (e.g., political analysis, news mining, etc.…). However, the potential of the cloud has yet to be fully utilized for natural resource management apart from climate change, atmospheric-oceanic computations, and a few biodiversity related programs (often via citizen science programs like eBird, described below). A brief review of some major textbooks used for teaching purposes in natural/life sciences demonstrates this to a convincing degree. (Table 18.1). Consequently, this means that as students come through life sciences programs, their digital fluency and knowledge of the cloud is limited, despite its powerful potential.

Table 18.1 Disciplines and textbooks for natural resource conservation and their use and embracement of 'the cloud' (see also Mordecai et al. 2010 for applications and potential with natural resource management)

Discipline	Textbook	Content discussing the cloud	Relevance
Wildlife management	Silva (2012)	Almost no	Very large, e.g. for North America
Ornithology	Gill (2007)	No	Large, e.g. for birds and their habitats
Conservation	Primack (2016; 6th edition)	No	Very large globally, any conservation of any species
Landscape ecology	Forman (1995), Gergel and Turner (2001)	No or very little	Very relevant, e.g. urban planning and design
Strategic conservation planning	Moilanen et al. (2009)	Somewhat	Very relevant, e.g. coral reefs and marine protected areas
Statistics and quantitative analysis	Zar (2010), Burnham and Anderson (2002)	No	Very relevant for all quantitative and model analysis

18.1.3   'The cloud': A Powerful Tool and its Implications

The cloud currently plays a major role in development and applications of technologies worldwide. We have referred to the cloud earlier as a sword that allows us to provide cutting-edge applications. Here we mean a double-edged sword that has major positive potential, but also some inherent dangers that must be considered (see Fig. 18.1).

While the cloud offers huge opportunities for computing, it is often sold to the general public who may not have the skills to use it properly or interpret outputs. Thus, it ends up being somewhat of a waste with respect to energy expenditure, Rapid development of cloud technologies means that the average person (who has an un-related job or hobbies) will likely be unable to keep up with the understanding, expertise and frameworks required to take full advantage of cloud services. Furthermore, the ethics behind these developments are not moving forward at the same pace, which then means that corporations are virtually unconstrained when it comes to how the cloud can be manipulated for their benefit. Before we elaborate deeper on this subject, we will show some cloud applications to present its power.

18.1.4   Specific Cloud Service Examples

Perhaps one of the most successful and accessible cloud computing and storage services is Amazon Web Services (https://aws.amazon.com). Amazon hosts many data centers around the world that are connected by a fast and redundant network. The large number of data servers allow for redundant back-ups but also enables users to access a virtually unlimited bandwidth for data storage and processing. Using virtual machines, users can run processes with their own custom algorithms and code, or more recently, they can use some of Amazon's deep learning tools to perform 'out of the box' analyses on large datasets. The ability to vary the size and power of virtual machines being used means that users can decrease speed of analyses and results drastically when working with large datasets. The downside to this, however, is that as users increase the power of the machines they use, the cost goes up exponentially and this could be a barrier to some applications. From a natural resource management perspective, this means that scientists and managers are now (and have been) able to run meaningful and accurate modeling scenarios without having to rely on local machines (see Che-Castaldo et al. 2017 where Amazon web services was used to run population models), reducing costs on local infrastructure. So far, examples for such work are still few in Natural Resource Management, but projects in Australia and ones employing Strategic Conservation Planning are currently working on such applications and to scale-them up more easily (F. Huettmann and J. Hanson pers. com.).

18.1.5   Specific Cloud Services: Supercomputing Centers

According to the TOP500 list (www.top500.org), supercomputing centers are distributed around the world but concentrated in the some of the more developed countries (primarily China and United States, but followed by Japan, Germany, France and the UK). Access to these centers varies depending on the organizations who own them. For example, the Trinity and Cielo supercomputers operated by the Los Alamos National Laboratory in the U.S.A., are only accessible by employees or direct collaborators with scientists from the lab. However, others have public access supercomputers that can be accessed for a fee (e.g., IBM's Watson), or free of charge (Zennet.sc). These operate somewhat differently from cloud systems as they are large computers designed for fast processing, with very little permanent memory. This means that users typically cannot store large data objects for long periods of time on them. A user is given access to a working "node" on a supercomputer, and then a virtual machine allows them to perform operations on that node. If nodes are busy on a supercomputer then the user must wait until that node is free, usually controlled by a queuing system. Whereas in the cloud, users have access to a near unlimited number of processes and space due to the ability to split data across multiple data centers. Furthermore, some supercomputers are operated by corporations that are prone to inherent market fluctuations. For example, "Silicon Graphics Inc.", which developed the graphics for movies like *Jurassic Park*, have gone under because the products they were offering were unable to keep up with rapid changes in the technology. In some cases, it is therefore possible that, like with some data portals (Costello et al. 2014), supercomputers are vulnerable to boom-bust cycles. The Arctic Center for Supercomputing at the University of Alaska Fairbanks is a classic example for that (https://en.wikipedia.org/wiki/Arctic_Region_Supercomputing_Center) due to their reliance on government grants to operate which can vary depending on political administrations. If using supercomputing for natural resource development, the longevity and sustainability of the system needs to be considered though to maintain longevity. We would guess that at least two human generations into the future should be considered as a good time frame here as it will allow for projects to be handed over in stable ways for them to make them viable in the long term.

18.1.6   'The cloud' and Mobile Devices

All cloud services discussed previously were initially based on stationary computing, which usually required a PC and an office with an authentication. However, many applications are now more flexible and require dynamic uploads, additions, batching and re-runs. This can now be achieved through mobile devices (e.g., smart phones and tablets) and remote log-ins. This has made the cloud even more dynamic, bringing it directly into the home, and adds many more applications. Because it can be used by virtually any user anywhere, the cloud has arguably become the most important computational infrastructure in the world! There are certain limitations in remote areas, for example, where users (even with mobile devices) are unable to access internet connections or reliable energy sources (though this is rapidly changing). Many parts in Alaska for instance qualify for those situations, others are found in China and North Korea where Google is not supported or allowed even. Regardless, the use of mobile devices for field work and field courses in the natural sciences is becoming 'part of the course' (Huettmann 2015a,b,c). There are also many implications of the use of mobile devices for natural resource management (Huettmann 2007a,b).

18.2 Current Uses of the Cloud in Wildlife Biology or Ecology

The cloud was primarily used for enhanced computational speed on large programming jobs. This was quickly modified for financial gain (e.g., speed trading and analysis on the stock market). However, this has changed, and the cloud is now used wider and more flexibly in a wide variety of tasks, including quantitative computations, visualizations, database storage, optimization problems, simulations, banking, gaming and spatial applications.

18.2.1 GBIF and the Rio Convention

The act of hosting and delivering biodiversity data in web portals falls under the purview of cloud storage and thus our definition of 'the cloud' for the purposes of this chapter. Biodiversity data tend to be 'huge' and need digital attention and conversion. On the web, they were initially stored on single servers requiring SSH (Secure Shell) or FTP (File Transfer Protocol) access, which limited their capabilities. These efforts were initiated by the Rio Convention and its Convention of Biodiversity (CBD) as early as 1992. It led to the creation of the Global Biodiversity Information Facility (GBIF.org), which houses occurrence records of species from around the world. However, due its international nature and set up, it evolved into a computational online architecture tailored for biodiversity. It involves data sharing and updating in-time from all its member nations. This resulted in the formation of a variety of federated hubs and servers that feed into one global database that is served via the www. GBIF went through several generations of data schemas and software to reach its goals, while the operating systems are changing and moving towards 'virtual standards'. This has pushed GBIF eventually into 'the cloud' and made it a 'cloud' computational service (details are found at gbif.org).

18.2.2 OBIS, OBIS–Seamap and Similar Sites

Due to the success of GBIF, many user communities and funders wanted to emulate this product or improve on their data delivery within the discipline. Many used similar concepts and developed their own schemes, such as presented in Table 18.2. However, this was and is often done on a smaller and more narrow scale, which has the unfortunate side effect of fragmenting global efforts. While this might have triggered some additional data fed into the public sphere, these were easier to fund as they are smaller programs (n.b. the convention on biological diversity is legally binding for signatories' parties and as such, all public data are meant to be in gbif. org, although this is not enforced). These applications became their own cloud services and are often intertwined and make use of similar data and computational platforms. As such in some cases, double-serving of data can be observed, making the value of metadata and ethical questions vital for these efforts.

Table 18.2 A selection of web-portals serving biodiversity data in 'the cloud'

Webportal name (selection) and URL for online data, GBIF-style	Topic	Goal	Year of initiation	Organization
OBIS (http://www.iobis.org/)	Marine biodiversity	Serve all marine biodiversity data funded by COML	2009	COML
SCAR-MarBIN (http://www.scarmarbin.be/)	Antarctic biodiversity	Serve Antarctic marine biodiversity data	2007	SCAR & Belgium Museums and Science Funders
ICES (http://ices.dk/marine-data/data-portals)	Fisheries related data from ICES	Serve all data funded by ICES	2010	ICES
eBird (ebird.org)	Public birding data	Provide birding data worldwide	2005	NSF U.S., Wolf Creek foundation, Leon levy foundation
iNaturalist (http://www.inaturalist.org/)	Naturalist data	Provide and improve naturalist data worldwide	2008	California Academy of Science
Copepod (http://www.st.nmfs.noaa.gov/copepod/)	Global plankton database	Provide best available global plankton data	2004	NOAA
MAPPPD (http://www.penguinmap.com)	Antarctic penguin population database	Open access source for all publicly accessible penguin abundance data	2016	Oceanites, Inc. and Stony Brook University

18.2.3   OpenModeller

OpenModeller is a true cloud-based computational infrastructure that allows a user to make in-near-real-time use of GBIF data (Muñoz et al. 2011). Users can upload their own data sources to the cloud servers used by OpenModeller and perform analyses without using resources on their local machines. This concept enables virtually any user in the world (with internet access) to use and apply predictions on topics of interest. This falls directly in the category of both cloud storage and computing. OpenModeller provides temporal and spatial predictions using an ensemble model algorithm set which can be tuned via several different settings to obtain the best possible predictions required by the global user community. Despite these great services and to assess best-available data and models, textbooks on natural resource management and related journals lack citations of publications employing such methods and concepts (as seen in Table 18.1). Arguably, while global destruction continues OpenModeller is widely underused, e.g. in conservation and wildlife management, court cases and for climate change predictions. A representative multiyear list of publications using OpenModeller is found here: http://openmodeller.sourceforge.net/publications.html

18.2.4   eBird

A large-scale citizen science initiative, eBird.org, is a global repository of bird occupancy data (submitted by bird-watchers). It is somewhat based on the long tradition of 'citizen science' (e.g. the U.S. American Rosalie Edge starting hawk watches in the 1920s; http://www.hawkmountain.org/) and generic environmental concern about global well-being and society, e.g. Carson's Silent Spring (http://www.rachelcarson.org/SilentSpring.aspx).

While it started in the U.S. and North America with bird observations (e.g. run by Cornell University's lab of Ornithology and its funders), eBird has become a truly global program (Sullivan et al. 2009) and has led to many 'spin-offs' in other disciplines and topics (see sections below). In terms of adoption and growth, this scheme has been a global success story and feeds its data to GBIF. eBird obtains app. 25,000 new records a day, and over nine million new records per year. Its data include geo-referenced surveys, opportunistic sightings and photos to document sightings. Due to the sheer amount and diversity of data in eBird, we argue that, so far, only machine learning and data mining techniques via cloud computing can be used for analysis and modeling exercises. This program has been taking advantage of cloud and machine learning techniques for several years now, with one of the major advances being in how citizen science records can be tested with machine learning techniques, with information feeding directly back into the model and to the user (Kelling et al. 2012). The visualizations (i.e., migration movements, observation density) that can be found on the eBird website are driven by cloud computing technologies, and the species distribution models have been created with a pseudo-machine learning algorithm based on adaptive trees (AdaSTEM; Fink et al. 2013). Further, eBird is now in the process of moving their services to Microsoft's AZURE cloud computing system, which will give way to opportunities to expand into more advanced machine learning technologies. However, some studies (e.g., Huettmann et al. 2011; Humphries and Huettmann 2014) have used machine learning techniques on eBird data downloads with good success. Considering the wealth of citizen science data, and despite the popular increase of its use, this resource may still be considered as somewhat under-analyzed and under-used. For example, once GBIF went online over a decade ago, there was an 'explosion' of publications that arose from the data (see https://www.gbif.org/science-review). Although there is a mass of citizen science data on eBird, we have yet to see the same effect outside of the core research group. A quick search through Alaskan publications for instance shows no use of eBird data to date and could be a good place for people to start.

18.2.5   iNaturalist

iNaturalist.org banks on the success of eBird but opens a much wider array of species to be included, e.g. plants, insects, mammals, reptiles and amphibians. Data collected and housed in iNaturalist are even more complex than in eBird and will eventually provide a larger amount and diversity of data. Crowd-sourced identifications and detections are now mainstream and offer completely new ways to conduct surveys and data collection, world-wide. As we argued in the previous section, its information can only truly be extracted and analyzed in a meaningful and effective way when advanced data mining and machine learning via cloud computing services are employed. The full potential of these programs and data are still to be explored and await heavier use.

18.2.6   MoveBank

The online web portal, Movebank, is free but not completely 'open access' (i.e., data are not immediately downloadable). The program opted for a data management model wherein data are protected via user name and password (which can be obtained by signing up to the web service upon approval). Data cannot be readily accessed and harvested thus the web portal is not truly transparent, which violates public trust and gives the impression of elitism towards a small circle of people who are funded. This could be appealing to some biologists who are not wanting to share exact locations of vulnerable species (i.e. to protect them from poachers), or those who are waiting to publish papers before making data available (see Textbox 18.1).

Textbox 18.1 A few reasons for withholding data and arguments against it

The reasons why data aren't made available are not public, making it difficult to comprehend the real issues at hand. Aside from protecting species from poachers (i.e. giving poachers exact locations in near real time of vulnerable species), there is not much justification for holding back data, in our opinion. Ecologists generally site 'scooping' as an issue, but if it happens, it is not recorded and thus no evidence exists to suggest that this is an actual problem. Perhaps in more competitive fields where millions of dollars of research funding are on the line (medical research for example), this would be expected to be a problem. In ecology, however, budgets and resources are very constrained and it just isn't feasible for another research group to simply come along and grab someone else's data (even online via GBIF) and just 'pop' out a paper. Furthermore, these issues were addressed in GBIF's data policy as per the RIO convention).

Carlson (2011) states that other arguments for holding back data include copyright claims, lack of appropriate budget for data management and potential financial losses. There are yet to be any truly scientific reasons for withholding data from the public.

The data content of Movebank (supported by public EU and German money) is primarily focused on movement of animals and somewhat in competition with GBIF (with regards to the storage of presence only data). It can be submitted and downloaded by anyone, as long as they have an approved username and login. Unfortunately, the data lack mandatory ISO compliant metadata to date, primarily since enforcements are missing from submissions coming from a variety of species and sources (e.g., telemetry, tracking, sightings; primarily from well-funded projects). Within this service, a contributor (usually agencies, NGOs and highly-funded project PIs who can afford such projects) can request limited or no access to their data at any time, which is why Movebank does not qualify as a truly open-access service (as otherwise required for RIO Convention signatory nations). Another arguable pitfall of the data is that if a contributor requests, public access to their data can be revoked even. As we mentioned, this leads to non-repeatable science and a lack of transparency, despite working on a public resource (migratory species and their habitats), with approved public permits, and all being funded by government sources (i.e. tax-payer's money and with a mandate to serve the wider public good).

Movebank acts as a cloud computing system however, as it offers behind its 'wall', a suite of analytical tools to temporally and spatially overlay environmental parameters with GPS tracks, and presents quick graphs which analyze movement behavior. It can serve as a platform of innovation and development. The tools can be helpful to individuals who lack skills in programming or ArcGIS (particularly for analyzing movement) but can be done in both of those platforms for a more repeatable / transparent analysis. Some of its capabilities are based on machine learning methods however, which is a good step forward in our opinion. Regardless of some of the dis-advantages, Movebank, as a cloud-based system, is a typical example where results look shiny and progressive, but certain underlying issues are yet to be included or documented. For instance, reporting of animal ethics concerns (e.g., impact of transmitters or tags on survival or behavior) is virtually not included, nor is relevant progress in conservation when many species are on the decline or their habitats are in a poor state (most shorebirds, seabirds, Arctic wildlife). The notion of international species, their ownership and best management remains to be addressed to satisfaction (a topic that essentially has not been resolved worldwide for over 100 years, e.g. straddling stock problem and Migratory Species Act 100 years ago). Statistical assumptions are widely left unaddressed and many users of Movebank still try to remove autocorrelation and apply self-designed outlier removal algorithms. A stronghold of these tools is to make descriptive data analysis, data cleaning and fancy visualizations (with appealing colors and in three-dimensions) easier for untrained practitioners, managers and users.

18.2.7   MAPPPD

The Mapping Application for Penguin Populations and Projected Dynamics (MAPPPD; https://www.penguinmap.com/) is a fully open access tool for downloading and exploring penguin population data for the Antarctic ecosystem (Humphries et al. 2017). The web application has a dynamic map linked to a database of historic counts of penguins and a Bayesian population model. As of writing this book chapter, MAPPPD was still under development, however all data can already be downloaded with ISO compliant metadata so that the user community can fully understand them. Furthermore, code for the model and database are available on Github, a website for hosting open source code (https://github.com), for public examination (n.b., at the time of publishing this book, the code was not available as it was under development, but links to this code will be found at www.penguinmap.com). This application makes use of cloud computing technologies available through Amazon Web Services, which allows for complex models to run just within a few days (Che-Castaldo et al. 2017). This means the model can be updated and added to "on the fly" as new data become available. We think this concept has a lot of merit and presents a role model how public resources, the Antarctic region and its penguins, can be analyzed and delivered to a global audience. MAPPPD is one example of a program that has taken advantage of cloud computing methods developed outside of ecology (e.g., economics).

18.2.8 Translating Economic Applications of 'the cloud' to Ecology

In the last decade or so, the cloud became rather popular in economic applications (outside of wildlife conservation). Worldwide currencies, and high-speed trading (stock markets) rely on a digital and online infrastructure, often supported by cloud applications due to the huge amounts of data. With that, the cloud is a highly strategic tool in support of national and global wealth affecting landscapes, watersheds and the atmosphere (Drew et al. 2010). The stock market applications are also behind the push to provide broadband internet connections linking the EU with Britain, the U.S. and then, Japan and other financial trading hotspots. Because every millisecond counts in the financial sector, cloud computing is fundamental there.

In addition, cloud applications play a role in modeling and predicting the behavior of business partners (e.g., to aid in negotiations; game theory mentioned below). Computing the gain in cost-benefit analysis is such an application. Applications with an environmental spin include cap-and-trade of carbon, price-setting in complex carbon markets or for the insurance risks of stochastic environmental events (tsunamis, fire, earthquakes). In addition, game theory is steeply on the rise in economics including 'tit for tat' reasoning. Similar concepts are also widely developed in gaming; a huge global industry. These approaches have now become of strategic interest in decision-making for finding an optimum in complex scenarios and with data, e.g. for climate change questions. Albeit driven by costs, in modern natural resource decision-making, these approaches can also be used and applied. They offer some conservation value but are not yet used and examined to their full potential for natural resource management. A wider use and assessment for the benefit of natural resources seems beneficial due to the global impacts, e.g. in CITES and with simple Tariffs of Trade, which happen to affect global biodiversity, habitats and the atmosphere in dramatic ways. One could imagine a "stock index" for wildlife that could be tracked in real time with existing stock markets. This would allow us to take advantage of advanced cloud techniques for analysis. Sadly, most banks would "short" wildlife stocks (i.e. bet on the decline).

18.2.9 Crowdsourcing/Storage

We feel it is pretty clear at this point that leveraging computational power and storage on the cloud is linked to the best available science and decision making in conservation ecology. If applied correctly, it can be used to advance conservation management (Cushman and Huettmann 2010; Drew et al. 2010). So beyond 'just' computational gains, 'the cloud' offers equally attractive services in relation to online storage to maintain long-term viability of data. Data storage and associated data management remain major schemes in science for wildlife conservation data (Zuckerberg et al. 2011). Data storage is likely the most common usage of 'cloud' services. 'The cloud' allows users to store data of any type and form in a dynamic fashion to process them and make them available as needed; often this can be done 'in-time'. With that, 'the cloud' has become part of a business strategy, e.g. for companies (book keeping) and when photos, videos and scanned images (or records) are involved. Examples for such a service applied to conservation can be seen in dSPACE (data storage at universities and repositories for self-archiving), as well as scanned specimen collections of museums and herbaria for instance (see the internationally linked Alaska Museum, ARCTOS database as an example: https://arctos.database.museum/home.cfm). Youtube, Facebook, Dropbox, Google and 'wetransfer' offer similar services and so it becomes a global commodity. There, machine learning applications are often provided in-time when high precision pattern and outlier recognitions is required. Facial or vocal recognition of people and individual pattern learning, and recognition are among classic examples for such applications. Just like Facebook might recognize the face of a person in a photo you post, for conservation purposes, cloud services like this could recognize images of animals with geographic tags.

18.2.10   Other Applications

As we have shown above, the cloud is an almost ideal platform for many citizen science projects. The cloud is 'cool', has a good public buy-in, is relatively cheap, nearly infinitely expandable and is globally accessible by anyone with an internet connection. It provides a good learning ground for society to aid in understanding online computing tools and infrastructure and machine learning. The cloud (and its effective use through machine learning) can be and already is part of modern society and could even play a major role in guiding democracy and natural resources.

As we have eluded to above, many international corporations such as Google, Microsoft, QQ (Chinese version of Skype), Facebook, Twitter, Instagram etc. heavily use the cloud. They are cloud users as well as providers and could not operate without it. While applications are too numerous to be mentioned here, many of them rely on machine learning algorithms. One might think that this is stand-alone and has no big implications for natural resource management, but we argue that it does. We see at least four applications when it comes to natural resource management:

1) Machine learning-supported cloud applications create their own reality and image of what nature is and how we relate to nature and its resources. Albeit this is initially indirect, it might well be the biggest impact.
2) Face and pattern recognition can help in enforcement cases, like with poaching and CITES and for species identification 'in time', virtually anywhere.
3) Generic information requests of any sort, for science and research will be greatly helped by 'the cloud' and machine learning.
4) Decision support tools that can provide risk analysis (e.g., illegal fishing, or poaching risks). However, potential is forming based on creativity which is opening many other possible applications.

18.3 How Does the Cloud Link Specifically with Machine Learning?

18.3.1 Mining 'the Cloud' before Analysis and Posting (Data Cleaning and Statistical Filtering)

A major use of data mining methods lies in the use of sorting and pre-screening data. Machine learning algorithms are used to assess and classify incoming data streams from a trained model, and then follow rules when certain outcomes are predicted. Another major application is pattern recognition where data and images are used for training detection of an object. Once sufficiently 'learned', new data or images are searched for the trained pattern(s). There are so many applications of this sort that it has become a growing discipline and business (Costello et al. 2014). More intelligent applications involve outlier detection. This is very powerful for finding a 'needle in a haystack' and for sorting out bad data in a constant stream of incoming data packages. Classic examples here are presented in eBird when a user enters an unusual record or bird number. Often these 'dubious' entries get detected in-near-real-time and produce an alert. Or, once entered, data get compared with a computed average record and then can help to pin-point such outliers in the existing vast pool of data which can then be verified by human eyes (perhaps in the future, machine learning will allow for less human intervention in these cases). eBird also makes aggressive use of the second approach to verify records post-hoc. A third application of machine learning and predictions lies in the use of extrapolating trends from existing data, e.g. for early warning assessments (tripwires) and pre-alerts or risk identification. In such cases, forecasted trends get reported based on existing trends. These trends can arguably be weak or incorrect even, but they invoke a prioritization method and what to check and assess for Quality Control questions, or where potentially risky situation could occur for action to be remedied before danger occurs. This should be part of any pre-cautionary management technique.

18.3.2 Mining the Cloud in a Smart Way, with Consistent Learning Feedback

Artificial Neural Network (ANN) analysis is part of machine learning and can be pretty powerful because it keeps a memory. This is in line with 'Deep Learning' (Mueller and Massaron 2016). Making the memory smarter and filling it with 'knowledge from the world' can make ANNs not only powerful, but it can provide the best and only way of making decisions (i.e. it has no competitor). It is relatively easy to digitize most libraries of the world, and extract knowledge and feed it into such an ANN (e.g. to provide it to politicians and think-tanks). Applications of this sort are already used in online marketing where the customer behavior of the past informs offers of the present, or even how discounts and incentives are offered. Applications to natural resource management are still lacking but can easily be envisioned. More recently, a company called Deepsense used deep learning to classify images of Right whales for individual recognition. However, an envisioned use would be for time series analysis of biodiversity changes over time based on past knowledge. It would be reasonable to include traditional ecological knowledge into the memory of a neural network to inform potential changes under varying scenarios.

18.3.3 'The Cloud' as the Source of Data Mining and for Inference

Data mining needs data to run on. It is 'the cloud' that collects and stores a near unfathomable number of records per second to be used, re-used and re-analyzed in context with the existing and newly forming information. Both, 'the cloud' and machine learning are interacting and evolving into a new entity, and to some extent becoming a nearly sentient process. We find the potential, dynamics and trends of such works are not yet fully explored. However, they are already recognized and make for big applications and opportunities. Rapid assessments and summaries come to mind; see for instance Huettmann and Ickert-Bond (2017). So far, managers of natural resources and their agencies and institutions have not fully embraced these steps. For instance, a nation-wide institutionalized and widely promoted policy to use parsimony for decision-making can only be harmful for making best use of all information available (Manly et al. 2002; Burnham and Anderson 2002).

18.4 Specific Applications and Issues of the Cloud with Machine Learning

18.4.1 Machine Learning with GARP (in OpenModeller), Genbank and Influenza Research Database (IRD)

For model predictions, the mentioned OpenModeller is primarily based on the GARP model concept but it still employs several algorithms. It is in the hand of the user to pick and choose whatever provides the best output (as assessed by model accuracy metrics).

Genbank (https://www.ncbi.nlm.nih.gov/genbank/) can employ machine learning in various ways. One is for sequence matching, where a new gene sequence gets compared (matched) with the existing set of genes. Due to the new generation sequencing methods, these applications are rapidly advancing. For applications like Genbank and IRD (fludb.org) these methods are often used as a work bench, making analysis runs and workflows easier to employ, to repeat and update.

18.4.2   'The Cloud' and 'R'

'R' is a statistical tool, coding and programing environment (https://cran.r-project.
org/). It is global and open source and it is increasingly applied to computing problems in natural resource management. The R-language makes extensive use of
online resources for package management and the language itself is fully available
for free download worldwide on multiple platforms. R uses national server mirrors
to serve the raw console and its packages. R (and other languages) can be used on
the cloud by logging into remote machines that have the R language installed, and
some R packages allow you to log directly into cloud applications e.g. https://
www.r-statistics.com/tag/cloud-computing/ for using the Amazon cloud. Using R in
the cloud means that we can run memory expensive tasks that would not run effectively on our local computers. This is a great advantage when paired with machine
learning packages in R. This is probably a major lead into the notion that frequentist
statistics are now moving directly into machine learning (see also https://normalde-
viate.wordpress.com/2013/04/13/data-science-the-end-of-statistics/).

18.4.3   Known Problems When Applying the Cloud

'The cloud' makes use of huge server farms to deliver all services, store data and
provide the computational backbone. This is a big business, which includes interest
in the real estate market due to specific site requirements. While many of them were
initially located in more developed nations (e.g. USA, Japan, Germany), real estate
and maintenance costs are relatively high. This results in a shift towards data centers
being moved to 'cheaper' nations which still have the infrastructure to provide these
services in similar conditions. However, this might remain a widely dubious concept, unless back-ups and pre-emptive security measures are in place (N.B., the
advantage of multiple data servers in various countries means that redundancies can
be set up in varying places to ensure no disruption of service. Amazon web services
and Google are companies that do this). Arguably, this safety net comes with a
heavy price of cyber security expenses. The actual failure or certain break down of
a server farm, or its maintenance and service, can have dramatic consequences for
'the cloud' service. While the full-scale loss is probably not so realistic, outages and
online security topics might play equal or worse roles.

   We discuss the issue of sustainability from an environmental perspective shortly,
but there are other issues around the business of server farms which are not clear.
Growth in the industry, right now, is very high, with many server farm providers
structuring around this increase over the next 20 years or so. However, the underlying business model as well as the long-term viability of the market remains unclear
and is heavily tied to the environment (i.e. rare earth mineral extraction, energy
usage, etc. References by Eichstaedt (2016) for the Congo and www.meltdowninti-
bet.com/ for the Tibetan Plateau already show us those problems very clearly). It is

reasonable to suspect that at some point data farms may even become obsolete when newer and sleeker technologies replace them. 'Moore's law' and the history of faster computing points towards such situations: the ever-increasing progress directly linked with smaller and faster circuits and transistors seems not to stop for the foreseeable future, thus far (https://en.wikipedia.org/wiki/Moore%27s_law).

The unknown future and security of the cloud remain key issues, which can make usage a risky endeavor. Further, many data ownership issues are not so clear when considering that data are often international, they frequently cross borders and occur simultaneously across various regimes. Copyright infringements and privacy issues remain a wide concern (some readers might find it interesting to know that Google for instance considered for a while to move its operation onto a ship operating in international waters! The real-world, legal and governance implications of operating servers and communications 'offshore' in international waters are not fully explored yet. But consider solar panels providing cheap energy etc.).

One way or another, the cloud is usually not very transparent for users. On the one hand, this has hindered many applications that otherwise could benefit from cloud applications due to some confusion on how to best use it. On the other hand, the steep rise in cloud applications shows those problems of being less of a concern to users, thus far. Virtually no legislation exists anywhere that truly protects users, customers and even operators of 'the cloud' (data use and web portal agreements to be signed online prior to use make for a good point in case). Arguably, cloud applications could easily invoke the constitutional questions such as those related to privacy, property and freedom of expression, liability, financial damage compensation and international agreements.

18.4.4 Lack of Sustainability of the Cloud (environmental footprint)

The ecological footprint of server farms is 'very high' due to a large amount of real estate space and energy requirements (Walsh 2013). The primary costs are financial, whereas the secondary costs are the resources and materials needed; thirdly, the required infrastructure, and fourthly, the energy that is emitted (usually heat). Social costs of excessive online usage are not yet even fathomed. Server farms need constant temperatures and cooling thus leading to massive energy requirements. Serious proposals have been suggested to put server farms in Alaska and cold/polar regions as cooling would be naturally less expensive in these areas. In 2018, Microsoft famously sunk a data center off the coast of Scotland in order to help mitigate the energy costs of cooling it. Implications of the wiring and geo-strategic connectivity have yet to be thoroughly explored.

Reliable assessments of the energy consumption of server farms are at its infancies. However, one might believe that over 10% of all energy, worldwide, is by now devoted to the internet (Mills 2013). The amount of energy required to run server farms is probably a large proportion within that. The consumption of such energy is on the rise, and with China being a huge builder and user of 'the cloud' as well as coal-fired and nuclear plants, the issues are abundantly clear. Secondly, servers usually need hard drives and transistor circuits. They need specific metals and rare metals and earth to run. They rely heavily on mining products, some of them are known to fuel civil war, rape, genocide and poverty on a grand scale, e.g. in Congo (Eichstaedt 2016) and in Tibet (www.meltdownintibet.com/), besides others. Either way, it is readily clear that server farms and their maintenance are now presenting a political issue; one that affects the global environment (Fig. 18.2).

18.4.5 Democracy and Control of 'the cloud'

The cloud is not just a tool of convenience as it fully supports globalization of businesses and industry and is therefore not likely to go away. However, the control of the cloud remains widely undefined. Net neutrality is a wide discussion item and is not set to change in the immediate future (at least within the U.S. and the EU). However, China and other nations (North Korea) actively censor services like Google or Facebook, which then poses questions about the internet, its players and its services (including the cloud). Current political circumstances in the USA and in Britain may cause them to move this direction as well, but other countries might remain on the forefront of ethical issues around technological advancement.

China and India are among the world's biggest providers of software, hardware and users. However, both carry notorious human rights records and have not shown accepted track records of a balanced and fair worldview (e.g. the extreme poverty rate in India's Hindu cast system is 20%; China runs officially on a single-party Marxist-Leninist government with poor democratic policies for its minorities). None of these approaches will truly scale up globally and with western human-rights in mind. North America and 'the western world' present just a small fraction of the world's citizens, whereas France, Mexico and Brazil for instance still promote an exclusive French, Spanish and Portuguese web respectively. The U.N. thus far, has not become engaged with internet issues, and thus, major decisions remain in the hands of commercial giants like Google, Microsoft, Facebook and Apple. It seriously affects The United Nations Environmental Program (UNEP; http://www.unep.org/) or Food and Agriculture Organization (http://www.fao.org/home/en/), its decision-making and leadership, as consequently natural resources, the world and future generations. From such a structure and setup, it is unlikely that the internet, software and hardware, as well as the cloud, are currently designed for fair and socialistic concepts helping the poor (app 1/3 of the world by now), not resulting into a better and more equal distribution of wealth and making the environment the constraining factor to protect and act upon. The international corporations simply do not care for those metrics and are not set up to handle such topics; they are not democratic. Under such a framework and governance though it is not really possible to stop the decay of ecological services and associated extinction of species and their habitats, nor to stop climate change.

Fig. 18.2 Four images showing some of the concepts we discuss in this chapter (**a**) Data storage: The Facebook mega data center in Prineville, Oregon. This center sits on 120 acres of land, and is over 300,000 square feet. This data center consists of two stages as of 2017 with a third planned. When completed, it will use nearly 80 MW of power. Some data centers have a power capacity of around 100 MW (e.g., Apple in North Carolina; which claims to be completely powered by renewable energy; http://www.apple.com/environment/report). (**b**) Power utilization of the internet: for comparison, the Datteln 2 coal power plant (built in the 1960s and decommissioned in 2014) had a power capacity of 86 MW (just enough to power a SINGLE data center!). Coal plants are more efficient now (upwards of 1000 MW), but a single data center can use 10% of this type of coal plant's power capacity. (**c**) Citizen science on the cloud: a map of the distribution of Baird's sandpiper (*Calidris bairdii*) made from open access data in eBird. (**d**) Facial/image recognition (i.e. automated species recognition from cameras): The endangered red panda (*Ailurus fulgens*), which has a rapidly declining wild population. Not only is it cute, but it is one of the many species potentially at risk from the digital sword because public awareness of its 'cuteness', as spread by the internet, fuels illegal trade, notwithstanding impacts of climate change partly due to energy consumption of web resources (Fig. 18.1). Some recent work has quantified its distribution using machine learning techniques (Kandel et al. 2015)

18.5   Where is the Cloud Going with Machine Learning?

18.5.1   Rise of the Relevance of the Cloud

The many applications discussed here are arguably just 'the tip of the iceberg' as many new applications are in development and still evolving. A first glimpse might come from S. Korea and Japan where electronic and online applications have already been fully integrated with home and private life (e.g., kitchen appliances, living rooms, heating, cars, gardens and even toilets, some of them designed to operate during earthquakes!). Some of the latest applications involve 'sheep & cattle herding' with robots that detect flock behavior by cameras and with online connection back to the owner, all done in-time and with insurance, e.g. using drones. This essentially presents a new form of land management driven by an effective cloud.

18.5.2   Teaching in the Cloud (e.g. MOOCs)

The first author of this chapter uses the cloud frequently in lectures and workshops for teaching on natural resource management topics with graphic and computational examples. Most software is now served online, and can be run 'online' (meaning, in the cloud). Becoming fluent with the cloud is probably a major shortcoming, for now (Huettmann 2007a, b; 2015b, c). Students, and therefore society as a whole, lacks awareness and expertise as to what the cloud is, what it and should, and what it should not do. Teaching the cloud is therefore an essential feature in virtually any syllabus and to educate future citizens in the Anthropocene.

18.5.3   Interacting and Synergetic Applications and Implications

The power of the cloud sits in its network, as well as in its sophisticated intelligence when connected with machine learning. Adding different applications together, and across nations or continents towards a coherent 'one', creates new entities that never existed before. For instance, business applications that are optimized from game theory adding climate data forecast models for best-possible biodiversity outcomes considering human behavior! On one hand, applications of this complexity create new options for solutions, presenting optimized systems for economy, ecology and sociology. On the other hand, it can easily be envisioned how bad leadership can have extremely negative results. The cloud and optimized with machine learning can act "like a machete in the hand of Neanderthal" when not driven by ethical control mechanisms to use it in the best-possible manner.

18.5.4 Ecology and the Cloud

Most applications of the cloud that are discussed here consider unlimited growth, and see no limits set by the environment. Carrying capacity, and that the earth is finite, is usually entirely left out of the discussion. A study demonstrating the connection between the ecosystem, global well-being and 'the cloud', has yet to be performed for all its impacts (e.g., energy use, space use of data centers, spread of mis-information, etc....). This means that to date, the fact that that 'the cloud' can easily destroy the ecosystem (taking up resources and energy) is ignored. The virtual world happens just off the screen but does not show its wider network worldwide and the footprint.

We find, this is a shortcoming and should be corrected. The ecological footprint of 'the cloud' was already discussed above, however, one must not be entirely opposed to a human footprint, as long as it is benign, serves a good meaning and can be sustained, e.g. done through renewable resources and/or with a defendable democratic impact. These things are possible, but again, widely left out of the debate. Deep Ecology, and Ecological Economics, provide good platforms to deal with those issues, and those should be included in any discussion about 'the cloud'.

18.5.5 Ethics in the Cloud?

Similar to the endless growth paradigm (Rosales 2008a, b for an assessment), and somewhat related, one will not find ethical issues discussed with 'the cloud' (but see O'Neil 2016 for so-called 'weapons of math destruction'. Questions like:

– who benefits from the cloud?
– what's the cloud really for?
– who pays for the cloud, and at what cost?
– what does the cloud really produce?
– what are the impacts from the cloud, and what is gained?

are rarely discussed. It is clear that many of these questions and topics are still in their infancy and will trickle down to issues in ecology regarding how or if the cloud should be used (i.e. is it beneficial for wildlife). However, while 'the cloud' is pushed and growing, the larger questions about ethics and society, are not resolved and hardly on the 'agenda'. We find, they must be part of a valid discussion about 'the cloud', specifically when the gains for ecology and nature are to be harvested for the wider public good, e.g. following concepts of Næss (1989), Cushman and Huettmann (2010), Zuckerberg et al. (2011), Huettmann (2015a, c).

18.5.6   Some Future trends of the Cloud

Nobody really knows what the future will look like. However, a few trends are pretty obvious. Unless we have a major turn-around, wilderness (as somewhat protected by U.S. and international law) is unlikely to survive as we know it. The speed of development, resource consumption and lack of awareness and governance is just too big on a global scale. We might be lucky if only a few small parcels remain (e.g. 2%). Already in the western world, habitat losses of 90% can easily be found, e.g. tall grass prairie, Canadian old growth forest on Vancouver Island, wetlands in California and Ontario or swamp forests in the Gulf of Mexico. Global access and human population and consumption rise will certainly drive this destructive progress, fueled by industrialization and globalization for years to come. The notion of national parks - as we know them- is unlikely to be upheld for the next 100 years. All signals for Antarctica for instance show just that, e.g. pressure to mine for uranium, even oil and gas drilling, besides the already intense pressures to fish and for tourism. The U.S. experience goes along the same way, e.g. pipelines to be built in National Parks and watersheds affected by grazing, hydro dams and fracking. The loss of tropical rainforests, e.g. in Amazonia, SE Asia and Congo is on that same trend (see Yen et al. 2005 for a tropical national park in Vietnam). Major urbanization and decay of ecological services are on the rise. The hope is that 'the cloud' and machine learning can detect impacts early, so we can make better decisions, and help to minimize impacts and global attitudes. We can hold out hope that perhaps, people can change the framework and attitude to achieve harmony with nature.

18.6   Conclusion: Ethical Non–Parsimony to the Rescue

Natural resource managers live in an exciting, yet tragic time. While experiencing a global species extinction event, they face their expertise being replaced and becoming often superfluous due to technology. However, in many instances those natural resource managers did not halt the problems and perhaps exacerbated them, e.g. when promoting reductionism (Silva 2012; Primack 2016), parsimony (Burnham and Anderson 2002), being by-standers of environmental destruction schemes, and locking up data without relevant ethics and metadata even (Movebank.org). Humans are now moving into a computationally driven society, where many decisions are made by cloud computing technologies. While the WWW is the empirical data platform for this, the algorithms - specifically machine learning employed as ensembles - sit at the core of that intelligence and trend (Cushman and Huettmann 2010; Drew et al. 2010). We find that currently this trend is virtually unaccounted for with managers of conservation and natural resources. This has the large implication that it indicates once more that natural resource management is widely behind. This profession is passive and conservative but not pro-active (see Anderson et al. 2003 for an assessment, and Silva 2012 and Primack 2016 for institutionalized forms as per wildlife & conservation management textbooks), just watching the decay of what they are trusted with: a global asset (this equals to watch and study and 'the deckchairs on the Titanic', instead of saving the entire ark). Secondly, natural resources and their conservation are likely to be marginalized further. Such as state of denial will not only harm natural resources, the world, but leads to a continued decay of human development (Cockburn 2013), society (Diamond 2005), vision of modernity and progress (Alexander 2013) and the decision-making process, while better options exist and await to be urgently employed.

Acknowledgements We acknowledge Springer Publisher for the opportunity to write this chapter. FH wishes to thank his colleagues, the EWHALE lab, S. Linke, I. Presse and many co-workers that helped shaping this manuscript. T. Marr,B. Barnes and the Canadian Wildlife Service helped us to show how not to set up, run and organize Bioinformatic and Survey Data Centers. This is EWHALE lab publication #181.

References

Alexander JC (2013) The dark side of modernity. Polity Press, Cambridge, p 187
Anderson DR, Cooch EG, Gutierrez RJ, Krebs CJ, Lindberg MS, Pollock KM, Ribic CA, Shenk TM (2003) Rigorous science: suggestions on how to raise the bar. Wildl Soc Bull 31:296–305
Armbrust M, Fox A, Griffith R, Joseph AD, Katz RH, Konwinski A, Lee G, Patterson DA, Rabkin A, Stoica I, Zaharia M (2009) Above the clouds: a Berkeley view of cloud computing. http://home.cse.ust.hk/~weiwa/teaching/Fall15-COMP6611B/reading_list/AboveTheClouds.pdf
Burnham KP, Anderson DR (2002) Model selection and multimodel inference: a practical information-theoretic approach. Springer Science and Business media, New York
Carlson D (2011) A lesson in sharing. Nature 469:293
Cockburn A (2013) A colossal wreck. Verso Publishers, London
Costello MJ, Appeltans W, Bailly N, Berendsohn WG, de Yong Y, Edwards M, Froese R, Huettmann F, Los W, Mees J (2014) Strategies for the sustainability of online open-access biodiversity databases. Biol Conserv 173:155–165
Cushman S, Huettmann F (2010) Spatial complexity, informatics and wildlife conservation. Springer, Tokyo, p 448
Che-Castaldo C, Jenouvrier S, Youngflesh C, Shoemaker KT, Humphries G, McDowall P, Landrum L, Holland MM, Li Y, Ji R, Lynch HJ (2017) Pan-Antarctic analysis aggregating spatial estimates of Adélie penguin abundance reveals robust dynamics despite stochastic noise. Nat Commun 8(1):832
Diamond J (2005) Collapse: how societies choose to fail or succeed. Viking Press, New York
Drew CA, Wiersma Y, Huettmann F (2010) Predictive species and habitat modeling in landscape ecology. Springer, New York, pp 45–70
Eichstaedt P (2016) Consuming the Congo: war and conflict minerals in the World's deadliest place. Chicago Review Press, Chicago
Fink D, Damoulas T, Dave J (2013) Adaptive Spatio-temporal exploratory models: hemisphere-wide species distributions from massively crowdsourced eBird data. Proceedings of the twenty-seventh AAAI conference on artificial intelligence. http://www.aaai.org/ocs/index.php/AAAI/AAAI13/paper/viewFile/6417/6852
Forman RTT (1995) Land mosaics: the ecology of landscapes and regions. Cambridge University Press, Cambridge
Gergel S, Turner MG (2001) Learning landscape ecology. Springer, New York
Gill FB (2007) Ornithology, 3rd edn. W. H. Freeman & Co., New York

Hochachka W, Caruana R, Fink D, Munson A, Riedewald M, Sorokina D, Kelling S (2007) Data mining for discovery of pattern and process in ecological systems. J Wildl Manag 71:2427–2437

Huettmann F (2007a) The digital teaching legacy of the international polar year (IPY): details of a present to the global village for achieving sustainability. In: Tjoa M, Wagner RR (eds) Proceedings 18th international workshop on Database and Expert Systems Applications (DEXA) 3–7 September 2007, Regensburg, Germany. IEEE Computer Society, Los Alamitos, pp 673–677

Huettmann F (2007b) Modern adaptive management: adding digital opportunities towards a sustainable world with new values. Forum on public policy: climate change and sustainable development. 3: 337–342

Huettmann F (2015a) On the relevance and moral impediment of digital data management, data sharing, and public open access and open source code in (tropical) research: the Rio convention revisited towards mega science and best professional research practices. In: Huettmann F (ed) Central American biodiversity: conservation, ecology, and a sustainable future. Springer, New York, pp 391–418

Huettmann F (2015b) Field schools and research stations in a global context: La Suerte (Costa Rica) and Ometepe (Nicaragua) in a wider perspective. In: Huettmann F (ed) Central American biodiversity: conservation, ecology, and a sustainable future. Springer, New York, pp 174–198

Huettmann F (2015c) Teaching (tropical) biodiversity with international field schools: a flexible success model in a time of "wireless" globalization. In: Huettmann F (ed) Central American biodiversity: conservation, ecology, and a sustainable future. Springer, New York, pp 215–245

Huettmann F, Ickert-Bond S (2017) On open access, data mining and plant conservation in the circumpolar north with an online data example of the herbarium, University of Alaska Museum of the north Arctic Science http://www.nrcresearchpress.com/toc/as/0/ja

Huettmann F, Artukhin Y, Gilg O, Humphries G (2011) Predictions of 27 Arctic pelagic seabird distributions using public environmental variables, assessed with colony data: a first digital IPY and GBIF open access synthesis platform. Mar Biodivers 41:141–179

Humphries GRW, Huettmann F (2014) Putting models to a good use: a rapid assessment of Arctic seabird biodiversity indicates potential conflicts with shipping lanes and human activity. Divers Distrib 20(4):478–490

Humphries GRW, Naveen R, Schwaller M, Che-Castaldo C, McDowall P, Schrimpf M, Lynch HJ (2017) Mapping application for penguin populations and projected dynamics (MAPPPD): data and tools for dynamic management and decision support. Polar Rec 53(2):160–166

Kandel K, Huettmann F, Suwal MK, Regmi GR, Nijman V, Nekaris KAI, Lama ST, Thapa A, Sharma HP, Subedi TR (2015) Rapid multi-nation distribution assessment of a charismatic conservation species using open access ensemble model GIS predictions: red panda (*Ailurus fulgens*) in the Hindu-Kush Himalaya region. Biol Conserv 181:150–161

Kelling S, Gerbracht J, Fink D, Lagoze C, Wong W-K, Yu J, Damoulas T, Gomes C (2012) eBird: a human/computer learning network for biodiversity conservation and research. Proceedings of the twenty-fourth innovative applications of artificial intelligence conference

Manly FJ, McDonald LL, Thomas DL, McDonald TL, Erickson WP (2002) Resource selection by animals: statistical design and analysis for field studies. Second edition. Kluwer Academic Publishers, Dordrecht, Netherlands

Mills MP (2013) The cloud begins with coal: big data, big networks, big infrastructure, and big power; an overview of the electricity used by the global digital ecosystem. Report: digitalpower group. National Mining Association. American Coalition for Clean Coal Electricity. Washington D.C. U.S. https://www.tech-pundit.com/wp-content/uploads/2013/07/Cloud_Begins_With_Coal.pdf?c761ac&c761ac

Moilanen A, Wilson KA, Possingham H (2009) Spatial conservation prioritization: quantitative methods and computational tools, 1st edn. Oxford University Press, Oxford

Mordecai R, Laurent E, Moore-Barnhill L, Huettmann F, Miller D, Sachs E, Tirpak J (2010) A field guide to web technology. Southeast Partners in Flight (SEPIF).http://sepif.org/content/view/62/1/

Mueller JP, Massaron L (2016) Machine learning for dummies. John Wiley & Sons, Hoboken, p 435

Muñoz MES, Giovanni R, Siqueira MF, Sutton T, Brewer P, Pereira RS, Canhos DAL, Canhos VP (2011) Open modeller: a generic approach to species' potential distribution modelling. GeoInformatica 15:111–135

Næss A (1989) Ecology, community and lifestyle: outline of an ecosophy (trans: Rothenberg D). Cambridge University Press, Cambridge

O'Neil C (2016) Weapons of math destruction. How big data increases inequality and threatens democracy. Crown Publisher, New York

Primack R (2016) Essentials of conservation biology, 6th edn. Sinauer Press, Bosto

Rosales V (2008a) Globalization and the new international trade environment. CEPAL Rev

Rosales J (2008b) Economic growth, climate change, biodiversity loss: distributive justice for the global north and south. Conserv Biol 22(6):1409–1417

Silva NJ (2012) The wildlife techniques manual: research & management, vol 2, 7th edn. The Johns Hopkins University Press, Baltimore

Sullivan BL, Wood CL, Iliff MJ, Bonney RE, Fink D, Kelling S (2009) eBird: a citizen-based bird observation network in the biological sciences. Biol Conserv 142:2282–2292

SYS-CON Media (2008) Twenty experts define cloud computing, http://cloudcomputing.sys-con. com/read/612375_p.htm

Walsh B (2013) The surprisingly large energy footprint of the digital economy [UPDATE]. TIME. Aug 14. http://science.time.com/2013/08/14/power-drain-the-digital-cloud-is-using-more-energy-than-you-think/

Yen P, Ziegler S, Huettmann F, Onyeahialam AI (2005) Change detection of forest and habitat resources from 1973 to 2001 in Bach Ma National Park, Vietnam, using remote sensing imagery. Int For Rev 7(1):1–8

Youseff L, Butrico M, Da Silva D (2008) Toward-a-Unified-Ontology-of-Cloud-Computing. Conference paper December 2008. DOI:https://doi.org/10.1109/GCE.2008.4738443Source: IEEE Xplore Conference: Grid Computing Environments Workshop, 2008. GCE '08. http://dosen.narotama.ac.id/wp-content/uploads/2012/01/Toward-a-Unified-Ontology-of-Cloud-Computing.pdf

Zar JH (2010) Biostatistical analysis, 5th edn. Prentice Hall, Upper Saddle River

Zuckerberg B, Huettmann F, Friar J (2011) Proper data management as a scientific foundation for reliable species distribution modeling, Chapter 3. In: Drew CA, Wiersma Y, Huettmann F (eds) Predictive species and habitat modeling in landscape ecology. Springer, New York, pp 45–70

Chapter 19
Assessment of Potential Risks from Renewable Energy Development and Other Anthropogenic Factors to Wintering Golden Eagles in the Western United States

Erica H. Craig, Mark R. Fuller, Tim H. Craig, and Falk Huettmann

19.1 Introduction

Golden Eagles occur in a variety of landscapes across a wide latitudinal range in North America (NA). As a result, they are subject to many factors that could have negative cumulative effects on geographic populations (Kochert et al. 2002). These include fire (Balch et al. 2013; Knick et al. 2005), wind and various forms of energy development (Watson 2010; Watson et al. 2014), changes in management of rangelands and forests, and urban sprawl that are occurring on a broad scale within Golden Eagle range (Hunt and Watson 2016; Hunt et al. 2017; Kochert and Steenhof 2002; Paprocki et al. 2017; Paprocki et al. 2015). Simultaneously, effects from climate change in the same regions (e.g., severe change in water run-off, drought conditions) are predicted to exacerbate the existing effects of altered fire regimes and invasive annual grasses (Dukes and Mooney 1999; Chapin III et al. 2000; Flannigan et al. 2009; Bradley 2010; Balch et al. 2013; Creutzburg et al. 2015). These factors will subsequently effect the distribution and abundance of avian and prey species (Bradley 2010; Knick et al. 2003; Knick and Dyer 1997; Kochert et al. 1999; Nielson et al.

E. H. Craig (✉) · T. H. Craig
Aquila Environmental, Fairbanks, AK, USA
e-mail: goea.rs@gmail.com

M. R. Fuller
Boise State University, Raptor Research Center, Boise, ID, USA

F. Huettmann
EWHALE Lab, Biology and Wildlife Department, Institute of Arctic Biology, University of Alaska-Fairbanks, Fairbanks, AK, USA
e-mail: fhuettmann@alaska.edu

© Springer Nature Switzerland AG 2018
G. R. W. Humphries et al. (eds.), *Machine Learning for Ecology and Sustainable Natural Resource Management*, https://doi.org/10.1007/978-3-319-96978-7_19

2016). However, managers often lack landscape scale analyses of the effect of these potentially harmful factors on eagle populations. It is particularly difficult to evaluate the effects on wintering birds because information about winter use and distribution patterns of Golden Eagles is often lacking over much of their range. In addition, wintering populations frequently comprise a mix of resident eagles, and regional and long-distance migrants, representing populations from a broad geographic area (Craig and Craig 1998; see e.g., Marzluff et al. 1997; Poessel et al. 2016; Watson et al. 2014). Published research specifically addressing the winter ecology and distribution of Golden Eagles in NA, and related risk factors, is limited. Further, data are frequently lacking on the geographic origin of eagles in wintering areas. All these factors constrain the ability of management decision makers to adequately address the geographic scale of potential impacts from anthropogenic activities on Golden Eagle populations.

Blade strikes by wind turbines at some wind-energy facilities can be a substantial source of mortality for raptors (Carrete et al. 2009; Dahl et al. 2012; De Lucas et al. 2008; Kikuchi 2008; Loss et al. 2013), including Golden Eagles (*Aquila chrysaetos*) (Pagel et al. 2013; Smallwood and Thelander 2008); rapid expansion of wind and solar energy development is occurring in shrub and grassland habitats (Copeland et al. 2011; Tack and Fedy 2015) within the range of the Golden Eagle in NA. At the same time, other changes, wildfires, invasive plants, drought, and climate change are altering or destroying native habitats at an unprecedented scale in the western United States (West; US; Abatzoglou and Kolden 2011; Copeland et al. 2011; Dennison et al. 2014; Tredennick et al. 2016); eagles are known to rely on these habitats (Kochert et al. 2002). Recent US Fish and Wildlife Service (FWS) models indicate that Golden Eagle populations in the western US may be declining slightly (USFWS 2016). Long-term datasets from migration counts and nest occupancy data from locations in the West also reveal evidence of local population declines (Hoffman and Smith 2003; Kochert and Steenhof 2002; Millsap et al. 2013). Other models indicate that a "floater" population of Golden Eagles, comprised of surplus non-breeders, may be limited (Hunt 1998; Millsap and Allen 2006; USFWS 2009).

The FWS has primary statutory authority for management of Golden Eagles in the US under the Bald and Golden Eagle Protection Act (16 U.S.C. 668-668d; Eagle Act). Consequently, the agency is developing strategies for eagle conservation (USFWS 2016; 2013), and in 2008 the FWS designated the Golden Eagle as a Species of Conservation Concern in much of its range in the West (USFWS 2008). As a result of the accelerated development of renewable energy projects in the US, the FWS has established new policy for permits to take Golden Eagles under the Eagle Act, "when necessary to protect certain public interests", in particular localities (e.g., wind energy projects). The permits are based on the ability of the eagle population to sustain permitted loss that is compatible with maintaining stable or increasing populations; permits are to be implemented using an adaptive management strategy (Dept. of Interior FWS: 50 CFR Parts 13 and 22; Eagle Permits; revisions to regulations for eagle incidental take and take of eagle nests: see Federal Register final rules: 81 FR 91494; Dec. 16, 2016 and 82 FR 7708; Jan. 23, 2017). Currently, this is interpreted to mean that

Golden Eagle populations can sustain "no net loss". The FWS proposes to manage populations for "no net loss" at the scale of Eagle Management Units (EMU's), but the boundaries of geographic Golden Eagle populations in NA are not well established, creating additional challenges for FWS management objectives (Brown et al. 2017; USFWS 2016). Current FWS analyses utilize the boundaries of existing NA Bird Conservation Initiative (NABCI), Bird Conservation Regions (BCR's; http://nabci-us.org/resources/bird-conservation-regions/),or possibly administrative migratory bird flyways (e.g., Pacific and Central Flyways; USFWS 2014, 2015, 2016); flyways are more likely to include the full annual cycle of migratory Golden Eagles (USFWS 2016), although see Brown et al. (2017).

Considerable research on NA Golden Eagles has been conducted in recent years (Brown et al. 2017; Domenech et al. 2015a, 2015b; McIntyre and Lewis 2016; Poessel et al. 2016; see e.g., Watson et al. 2014), and is currently underway. However, there is still uncertainty regarding aspects of eagle population biology and status. In particular, gaps in the knowledge base on key factors that are necessary to manage stable or increasing populations challenge the ability of the FWS to prevent further population declines under the new 30-year permit rule for Golden Eagles (81 FR 91494; Dec. 16, 2016 and 82 FR 7708; Jan. 23, 2017). There is currently inadequate information to develop a management plan that sets management objectives with strategic habitat conservation goals for wintering Golden Eagles over a broad geographic range, to identify and assess the cumulative effects of risks within eagle winter range to the population in the western US and Canada, or to identify key areas for habitat conservation and eagle monitoring.

Our research objectives were to: (1) predict potentially important Golden Eagle winter use areas in Idaho, Utah, Nevada and eastern Oregon, by applying a machine learning predictive model developed from wintering eagles in Idaho and Montana (The Authors, Unpublished Data); (2) describe the characteristics of predicted winter range that are of particular relevance for the management and conservation of eagles within the study area; (3) relate predicted winter distribution to known risk factors (stressors) and evaluate potential for cumulative risks to wintering eagles; and (4) highlight additional research needs at other spatial scales.

19.2 Methods

19.2.1 Study Area

The study area in the western US encompasses Idaho, Utah, Nevada and eastern Oregon. It falls within five Bird Conservation Regions (BCR's; 9 [Great Basin], 10 [Northern Rockies], 15 [Sierra Nevada], 16 [Southern Rockies/Colorado Pleateau] and 33 [Sonoran and Mojave Deserts]; Fig. 19.1) and lies entirely within the Pacific Flyway (see USFWS 2016). This area supports wintering eagles that are resident in the study area, as well as, migrants primarily from the Pacific Flyway, and from at

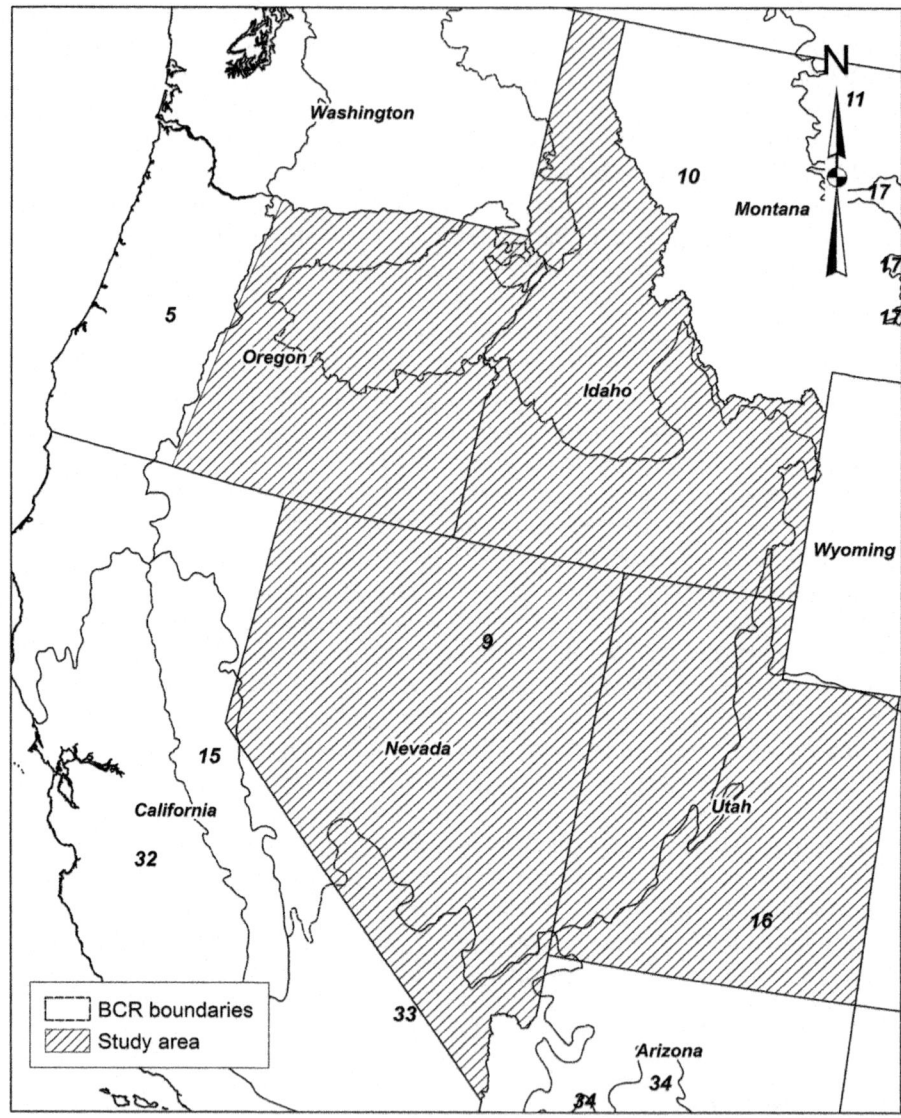

Fig. 19.1 The four state study area in western North America (indicated with diagonal lines) is overlain with Bird Conservation Region (BCR) boundaries (http://nabci-us.org/resources/bird-conservation-regions/; BCR reference numbers are shown within each boundary). The entire study area lies within the Pacific Flyway (see Fig. 19.1 in USFWS 2016)

least two other BCR's: 4 (Northwestern Interior Forest) and 5 (Northern Pacific Rainforest) in Canada and Alaska (USGS data; Craig and Craig 1998). We selected this region because: (1) it is within the known range of the Golden Eagle in NA; (2) digital layers are available for the study area that were similar to the environmental layers used to develop the original model (The Authors, Unpublished Data), (3) there are digital layers available of known potential threats in the region (e.g., energy development, wildfires) that might affect Golden Eagle population dynamics and status; and (4) inferences from our research can be applied to this region for management planning.

19.2.2 The Model

Prediction To estimate potential Golden Eagle winter distribution, we used a predictive model derived from a 1990's dataset of satellite-tracked eagles wintering in shrub-steppe habitat associations in Idaho and Montana and applied it to the larger geographic four state area. We derived the original distribution model using the machine learning software, TreeNet (The Authors, Unpublished Data; Salford Systems, Inc., San Diego, CA), a stochastic gradient boosting algorithm that is known for highly accurate predictions (Craig and Huettmann 2009; Cutler et al. 2007; Elith et al. 2006; Prasad et al. 2006). We found it to be accurate for predicting winter distribution of Golden Eagles across broad temporal and spatial scales in the original application in Idaho and Montana (Precision: 98.7%, Sensitivity: 89.1%, Specificity: 99.7%, F1 Score: 93.7%; see Shalev-Shwartz and Ben-David 2014 for explanation of statistical terms). The original model also had 89.4% accuracy (The Authors, Unpublished Data) when applied to an independent, external sample of 770 eagle winter locations from eBird (eBird 2013; Munson et al. 2011; Sullivan et al. 2009) in Idaho and Montana. The model predicts the distribution of eagles at the scale of the wintering area that an eagle has already selected from the broader landscape. It represents only one of the many hierarchical choices made by wintering Golden Eagles (e.g., how to use the landscape within an already selected winter area) in their selection and use of resources in the region (see e.g., Marzluff et al. 1997). For the analyses presented here, we applied the original model, which overlaps a portion of the current study area, to the four state area.

Application All spatial interpretations and analyses were performed in a geographic information system (GIS) using ArcMap 10.1 (ESRI, Redlands, CA), or in Geospatial Modeling Environment (GME; Beyer 2012). To transfer the outputs from the original model to a new geographic area, we generated a lattice of approximately 900,000 regularly spaced points across the four state study area (1 km x 1 km spacing). We extracted data to each point from 11 digital environmental layers representing the same variables that were used to construct the original model (see Appendix Table A19.1 for digital layer details). We then applied the TreeNet output (grove file) from the original model to the resultant gridded point dataset; the grove file retains the relationships among the response (satellite derived location estimates of wintering eagles) and 11 predictor variables. This provided a relative index of occurrence (RIO), on a scale from 0.00–1.00 of the suitability for eagle winter habitat at each gridded point. The RIO does not provide a frequency statistic (Gaussian) probability; however, it does produce a testable model that allows a comparison ranking among areas. We use the model-produced RIO to predict potential eagle wintering areas in the study area; we also refer to these as wintering grounds or habitat in the text.

We used inverse distance weighting (IDW; similar to Booms et al. 2010) for visual representation of predicted wintering areas; IDW is an interpolation method that considers the weight of the closest measured values more heavily than those further away (see Lu and Wong 2008). For clarity of display, we converted the RIO into percentages and classified them into three equal intervals using the following arbitrary thresholds: (1) < 33%: low suitability (interpreted for our purposes as unsuitable winter habitat and a conservative estimate at 0.10 above prevalence; see Liu et al. 2005), (2) 33% - 67%: medium suitability and (3) > 67%: high suitability for eagle winter areas. The arbitrary selection of the threshold value for prediction has the potential to influence accuracy assessment of the model (Manel et al. 2001). However, this metric still provides a basis for comparing the suitability of locations within the study area for eagle wintering habitat, allows for good classifiers overall (Anderson et al. 2003) and is testable using external data for validation.

19.3   Winter Range Characteristics and Potential Risk Factors

We used digital GIS layers for spatial overlay of predicted eagle winter distribution with features of particular relevance for management and conservation of Golden Eagles (Kochert et al. 2002; e.g., shrub/grassland vegetation associations, land ownership categories; Marzluff et al. 1997), including potential risks to wintering eagles. We selected layers of potential risks to eagles based on our knowledge of Golden Eagle biology, the published literature, and the availability of digital risk datasets that spanned the entire study area. We use simple spatial assessment (see Suter II 2016) to describe where predicted winter areas overlap risks. These do not necessarily represent all risks to Golden Eagles and because changes are occurring rapidly in the West, our digital layers contain some data that are already out of date. Risks included in our analyses include: anthropogenic fragmentation, human footprint intensity, potential wind and solar commercial development areas, oil and gas well distribution and density, invasive plant species risk, and fire history within the last decade (see Appendix Table A19.2 for additional layer details). We report human footprint intensity (from Leu et al. 2008) separately from our map of cumulative risks to eagles. It is a conglomerate that estimates the combined physical and ecological effects of many anthropogenic influences (Leu et al. 2008), thereby duplicating some risk factors presented here, and would artificially inflate our results.

We extracted data from the digital risk layers to each of the gridded points from the predictive surface. For risk layers represented by point data, we used the '*select by location*' option with a spatial selection of 1500 meters to determine the model predicted RIO value for the closest gridded point. To map the risks to wintering Golden Eagles, each risk factor at a given gridded point was assigned a numerical rank of 0 for no, or low risk, and a 1 for risk; all risks were weighted equally and then summed at each gridded point within predicted winter range. The number of potential risks (range: 1–7) at any one site does not account for the dissimilarities among risks to eagles (e.g., direct mortality vs. displacement of prey; also see Carrete et al. 2009). It does provide a first step in spatial assessment of some of the risks to wintering eagles across the study area (see, e.g., Suter II 2016 for an overview of ecological risk assessments).

Evaluation Using external, independent, spatially explicit data can be informative for testing model assumptions and also provide one of the most realistic evaluations of model accuracy (Booms et al. 2010; Hernandez et al. 2006; Zorn 2012). We assessed the reliability of model predicted eagle wintering areas in the four-state region by employing subsets of the publicly available eBird data (eBird 2013; Munson et al. 2011) and results from Christmas Bird Counts (CBC; Audubon 2010), which include winter sightings of Golden Eagles across the study area. We used eBird records from 1979–2011 (n = 2012; Munson et al. 2011) that were collected during the same winter months we used for the original model (1 Nov. – 30 March; The Authors, Unpublished Data), but excluded any that were classified as not valid by the eBird review process (Sullivan et al. 2009). We overlaid the eBird eagle winter locations with a raster layer of mean RIO values in the study area (4.1 x 4.1 km grid cell) to determine if the winter sightings fell within model-predicted suitable wintering areas for Golden Eagles. We then compared the difference in the observed number of eagle locations in suitable and unsuitable habitat with the expected frequency of an equal number of aspatial points (based on the proportion of the study area predicted as suitable or unsuitable; chi-square goodness of fit test; R Core Team 2013). Additionally, we report the percent of buffered eBird locations (1500 m radius) that contain predicted suitable winter habitat vs. those that did not. We buffered the eBird sightings to account for potential errors in location accuracy and selected 1500 m because that represents the maximum error of the satellite derived eagle locations used to develop the original model (The Authors, Unpublished Data), and because eagles are highly mobile and can easily move this distance during their daily winter movements (EHC, THC, MRF pers. observation). For further validation, we used data collected from 39 Christmas Bird Count (CBC) circles representing survey routes where winter sightings of eagles were made (1979–2010; www.audubon.org, www.christmasbirdcount.org). Any CBC circle that contained ≥1 Golden Eagle sighting was treated as one eagle survey unit for our analysis, regardless of the number of eagles observed in it, or number of years in which sightings were made; most CBC circles were surveyed multiple years but eagles were not necessarily sighted every year. CBC's do not record the specific locations of birds observed within the 24 km (15 mi) diameter circle but simply the number and species of birds sighted. Because CBC circles encompass a large area, we could not directly equate the predicted suitability for wintering eagles with a specific location. We calculated the proportion of predicted high and medium suitability and unsuitable eagle winter habitat within the CBC circles. We then compared the distribution of the categories within the CBC circles in which eagles were sighted, with the distribution within all 100 CBC circles in the study area (chi-square goodness of fit; R Core Team 2013). When centers of circles from routes changed from year-to-year, but still overlapped, we counted the overlapping routes as a single route for our analysis. Although CBC and eBird citizen science datasets may be somewhat biased toward population centers (Bonney et al. 2014; Dickinson et al. 2010), the extent of the data is widespread and reasonably represents the study area. Therefore, they were the best independent data available for evaluating our predictive model when applied to the study area.

19.4 Results

19.4.1 Model Evaluation

Our predictive model correctly classified the 2012 eBird Golden Eagle winter sightings as occurring in suitable winter habitat significantly more often than expected, based on the predicted winter area available ($\chi^2 = 19.1417$, df $= 1$, $p < 0.001$; see the chi-square residuals Appendix Table A19.3). In addition, most ($> 94.0\%$) of the buffered (1500 m radius) 2012 eBird eagle winter locations contained predicted suitable wintering areas. Similarly, there were significantly more predicted Golden Eagle wintering areas within the 39 CBC circles where eagles were sighted than were observed (available) in all 100 CBC circles in the study area ($\chi^2 = 25.7315$, df $= 2$, $p < 0.001$). Chi-square residuals indicate that more than expected winter habitat classified as highly suitable was found in the CBC circles where eagles were sighted (see Appendix, Table A19.4).

19.4.2 Winter Range Characteristics and Risk Factors

Model predicted wintering areas that are potentially suitable for Golden Eagles comprise 44.9% (31.5% high and 13.4% mid suitability) of Idaho, Utah, Nevada and eastern Oregon and are widespread across the study area (Fig. 19.2). Public land predominates in the region and comprises most of predicted eagle wintering areas (75.2%: Fig. 19.3). Almost half of the predicted wintering areas are on BLM land and 62.3% encompass shrub/grassland vegetation communities (Fig. 19.3). Sagebrush encompasses about a third of predicted eagle wintering areas (Fig 19.4a) and digital overlays of predicted eagle winter distribution and mapped distribution of sagebrush correspond closely in many regions of the study area (Fig 19.4b).

All of model predicted Golden Eagle wintering areas overlapped one risk factor (potential for commercial solar development) and 39.4% overlapped two or more

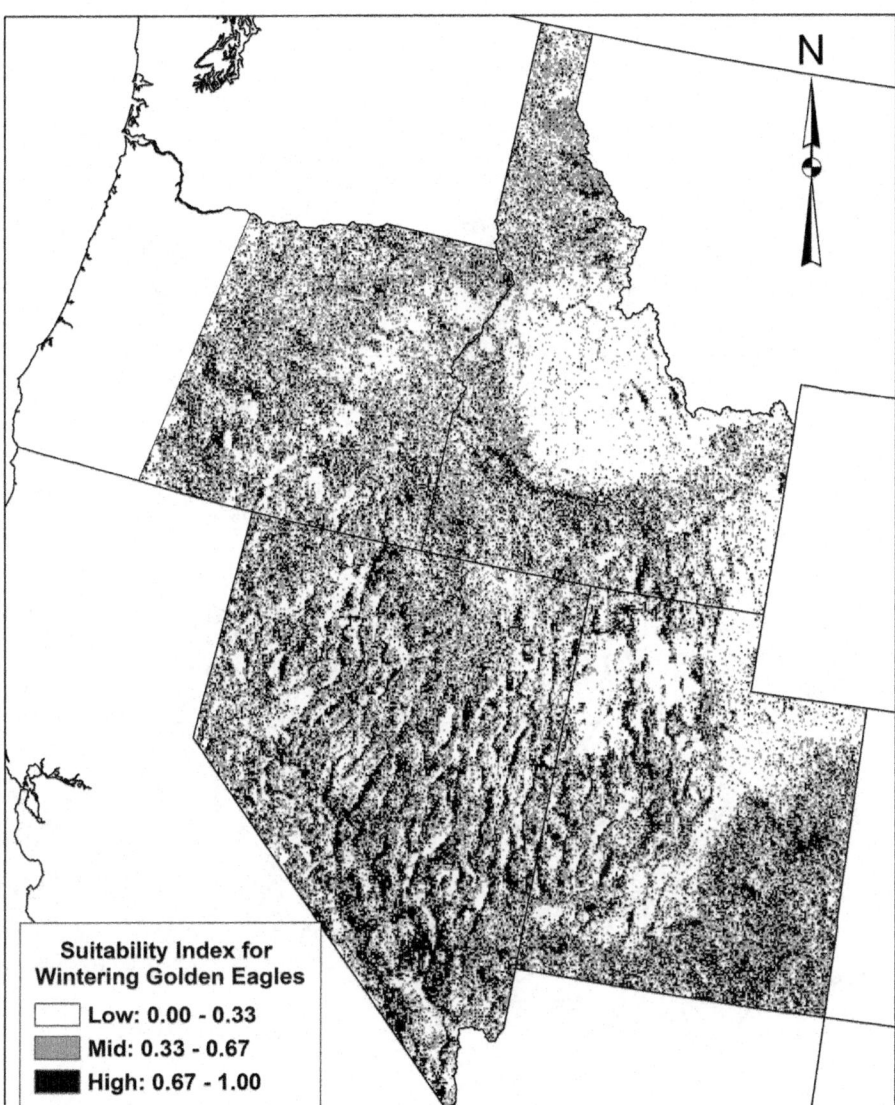

Fig. 19.2 Model predicted areas potentially suitable for wintering Golden Eagles in the states of Idaho, Nevada, Utah and eastern Oregon in western North America

potential risks (Fig. 19.5). About a quarter (25.6%) of predicted wintering areas are at risk to invasive weed species because of proximity to roads, human population centers and agricultural areas (risk classified as greater than or equal to 2 in Leu et al. 2008) or because of wildfires and their relationship to invasive weed encroachment into burned areas (Figs 19.6, 19.7; Brooks and Pyke 2001; Keeley 2006;

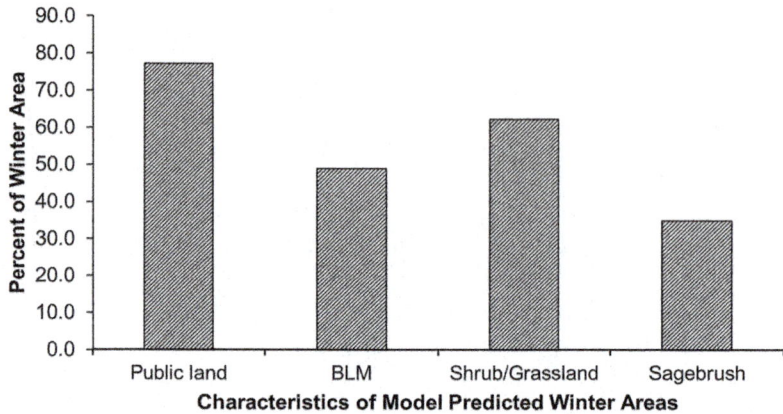

Fig. 19.3 Characteristics of predicted Golden Eagle winter range (classified as mid to high suitability), that may have significance for conservation of eagles in the four state study area of western North America (shown in percent--categories may overlap within the wintering area)

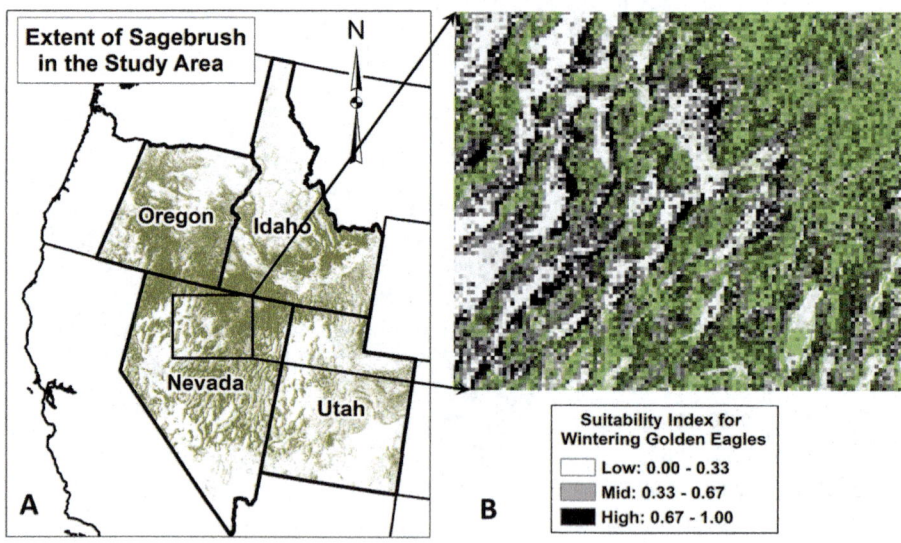

Fig. 19.4 (a) Distribution of sagebrush vegetation communities in the four state study area of western North America. (b) Close-up of the sagebrush habitats overlain with a portion of model predicted winter areas, showing the close association between the two (original sagebrush layer from Knick and Connelly 2011)

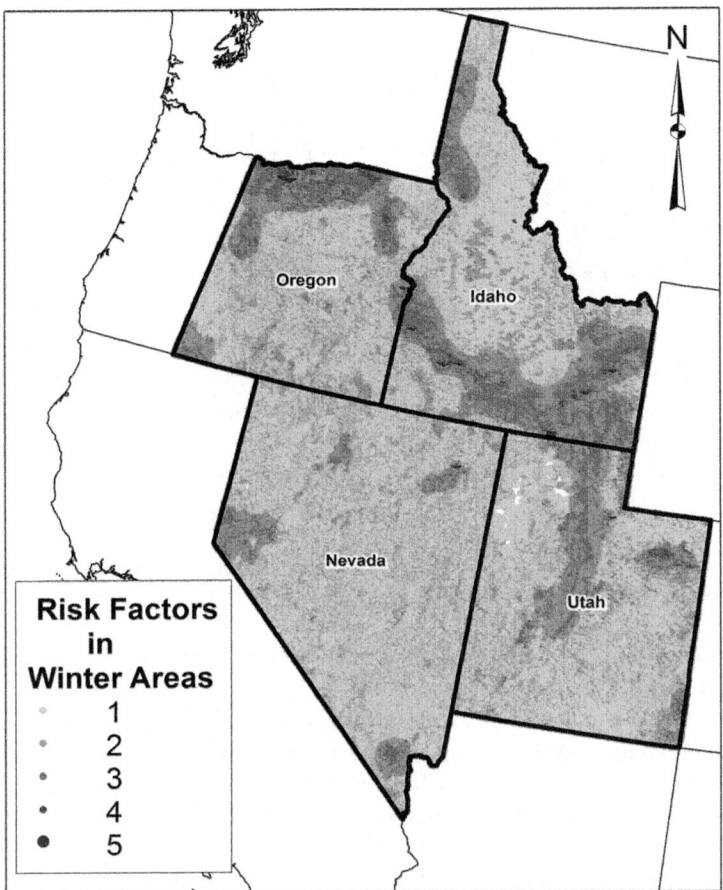

Fig. 19.5 Distribution and number of potential risks analyzed, that occur in model predicted Golden Eagle wintering areas in the four state study area of western North America

Abatzoglou and Kolden 2011) and subsequent influence on prey availability (Knick and Dyer 1997; Kochert and Steenhof 2002).

Significantly more than expected of the predicted wintering areas (59.4%) are in the least fragmented landscapes (<10%) while less than expected (3.3%) of model predicted winter areas are in landscapes fragmented ≥50% (χ^2 = 977.8901, df = 5, p < 0.001; and see chi-square residuals Appendix Table A19.5; Fig. 19.8). Predicted Golden Eagle winter habitat occurred across the spectrum of human footprint intensity (1–10; calculated from the original data layer created by Leu et al. 2008). However, there is a significant difference in the distribution of the intensity of the human footprint in suitable vs. unsuitable winter habitat; more than expected potential eagle winter areas occurred in landscapes with the least human footprint (< 2: from Leu et al. 2008) and less than expected in landscapes where the human

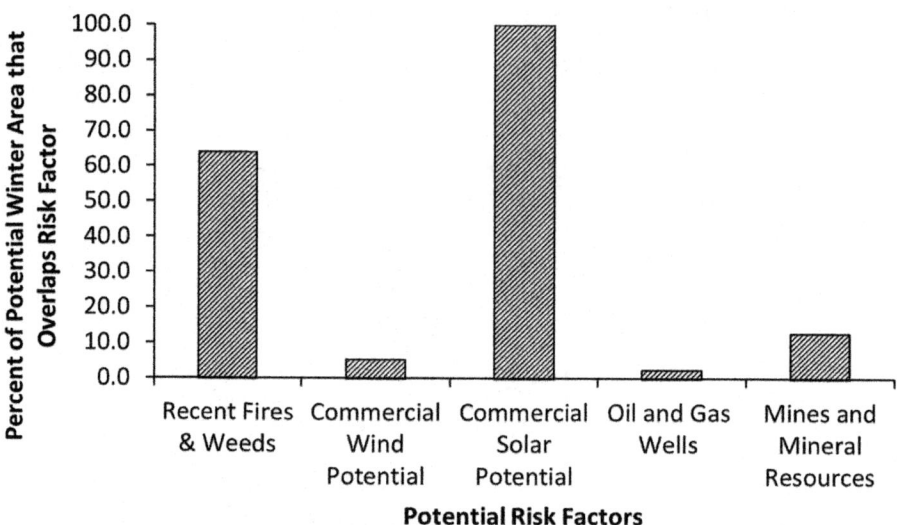

Fig. 19.6 Extent of potential or known risk factors to Golden Eagles that overlap model predicted eagle wintering areas in a four state region of western North America

footprint was ≥3 (χ^2 = 826.8963, df = 5, p < 0.001; and see chi-square residuals Appendix Table A19.6).

All the model predicted winter areas overlap potential sites for commercial solar development and < 5% overlap potential sites for wind energy development (Fig. 19.6). However, about half of BLM approved wind development sites (47.2%), areas with potential for utility grade wind (42.2%; Fig. 19.9a) and solar (44.9%; Fig. 19.9b) development in the four state study area, occur in predicted Golden Eagle winter habitat. Approved oil and gas wells and development sites, mining claims, active mines and other mineral resources overlap about 15.0% of predicted winter areas in the four state region (Fig. 19.6), but most that do occur are within predicted eagle winter areas (Figs. 19.10a, 19.10b and 19.11).

19.5 Discussion

Our results are the first to predict Golden Eagle wintering areas across a multi-state region of the western US, by applying the outcome of a previous predictive model (The Authors, Unpublished Data) to a new geographic area. The model was derived using TreeNet, a machine learning algorithm, and represents broad-scale ecological factors associated with eagles on their winter grounds. It was developed using publicly available digital layers. The same ecological factors used to construct the original model for Idaho and Montana were used as a basis for extrapolating Golden Eagle wintering areas across Idaho, Utah, Nevada and eastern Oregon. Based on the accuracy of our predictions using independent observations from citizen science

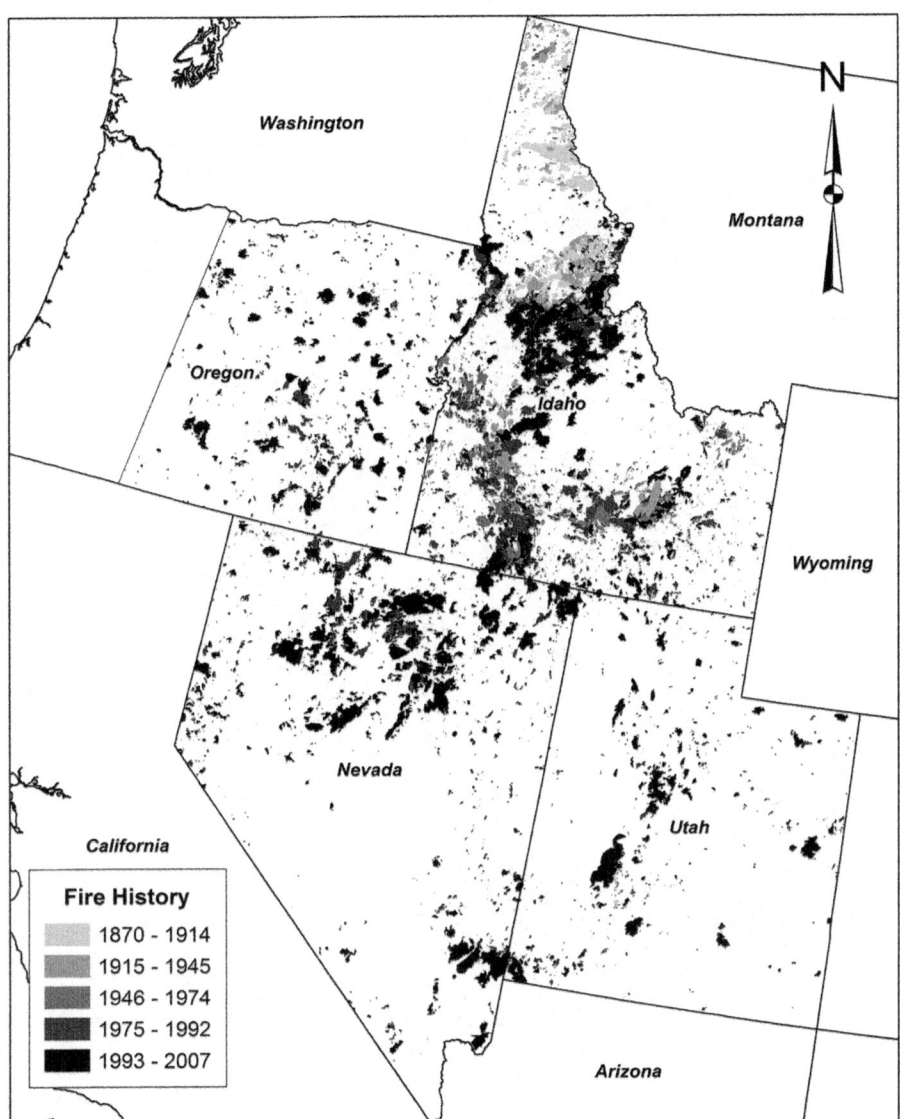

Fig. 19.7 Distribution of burns and years in which areas burned within the four state study area of western North America (original layer from US Department of Interior)

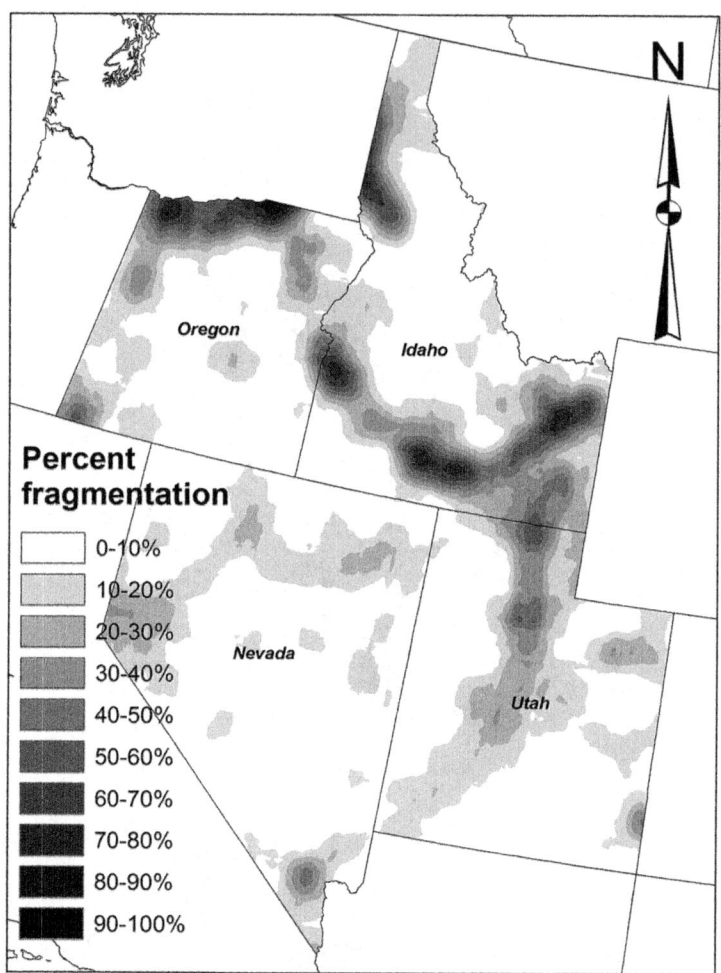

Fig. 19.8 Percent fragmentation of the landscape in the four state study area in western North America (original layer from Leu et al. 2008)

datasets (CBC: www.christmasbirdcount.org, Audubon 2010; eBird: Sullivan et al. 2009; eBird 2013), we believe that our model is sufficiently accurate to be useful for predicting suitable eagle wintering areas, their characteristics and the co-occurrence of potential risk factors within Golden Eagle winter range in the four-state region. Our model results were further validated by agreement with published findings that wintering eagles in the West are often associated with sagebrush and other shrub/grassland vegetation associations (see e.g., Marzluff et al. 1997; Poessel et al. 2016) and regions where habitat fragmentation and anthropogenic influences are less evident (Craig et al. 1986; Domenech et al. 2015a; see e.g., Fischer et al. 1984; Marzluff et al. 1997; Nielson et al. 2016). A visual comparison also shows considerable

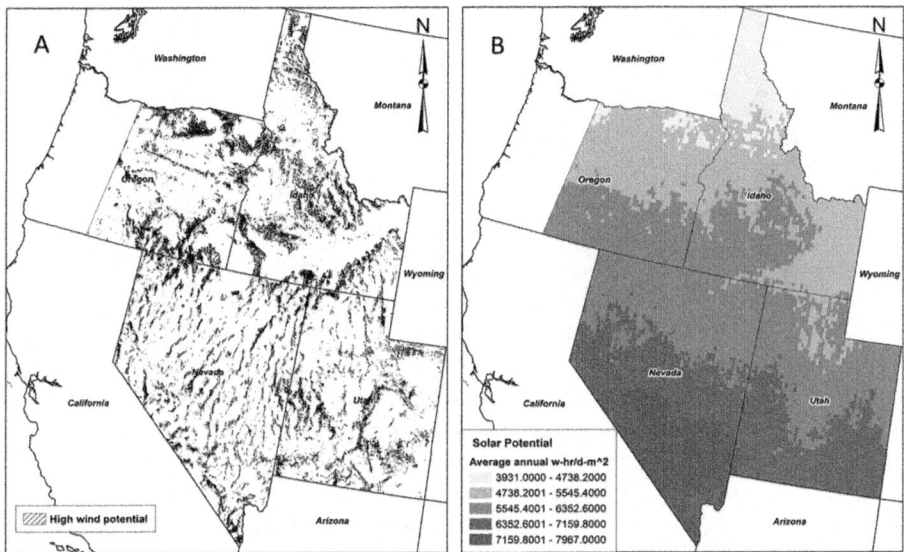

Fig. 19.9 (**a**) Extent of potential utility grade wind (original layer from National Renewable Energy Laboratory) and (**b**) solar areas (Perez et al. 2002) suitable for development within the four state study area in western North America

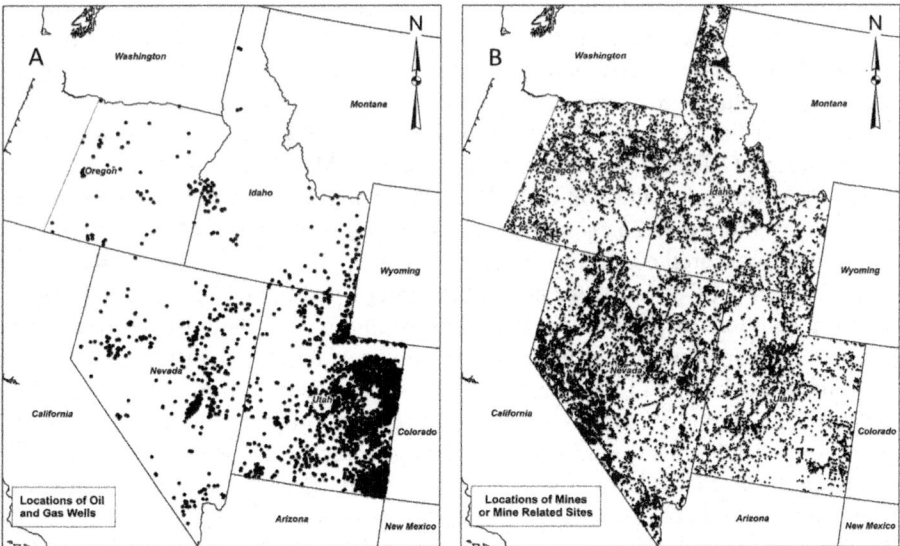

Fig. 19.10 (**a**) Distribution of oil and gas wells (layer original source: USGS National Oil and Gas Assessment website) and (**b**) mineral resource sites (Fernette et al. 2016) within the four state study area in western North America

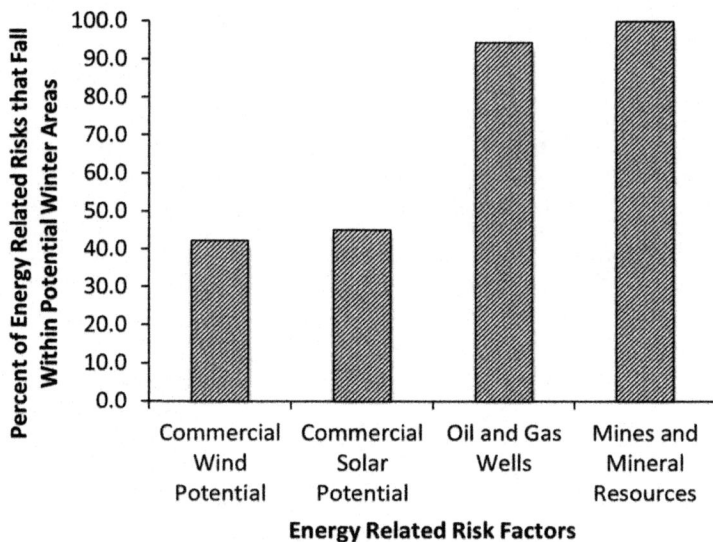

Fig. 19.11 Percent of energy related risk factors in the four-state study area in western North America that fall within model predicted Golden Eagle wintering areas

overlap of our predicted winter areas with Nielson et al.'s (2016) model of the intensity of Golden Eagle use in the western US during late summer.

Identifying areas that are suitable for wintering eagles is an essential first step for determining their distribution, the potential extent of risks to the population, and for informing future management actions. Our results predict that eagle wintering areas extend in a mosaic pattern across all the current FWS EMU's (BCR's; USFWS 2016) in the study area. Eagles that winter there are known to represent breeding populations from at least 5 BCR's within the Pacific Flyway, which extend from the southern tip of Nevada, north to Alaska (see e.g., Marzluff et al. 1997; McIntyre et al. 2008; McKinley and Mattox 2010). During migration, eagles could potentially cross any of more than 25 ecoregions (see Level III ecoregions of NA; https://www.epa.gov/eco-research/ecoregions-north-america),   5   Landscape   Conservation Cooperatives (LCC's; https://lccnetwork.org/map), and 5 Joint Ventures (JV; https://iwjv.org/resource/map-north-american-joint-ventures) as they travel from their summer areas to wintering grounds in the study area. A cluster analysis of the yearly movement tracks of 571 Golden Eagles in North America from 1992–2016 similarly indicated that different clusters of eagles, representing both resident and migratory populations, often used the same geographic units (e.g., BCR or LCC; Brown et al. 2017). Brown et al. (2017) also found that the clusters of eagle data they analyzed did not adequately conform to the geographic boundaries of the four mapping systems they examined in western NA (BCR, Flyway, LCC, JV). This emphasizes the challenges faced by managers as they attempt to identify geographic populations of Golden Eagles in order to estimate the impacts of local risks on long-term population stability. The numbers of wintering eagles, their distribution in an

EMU (BCR), and the extent to which different breeding populations are represented in specific winter areas within the four state region are unknown. There is a need for eagle surveys across the species' winter range and for identifying the composition of the wintering population in order to identify important use areas and to gather data for advancing population models (USFWS 2016).

Our predictions of Golden Eagle winter habitat in Idaho, Utah, Nevada and eastern Oregon are based on eagle locations and habitat characteristics in southwestern Idaho and the east-central Idaho/west-central Montana region. Shrub/grassland vegetation associations predominated in the winter areas of those satellite-tracked eagles (The Authors, Unpublished Data) and shrub/grasslands predominate the results of the predicted eagle winter areas in the four-state study region. It is interesting to note that although we did not use a specific layer representing sagebrush to construct the original model, there was remarkable overlap in portions of the predicted winter areas and the digital sagebrush layer created by Knick and Connelly (2011). These results corroborate the importance of sagebrush habitats to Golden Eagles in the study area during winter and agree with observations in the field (EHC, THC, MRF; Marzluff et al. 1997). They further support the general assumption that Golden Eagles often rely on these habitats in much of the US intermountain west (Kochert et al. 2002; Marzluff et al. 1997). Future models could be further refined by using sagebrush distribution maps overlain with predicted winter areas in western US landscapes where shrub-steppe vegetation cover prevails.

Of particular interest is that most of predicted wintering areas in the four state region occur on sparsely populated public lands, and almost half occurs on BLM managed lands. These include much of the shrub-steppe habitat in the western US and many of the areas targeted for rapid expansion of energy development projects (Copeland et al. 2011). As a result, public land management decisions, particularly those involving permits for development on BLM managed lands, and the transfer of public lands to private ownership (see e.g., 1 March 2018, Bill H.R. 5133, which allows, "...for the sale or exchange of public land..."), will play a vital role in the future conservation of Golden Eagles in the West.

We found considerable overlap in our model predicted eagle wintering areas and sites with high potential for renewable energy development. Nielson et al. (2016) report similar results for their modeled density distribution of eagles during late summer. Direct mortality from wind farms (Katzner et al. 2017; Loss et al. 2013; Pagel et al. 2013; Smallwood and Thelander 2008) is just one of the energy development related threats to Golden Eagles. Copeland et al. (2011) estimate that some of the greatest effects to ecosystems in western NA from energy development will occur in shrublands (>40,000,000 h), and that >75,000,000 h of grasslands are also vulnerable because of their occurrence in areas with high oil and gas reserves. They further speculated that, if development in the US occurs at a maximum rate, the future effects of commercial wind and solar development are greatly underestimated, and could have major impacts on shrub and grassland ecosystems (Copeland et al. 2011). At the same time, climate change, wildfires, and invasive weeds are also predicted to further transform ecosystems in the western US and present additional major challenges for conserving sagebrush and other native shrub-steppe vegetation communities and the species reliant upon them (Abatzoglou and Kolden 2011; Balch et al. 2013; Bradley 2010; Chambers et al. 2014; Creutzburg et al. 2015). More than half of predicted winter areas in the four state region have already been altered by fires and/or invasive weed species, emphasizing the importance of monitoring the cumulative effects of these multiple threats on eagle populations in the West.

Currently, little of our model predicted eagle wintering areas have active development in them, although ~40% contain at least two risk factors. Because winter areas are scattered across much of Idaho, Utah, Nevada and eastern Oregon, it is likely that many of the individual eagles wintering there are exposed to some of the risks identified in our analysis. Potential sites for renewable energy development are the most widespread of the risks we examined, but almost all oil and gas development and other mineral resources occur in potential eagle wintering areas. Identifying critical habitat for wintering eagles and the density and composition (breeding populations represented) of wintering populations is essential for effective conservation planning. It is particularly important for assessing the population level impacts of potential future development and other risk factors in eagle winter areas. Our model does not provide an estimate of eagle density, but we speculate that the most suitable habitats will likely support higher numbers of eagles; additional investigation is needed in this area.

Predicted winter areas occur disproportionately where the intensity of human activity is low, and where habitat fragmentation is also relatively low. We assume these areas reflect more intensive and consistent winter use by Golden Eagles. Similarly, Domenech et al. (2015a) reported wintering eagles to avoid urban areas and Nielson et al. (2016) found a negative relationship between the intensity of Golden Eagle use during late summer in the western US and the proportion of developed landscape. These results imply that the potential for adverse effects on eagles in western NA is high as a result of expanding anthropogenic influences. Generalist apex predators [such as the Golden Eagle] that have low population densities and range over large areas are particularly at risk from habitat fragmentation, partly because it may cause them to move more widely or relocate, where they may also encounter other threats (Terborgh et al. 2011). Similarly, the distribution of threats we analyzed provides evidence of the extent of some of the stressors that resident and migrant eagles face in the study area during winter.

Many risks we identified are concentrated near population centers and where habitat fragmentation is high, both, areas where fewer wintering eagles are likely to occur (this study; Domenech et al. 2015a; Nielson et al. 2016). However, we caution that our map of risks to eagles should be considered a minimum estimate because changes are occurring rapidly in the West and some of the digital layers we used contain data that are out of date; in addition, we weighted all risks equally. Our analysis makes no attempt to take into account the varying demographic consequences of individual risks to Golden Eagles such as direct mortality from blade strikes at wind farms (Pagel et al. 2013) vs. effects of wildfires and invasive weed encroachment in native habitats (Balch et al. 2013; Bradley 2010) that may affect prey distribution and availability (Knick and Dyer 1997; Kochert et al. 1999).

Further, our results represent only a portion of risks to eagles. The widespread extent and level of exposure to environmental contaminants (e.g., heavy metals or polychlorinated biphenyls; Herring et al. 2017) remain largely unmapped and unavailable as digital data layers. Still other risks to eagles may be unknown. This challenges resource managers attempting to quantify and estimate local risk factors on the long-term stability of widespread geographic populations of Golden Eagles. The problem is further exacerbated because, as evidenced by Brown et al.'s (2017) research on the movements of hundreds of eagles and recognized by USFWS (2016) analyses, it is difficult to identify the appropriate boundaries of an EMU. Further, the proportional extent to which different breeding populations are represented in specific winter areas within the four-state region is still unknown. Preliminary estimates of wintering Golden Eagle density at the larger, landscape scale have been highly variable and challenge interpretation (USFWS 2016).

Some may assume that alterations in habitats have less impact on wintering eagles than to birds during the breeding season because wintering eagles are not tied to a nest site and can forage elsewhere when winter habitats are altered or lost. However, recent evidence indicates that migrants, like resident Golden Eagles, show fidelity to wintering areas, particularly adult birds (The Authors, Unpublished Data; McKinley and Mattox 2010; Watson et al. 2014; Domenech et al. 2015a). Site fidelity during winter indicates the importance of consistently used areas and emphasizes possible negative impacts from deterioration of winter habitats. There is a need to identify these areas *ante factum* as an essential component of informing appropriate management decisions relative to population level impacts of perturbations on eagle winter grounds. Our findings are an important first step toward that end. However, we emphasize that our model predicted winter areas do not necessarily represent the actual winter distribution of eagles, but their potential distribution in the study area. Additional surveys or other assessment of Golden Eagle presence at the local level are necessary to refine this broad-scale predictive model. Further, until Golden Eagle winter distribution and resource use are studied and more fully understood across landscapes, it will continue to be difficult to determine how interacting and multiple risk factors affect eagle population dynamics. This is particularly true if rapidly expanding energy development occurs where migrants and residents concentrate.

Our results and distribution maps describe characteristics of, and risks associated with predicted Golden Eagle wintering areas in the four state region and potentially identify where management can best be implemented toward conserving a stable or increasing Golden Eagle population. Our results can also provide geographic focus for prioritizing surveys to help identify important eagle use areas in western US landscapes, where the greatest potential for conflicts between development projects and wintering eagles occurs, and where risks from degradation of habitat, disturbance, or take is likely to affect eagles (USFWS 2013). They can additionally provide geographic input on relevant anthropogenic stressors for inclusion in more complex ecological risk assessments (Suter II 2016) such as spatial demographic models that estimate population level responses to risks at smaller geographic scales, (see, e.g., Wiens et al. 2017).

19.5.1 Summary

Our results indicate that wintering Golden Eagles are predicted to occur in a continuum of suitable habitat across much of the four state region, but are likely exposed to a variety of risk factors on their wintering grounds. The demographic consequences of wintering in habitats of varying suitability are unknown, as are the effects of exposure to various risk factors or multiple risks. We believe that managing for a stable or increasing population at the continental or regional scale by permitting and mitigating take at the local site or EMU scale (USFWS 2016) can benefit from our predictions of potential winter areas. Using the predictions and our documented distribution of risk factors, biologists can stratify surveys to locate eagle use areas and document eagle resource use responses to risk factors. Management of wintering areas to minimize and mitigate adverse effects or to maximize conservation is likely to be especially important in areas of suitable habitat that are consistently used by resident and migrant eagles (The Authors, Unpublished Data; Marzluff et al. 1997; Domenech et al. 2015a). Comparatively dense concentrations of eagles at risk could result in greater demographic consequences. Also, if habitat loss or eagle take affects individuals that would have dispersed to or returned to breeding season range in a different EMU, a decline in eagles detected in a distant area will be difficult to understand and manage without more complete knowledge of Golden Eagle winter ecology.

Although a relatively small percent of our model predicted eagle wintering areas have active development in them, it is important to note that: (1) mortality at the localized scale can have widespread impacts to eagle populations; (2) a high percentage of development projects or areas with potential for development, occur within predicted winter areas; (3) our map of risks to wintering Golden Eagles is appropriate for initial, broad-scale project level planning but it does not represent all risks that occur in the study area; (4) our model of winter areas reflects only one of a series of hierarchical choices that eagles make when selecting and using a wintering area; (5) our model of predicted winter areas is an estimate and does not necessarily represent the actual winter distribution of wintering eagles in the study area; (6) the resolution of the digital layers we used and error associated with the original input datasets also determine the scale at which our models can be interpreted; and (7) further field validation is essential for fine-scale project level risk assessment.

Conflicts with Golden Eagle winter populations could be greatly reduced by encouraging renewable energy development in fragmented landscapes near population centers that are already heavily impacted by the human footprint and where current research indicates that fewer wintering Golden Eagles are likely to occur (this study; Nielson et al. 2016). Further, solar and wind development at already impacted population centers and by individual home owners and businesses are more energy efficient and cost effective because they provide power at or near the site where it is used; energy loss due to transportation from remote locations is avoided (Bialek 1996; Rao et al. 2013). In addition, energy development that occurs in disturbed areas has been shown to result in fewer ecological impacts overall

(Jones and Pejchar 2013). Privately developed solar projects in some states have been highly successful and occur in impacted environments where Golden Eagles are not likely to occur (Aylett 2013; Cusick 2015). An avoidance approach for development could be a useful management strategy even while there is still uncertainty about aspects of eagle population biology.

Acknowledgements Brian Millsap, Diana Whittington and other colleagues have collaborated on development and prioritization of information needs relative to Golden Eagle management. We would like to thank H. McFarland and J. & R. Craig, for their considerable contributions in earlier phases of this research and L. Dunn and L. Schueck for help with digital layers. FH acknowledges S. & J. Linke, E. Bruenning and L. Strecker for contributions while developing some of the research leading to this publication. We acknowledge the value of the publicly available GIS datasets used in this analysis and the contribution of data, access and permission for use of eBird data by the Cornell Laboratory of Ornithology and the National Audubon Society. CBC Data is provided by National Audubon Society and through the generous efforts of Bird Studies Canada and countless volunteers across the western hemisphere. This project was funded through the Science Support Partnership program of the US Geological Survey, Aquila Environmental and the Bureau of Land Management. Any use of trade, product, or firm names is for descriptive purposes only and does not imply endorsement by the US Government.

Appendix

Table A19.1 Descriptions and sources for environmental layers used to apply the grove file from the original wintering Golden Eagle model in Idaho and Montana (The Authors, Unpublished Data) to the geographic area of Idaho, Nevada, Utah and eastern Oregon[1]

Environmental layer	Description (labels)	Time period	Data source
Climate	30 year averages used to interpolate minimum monthly temperature during winter; raster, cell: 1 km.	Nov–March 1980–1997	PRISM, http://prism.oregonstate.edu
	30 year averages used to interpolate monthly precipitation during winter; raster, cell: 1 km.	Nov–March 1980–1997	PRISM, http://prism.oregonstate.edu
	Average Hillshade[2], a relative measure of solar illumination (value during winter on the 15th of each winter month at 10:00 hr). Derived in ArcMap; raster, cell: 1 km.	Nov–March, winter	Derived in ArcMap (Azimuth and Altitude calculated using NOAA's solar calculator
Topography	Digital elevation models for the 4-state area; datasets were merged for our purposes; raster, cell: 90 m	NA	https://catalog.data.gov/dataset/usgs-national-elevation-dataset-ned
	Slope: Derived from Digital Elevation Model;raster, cell: 90 m	NA	Derived in ArcMap

Environmental layer	Description (labels)	Time period	Data source
	Aspect[2]: Degrees derived from Digital Elevation Model and then classified;raster, cell: 90 m	NA	Derived in ArcMap
Water	Distance[3] to rivers and streams derived from original hydrography layer (https://nhd.usgs.gov/); raster, cell: 1 km.	2002	Derived in ArcMap
Highways and roads	Distance[3] to highways or roads derived from original Tiger transportation layers; raster, cell: 1 km.	2000	Derived in ArcMap
Land Cover	Vegetation cover[2], reclassified to 9 classes from the original GAP layers (US Geological Survey 2011); raster, cell: 30m)	2006	https://gapanalysis.usgs. gov/gaplandcover/data/ Version 2
Land Ownership	Land ownership[2] clipped to the four states' boundaries from original, westNA_own.shp; polygon.	1992–2003	http:\\sagemap.wr.usgs. gov
Human population density	Human population density from Census Blocks 2000 with Associated Data; polygon.	2000	US Census Bureau Tiger Files http://sagemap.wr. usgs.gov/ftp/regional/ usgs/us_ population_1990-2000_ sgca.zip

[1]Spatial Reference Information: Projection: North American Datum Albers 1983, False Easting: 0.000000, False Northing: 0.000000, Longitude of Central Meridian: -96.000000, Latitude of Origin: 23.000000, Standard Parallel 1: 29.500000 and 2: 45.500000

[2]*Additional GIS Layer Information*

Hillshade was calculated in ArcMap but input parameters for azimuth and altitude were calculated using an online solar calculator (NOAA Research, http://www.esrl.noaa.gov/gmd/grad/solcalc/)

We reclassified the vegetation classes into 9 general categories from the original GAP layers (Homer, 1998; Redmond et al., 1998): 1 = urban or developed land, 2 = agricultural land, 3 = grasslands and shrublands, 4 = forest uplands, 5 = water, 6 = riparian and wetland areas, 7 = barren land (exposed rock, sand dunes, shorelines and gravel bars), 8 = alpine meadow, 9 = snow, ice, cloud or cloud shadow

For Land Ownership we retained the original 16 categories described by Finn (2003): BIA=Bureau of Indian Affairs, BLM=Bureau of Land Management, BOR=Bureau of Reclamation, DOD=Department of Defense, DOE=Department of Energy, LOCAL=local ownership, MISC_ FED=miscellaneous federal lands, NPS=National Park Service, PRIVATE=private land, STATE=state land, TNC=the Nature Conservancy, UNKNOWN, USFS=U.S. Forest Service, USFWS=U.S. Fish and Wildlife Service, WATER=bodies of water, USFS/BLM=U.S. Forest Service or Bureau of Land Management. The combined USFS/BLM category comprised a very small portion of the land ownership and was included by Finn (2003) for the small percentage of lands in which the exact federal ownership was undetermined

[3]All distance layers were derived using the Euclidian Distance tool in the Spatial Analyst extension of ArcMap 10.1

Table A19.2 Descriptions and source information for layers included in the analysis of risks to Golden Eagles in their predicted winter areas in Idaho, Utah, Nevada and eastern Oregon. Original source layers were clipped to the study area boundaries and reprojected, if necessary. Spatial Reference: North American Datum Albers 1983, False Easting: 0.000000, False Northing: 0.000000, Longitude of Central Meridian: -96.000000, Latitude of Origin: 23.000000, Standard Parallel 1: 29.500000 and 2: 45.500000

Risk layer	Description (labels)	Time period	Data source
Prospect and mine-related features	Prospect- and Mine-Related Features on USGS Topographic Maps of the Western United States (point; Fernette et al. 2016)	1888–2006	https://www.sciencebase.gov/catalog/item/57962314e4b007df0739fede
Fire history	Historical fire record for the western United States (polygon; Western Fire Map 1870-2007; US Dept. of Interior);	1870–2007	https://catalog.data.gov/dataset/western-range-fires-1870-2007
Anthropogenic fragmentation	Estimates the degree of fragmentation caused by development of human features on the landscape (raster, cell: 180 m; Leu et al. 2008).	2007	USGS SAGEMAP: http://sagemap.wr.usgs.gov
The human footprint in the West	Model of the influence of anthropogenic disturbance in the western United States. It contains 10 human footprint classes based on input from 14 anthropogenic related features (raster layer, cell size: 180 m; Leu et al. 2008).	2007	USGS SAGEMAP: http://sagemap.wr.usgs.gov
Exotic plant invasion risk in the western United States	Model of the risk of invasion by exotic plant species (raster, cell: 180;Leu et al. 2008).	2008	USGS SAGEMAP: http://sagemap.wr.usgs.gov
Oil and gas well distribution	Displays the oil and gas well distribution in the western United States (points; all_wells_wus_1899-2007). Original layer from USGS National Oil and Gas Assessment website	1899–2007	USGS SAGEMAP: http://sagemap.wr.usgs.gov; http://energy.cr.usgs.gov/oilgas/noga/
Oil and gas well density	Density of oil and gas wells in the western United States (raster, cell: 180; Original layer from USGS, National Oil and Gas Assessment)	2004	http://energy.cr.usgs.gov/oilgas/noga/

Risk layer	Description (labels)	Time period	Data source
Western US wind resource at 50 meters above ground level	Identifies areas of high wind resource potential in the western US on a scale of 1-7, with 7 having the highest potential (polygon; original source: National Renewable Energy Laboratory)	2002	USGS SAGEMAP: http://sagemap.wr.usgs.gov/ftp/sab/wus_50mwind.zip
Utility grade solar development potential	Monthly and annual average solar resource potential for 48 Contiguous United States. (polygon; Perez et al. 2002)	1998–2005	Created by the National Renewable Energy Laboratory which is operated by the Alliance for Sustainable Energy, LLC for the U.S. Department of Energy
Sagebrush in the Western US	Sagebrush habitat in the western US (raster, cell size: 90 m; Knick and Connelly 2011)	2006	USGS SAGEMAP: http://sagemap.wr.usgs.gov

Table A19.3 Chi square residuals from a comparison of 2,012 independent Golden Eagle (eBird) winter sightings that were located in predicted suitable and unsuitable Golden Eagle wintering areas (observed), with the expected distribution, based on the the percent of predicted suitable vs. unsuitable winter areas available in the Idaho, Utah, Nevada and eastern Oregon study area (χ^2 = 19.1417, df = 1, p <0.001)

Suitability for Golden Eagle winter habitat	Low (unsuitable): ≤ 0.33 RIO	Mid and High: ≥ 0.33 RIO
Chi-square Residual	−2.932	3.2476

Table A19.4 Comparison of the distribution of predicted suitable Golden Eagle winter habitat in CBC circles where eagles were sighted (observed), with the expected distribution in all CBC circles in the study area. Residuals of the chi-square test are shown for each category of habitat from low (unsuitable) to high suitability for winter habitat (χ^2 = 25.7315, df = 2, p <0.001)

Suitability for Golden Eagle winter habitat	Low (unsuitable): ≤ 0.33 RIO	Mid: > 0.33 and ≤ 0.67 RIO	High: > 0.67 RIO
Chi-square Residual	−2.6282	−0.7986	4.2645

Table A19.5 Chi square residuals for a comparison of the degree of fragmentation in predicted suitable Golden Eagle wintering areas (observed), with fragmentation of all the available (expected) area in the Idaho, Utah, Nevada and eastern Oregon study area (χ^2 = 977.8901, df = 5, p <0.001)

Percent of the habitat that is fragmented	< 10	10 – 20	20 – 30	30 – 40	40 – 50	≥ 50
Chi-square Residual	20.013849	−13.972032	−10.307536	−7.616126	−9.877737	−10.968042

Table A19.6 Chi square residuals for a comparison of the intensity of the human footprint (scale 1 – 10) in predicted suitable Golden Eagle wintering areas (observed), with the intensity of the human footprint in all the available (expected) area in the Idaho, Utah, Nevada and eastern Oregon study area. Predicted winter areas were more likely to be in habitats with the lowest intensity human footprint (1 and 2) and less likely than expected in areas with greater human footprint (>3; $\chi^2 = 826.8963$, df = 5, p <0.001)

Intensity of human footprint	1(least)	2	3	4	5	> 5 (most)
Chi-square Residual	25.1540	3.5628	−2.8829	−4.0488	−4.7643	−11.5790

References

Abatzoglou JT, Kolden CA (2011) Climate change in western US deserts: potential for increased wildfire and invasive annual grasses. Rangel Ecol Manag 64:471–478

Anderson RP, Lew D, Peterson AT (2003) Evaluating predictive models of species' distributions: criteria for selecting optimal models. Ecol Model 162:211–232

Audubon (2010) The Christmas bird count historical results. National Audubon Society, New York http://www.christmasbirdcount.org

Aylett A (2013) Networked urban climate governance: neighborhood-scale residential solar energy systems and the example of solarize Portland. Environ Plan C Gov Policy 31:858–875

Balch JK, Bradley BA, D'antonio CM, Gómez-Dans J (2013) Introduced annual grass increases regional fire activity across the arid western USA (1980–2009). Glob Change Biol 19:173–183

Beyer HL (2012) Geospatial modelling environment. [WWW Document]. URL http://www.spatialecology.com/gme/ (Accessed 21 May 2016)

Bialek J (1996) Tracing the flow of electricity. IEE Proc-Gener Transm Distrib 143:313–320

Bonney R, Shirk JL, Phillips TB, Wiggins A, Ballard HL, Miller-Rushing AJ, Parrish JK (2014) Next steps for citizen science. Science 343:1436–1437

Booms TL, Huettmann F, Schempf PF (2010) Gyrfalcon nest distribution in Alaska based on a predictive GIS model. Polar Biol 33:347–358

Bradley BA (2010) Assessing ecosystem threats from global and regional change: hierarchical modeling of risk to sagebrush ecosystems from climate change, land use and invasive species in Nevada, USA. Ecography 33:198–208

Brooks ML, Pyke DA (2001) Invasive plants and fire in the deserts of North America. In: KEM G, Wilson T (eds) Proceedings of the invasive species workshop: the role of fire in the control and spread of invasive species. Presented at the fire conference 2000: the first National Congress on fire ecology, prevention, and management. Tall Timbers Research Station, Tallahassee, pp 1–14

Brown JL, Bedrosian B, Bell DA, Braham MA, Cooper J, Crandall RH, Didonato J, Domenech R, Duerr AE, Katzner TE (2017) Patterns of spatial distribution of golden eagles across North America: how do they fit into existing landscape-scale mapping systems? J Raptor Res. 51:197–215

Carrete M, Sánchez-Zapata JA, Benítez JR, Lobón M, Donázar JA (2009) Large scale risk-assessment of wind-farms on population viability of a globally endangered long-lived raptor. Biol Conserv 142:2954–2961

Chambers JC, Bradley BA, Brown CS, D'Antonio C, Germino MJ, Grace JB, Hardegree SP, Miller RF, Pyke DA (2014) Resilience to stress and disturbance, and resistance to Bromus tectorum L. invasion in cold desert shrublands of western North America. Ecosystems 17:360–375

Chapin FS III, Zavaleta ES, Eviner VT, Naylor RL, Vitousek PM, Reynolds HL, Hooper DU, Lavorel S, Sala OE, Hobbie SE (2000) Consequences of changing biodiversity. Nature 405:234–242

Copeland HE, Pocewicz A, Kiesecker JM (2011) Geography of energy development in western North America: potential impacts on terrestrial ecosystems, Chapter 2. In: Naugle DE (ed) Energy development and wildlife conservation in western North America. Island Press, Washington, DC, pp 7–22

Craig EH, Craig TH (1998) Lead and Mercury levels in Golden and Bald Eagles and annual movements of Golden Eagles wintering in East Central Idaho, 1990–1997, Technical Bulletin No. 98–12. Idaho Bureau of Land Management, Boise

Craig E, Huettmann F (2009) Using "Blackbox" algorithms such as TreeNet and random forests for data-mining and for finding meaningful patterns, relationships and outliers in complex ecological data: an overview, an example using golden eagle satellite data and an outlook for a promising future. In: Wang H-F (ed) Intelligent data analysis: developing new methodologies through pattern discovery and recovery. IGI Global, Hershey, pp 65–84

Craig EH, Craig TH, Powers LR (1986) Habitat use by wintering golden eagles and rough-legged hawks in southeastern Idaho. J Raptor Res. 20:69–71

Creutzburg MK, Halofsky JE, Halofsky JS, Christopher TA (2015) Climate change and land management in the rangelands of Central Oregon. Environ Manag 55:43–55

Cusick D (2015) Solar power sees unprecedented boom in US, Climatewire. Environment & Energy Publishing, LLC, Washington, DC. https://www.eenews.net/climatewire/stories/1060014730/search?keyword=solar (Accessed 27 July 2017)

Cutler DR, Edwards TC, Beard KH, Cutler A, Hess KT, Gibson J, Lawler JJ (2007) Random forests for classification in ecology. Ecology 88:2783–2792

Dahl EL, Bevanger K, Nygard T, Røskaft E, Stokke BG (2012) Reduced breeding success in white-tailed eagles at Smøla windfarm, western Norway, is caused by mortality and displacement. Biol Conserv 145:79–85

De Lucas M, Janss GF, Whitfield DP, Ferrer M (2008) Collision fatality of raptors in wind farms does not depend on raptor abundance. J Appl Ecol 45:1695–1703

Dennison PE, Brewer SC, Arnold JD, Moritz MA (2014) Large wildfire trends in the western United States, 1984–2011. Geophys Res Lett 41:2928–2933

Dickinson JL, Zuckerberg B, Bonter DN (2010) Citizen science as an ecological research tool: challenges and benefits. Annu Rev Ecol Evol Syst 41:149–172

Domenech R, Bedrosian BE, Crandall RH, Slabe VA (2015a) Space use and habitat selection by adult migrant golden eagles wintering in the western United States. J Raptor Res 49:429–440

Domenech R, Pitz T, Gray K, Smith M (2015b) Estimating natal origins of migratory juvenile golden eagles using stable hydrogen isotopes. J Raptor Res 49:308–315

Dukes JS, Mooney HA (1999) Does global change increase the success of biological invaders? Trends Ecol Evol 14:135–139

eBird (2013) eBird: an online database of bird distribution and abundance. Cornell Lab of Ornithology, Ithaca. http://www.ebird.org

Elith J, Graham CH, Anderson RP, Dudík M, Ferrier S, Guisan A, Hijmans RJ, Huettmann F, Leathwick JR, Lehmann A, Li J, Lohmann LG, Loiselle BA, Manion G, Moritz C, Nakamura M, Nakazawa Y, Overton JMCM, Townsend Peterson A, Phillips SJ, Richardson K, Scachetti-Pereira R, Schapire RE, Soberón JR, Williams S, Wisz MS, Zimmermann NE (2006) Novel methods improve prediction of species' distributions from occurrence data. Ecography 29:129–151. https://doi.org/10.1111/j.2006.0906-7590.04596.x

Fernette GL, Horton JD, King Z, San Juan CA, Schweitzer PN (2016) Prospect- and Mine-Related Features from U.S. Geological Survey 7.5- and 15-Minute Topographic Quadrangle Maps of the Western United States: U.S. Geological Survey data release. U.S. Geological Survey data release. https://doi.org/10.5066/F7JD4TWT

Finn S (2003) Sage-grouse range wide conservation assessment project. USGS, Snake River Field Station. http://sagemap.wr.usgs.gov

Fischer DL, Ellis KL, Meese RJ (1984) Winter habitat selection of diurnal raptors in Central Utah. J Raptor Res 18:98–102

Flannigan M, Stocks B, Turetsky M, Wotton M (2009) Impacts of climate change on fire activity and fire management in the circumboreal forest. Glob Change Biol 15:549–560

Hernandez PA, Graham CH, Master LL, Albert DL (2006) The effect of sample size and species characteristics on performance of different species distribution modeling methods. Ecography 29:773–785

Herring G, Eagles-Smith CA, Buck J (2017) Characterizing golden eagle risk to lead and anticoagulant rodenticide exposure: a review. J. Raptor Res. 51:273–292. https://doi.org/10.3356/JRR-16-19.1

Hoffman SW, Smith JP (2003) Population trends of migratory raptors in western North America, 1977-2001. Condor 105:397–419

Homer CG (1998) Idaho/western Wyoming landcover classification. Utah State University, Remote Sensing/GIS Laboratories, Logan

Hunt WG (1998) Raptor floaters at Moffat's equilibrium. Oikos 82:191–197

Hunt GW, Watson JW (2016) Addressing the factors that juxtapose raptors and wind turbines. J Raptor Res 50:92–96

Hunt WG, Wiens JD, Law PR, Fuller MR, Hunt TL, Driscoll DE, Jackman RE (2017) Quantifying the demographic cost of human-related mortality to a raptor population. PLoS One 12:e0172232

Jones NF, Pejchar L (2013) Comparing the ecological impacts of wind and oil & gas development: a landscape scale assessment. PLoS One 8:e81391

Katzner TE, Nelson DM, Braham MA, Doyle JM, Fernandez NB, Duerr AE, Bloom PH, Fitzpatrick MC, Miller TA, Culver RC (2017) Golden eagle fatalities and the continental-scale consequences of local wind-energy generation. Conserv. Biol J Soc Conserv Biol 31:406–415

Keeley JE (2006) Fire management impacts on invasive plants in the western United States. Conserv Biol 20:375–384

Kikuchi R (2008) Adverse impacts of wind power generation on collision behaviour of birds and anti-predator behaviour of squirrels. J Nat Conserv 16:44–55

Knick S, Connelly JW (eds) (2011) Greater sage-grouse: ecology and conservation of a landscape species and its habitats. Univ of California Press, Berkeley

Knick ST, Dyer DL (1997) Distribution of black-tailed jackrabbit habitat determined by GIS in Southwestern Idaho. J Wildl Manag 61:75–85

Knick ST, Dobkin DS, Rotenberry JT, Schroeder MA, Vander Haegen WM (2003) Teetering on the edge or too late? Conservation and research issues for avifauna of sagebrush habitats. Condor 105:611–634

Knick ST, Holmes AL, Miller RF (2005) The role of fire in structuring sagebrush habitats and bird communities. Stud Avian Biol 30:63

Kochert MN, Steenhof K (2002) Golden eagles in the US and Canada: status, trends, and conservation challenges. J Raptor Res 36:32–40

Kochert MN, Steenhof K, Carpenter LB, Marzluff JM (1999) Effects of fire on golden eagle territory occupancy and reproductive success. J Wildl Manag 63:773–780

Kochert MN, Steenhof KM, McIntyre CL, Craig EH (2002) Golden eagle (Aquila chrysaetos), version 2.0. In: Poole AF, Gill FB (eds) The birds of North America. Cornell Lab of Ornithology, Ithaca. https://doi.org/10.2173/bna.684

Leu M, Hanser SE, Knick ST (2008) The human footprint in the west: a large-scale analysis of anthropogenic impacts. Ecol Appl 18:1119–1139

Liu C, Berry PM, Dawson TP, Pearson RG (2005) Selecting thresholds of occurrence in the prediction of species distributions. Ecography 28:385–393

Loss SR, Will T, Marra PP (2013) Estimates of bird collision mortality at wind facilities in the contiguous United States. Biol Conserv 168:201–209

Lu GY, Wong DW (2008) An adaptive inverse-distance weighting spatial interpolation technique. Comput Geosci 34:1044–1055

Manel S, Williams HC, Ormerod SJ (2001) Evaluating presence–absence models in ecology: the need to account for prevalence. J Appl Ecol 38:921–931

Marzluff JM, Knick ST, Vekasy MS, Schueck LS, Zarriello TJ (1997) Spatial use and habitat selection of golden eagles in southwestern Idaho. Auk 114:673–687

McIntyre CL, Lewis SB (2016) Observations of migrating golden eagles (*Aquila chrysaetos*) in eastern interior Alaska offer insights on population size and migration monitoring. J Raptor Res 50:254–264

McIntyre CL, Douglas DC, Collopy MW (2008) Movements of golden eagles (*Aquila chrysaetos*) from interior Alaska during their first year of independence. Auk 125:214–224

McKinley JO, Mattox B (2010) Winter site fidelity of migratory raptors in southwestern Idaho. J. Raptor Res. 44:240–243

Millsap BA, Allen GT (2006) Effects of falconry harvest on wild raptor populations in the United States: theoretical considerations and management recommendations. Wildl Soc Bull 34:1392–1400

Millsap BA, Zimmerman GS, Sauer JR, Nielson RM, Otto M, Bjerre E, Murphy R (2013) Golden eagle population trends in the western United States: 1968–2010. J Wildl Manag 77:1436–1448. https://doi.org/10.1002/jwmg.588

Munson MA, Webb K, Sheldon D, Fink D, Hochachka WM, Ili M, Riedewald M, Sorokina D, Sullivan B, Wood C, Kelling S (2011) The eBird reference dataset, Version 3.0. Cornell Lab of Ornithology and National Audubon Society, Ithaca

Nielson RM, Murphy RK, Millsap BA, Howe WH, Gardner G (2016) Modeling late-summer distribution of golden eagles (*Aquila chrysaetos*) in the western United States. PLoS One 11:e0159271

Pagel JE, Kritz KJ, Millsap BA, Murphy RK, Kershner EL, Covington S (2013) Bald eagle and golden eagle mortalities at wind energy facilities in the contiguous United States. J Raptor Res. 47:311–315

Paprocki N, Glenn NF, Atkinson EC, Strickler KM, Watson C, Heath JA (2015) Changing habitat use associated with distributional shifts of wintering raptors. J Wildl Manag 79:402–412

Paprocki N, Oleyar D, Brandes D, Goodrich L, Crewe T, Hoffman SW (2017) Combining migration and wintering counts to enhance understanding of population change in a generalist raptor species, the north American red-tailed hawk. Condor 119:98–107

Perez R, Ineichen P, Moore K, Kmiecik M, Chain C, George R, Vignola F (2002) A new operational model for satellite-derived irradiances: description and validation. Sol Energy 73:307–317

Poessel SA, Bloom PH, Braham MA, Katzner TE (2016) Age- and season-specific variation in local and long-distance movement behavior of golden eagles. Eur J Wildl Res 62:1–17. https://doi.org/10.1007/s10344-016-1010-4

Prasad AM, Iverson LR, Liaw A (2006) Newer classification and regression tree techniques: bagging and random forests for ecological prediction. Ecosystems 9:181–199

R Core Team (2013) R: A language and environment for statistical computing. R Foundation for Statistical Computing, Vienna. http://www.R-project.org/

Rao RS, Ravindra K, Satish K, Narasimham SVL (2013) Power loss minimization in distribution system using network reconfiguration in the presence of distributed generation. IEEE Trans Power Syst 28:317–325

Redmond RL, Hart MM, Winne JC, Williams WA, Thornton PC, Ma Z, Tobalske CM, Thornton MM, McLaughlin KP, Tady TP, Fisher FB et al (1998) The Montana gap analysis project: final report. Montana cooperative wildlife research unit. The University of Montana, Missoula. zii + 136 pp. + appendices

Shalev-Shwartz S, Ben-David S (2014) Understanding machine learning: from theory to algorithms. Cambridge University Press, New York

Smallwood KS, Thelander C (2008) Bird mortality in the Altamont pass wind resource area, California. J Wildl Manag 72:215–223

Sullivan BL, Wood CL, Iliff MJ, Bonney RE, Fink D, Kelling S (2009) eBird: a citizen-based bird observation network in the biological sciences. Biol Conserv 142:2282–2292

Suter GW II (2016) Ecological risk assessment. CRC press, Boca Raton

Tack JD, Fedy BC (2015) Landscapes for energy and wildlife: conservation prioritization for golden eagles across large spatial scales. PLoS One 10:e0134781

Terborgh J, Holt RD, Estes JA (2011) Trophic cascades: What they are and why they matter. Chapter 1. In: Terborgh J, Estes JA (eds) Trophic cascades: predators, prey, and the changing dynamics of nature. Island Press, Washington, DC, pp 1–20

Tredennick AT, Hooten MB, Aldridge CL, Homer CG, Kleinhesselink AR, Adler PB (2016) Forecasting climate change impacts on plant populations over large spatial extents. Ecosphere 7(10):e01525. https://doi.org/10.1002/ecs2.1525

USFWS (2008) Birds of Conservation Concern 2008. United States Department of Interior/Fish and Wildlife Service/Division of Migratory Bird Management, Arlington

USFWS (2009) Final environmental assessment; proposal to permit take as provided under the bald and golden eagle protection act. Division of Migratory Bird Management/U.S, Fish and Wildlife Service, Washington, DC

USFWS (2013) Eagle conservation plan guidance. Module 1 land-based wind energy, version 2. Division of Migratory Bird Management, Washington, DC

USFWS (2014) Migratory bird hunting; final frameworks for late-season migratory bird hunting regulations. Fed Regist 79:56864–56890

USFWS (2015) USFWS administrative waterfowl flyway boundaries. Vector digital data, USFWS, Falls Church

USFWS (2016) Bald and golden eagles: population demographics and estimation of sustainable take in the United States, 2016 update USFWS. Division of Migratory Bird Management, Washington, DC

US Geological Survey, Gap Analysis Program (GAP) (2011) May 2011 National Land Cover, Version 2. https://gapanalysis.usgs.gov/gaplandcover/data/Version2

Watson J (2010) The golden eagle, 2nd edn. T & AD Poyser, London

Watson JW, Duff AA, Davies RW (2014) Home range and resource selection by GPS-monitored adult golden eagles in the Columbia plateau ecoregion: implications for wind power development. J Wildl Manag 78:1012–1021

Wiens JD, Schumaker NH, Inman RD, Esque TC, Longshore KM, Nussear KE (2017) Spatial demographic models to inform conservation planning of golden eagles in renewable energy landscapes. J Raptor Res 51:234–257. https://doi.org/10.3356/JRR-16-77.1

Zorn PA (2012) Developing a "severe test" of species distribution modelling for conservation planning. Carleton University, Ottawa

Part VI
Conclusions

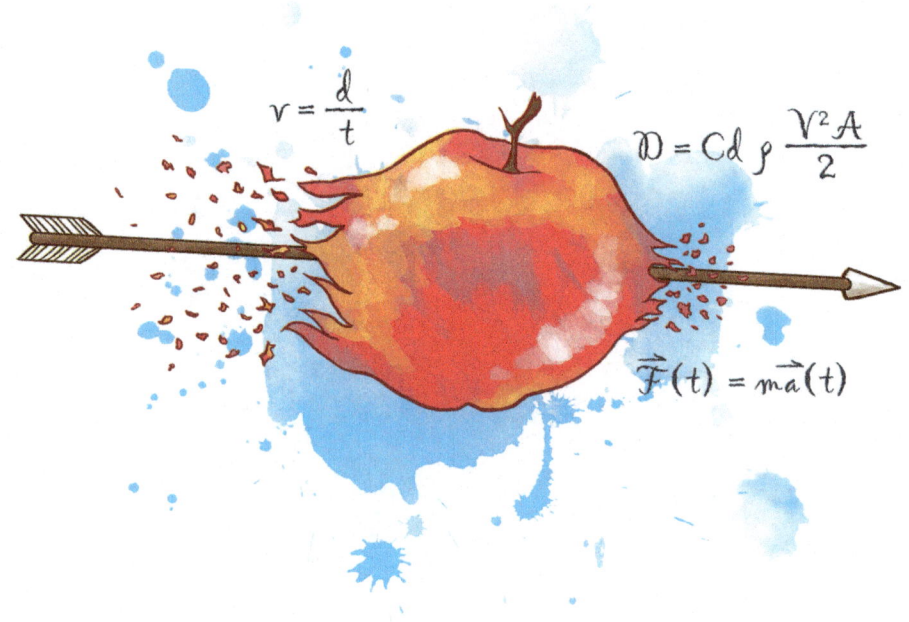

$$v = \frac{d}{t}$$

$$\mathcal{D} = Cd \, \rho \, \frac{V^2 A}{2}$$

$$\vec{\mathcal{F}}(t) = m\vec{a}(t)$$

Artwork by Andrea Price and Catherine Humphries

"In God we trust. All others must bring data."
– W. Edwards Deming

Chapter 20
A Perspective on the Future of Machine Learning: Moving Away from '*Business as Usual*' and Towards a Holistic Approach of Global Conservation

Grant R. W. Humphries and Falk Huettmann

20.1 The Future is Here

The year 2100 seems so far away for us that it's almost inconceivable to picture what life might be like then. In fact, very few people who were born before this book was published, will survive to that year. For some of us, the year 2100 may as well be 3000 years or even eons from now (i.e. when our sun is predicted to consume the planet). Therefore, it is hard for those of us currently alive to care about it. By the time 2100 arrives, three more generations of humans (assuming a generation is approximately 25 years) will come to be. Scientists have arbitrarily chosen 2100 as a benchmark for climate change model predictions (e.g., www.ippc.ch, Friedman 2010). Thus, natural resource managers and ecologists predicting distributions and populations into the future in relation to climate data are limited to using the same benchmark year.

We draw your attention to the fact that 2100 is a benchmark year for ecologists to highlight that technological advances, particularly those in machine learning, are going to be well beyond our current imagination by the time this comes around. This almost need not be said, but if you consider the fact that the first computer was created in the 1940s (just ~78 years before the publication of this book), and there are still ~82 years until 2100, the possibilities are actually mind-boggling. Ever since machine learning started becoming mainstream in the early 2000s, we have

G. R. W. Humphries (✉)
Black Bawks Data Science Ltd., Fort Augustus, Scotland
e-mail: grwhumphries@blackbawks.net

F. Huettmann
EWHALE Lab, Biology and Wildlife Department, Institute of Arctic Biology,
University of Alaska-Fairbanks, Fairbanks, AK, USA
e-mail: fhuettmann@alaska.edu

© Springer Nature Switzerland AG 2018
G. R. W. Humphries et al. (eds.), *Machine Learning for Ecology and Sustainable Natural Resource Management*, https://doi.org/10.1007/978-3-319-96978-7_20

developed software that can accurately classify voices, faces, and even determine which Harry Potter character we most resemble based on our social media posts. Consider also the use of robots, data sensors and drones which have massive implications not only for society, but also in ecological studies and subsequently, conservation management. These technologies rely on machine learning algorithms that are developing faster than we can keep up with. So where will that take us in the next 10, 50 or 100 years? Will humanity (and society) and landscapes still be the same? Having a concept of 'the future' in mind is essential.

Beyond these technical advances, which obviously have implications in how we do science, we have to ask a larger question: *What is the future of machine learning in ecology*? More specifically, what will its impact be on our understanding and conservation of 'Mother Earth', and its various systems?

As with nature and earth issues in general, what we don't know vastly outweighs what we do. We do know that the virtual economy (i.e. economic growth driven by technological change) and related concepts (Romer 1990) are likely not sustainable due to resource limitations (Daly and Farley 2010). We also know that humanity has become interwoven with technologies (driven by machine learning), which impacts every aspect of our lives, even in remote areas thanks to the advent of smart phones and wireless technologies. This means that virtually every landscape, its flora, fauna and indigenous communities are directly impacted (Huettmann 2015, *sensu* Czech 2000).

Arguably, massive changes on all fronts (hardware, software, human society, and the earth) may be expected when creating these new interactions and synergies. But we simply cannot predict what those interactions will look like for 'Mother Earth', or how they will affect her. While parsimony will only lead us to a single predictor, machine learning technologies will be central to how we study this moving forward due to their ability to learn and predict patterns as data evolve. But as we just previously mentioned, these technologies are limited by natural resources (e.g., rare earth metals) and their extraction. So, despite various claims (Romer 1990), the 'Information Economy' cannot grow forever and is not decoupled from the planet or her resources and ecology (Czech 2000). This is something that must be borne in mind as we move forward as a society and species. Work by Eichstaedt (2016) for the Congo, Buckley (2014) for the Tibetan Plateau, or Henton and Flower (2007) for Papua New Guinea demonstrate the impacts of the 'Information Economy' on human environments at multiple scales.

We are making great strides in using machine learning technologies, yet there are large conceptual and ideological bottlenecks to progress in ecology that cannot be ignored. The rise of nationalism, for example, which tends to lead to decreased interest in advancing ecological studies (i.e. through federal funding cuts), could lead to situations where ecologists fall too far behind the technology to be able to fully understand the potential impacts of machine learning on the planet. It is vitally important to overcome these bottlenecks through creative solutions to keep abreast of future advancements.

Below is a short list of research fields and technologies that are currently advancing thanks to machine learning, and which will have implications in ecology. We also provide examples of a few hypothetical possibilities for the future.

1. DNA research: disease screening is much faster, and entire genomes can now be analysed using big data techniques. Perhaps this, combined with new lab techniques, could lead to bringing back extinct species or at least provide early warning signs for human disease. If performed correctly, linking DNA information with spatial information (e.g. Landscape Genetics) can allow for a more inclusive approach when using DNA.
2. Taxonomy: related to DNA research, the study of entire genomes with machine learning algorithms (e.g. intelligent clustering) could lead to the re-definition of species and underlying concepts (Fernández et al. 2010).
3. Drones: the peaceful application and automation of drones is already advancing rapidly. As they become cheaper and automated algorithms for detecting animals are integrated, it could drastically improve how we perform wildlife surveys worldwide.
4. Robotics: related to drones to some degree, with automation using machine learning algorithms imminently due to take over certain activities normally carried out by humans (e.g., expert scientists, factory workers, ship operators, etc....). These could be ground-based robots that take samples of plants, or perhaps marine-based UAVs for sampling organisms in the ocean.
5. Databases: although databases have been used for many decades, machine learning has helped to build on, and add to, these technologies. Algorithms can query and populate databases in near-real time (i.e., data are uploaded quickly and can be accessed instantaneously by other after uploading), which could facilitate running near-real time species distribution models. Those databases can be complex and very big (i.e., 'Big' Data) but it works equally well for data gaps and small data sets. eBird and its machine learning approach is a good example of how these algorithms can be used to improve databases and inferences (Kelling et al. 2012).
6. Risk assessment: commonly used by insurance agencies and economics, these methods are getting more efficient and accurate. We see many applications applicable to ecology with outlier / anomaly detection as a showpiece, helping us to focus our conservation efforts.
7. Forecasting: Time series forecasting or forecasting of species distributions and densities is now common in ecology, but machine learning has greatly advanced our ability to do this accurately (Craig and Huettmann 2008; Taylor and Letham 2017). This is bound to improve with new statistical methods, and machine learning is likely to be central to this.

There are other social movements that are predicted to impact machine learning technologies as well, including: global availability of the internet (e.g., public WIFI), cheap computers, smaller microchips, faster internet connections (4G, 5G, etc....), changes in how we produce energy, and the push to reduce carbon emissions. We have to be aware of these developments, movements, and changes as they occur in order to ensure that we fully take into account our impacts on the environment into the future.

With this perspective in mind, there are still facets of wildlife and natural resource management that seem unprepared for these changes; getting stuck in and re-embracing older methodologies and technologies in the hope that the system and society will change to adopt their paradigms. So where do we go from such situations? We believe that the key is to re-visit some of the short-comings of where we are right now in order to determine the best strategies for moving forward. We are usually optimistic of the future, but to even begin to understand where we are going, we must know where we are and what is causing delays in moving towards new strategies.

20.2   Natural Resource Management (and Species Distribution Models)

For over five decades habitat selection studies have been the cornerstone for wildlife research, and natural resource management, particularly in the United States. A quick search on 'white-tailed deer *Odocoileus virginianus* and habitat studies' online will reveal a bewildering amount of literature on this topic (41,500 hits on Google Scholar; bearing in mind that not all of these will be on this exact subject, but will be somewhat related). This has the unfortunate consequence of terminology being blurred (e.g., Jones 2001, Silvy 2012, Kirk et al. 2018), or potentially conflicting arguments. Considering burgeoning white-tailed deer numbers, we have to question how effective the current habitat selection function techniques have been for wildlife management. Sophisticated research designs and analysis are still treated like the 'holy grail' (Johnson 1980 for habitat preference; Manly et al. 2002 for the original theory and quantitative methods at the time). Meanwhile, efforts by citizen science have vastly changed the playing field (e.g. Hochachka et al. 2007).

Habitat suitability indices (HSIs) were dominant in the literature and quite cutting edge in the absence of laptop or personal computers in the 1970s and 1980s. The use of HSIs attracted a lot of attention (Brooks 1997; Silvy 2012) and funding, leading to many updates in the methodology (e.g., Verner et al. 1986; Manly et al. 2002; Elith et al. 2006). But, by their very nature, HSIs are considered reductionist, and do not take the entire ecosystem into account. A continuation of this reductionistic thinking then came with Manly et al. (2002) who widely formulated and promoted the use of Resource Selection Functions (RSF)s, all based on (basic) computing and a controlled research design (see Eq. 1 below for an example RSF). Linked with GIS and computational statistics, in some governmental and federal contractor corners this overly simplistic and reductionistic approach then became the 'latest cry' in their species conservation management plans and for their solutions presented to the public audience on how to best manage wildlife. It was relatively easy to do in an environment that is neither in favor of technology, nor computing and certainly not of math. The North American performance metrics on basic mathematical knowledge express that clearly, and thus, simple formulas carry great weight in a public discussion.

$w(x) = e^{(\beta 1x1 + \beta 2x2 + ...\beta kxk)}$	(Manly et al. 2002; Johnson et al. 2004; Silvy 2012)	Eq. 1

Linked with the information-theoretic approach to decision-making (Burnham and Anderson 2002) RSFs became virtually mandated for federal agencies. It became a recommended method for charismatic megafauna like polar and grizzly bears, whales, etc.... (Manly et al. 2002 for examples). However, early work by Ferrier (2011 for review), Peterson et al. (2002) and Guisan and Zimmermann (2000) sowed the seeds of change in this culture. This change was likely slow perhaps because the initial works linking machine learning to spatial ecology were rooted in botany and not in charismatic species (see Huettmann and Diamond 2001 for an exception). However, this has been improving since that time (e.g., Tobeña et al. 2016).

It was left to a study within an international working group of the National Center for Ecological Analysis and Synthesis (NCEAS) in St. Barbara, California, by Elith et al. (2006) which probably focused the ecological world's attention on the use of advanced modeling methodologies. Still, the role of earlier model work by D.R.B. Stockwell, A.T. Peterson, M. Burgman, S. Ferrier, C. Moritz, A. Hirzel and the open access museum community should really be emphasized first, strongly leading to NCEAS and its studies and concepts. Judging by the many citations of this NCEAS effort and its related papers and outcome, Elith et al. (2006 and many other publications coming from it) truly represents a big leap forward demonstrating that machine learning methods were becoming the method of choice for wildlife-habitat investigations. It remains a seminal study to this very day and the concept is continually growing.

Machine learning methods like Maxent (maximum entropy; Phillips et al. 2006) were now on the rise, geared towards species distribution models (SDMs; Guisan and Zimmermann 2000; Guisan and Thuiller 2005). Maxent for SDMs became very popular due to it being a 'point & click' platform, which led to somewhat of a 'me too' type of approach to ecology (this type of 'me too' science was critiqued by O'Connor 2000). However, despite their widespread use, these point and click methods typically did not have code that could be shared, making them somewhat opaque, scientifically. Also, it is somewhat difficult to report on all the settings used within the graphical user interface, especially during the initial project phases when most scientists forget to record such details (this is a reference to poor data management skills as well, commonly found in ecology). This is changing though as programming languages become more commonly used by ecologists (see https://github.com/julienvollering/MIAmaxent for an open source R implementation of maximum entropy modeling, and Phillips et al. 2017 for the now open sourced code for Maxent). However, to date there is still a needed change of culture away from the typical SDM template which do not adhere to true machine learning philosophies (see Breiman 2001a).

We think that the lessons from the seminal Breiman (2001a) statistical culture publication are not yet strictly adhered to in SDMs because:

a) the specific compiled input data are not always shared open access,
b) the specific code that was run are not shared in relevant detail,
c) the output prediction layer is rarely made available for an assessment and use, and worst of all
d) the study is rarely linked to effective conservation management in that machine learning is not referred to in most policy or legal (court) decisions.
e) these studies widely lack a reflective component which advances ethical and societal questions

Despite this, machine learning (often linked with Artificial Intelligence), has become a much more commonly referenced term in the scientific literature. However, there is much more work to be done to fully incorporate creative and scientifically sound applications of machine learning. For example, examining new and upcoming algorithms (e.g., Prophet; Taylor and Letham 2017, or convolutional neural nets; LeCun and Bengio 1995) could lead to new insights into ecosystem scale (i.e. holistic) modeling.

20.3   Falling 'out of love' with your Model to Help Develop New Methodologies and Insights

Many modelers (and programmers) tend to fall in love with their models, algorithms, code, and insights, even in cases where they are not relevant for conservation (McArdle 1988; Keating and Cherry 2004; Guthery 2008; Stephens et al. 2007; Arnold 2010; Faraway 2016). Meanwhile, the destruction of nature continues unabated in many places, especially hotspots of biodiversity like Asia, Africa, and South America (Mace et al. 2010; Pimm et al. 2014). Falling in love with a method can potentially lead to costly, drawn out debates in public fora and elsewhere (e.g., published rebuttals). Science and progress entails debate. It means that, apart from being teachers, mentors, researchers, programmers, and writers, ecologists must also be part- time debaters (see Silvy 2012). This can be exhausting and demoralizing and can often end up with one side simply conceding to the fact that "nothing can be done". Although scientific debate is important, and certainly there is merit in being able to defend a methodology, oftentimes, debates and arguments lie around issues with statistical assumptions, or problems with parametric approaches. Issues like significance (p-values; which are beginning to fall out of favor in the ecological community due to their various drawbacks), linear model fits, data distributions (e.g., the assumption of normal distributions which are hardly ever found in ecological data; McArdle 1988; Breiman 2001a) and localized optima problems can drastically sideline projects and publication. This is where machine learning techniques can be of great value and provide progress.

As you have learned throughout this book, machine learning algorithms have the advantage of being flexible across a range of different datasets while providing optimal results against traditional parametric statistics. They are also independent of issues with *a priori* statistical assumptions about the parameters in the model. Not having to worry about how you have transformed your data (and whether you should have done so in the first place), as well as eliminating problems around significance and information criterion model selection, means that the ecologist can spend significantly more time on inference and applying his or her findings for conservation (if applicable). However, after spending time on a project that leads to a publication (or two), it is understandable that one might get attached to the process and then continue to apply it. Unfortunately, this attitude leads to several arguments against using new methods (like machine learning). Below we list briefly some of these arguments, and our rebuttals to them (Table 20.1).

Table 20.1 Some brief arguments made against using machine learning algorithms in our experience, and our brief rebuttals

Argument	Rebuttal
"This is too much of a 'black box'"	All the sub-algorithms that make up algorithms like random forests and boosted regression trees are well described in the literature. We would argue that machine learning algorithms are transparent if you spend the time to dig into the methods.
"The methods are very new and untested"	Machine learning methods have been around since Alan Turing in the 1950s, and basic implementations of algorithms have been used in search engines since the early 2000s.
"My supervisor does not want it"	Graduate students are important for pushing the boundaries of science, and if they are keen to use machine learning methods it behooves them to try the methods and convince their supervisors of the merits.
"We lack the software"	Free software exists online, and some paid-for software have free (and sufficient) trial periods, and most algorithms can be implemented via Python or R with great resources online.
"Machine learning is not sufficient for inference"	As we have demonstrated in this book, and as seen throughout the machine learning literature, inference is possible, and can be powerful (i.e., it can be more holistic than other methods).
"My agency does not allow using machine learning"	This is an unfortunate consequence of policy legislation that has yet to be updated. This falls on government scientists to be willing to put in ground work to demonstrate the power of machine learning.
"Predictions are not our goal"	Generalizations are best achieved through strong predictions. See Breiman (2001a) for discussions on how to perform the best possible inferences. And as mentioned above, powerful inference can be made through machine learning.
"We just want to stay with the mainstream analysis methods"	Machine learning is mainstream in many disciplines and is a cornerstone for many industries (e.g., social media, advertising, bioinformatics).
"Our clients asked us to use commonly applied methods"	This relates to legislation and is the responsibility of government scientists to demonstrate the benefits of machine learning. This also falls to analysis providers to demonstrate potential power of machine learning algorithms.

So, how does one go about 'falling out of love' with a preferred method? In our opinion, the best way is to try several methods yourself and then assess their performance. There are many good tutorials on the subject that can be found on the internet with a simple search. The programming language, R, is constantly improving and making it simpler to apply machine learning techniques to datasets, and most universities offer courses or have tutors willing to help with such matters. Other programming languages like Python or java will be equally insightful. Benchmarking methods (i.e., modeling platforms) against each other is the best way to convince yourself of the merits of particular techniques and helps to advance scientific processes and innovation.

20.4 Innovations in Machine Learning and Innovative Uses that can Benefit Ecological Studies

The true depth of how machine learning affects our daily lives is not well understood by the average person on the street. However, machine learning is used so frequently that essentially everyone in the world is affected by it daily and will be from now onwards (or at least until the downfall of human civilization and computing). The massive push of machine learning has been so successful (due to the power of these algorithms for prediction and inference), that an entire industry and even sub-industries have formed around it. This means that there are billions of dollars (or Euros, Yens or Pounds sterling) being thrown into the industry which is driving innovation in ways which we are probably not yet prepared for as a society. Unfortunately, wildlife science (and several other fields) are falling behind the curve when it comes to innovative uses of these technologies. In Table 20.2, we list some machine learning concepts and references that ecologists could find handy.

Despite the plethora of machine learning tools that have shown great promise in other fields, wildlife and ecology textbooks still lack any relevant mention of these methods (Caughley and Sinclair 1994; Krausman 2002; Gergel and Turner 2001; Moyes and Schulte 2007; Primack 2010; Silvy 2012). Furthermore, there is a serious disconnect between ecology and the latest computational techniques. Although this is changing to some degree with some quantitative ecology courses being taught in R or Python, there is still much work to be done to bring the field up to speed. Unless the new generation embraces some of these new techniques, things are unlikely to change when it comes to natural resource management, which is arguably failing, e.g. Mace et al. (2010).

We have presented several machine learning concepts in this book that are somewhat 'new' to ecology (within the last decade or so) that could offer some new insights into natural resource management and conservation (e.g., image recognition; Chap. 15, classification of species by call; Chap. 16, ensembles of ensembles; Chaps. 5 and 6). However, these types of analyses are commonplace outside of ecology, but we expect them to be used more and more in this field in the near future

Table 20.2 List of relevant machine learning developments that could be useful for Ecological studies; many of them are referred to and applied in this publication

ML Development	Reference	Description	Impact	Comment
Automated linear regression analysis	Harrel (2001), Mac Nally (2000)	LM and GLM	Modern linear statistics	Those methods are tempting but are not in line with machine learning philosophies
	Friedman (1991) Salford Systems Ltd. (2017)	MARS, LASSO	More advanced analysis methods	SPLINES can outcompete linear regressions and are easy to interpret
	Elder IV (2003)	Ensembles	Comparative and context	Probably one of the biggest steps forward in machine learning
Neural network	Rumelhart et al. (1986)	Backward feed, back-propagation	NN became an accepted standard tool	The standard for neural networks. Still rarely used in wildlife studies
	Cortes and Vapnik (1995)	Support vector machines	Powerful use of predictions	Powerful but not widely used in wildlife studies
Recursive partitioning	Breiman et al. (1984)	CART	Introduction to a new form of inference	The basis for both random forests and boosted regression trees
	Antipov and Pokryshevskaya (2010)	Multi-path split CARTs (CHAID)	An alternative to binary CARTs	Not widely used but intriguing approach.
Boosting	Freund and Schapire (1997)	ADABOOST	Introduction to a new form of inference	Can be used to boost a variety of modeling approaches
	Drucker (1997)	Boosted neural networks	An intriguing combination of concepts toward powerful outcomes	Virtually unused in wildlife studies
	Friedman (2002)	TreeNet	Powerful use of predictions	TreeNet invokes boosting and (basic) bagging concepts

(continued)

Table 20.2 (continued)

ML Development	Reference	Description	Impact	Comment
Bagging	Breiman (2001b)	Random forests	Introduction to a new form of inference	An ensemble of CART trees and shown to be quickly becoming a favorite in ecological studies
	YAIMPUTE (Crookston and Finley 2008)	Imputation	Allows innovative approaches to multi-response and gap imputations	Method that can suggest or fill data gaps are in big demand
Entropy	Shannon (1949)	Entropy	Introduction to a new form of inference	A basic and early concept for quantification, information theory and machine learning
	Phillips et al. (2006)	Maxent	Powerful use of predictions	One of the most popular algorithms in SDMs, thus far
Ensembles	BIOMOD package (Thuiller et al. 2009)	Ensembles	An R code that combines several model (standard) algorithms	A standard package that performs 'standard' models. (lack of fine-tuning). Difficult to use due to documentation.
	Lieske et al. (Chap. 5), Salford Systems Ltd. (2017)	Ensembles of ensembles	Ensembles can be run and combined with many algorithms	An open and developing field, very close to deep learning and holistic analysis
	Coulouris et al. (2011)	Distributed computing	Ensembles linked in the cloud	Probably one of the most powerful and global applications, e.g., when including sensors. Various options exist.

(the general concepts here are in agreement with conclusions made in Cushman and Huettmann 2010; Drew et al. 2011; and Hochachka et al. 2007). In Table 20.3, we list a few machine learning innovations that will likely play a vital role in upcoming uses of these algorithms.

Implementing many of these methods offer ecologists an advantage because it shifts many of the issues away from the bottleneck of quantitative debate (e.g., *a priori* statistical assumptions), and towards the policy and applications of findings. Combining these with concepts of holistic ecological modeling could lead to real advancement in this field and to new information.

Table 20.3 Some innovations in machine learning that will likely become more mainstream in ecology

Innovation	Reference	Impact	Comment
Weighting	Salford Systems Ltd. (2017). 'Ranger' (Wright and Ziegler 2015)	Unbalanced data allowing users to find 'outliers'	This appears like a small innovation but will allow users to be more flexible in defining classes for prediction
Batteries	Salford Systems Ltd. (2017), Huettmann (Chap. 8)	Allows one to test for best predictors, an alternative to model selection	Virtually unused in wildlife conservation (see Chap. 12 for an example)
Swapping	Salford Systems Ltd. (2017)	Removes uncertainty in classic approaches (LMs, AIC) and allows one to truly test for robust predictors (=model selection problem)	A subset of batteries; It's a massive change in how we understand and infer on predictors, predictions and multiple regressions
Ensembles of ensembles	Lieske et al. (Chap. 5), Salford Systems Ltd. (2017), BIOMOD (Thuiller et al. 2009) and 'caret' (Kuhn 2008)	Automated and autonomous optimizations for classifications, predictions and inference	Probably the latest 'cry' in predictions and for inference. Likely, a research scheme to come for the next decade
Distributed computing	Coulouris et al. (2011)	Ensembles linked in the cloud	Probably one of the most powerful and global applications, e.g. when sensors are included.

20.5 Machine Learning and 'Mother Earth': Towards a holistic Understanding of Ecosystems for Better Conservation and Ethics

Ecology is complicated and convoluted, which makes quantifying systems a challenging task. We, as ecologists, generally agree that the reason for this is because environmental factors interact in ways that we cannot predict or quantify well, or where this concept can easily be misleading. This should go beyond just environmental factors, however, with anthropogenic forces affecting these natural phenomena and their interactions. Basically, this equates to something like the 'gaia' hypothesis and 'deep ecology', which states that all things in the natural environment are connected, and that once life exists it will try to remain (Lovelock and Margulis 1974; Naess 1989). Even if you don't accept this ideology, there is no doubt that there are still many connections/relationships in ecological systems that are too complex for standard linear or parametric techniques. Machine learning can help here by allowing us to make predictions and inference in complex systems; a more holistic form of modeling. It further provides progress on the unresolved questions of gaia and of 'Mother Earth'.

Machine learning is generally at its strongest when applied using the philosophies of the originators of some of the more frequently used techniques (e.g. *"many weak learners provide for a strong learner"* Freund and Schapire 1997; Breiman 1998;

Friedman 2002). The basic idea behind this is that instead of limiting ourselves to a few covariates in the model (in marine sciences these tend to be predictors like sea surface temperature, salinity, bathymetry and chlorophyll a), we should include a plethora of variables that may or may not be related to what we "know" about a system. We demonstrate a few examples of including many variables (more than what would be typically included in parametric methods) into a model in this book (Chaps 10, 12, 13, and 16). As ecologists, this might seem counterintuitive due to our "obsession" with mechanisms, as taught at good academic institutions. However, the approach we propose can lead to more powerful and holistic predictions (by considering those complex relationships between parameters), and drive hypothesis generation. For example, finding a relationship that was unexpected could lead to interesting and fundamental scientific discussions, and satisfy some of our desire to understand mechanistic relationships. However, this powerful aspect is still widely ignored (see Guisan and Zimmermann 2000; Guisan and Thuiller 2005; or Franklin and Miller 2009). Being all inclusive can play a powerful role in understanding whole ecosystems, but for now, no ethical guidance or 'best professional' practices really exists for machine learning applications, certainly not in the context of natural resource conservation management (see Huettmann 2007 and Zuckerberg et al. 2011 for data and metadata) (Fig. 20.1).

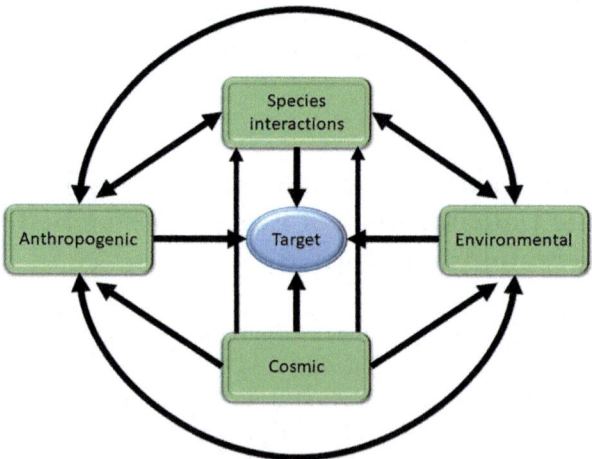

Fig. 20.1 Four broad categories of factors that are interconnected and that deserve incorporation into predictions of a target problem. Anthropogenic (socio-economic) forces include things like infrastructure (roads, buildings), pollution (e.g., sulfates, plastics) and burning of fossil fuels; species interactions include competition and predator/prey dynamics (or plants for herbivores); environmental factors include available habitat, weather conditions, and climate; cosmic forces include those often termed "environmental", but are from beyond our planet (e.g., solar phase, proximity of the moon, solar storms, meteor strikes, day phase, season). Those are very relevant, and one could easily argue that many environmental factors are actually very reflective of cosmic forces, thus leading us to not need inclusion of such parameters. However, in the absence of data on certain environmental forces, including well measured cosmic parameters might act as a good alternative (including solar cycles, moon stages, seasons and multi-annual cycles will often create fascinating and very relevant discussions and predictors for instance)

From a conceptual perspective, we believe that machine learning can be applied to whole ecosystems by first focusing on older data sets that may exist in the public domain that are often deemed 'too messy' to be included in traditional analyses. Secondly, trying out what some might call "the shotgun approach" (a.k.a. the 'kitchen sink' model) within a machine learning framework and using cross-validation to determine methods that create the best predictions (including ensemble models). Thirdly, assessing potential interactions, relationships and inferences from complex covariate structures to develop new and interesting hypotheses that could be tested (experimentally or otherwise). These techniques have the potential to deal with the fact that in the real world, data are messy, research designs are almost always incomplete while results are time-sensitive (concepts explained in Breiman 2001a, with examples in Herrick et al. 2013 and Wei et al. 2010), which will help move forward with strong conservation measures.

20.6   Where to Go from Here? Integrating Machine Learning and Good Ethics into your Scientific Process and Sustainability Policy

As shown throughout our book, the powerful predictions can actually be used to make inferences and generalize mechanistic understanding of systems (as per Breiman 2001a). We firmly believe that valid inferences are best made when using models that achieve powerful predictions on independent datasets. Machine learning methods can fill this niche, but those styles of research designs and techniques have yet to be fully adopted in natural resource conservation management and as 'best professional practices' (Silvy 2012). The best way forward here is simply to encourage the use of open access code, data and research to make these methods more transparent for other scientists.

First, we emphasize that predictions can be created from virtually any data and any research design, and the easiest covariates to obtain are those that exist in the public domain. This means that once a research question is identified, a next step would be to download (and document) variables from the internet. Many datasets exist, and we leave this up to the reader to decide where best to obtain them. For environmental variables, NOAA and NASA have some open access layers that marine scientists anywhere in the world might find useful. The European Center for Medium-Range Weather Forecasting (ECMWF) also hosts open access atmospheric data that might of interest to ecologists. Other governments (e.g., Australia and New Zealand) provide similar datasets relevant to their regions. If your readers know where your data come from, any modeling experiment can be replicated (or critiqued if needed).

Once data grooming is completed (i.e. getting rid of errors and potentially anomalous values, etc...) then the modeling process can begin. Unlike in parametric models (where, as we have noted, much time spent on dealing with statistical assumptions), we can immediately begin the modeling process in most cases.

Typically, an independent subset of the data are held back to predict onto in order to get a benchmark model. Keeping as many predictor variables as possible in this solution is a good way to do this as variables can be pared away (if desired, though not recommended if working within the Breiman 2001a philosophy and approach). From here, it is a matter of trying other settings, combining models, and in the end, deciding on which performed best against the independent test set. That is the science we promote. This differs from common parametric techniques because there are no "goodness of fit" tests really because no data are fit in the explicit statistical sense for model-based inference. Furthermore, it frees us from the shackles of p- and AIC values and gives us the ability to start thinking about hypotheses and better inference while at the same time having the best possible predictions handy.

In the classical sense, hypotheses should not ignore relevant concepts and biases, e.g. for ecology and conservation see Naess (1989), Bandura (2007) and Rosales (2008). Further, hypotheses can and should be predicted to test if they are generalizable and thus, supported (as per approach expressed by Leo Breiman 2001a). Predictions assessed against independent data, give us a much better idea of the accuracy of a model. This differs from the more traditionally used p-values, or AIC. This tiny detail pushes us into a new frontier in ecology though, where we can rigorously test our hypotheses by pitting them against many other ideas derived from machine learning applications.

In our work, we specifically promote the use of machine learning methods as a rigorous, often automated, prediction machine (e.g., Kandel et al. 2015 for an example). It can be done as a black-box, or otherwise, as a coded (from scratch) concept. Once predictions achieved then the process can move forward and further test mechanisms, e.g. in the field or with experiments and hypothesis-testing, and one can even incorporate p-values and AIC later, if one really wishes to do so. In times where conservation is time-sensitive, machine learning easily gets us a first rapid assessment. So why not use it?

Following this logic is the basic concept behind machine learning that we promote, where we know little or nothing of the system before we begin, find the best predictive model, and then make inferences from that (see Kandel et al. 2015 for a workflow). Any algorithm can be used to do so, as long as it predicts well (*sensu* Breiman 2001a, Elith et al. 2006). The basic concepts of stiff, stringent and untested parametric assumptions (Zar 2010) are replaced with a flexible prediction machine, regardless of r^2 values (Hastie et al. 2009). It is here where the power really lies for ecology.

How realistic is it to have all the basic statistical procedures following Zar (2010) and Burnham and Anderson (2002) replaced with open source machine learning code? Or, and based on input and output data that are predominately published open access in public repositories and web-portals? We believe we are fully heading in that direction. A culture change is coming in ecology, where the *p*-value is likely to become less important. And fortunately, we feel that this will happen quickly as a new generation of ecologists have started embracing machine learning concepts. A change of the global framework and 'how things are done' is easily possible in conservation. This has happened before, as the discipline of ecology itself frequently shows, e.g. Naess 1989, Begon et al. (2005).

20.7 Some Final Thoughts

Globalization in the Anthropocene, to the detriment of wildlife and its habitat, has been in full swing for four decades (arguably more) and it is clear by now that we have entered a widely acknowledged crisis-state (Rosales 2008; Mace et al. 2010). Solutions to global issues are time-sensitive, and complex, and we need techniques that can adapt to a fast-changing global ecosystem. Marginalizing these problems is the last thing that should be done.

Throughout our book, we have guided you, the reader, through a number of useful, practical applications of common ML techniques that are being adopted in the ecological community. Many of these examples revolve around using ML techniques to analyse patterns in space and time which are vital to global conservation efforts. There are two reasons for this: (1) Spatial ecology is the field that most commonly uses ML algorithms for prediction and is thus well developed in this regard. and (2) Holistic approaches to conservation must include spatial information whenever possible (hence why it's termed a 'holistic' approach). Despite this, however, there are many other useful applications of machine learning that are only recently becoming known to the ecological community. For example, deep learning techniques (deriving from machine learning) are now advancing and allowing ecologists to identify individual animals (e.g. right whales; *Eubalaena glacialis*; https://www.kaggle.com/c/noaa-right-whale-recognition). We also demonstrate how we can use the machine learning technique that we are most familiar with (e.g., random forests; Breiman 2001b, and boosted regression trees; Friedman 2002) to derive hypotheses, and how we can integrate all these techniques with 'the cloud' and web-based applications (see Sect. 20.5).

In all we are hopeful for the future of ecology as technology is advancing and scientists are learning more about new applications and helping to redefine ecology as a discipline. Although there is a lot of work to be done before ML techniques are fully accepted, we urge caution moving forward. The digital sword that we reference in Chap. 17 is a powerful weapon that can be easily misused to the detriment of the environment if the techniques are not properly applied, and over-used (e.g. massive data centers, etc...which could lead to issues around energy development in the future). But if used carefully, with a holistic approach in mind for conservation purposes, ML algorithms have great potential. Ecologists can look beyond a single study system and start applying ecosystem scale management and precautionary measures which will have massive implications on how we leave the planet for future generations.

References

Antipov E, Pokryshevskaya E (2010) Applying CHAID for logistic regression diagnostics and classification accuracy improvement. J Target Meas Anal Mark 18:109–117

Arnold TW (2010) Uninformative parameters and model selection using Akaike's information criterion. J Wildl Manag 74(6):1175–1178

Bandura A (2007) Impeding ecological sustainability through selective moral disengagement. Int J Innovation Sust Dev 2:8–35

Begon MC, Townsend R, Harper JL (2005) Ecology: from individuals to ecosystems, 4th edn. Wiley-Blackwell Publishers, Oxford

Breiman L (1998) Arcing classifier (with discussion and a rejoinder by the author). Ann Stat 26:801–849

Breiman L (2001a) Statistical modeling: the two cultures (with comments and a rejoinder by the author). Stat Sci 16:199–231

Breiman L (2001b) Random forests. Mach Learn 45:5–32

Breiman L, Friedman J, Stone CJ, Olshen RA (1984) Classification and regression trees. Taylor & Francis, New York

Brooks RP (1997) Improving habitat suitability index models. Wildl Soc Bull 25:163–167

Buckley M (2014) Meltdown in Tibet: China's reckless destruction of ecosystems from the highlands of Tibet to the deltas of Asia. St Martin's Press, Vancouver

Burnham KP, Anderson DR (2002) Model selection and multimodel inference: a practical information-theoretic approach. Springer, New York

Caughley G, Sinclair ARE (1994) Wildlife ecology and management. Blackwell Science, Oxford/Boston

Cortes C, Vapnik V (1995) Support-vector networks. Mach Learn 20:273–297

Coulouris GF, Dollimore J, Kindberg T, Blair G (2011) Distributed systems: concepts and design, 5th edn. Addison-Wesley Boston, Boston

Craig E, Huettmann F (2008) Craig E, Huettmann F (2008) In: Wang H-f (ed) Using "blackbox" algorithms such as TreeNet and random forests for data-mining and for finding meaningful patterns, relationships and outliers in complex ecological data: an overview, an example using golden eagle satellite/developing new methodologies through pattern discovery and recovery. IGI Global, Hershey, pp 65–83

Crookston NL, Finley AO (2008) yaImpute: an R package for kNN imputation. J Stat Softw 23:1–16

Cushman S, Huettmann F (2010) Future and outlook: Where are we, and where will the spatial ecology be in 50 years from now ? In: Cushman S, Huettmann F (eds) Spatial complexity, informatics and wildlife conservation. Springer, Tokyo, pp 445–450

Czech B (2000) Shoveling fuel for a runaway train: errant economists shameful spenders, and a plan to stop them all. University of California Press, Berkeley/California

Daly HE, Farley J (2010) Ecological economics: principles and applications. Island Press

Drew CAYW, Huettmann F (eds) (2011) Predictive species and habitat modeling in landscape ecology. Springer, New York, pp 45–70

Drucker H (1997) Improving regressors using boosting techniques. In: Fisher DH Jr (ed) Proceedings of the Fourteenth international conference on machine learning. Morgan-IGmfmunn, pp 107–I 15

Eichstaedt P (2016) Consuming the Congo: war and conflict minerals in the world's deadliest place. Chicago Review Press, Chicago

Elder JF IV (2003) The generalization paradox of ensembles. J Comput Graph Stat 12:853–864

Elith J, Graham CH, Anderson RP, Dudík M, Ferrier S, Guisan A, Hijmans RJ, Huettmann F, Leathwick JR, Lehmann A, Li J, Lohmann LG, Loiselle BA, Manion G, Moritz C, Nakamura M, Nakazawa Y, Overton JMM, Peterson AT, Phillips SJ, Richardson K, Scachetti-Pereira R, Schapire RE, Soberón-Mainero J, Williams S, Zimmermann NE (2006) Novel methods improve prediction of species' distributions from occurrence data. Ecography 29:129–151

Faraway JJ (2016) Extending the linear model with R: generalized linear, mixed effects and nonparametric regression models, vol 124. CRC press, Boca Raton

Fernández A, García S, Luengo J, Bernadó-Mansilla E, Herrera F (2010) Genetics-based machine learning for rule induction: state of the art, taxonomy, and comparative study. IEEE Trans Evol Comput 14(6):913–941

Ferrier S (2011) Extracting more value from biodiversity observations through integrated modeling. Bioscience 61:96–97

Franklin J, Miller JA (2009) Mapping species distributions: spatial inference and predictions. Cambridge University Press, Cambridge

Freund Y, Schapire RE (1997) A decision-theoretic generalization of on-line learning and an application to boosting. J Comput Syst Sci 55(1):119–139

Friedman JH (1991) Multivariate adaptive regression splines. Ann Stat 19:1–67

Friedman JH (2002) Stochastic gradient boosting. Comput Stat Data Anal 38:367–378

Friedman G (2010) The next 100 years: a forecast for the 21st century. Anchor Books, Random House, New York

Gergel S, Turner MG (2001) Learning landscape ecology. Springer, New York

Guisan A, Thuiller W (2005) Predicting species distribution: offering more than simple habitat models. Ecol Lett 10:993–1009

Guisan A, Zimmermann NE (2000) Predictive habitat distribution models in ecology. Ecol Model 135:147–186

Guthery FS (2008) Statistical ritual; versus knowledge accrual in wildlife science. J Wildl Manag 72:1872–1875

Harrell FE Jr (2001) Regression modeling strategies: with applications to linear models, logistic regression, and survival analysis. Springer, New York

Hastie T, Tibshirani R, Friedman J (2009) The elements of statistical learning: data mining, inference, and prediction. New York, Springer Series in Statistics

Henton D, Flower A (2007) Mount Kare gold rush: Papua New Guinea 1988–1994, as told by Andi Flower. Mt Kare Gold Rush, Queensland

Herrick KA, Huettmann F, Lindgren MA (2013) A global model of avian influenza prediction in wild birds: the importance of northern regions. Vet Res 44(1):42

Hochachka W, Caruana R, Fink D, Munson A, Riedewald M, Sorokina D, Kelling S (2007) Data mining for discovery of pattern and process in ecological systems. J Wildl Manag 71:2427–2437

Huettmann F (2007) Modern adaptive management: adding digital opportunities towards a sustainable world with new values. Forum Pub Policy: Clim Change Sustain Dev 3:337–342

Huettmann F (2015) On the relevance and moral impediment of digital data management, data sharing, and public open access and open source code in (tropical) research: the Rio convention revisited towards mega science and best professional research practices. In: Huettmann F (ed) Central American biodiversity: conservation, ecology, and a sustainable future. Springer, New York, pp 391–418

Huettmann F, Diamond AW (2001) Seabird colony locations and environmental determination of seabird distribution: a spatially explicit seabird breeding model in the Northwest Atlantic. Ecol Model 141:261–298

Johnson DH (1980) The comparison of usage and availability measurements for evaluating resource preference. Ecology 61:65–71

Johnson CJ, Seip DR, Boyce MS (2004) A quantitative approach to conservation planning: using resource selection functions to map the distribution of mountain caribou at multiple spatial scales. J Appl Ecol 41:238–251

Jones J (2001) Habitat selection studies in avian ecology: a critical review. Auk 118:557–562

Kandel K, Huettmann F, Suwal MK, Regmi GR, Nijman V, Nekaris KAI, Lama ST, Thapa A, Sharma HP, Subedi TR (2015) Rapid multi-nation distribution assessment of a charismatic conservation species using open access ensemble model GIS predictions: red panda (*Ailurus fulgens*) in the Hindu-Kush Himalaya region. Biol Conserv 181:150–161

Keating KA, Cherry S, Lubow (2004) Use and interpretation of logistic regression in habitat-selection studies. J Wildl Manag 68(4):774–789

Kelling S, Gerbracht J, Fink D, Lagoze C, Wong WK, Yu J, Damoulas T, Gomes C (2012) A human/computer learning network to improve biodiversity conservation and research. AI Mag 34(1):10

Kirk DA, Park AC, Smith AC, Howes BJ, Prouse BK, Kyssa NG, Fairhurst EN, Prior KA (2018) Our use, misuse, and abandonment of a concept: whither habitat? Ecol Evol. https://doi.org/10.1002/ece30.3812

Krausman PR (2002) Introduction to wildlife management: the basics. Prentice Hall, Upper Saddle River

Kuhn M (2008) Caret package. J Stat Softw 28(5):1–26

LeCun Y, Bengio Y (1995) Convolutional networks for images, speech, and time series. The handbook of brain theory and neural networks 3361(10):1995

Lovelock JE, Margulis L (1974) Atmospheric homeostasis by and for the biosphere: the Gaia hypothesis. Tellus 26:2–9

Mac Nally R (2000) Regression and model-building in conservation biology, biogeography and ecology: the distinction between – and reconciliation of – 'predictive' and 'explanatory' models. Biodivers Conserv 6:655–671

Mace G, Cramer W, Diaz S, Faith DP, Larigauderie A, Le Prestre P, Palmer M, Perrings C, Scholes RJ, Walpole M, Walter BA, Watson JEM, Mooney HA (2010) Biodiversity targets after 2010. Curr Opin Environ Sustain 2:3–8

Manly FJ, McDonald LL, Thomas DL, McDonald TL, Erickson WP (2002) Resource selection by animals: statistical design and analysis for field studies, 2nd edn. Kluwer Academic Publishers, Dordrecht

McArdle BH (1988) The structural relationship: regression in biology. Can J Zool 66(11):2329–2339

Moyes CD, Schulte PM (2007) Principles of animal physiology, 2nd edn. Pearson Publishers

Næss A (1989) Ecology, community and lifestyle. Cambridge University Press, Cambridge

O'Connor R (2000) Why ecology lags behind biology. The Scientist 14:35–36

Peterson AT, Ortega-Huerta MA, Bartley J, Sánchez-Cordero V, Soberón J, Buddemeier RH, Stockwell DRB (2002) Future projections for Mexican faunas under global climate change scenarios. Nature 416:626–629

Phillips SA, Anderson RP, Schapire RE (2006) Maximum entropy modeling of species geographic distributions. Ecol Model 190:231–259

Phillips SJ, Anderson RP, Dudík M, Schapire RE, Blair ME (2017) Opening the black box: an open-source release of Maxent. Ecography 40:887–893

Pimm SL, Jenkins CN, Abell R, Brooks TM, Gittleman JL, Joppa LN, Raven PH, Roberts CM, Sexton JO (2014) The biodiversity of species and their rates of extinction, distribution, and protection. Science 344(6187):1246752

Primack R (2010) Essentials of conservation biology, 5th edn. Sinauer Associates Inc., Sunderland

Romer P (1990) Endogenous technological change. J Pol Econ 98:S71–S102

Rosales J (2008) Economic growth, climate change, biodiversity loss: distributive justice for the global north and south. Cons Biol 22:1409–1417

Rumelhart DE, Hinton GE, Williams RJ (1986) Learning representations by back-propagating errors. Nature 323(6088):533–536

Salford Systems Ltd (2017) Salford predictive modeler [version 8.2] https://www.salford-systems.com/

Shannon CE (1949) Communication in the presence of noise. Proc IRE 37(1):10–21

Silvy NJ (2012) The wildlife techniques manual: research & management, vol 2, 7th edn. The Johns Hopkins University Press, Baltimore/Maryland

Stephens PA, Buskirk SW, Hayward GW, Martinez del Rio C (2007) A call for statistical pluralism answered. J Appl Ecol 44:461–463

Taylor SJ, Letham B (2017) Forecasting at scale. The American Statistician

Thuiller W, Lafourcade B, Engler R, Araújo MB (2009) BIOMOD–a platform for ensemble forecasting of species distributions. Ecography 32(3):369–373

Tobeña M, Prieto R, Machete M, Silva MA (2016) Modeling the potential distribution and richness of cetaceans in the Azores from fisheries observer program data. Front Mar Sci 3:202

Verner J, Morrison ML, Ralph CJ (1986) Wildlife 2000. Modeling habitat relationships of terrestrial vertebrates. University of Wisconsin Press, Madison

Wei C-L, Rowe GT, Escobar-Briones E, Boetius A, Thomas S, Julian Caley M, Soliman Y, Huettmann F, Fangyuan Q, Yu Z, Roland Pitcher C, Haedrich RL, Wicksten MK, Rex MA, Baguley JG, Sharma J, Danovaro R, MacDonald IR, Nunnally CC, Deming JW, Montagna P,

Lévesque M, Weslawski JM, Wlodarska-Kowalczuk M, Ingole BS, Bett BJ, Billett DSM, Yool A, Bluhm BA, Iken K, Narayanaswamy BE, Romanuk TN (2010) Global patterns and predictions of seafloor biomass using random forests. PLoS One 5(12):e15323

Wright MN, Ziegler A (2015) ranger: a fast implementation of random forests for high dimensional data in C++ and R. arXiv preprint arXiv:1508.04409

Zar JH (2010) Biostatistical analysis, 5th edn. Prentice Hall, Upper Saddle River

Zuckerberg B, Huettmann F, Friar J (2011) Proper data management as a scientific foundation for reliable species distribution modeling. Chapter 3. In: Drew CA, Wiersma Y, Huettmann F (eds) Predictive species and habitat modeling in landscape ecology. Springer, New York, pp 45–70

This piece by Gunnar Brehm shows just a tiny fraction of the incredible life that is at risk on our planet. We propose that machine learning approaches applied to ecology and natural resource management as shown throughout our book can be used to promote and maintain the natural beauty of our world for the sake of our global well-being and happiness as well as world peace and sustainability... for the only place in the universe that is currently known to carry life

Index

A

Aboveground forest biomass
 biomass data, 144
 boreal forest (*see* Boreal forest)
 boreal region, 141
 carbon exchange, 141
 calibration and validation
  datasets, 147
 dry mass portion of live trees, 141
 ecologically and economically, 143
 energy demands, 142
 environmental factors, 145–147, 152
 FIA, 143
 forest inventory data, 143
 investigations, 141
 maps, 143
 NDVI, 155, 156
 non-parametric models, 143
 predicted patterns, 154, 155
 prediction, 154
 predictive mapping, 148, 149, 157
 RF model, 150
 RTA, 153
 spatial dependency, 153–154
 spatial interpolation techniques, 143
 statistical methods, 147–149
 variable selection processes,
  149, 150
 variation, 143, 156
ADABOOST, 67
Advanced difference tests, 89
Advanced Very High Resolution Radiometer
  (AVHRR) data, 145
Akaike Information Criterion (AIC), 30

Alaska
 boreal forest (*see* Alaskan boreal forest)
Alaska Geospatial Data Clearinghouse
  (AGDC), 145
Alaskan boreal forest
 estimate and map, 144
 geographic distribution, 142
 ground measured inventory plots, 144
 landscape, 154
 mixed poplar/birch, 142
 mixed spruce, 142
 RF, 154
 woody biomass, 151
Algorithmic models, 27
Algorithms
 Bayesian analytics, 17
 CART, 8
 categories, 17
 complex behavior, 7
 and computational techniques, 4
 in ecology and wildlife biology, 17–19
 maxent, 13
Amazon, 342
Amazon Web Services, 356, 363, 368
American marten (*Martes americana*)
 AUC measures, 199
 CART, 200
 complete model predictor set, 209–210
 continuous sub-model, 215
 deforestation, 188
 ENMs, 207–208
 forest management, 185
 fragmentation, 186
 full model, 212–215

© Springer Nature Switzerland AG 2018
G. R. W. Humphries et al. (eds.), *Machine Learning for Ecology and Sustainable Natural Resource Management*, https://doi.org/10.1007/978-3-319-96978-7

American marten (*Martes americana*) (*cont.*)
    genetic analysis, 188
    GLM modeling, 201
    2012 habitat model, 189
    habitat suitability, 199
    HLC predictors, 220–221
    interaction effects, 221, 222
    investigations, logistic regression, 186
    landscape predictions, 211
    low-fragmentation landscapes, 186, 199
    marten habitat suitability, 200
    modeling approaches, 190, 191
    multi-scale modeling (*see* Multi-scale
        modeling)
    network analysis, 210–211, 214
    occurrence records, in Alaska, 208
    outcomes
        differences, spatial prediction, 197
        model performance, 198, 199
        qualitative interpretation, 196
        RF model, 197
        RF multivariate model, 192, 195
        RF univariate scaling, 192, 194
        sawtimber forest, 197
    parsimonious *vs.* interactive models, 220
    patterns of scale-dependency, 200
    performance of predictions, 191, 192
    perturbations, 185
    predicted marten distribution, 222–223
    prediction, 186
    predictive power and ecological
        interpretation, 186
    predictor variables, 189, 190
    recommendations, ENM models
        development, 223
    scale dependent fashion, 185
    scale-dependent habitat selection, 199
    Small Mammal sub-model, 215, 216
    spatial models, 217–219
    spatial predictions, 199
    Spearman correlation varclus analysis,
        210, 212
    study area, 187, 188
    timber harvest, 185
    training dataset, 208
    TreeNet, 210
    varclus analysis, 211
    variables, full model, 212, 213
Analysis of complex ecological systems,
    30–32
Analysis rules, 90, 92
Aquila chrysaetos, *see* Golden eagles
ArcGIS (remote sensing software), 291
ARCTOS database, 365

Artificial intelligence (AI), 79
Artificial Neural Network (ANN), 366
Autocorrelation, 89
Automatic identification system (AIS), 111, 112
AVGMOD, 131, 132
Avian vocalizations, 296

B
Bagging, 126
    applications, 76
    binary re-cursive partitioning, 68
    vs. boosting, 68
    bootstrapping, 69
    experiences, 76
    Leach's storm-petrel, 302
    linear regressions, 69
    parameters, 67, 68
    pseudo- r^2, 69
    re-sampling, 69
    RF algorithms, 69, 70
    ROC/AUC, 69
    trick, 69
Base learners, 125, 127, 128
Batteries
    available, in SPM, 173–174
    description, 165
    kitchen sink model, 168, 170
    for non-parsimonious solutions, 171
    predictive performance metrics, 169
    for Siberian crane, 167
    shaving, 165, 172
    styles, in SPM, 165
    on TreeNet (SPM7), 166–168
    visual assessment, 168–170
    in wildlife conservation, 172
Bayesian approaches, 63
Binary re-cursive partitioning, 66
Biodiversity, 13
BIOMOD R package, 52
Bird Conservation Regions (BCR's), 381
BirdVis program, 348
Blood lead levels (BLL), 244, 246, 249
Boosted regression trees (BRT), 110–112,
    114–119, 267, 302
Boosting, 110, 127, 128
    ADABOOST, 67
    applications, 75, 76
    binary re-cursive partitioning, 66
    CARTs, 66
    concept, 66
    correlations, 66
    description, 66
    experiences, 75, 76

fielding, 67
  Leach's storm-petrel, 302
  linear regressions, 68
  machine, 67
  non-parametric methods, 66
  parameters, 67
  sequence of algorithms, 66
  stochastic, 66
  testing and internal assessment data, 66
  tree-algorithms, 66
  tree-based methods, 68
Bootstrapping, 68, 69, 114
Boreal forest
  Alaska (*see* Alaskan boreal forest)
  ground measured inventory plots, 144
  terrestrial biome, 141
  types, 142
Boreal forest biome, 222, 223
Boruta algorithm, 149, 150
Buffer fuzzy data, 42

C
Camera traps, 285–287
Capture-mark-recapture models, 51
CAVGMOD, 131–134
Cheer pheasant (*Catreus wallichii*), 14, 15
Chi-square residuals, Golden eagles, 386, 402
Christmas Bird Count (CBC) circles, 385, 386
Citizen science, 22, 337–338
Classification and regression trees (CARTs), 5,
  6, 8–11, 18, 31, 64, 65, 190, 200
Climate change models, 48, 49
Climate models
  for Alaska, 228, 236
  climate predictions, 230
  comparisons, 232, 233
  data mining, 231
  expected temperature, 228
  generic 'issues, 229
  inference, 227, 230
  information source, 227
  IPCC models, 228
  methodological questions and problems, 230
  pixels, 232, 234, 235
  regional, 228
  on smaller scale, 228
Climate Research Unit (CRU) data, 145
Climate-scapes, 228, 230, 237
The cloud
  Amazon, 356
  China and India, 370
  citizen science, 371
  computing organizations, 354
  control of, 370
  data storage, 371
  democracy, 370
  and ecology, 373
  ethics, 373, 374
  facial/image recognition, 371
  future trends, 374
  in guiding democracy and natural
    resources, 365
  interacting and synergetic applications, 372
  internet, power utilization, 371
  and mobile devices, 357–358
  on programming jobs, 358
  role and its implications, 356
  'sheep & cattle herding' with robots, 372
  as a strategic sword, 354
  and storage services, 356
  Supercomputing Centers, 357
  tasks, 358
  teaching, 372
  technical definition, 353
  usage, 353
  web-portals serving biodiversity
    data, 359
  in wildlife biology/ecology
    Crowdsourcing/storage, 364–365
    eBird, 360–361
    economic applications, 364
    GBIF and the Rio Convention, 358
    iNaturalist, 361
    MAPPPD, 363
    Movebank, 361–363
    natural resource management, 365
    OBIS and OBIS-Seamap, 359
    OpenModeller, 360
  with ML
    ANN analysis, 366
    business, 368
    as data mining and inference, 367
    data mining before analysis and
      posting, 366
    future and security, 369
    GARP model, 367
    Genbank, 367
    Influenza Research Database (IRD), 367
    and 'R', 368
    server farms, 369–370
  and www, 355
Committee averaging, 111
Complex data, 102, 104
Confidence trick, 89
Consensus, 110, 111, 119
Contaminants, environmental, 243
Convolutional Neural Networks (CNN), 291

Cooperative Alaska Forest Inventory Database
        (CAFI), 144, 154, 155
Cosmic model covariates, 422, 423
Curse of dimensionality, 207

D
Databases, 413
    Birdlife International Seabird Tracking, 336
    functionality, 339
    hosting, 339
    integration, 339
    MySQL, 339
    PostgreSQL, 339
    scalability, 339
    security, 339
    selection, 339
    software/application development, 338
    usability, 339
Data cloning, 231
Data errors, 99
Data issues
    and availability, 17, 20, 21
Data mining
    best professional practices, 103
    on climate models, 231, 236
    data cube, 90
    education and culture, 103
    evidence-based analysis, 92, 95–97
    good practice, 87
    ignorance, 103
    inferences, 91–94, 101, 102
    misconduct, 103
    and ML, 17, 20, 21
    parametric ones, 90
    Pb activity, 253, 254
    and predictions, 91–94, 101
    professional bias, 103
    professional societies, 103
    real life applications, 90
    real-world data and analysis, 97–100
    simplicity/parsimony, 90
    standard rules, 90
    statistical tests, 87–89, 91–94
    suggestion and justification, 98–100
    valid generalization, 90
    vast datasets, 90
Data sources, 113
Decision support tools, 349
    See also Web-based ML applications
Decision trees, ENS
    bagging, 126
    base learners, 125
    boosting, 127, 128

    component, 126
    ensemble learning, 125
    ERF, 127
    high variance, 126
    intuitive, 126
    MSE, 126
    poor prediction ability, 126
    RF, 126, 127
    RuleFit, 128
Decorrelating, 114
Deep learning, 66, 79, 425
Diameter at breast height (DBH), 144
Digital sword, 354, 371, 425
    See also The cloud
Distance to communities (DTC), 150
Django framework, 344, 346
DNA research, 413
Drones, 413

E
Eagle Act, 380
Eastern population Siberian crane
                (*Leucogeranus leucogeranus*), 164
eBird, 336, 337, 348, 360–361
Eco-evolutionary processes, 124
Ecognition, 291
Ecological models, 27–53
    CARTs, 31
    complexities, 28
    habitat-species/biodiversity relationship
            modeling, 30
    maximum entropy, 31
    principal, 33
Ecological niche factor analysis (ENFA), 31, 32
Ecological niche models (ENMs)
    for American marten, 207
    macroscale (*see* Macroscale ENMs)
Ecology
    Cheer pheasant in Hindu-Kush Himalaya,
            13–15
    and natural resource managers, 12
    quantitative, 12
    and science-based conservation
            management, 13
    in scientific literature, 12
    SDM, 13
Ecology and Natural Resource Management, 92
El Niño Southern Oscillation (ENSO), 263,
        264, 280
Elevated blood lead levels (eBLL), 244, 247,
        248, 253–256
Ensemble (ENS) models
    accuracy assessment, 112

advantages, 74, 75
algorithms, 73
applications, 76
Biomod2 and SPM8, 74
classifiers and predictors, 74
climate and financial models, 74
and consensus, 110, 111
definition, 73
experiences, 76
GIGO ensemble models, 74
BRT (*see* Boosted regression trees (BRT))
committee averaging, 111
data set, 111, 112
decision trees (*see* Decision trees, ENS)
forecasting, 110
modeling algorithms, 114, 115
outcome, 115–119
overfitting, 110
prediction, 110
predictive, 109
RF (*see* Random forests (RF))
tree-based classifiers, 110
tree-based ML techniques, 110
Environmental data, 111
Ethics, in the cloud, 373, 374
Evidence-based analysis, 92, 95–97
Explanatory variables (predictors), 130
Extremely randomized trees (ERF), 127

F
Fielding, 67
Fishing traffic dataset, 115
Fishing vessels, 111, 112
Fixed effect approach, 42
FLIRT program, 347
Forecasting, 110, 413
Forest Inventory and Analysis (FIA), 123, 143
FRAGSTATS metrics, 190
Free open-access robust softwares, 79
Frequentist statistics, 7
Funding organizations, 336

G
Garbage in garbage out (GIGO) ensemble
        models, 74
Gaussian distribution, 45
Genbank, 367
General additive models (GAMs), 91
General circulation models (GCMs), 124, 130
Generalized linear models (GLMs), 46, 47, 91
Geographic Information Network of Alaska
        database (GINA), 145

Geographic Information System (GIS), 33, 50,
        52, 174–175, 320, 321
Geomorphometry and Gradient Metrics
        Toolbox, 189
The Global Biodiversity Information Facility
        (GBIF.org), 358–361
Global sustainability culture and governance, 78
Global sustainability, the cloud, 369–370
Golden eagles
    age classes, 246
    binary logistic approach, 246
    bivariate partial dependence plots, 248,
        250–251
    capture month and gender interactions, 248
    CBC circles, for validation, 385, 386, 402
    chi-square residuals, 386, 402
    cumulative effects, on geographic
        populations, 379
    distribution and abundance, 379
    eBird review process, 385
    eBLL, 244, 255
    environmental layers, 399–400
    evaluations, model accuracy, 385
    FWS management, 380, 381
    gender differences, 255
    habitats, 380
    in Idaho, 245
    management, 380
    migratory status, 246
    populations, in western US, 380
    predictive models (*see* Predictive models,
        Golden eagles)
    predictor variable, 247–249
    research objectives, 381
    research, on NA Golden eagles, 381
    risk Layer, 401–402
    shrub/grassland vegetation associations, 395
    stochastic gradient boosting, 244
    study area, 381–383
    in Sweden, 253
    threats, 395, 396
    topography, 244
    TreeNet, 246, 253
    weather conditions, 254
    winter distribution (*see* Winter distribution,
        Golden eagles)
    winter use and distribution patterns, 380
Google, 342

H
H2o implementation, random forests, 303, 304
Habitat suitability indices (HSIs), 414
Habitat suitability, white oak, 129, 130

Habitat-species/biodiversity relationship
        modeling, 30
Heteroscedasticity, 89
High-level categorical (HLC) predictors,
        220–221
Holistic management, 421–423
Hurdle models, 111
Hypothetical habitat factors, 45

I
Image recognition
    camera traps, 285, 286
    counting animals, for population census,
            290–292
    individual penguins, 288
    internet search engines, 286
    pattern recognition techniques, 291
    Snapshot Serengeti, 287
    species identification, 286, 287
    success and error rates, 289
    tracking individuals, 287, 288
Importance value (IV), 129, 131–135
iNaturalist, 361
Inductively coupled plasma-mass spectroscopy
        (ICP-MS), 246
Inference, 88, 89, 91–95, 97–103
Influenza Research Database (IRD), 367
Intergovernmental Panel on Climate Change
        (IPCC), 48
Internet, 335
    users, 335
    web-based mapping application, 335
Interval/ratio data, 88

K
Kappa statistic, 191
Kitchen sink model, 168–171, 178
Known-fate' datasets, 51

L
Landscape-scale conservation, 349
Lasso penalty, 128
Leach's storm-petrel (*Hydrobates leucorhous*)
    Atlantic and Pacific populations,
            298, 299
    bagging, 302
    boosting, 302
    call recording technique, 300
    chatter call, 298
    comparing modeling algorithms, 302, 303
    computational comparison, 303–305

    description, 297
    discriminant function analysis, 299
    environmental *vs.* genetic influences,
            307, 308
    field recordings, 300
    geographic variation, 296
    measurement, chatter calls, 301
    ML algorithms, 308–309
    predictive comparison, 305
    reproducibility, 303
    sexing birds with ML, 307–308
    study sites, 300
    taxonomic classification, 297
    variable ranking, 305
Lead (Pb)
    in blood of birds, 254
    eagle-Pb contamination, 254
    exposure, 255
    game carcasses, 244
    sublethal Pb contaminant, 244
Linear models (LMs), 52, 64, 65
Linear regression
    data *vs.* machine learning
            summary, 95
Local climate, 229
Locally weighted scatterplot smoothing
        (LOWESS), 192, 195, 196
Location only data, 32
Logistic regression model
    areas, 198
    AUC, 187, 198
    canopy, 197
    complex non-linearity and non-
            monotonicity, 200
    differences, 200
    ecological interpretation, 186
    effects, broad scales, 200
    forest-dependent mammal
            species, 186
    fragmentation signal, 197
    genetic samples, 186
    GLM, 187
    LOWESS splines, 200
    marten habitat, 197
    marten occurrence, 201
    and occurrence data, 188, 189
    original, 190
    performance of, 186
    predictions, 197, 198
    recommendations, 199
    and RF, 191
    spatial predictions, 199
    TOC curve, 192
    variables and scales of influence, 201

M
Machine learning (ML)
  accessibility, 3
  adoption, 3
  algorithms, 4, 7, 8, 17–19, 295, 412, 416,
      417, 425
  anthropogenic (socio-economic) forces, 422
  applications, 14, 15, 30, 32–34
  autocorrelation, 41, 42
  batteries (*see* Batteries)
  black box, 37
  citizen science, 22
  in climate change models, 48, 49
  and 'the cloud (*see* The cloud)
  community, 3
  criticism and misunderstanding, 78, 79
  databases, 413
  data grooming, 423
  and data mining, 17, 20, 21
  decades in disciplines, 30
  decision based on empirical data, 38
  definitions, 4–7
  developments, 32, 418–420
  disciplines, 3, 7
  DNA research, 413
  drones, 413
  ecological models (*see* Ecological models)
  in ecology (*see* Ecology)
  enhancement, 32
  and ethics, 423–424
  forecasting, 413
  frequentist statistics, 7
  future, 411, 412
  humanity, 412
  hypothesis testing, 7
  information economy, 412
  image recognition (*see* Image recognition)
  innovations, 418–421
  macroscale ecology (*see* Macroscale
      ENMs)
  metrics, 40, 41
  and Mother Earth, 421–423
  nationalism, 412
  natural resource management, 3
  neural networks, 7
  overfitting, 38, 39
  parsimonious, 39, 40
  pattern, 7
  pattern detection (*see* Pattern detection)
  predictions, on independent datasets,
      423, 424
  presence only data (*see* Presence only data)
  principle, 28
  programming language, R, 418

  pseudoreplication, 42, 43
  rapid expansion, 31
  regression coefficients, 43, 44
  response index, 43
  risk assessment, 413
  robotics, 413
  Salford systems, 8–11
  sample code, implementation, 304
  sexing birds, 307–308
  stochastic gradient boosting, 205–206
  strategic conservation and planning
      (*see* Strategic conservation
      planning)
  strengths, 34, 38
  taxonomy, 413
  TensorFlow, 32
  terms/concepts, 4–7
  trained in, 3
  tree-based models, 44
  Turing test, 7
  understand and visualize, 44
  unsupervised techniques, 295
  variable interactions, 206–207
  weaknesses, 38
  workflows, 21
Macroscale ENMs
  abundance response, 124
  algorithms, 123
  biotic interactions, 124
  classification, 123
  controlling bias, 125
  datasets, 124
  eco-evolutionary processes, 124
  ENS (*see* Ensemble (ENS) models)
  explanatory variables (predictors), 130
  GCMs, 124
  habitat quality, 133, 134
  habitat suitability, 129, 130
  model reliability, 133
  multi-model ensemble approach, 131
  non-parametric methods, 124
  parametric analysis, 124
  prediction confidence, 133, 134
  predictor importance, 134, 135
  regression, 123
  statistical analysis, 123
  variance, 125
Mapping Application for Penguin Populations
    and Projected Dynamics
    (MAPPPD), 348, 363
Marine protected area (MPA) networks, 316,
    318, 328
Marine traffic data, 111
Mark-recapture, 287–289

MARXAN, 316–318, 320
Matanuska-Susitna valleys, 154
Mature average diameter (MAD), 129,
    131–135
Mature number of trees (MNT), 129,
    131–135
Maxent, 31, 32, 34, 38, 44, 48, 52
Maximum entropy, 31
Mean absolute error (MAE), 149
Mean error rate, 115
Mean square error (MSE), 112, 115–117,
    125–127, 148
Metadata, 337
Mixed models, 91
Mobile devices, 357, 358
Model comparison, 115, 119
Model improvement ratio (MIR), 191, 192,
    194–197
Model reliability, 133
Moran's I statistic, 46
Movebank, 348, 361–363
Multi-model ensemble approach, 131
Multiple adaptive regression splines (MARS),
    8, 10, 118
Multiple comparison tests, 89
Multi-scale habitat modeling, 186
Multi-scale modeling
    habitat, 186
    logistic regression (*see* Logistic regression
        model)
MySQL databases, 339

N
National Center for Ecological Analysis and
        Synthesis (NCEAS), 415
National Land Cover Database
        (NLCD), 145
Natural Resource Conservation, 355
Natural resource management, 12
    HSIs, 414
    machine learning methods, 415
    RSFs, 414, 415
    SDMs, 415, 416
    white-tailed deer, 414
Negative binomial regression, 111
Nesting areas, Siberian crane, 165, 170
Neural networks, 7
Non-parametric models, 124, 143
Non-spatial databases, 33
Non-stocked clear-cuts, 186
Normal distribution, 88
Normalized Difference Vegetation Index
        (NDVI), 145, 150, 155, 156

O
Open access data, 14
OpenModeller, 360
Oracle RDBMS, 339
Out-of-bag (OOB), 112, 114, 118, 127, 133,
    147, 148

P
Parameter space, 119
Parametric ones, 90
Parsimony, 30, 39, 64, 65, 90
Pattern detection
    classification, 28
    data and making predictions, 28
Pb contamination, golden eagles, 244, 253, 256
Pearson's product-moment correlation
        coefficients, 149
"Penguin Watch", 337
Permanent sample plots (PSPs), 144, 145
Poisson distribution, 45
Polynomial regressions, 91
Population census, 290–292
Population viability analysis (PVA), 321
PostgreSQL, 339
Predictions, 90–96, 100–102, 104, 110, 424
    climate models, 230
    modeling methods, 227
Predictive mapping, 148, 149, 157
Predictive models, 27–53, 109
    spatial variation, 45
Predictive models, Golden eagles
    active development, 398
    distribution of threats, 396
    human activity, 396
    potential/known risk factors, 390
    results and distribution maps, 397
    shrub-steppe habitat, 395
    solar energy development, 390, 395
    visual comparison, 392
    wind energy development, 390
    wintering areas
        characteristics, 386, 388
        habitat, 389
        Idaho, Utah, Nevada and eastern
            Oregon, 386, 387
        potential risks, 386, 389
        sagebrush vegetation communities,
            386, 388
Predictors, in TreeNet model, 235
Presence only data, 27–53
    binary model data, 39
    biological sciences, 34
    and location only, 32

vs. random, 39
wildlife, biodiversity and habitat
analysis, 28
Pro-Active Management Principles, 77
Professional societies, 103
Pseudoreplication, 42, 43
Pseudo-residuals, 127
Pseudo-r² metric, 40
Public access, 338
Python, 69
Python language, 341–344

Q
Quantify complex systems, 30

R
R computing environment, 31
R packages, 32, 51
Random forests (RF), 31, 32, 38, 39, 42,
46–48, 50, 52, 110–119, 126, 127,
143, 147, 148, 150, 151, 153–156
applications, 186
AUC, 198
CART, 190
differences, spatial prediction, 197
and GLM, 187, 191
fine scale effect, canopy cover, 197
heterogeneity, predictions, 198
investigations, 186
Kappa statistic, 191
landscape condition, 191
and logistic regression, 191, 197, 199
MIR, 191, 197
multi-scale optimization approach, 187
multivariate model, 192, 195
outperforms, 199
performance, 192
probability distribution, 191
qualitative interpretation, 196
study landscape, 190
univariate scaling, 192, 194
Random Survival Forest, 31
R-code, 112
Real-world data, 97–100
Recursive partitioning, 34
Red Panda, 28
Regression coefficients, 43, 44
Regression slope, 88
Regression tree analysis (RTA), 149, 153
Remote-sensing data, 143
Renewable energy development, 380, 395,
396, 398

Resource conservation management
precautionary, 77, 78
predictive models, 77, 78
pro-active, 77, 78
Resource selection functions (RSFs), 33,
414, 415
Response index, 43
Rigorous fashion, 91
The Rio Convention, 358
Risk assessment, 413
R-language, 368
Robotics, 413
Root-mean-square error (RMSE), 41, 46,
47, 149
RuleFit, 128

S
Salford systems, 8–11, 114
Satellite-derived marine vessel traffic
information, 111
scCombined, 111
Scenarios Network for Alaska Planning
(SNAP), 145
Science-based natural resource conservation
management, 64
Science-based wildlife management, 28, 34,
37, 38, 53
Science-driven management, 79
Self-fulfilling prophecies, 28
Siberian crane (*Leucogeranus leucogeranus*)
'batteries (*see* Batteries)
beneficial use, ML, 165
climate change, 164
conservation measures, 164, 165
eastern population, 164
ecological niche, 170
elegant appearance, 164
GIS predictors, 166
nesting areas, 165, 170
populations, in Asia, 163
presence and absence points, 165–166
rivers and lake systems, 164
in Russian high arctic, 163, 164, 172
western population, 163
Simulated annealing algorithm, 316, 321
Slow learning technique, 114
Snapshot Serengeti project, 286, 287
Sooty shearwaters
chick harvesting, 264
ECMWF data, 265, 266, 276
eggs hatching, 266
ENSO, 264
environmental data, 265

Sooty shearwaters (*cont.*)
    ocean variables *vs.* SOI, 268
    oceanographic habitats, 264
    SOI, 264, 268 (*see also* Southern
        Oscillation Index (SOI))
    spatial generalized boosted regression
        models, 269
    spatial models, SOI, 267–268
    spatial SOI model results, 269–270
    study region, 265
Soundscapes, 297
Southern Ocean, 265, 278
Southern Oscillation Index (SOI)
    chick quantity and size, 264
    chick size and oceanography, 278–279
    chick size and upcoming shifts, 264
    gradient boosted regression models,
        272, 274
    lags, 268
    and oceanographic factors, 277–278
    oceanographic parameter(s), 264
    and ocean parameters, 270–271
    vs. ocean variables, 268
    shifts, 279
    spatial generalized boosted regression
        models, 269, 270
    spatial models, 267–268, 277
    spearman correlations, 270, 272
Spatial autocorrelation, 143, 149, 154
Spatial interpolation techniques, 143
Spatial population viability analysis (sPVA),
        321, 322
Spatially-explicit mark-recapture models, 51
Spearman correlation varclus analysis, 210, 212
Species distribution model (SDM), 13, 17, 46,
        319–321
Statistical tests, 91–94
Stochastic boosting, 66
Stochastic data models, 30
Stochastic gradient boosting, 205–208
Stock markets, 7
Strategic Conservation Planning
    ad hoc decisions, 317
    and decision-making, 329
    global applications, 317
    Great Barrier Reef, Australia, 317
    improvements, 327–328
    known failures, 328
    MARXAN, 317
    MPA networks, 316
    National Park concepts, 326
    planning units (PUs), 317
    and protected area design, principles, 322
    SDM approaches, 319, 320

simulated annealing algorithm, 316
sPVA, 321, 322
techno-fix, 326–329
use of scenarios, 322
'zones' subunits, 316
Supercomputing centers, 357
Support-vector machines, 31
System's Predictive Modeler (SPM), 165–169,
        172–174

T
Taxonomy, DNA research, 413
TensorFlow, 32
Tensorflow computer vision model, 287
Topography, 244
Total operating characteristic (TOC) curve,
        192, 198
Tree-based ML methods, 42, 110
Tree-based models, 44
TreeNet, 38, 166, 167, 170–172, 175–176,
        383, 390
    climate models, 233
    Pb, in golden eagles, 244, 246–248, 256
    predictors, ranking, 235
Turing test, 7

U
US Fish and Wildlife Service (FWS)
        models, 244

V
Variables, 145–147, 189, 193
Vocalizations
    automated recording systems, 296
    individual recognition, 296
    Leach's storm-petrel (*see* Leach's
        storm-petrel (*Hydrobates
        leucorhous*))
    notes and syllables, 296
    western grebe, 296
Votes-matrix, 191

W
Wainwright Forest Inventory Database
        (WAIN), 144, 154, 155
Web-based ML applications
    advantages, 337
    BirdVis program, 348
    citizen science, 337–338
    databases, 339–340 (*see also* Databases)

delivery methods, web tools, 336
design skills, 344
FLIRT, 347
for landscape-scale conservation, 349
good web hosts, 346, 347
machine learning algorithms, 342
mapping application, 335
MAPPPD, 348
metadata, 337
Movebank, 348
open access, 336–337
painless programming framework,
      340–341
public access, 338
Python, 341, 342
Python/Django framework, 344
R Statistical programming language, 341
software/application development, 338
steps, for modeling, 342–344
wildlife management, 335
Web frameworks
   PostgreSQL, 339
   programming, 340–341
   Python/Django framework, 344
White oak (*Quercus alba*), 129–136
Wildlife conservation, 28, 34, 37, 42, 49

Wildlife management, 38, 53
Wind energy, 390
Winter distribution, Golden eagles
   breeding populations, 395
   geographic information system (GIS), 383
   inverse distance weighting (IDW), 384
   potential risks, 384
   predictive model, 383
   relative index of occurrence (RIO), 383
   and resource, 397
   of sagebrush, 386
   'select by location' option, 384
   spatial assessment, 384, 385
   winter range, 384–386
Woody biomass, 141, 142
World-wide web (www)
   and the cloud, 353, 355–356
   TOP500 list, 357

X
xgboost, 128, 131

Y
YAIMPUTE, 48

Printed by Printforce, the Netherlands